# ENVIRONMENTAL MODELING

## A Practical Introduction

# ENVIRONMENTAL MODELING

## A Practical Introduction

### Michael J. Barnsley

**CRC Press**
Taylor & Francis Group
Boca Raton   London   New York

CRC Press is an imprint of the
Taylor & Francis Group, an informa business

CRC Press
Taylor & Francis Group
6000 Broken Sound Parkway NW, Suite 300
Boca Raton, FL 33487-2742

© 2007 by Taylor & Francis Group, LLC
CRC Press is an imprint of Taylor & Francis Group, an Informa business

No claim to original U.S. Government works
Printed in the United States of America on acid-free paper
10 9 8 7 6 5 4 3 2 1

International Standard Book Number-10: 0-415-30054-1 (Hardcover)
International Standard Book Number-13: 978-0-415-30054-4 (Hardcover)

**Library of Congress Cataloging-in-Publication Data**

Barnsley, Michael J. (Michael John)
    Environmental modeling : a practical introduction / Michael John Barnsley.
        p. cm.
    Includes bibliographical references and index.
    ISBN-13: 978-0-415-30054-4 (alk. paper)
    ISBN-10: 0-415-30054-1 (alk. paper)
    1. Environmental sciences--Mathematical models. I. Title.

GE45.M37B37 2007
577.01'5118--dc22                                                     2006030555

**Visit the Taylor & Francis Web site at**
**http://www.taylorandfrancis.com**

**and the CRC Press Web site at**
**http://www.crcpress.com**

# Contents

# List of Figures

# List of Tables

# List of Programs

# Preface

The initial motivation to write this book stemmed from the perception of a growing divide between the increasingly sophisticated computer-based models that are being developed to represent various aspects of Earth's environmental systems (including those pertaining to its climate, ecosystems, biogeochemical cycles and hydrological processes) and the ability of many undergraduate students, and even some graduate students, of the environmental sciences to engage constructively with these models. The aim of this book, therefore, is to provide a practical introduction to the various methods, techniques and skills involved in computerized environmental modeling, including (i) representing an environmental problem in conceptual terms (i.e., developing a conceptual model), (ii) formalizing the conceptual model using mathematical expressions (i.e., formulating a mathematical model), (iii) converting the mathematical model into a program that can be run on a desktop or a laptop computer (i.e., implementing a computational model) and (iv) examining the results produced by the computational model (i.e., visualizing the output from a model and checking the model's validity in comparison with observations of the target system).

The contents of this book are based on a course that I have taught for many years to honors degree undergraduate students at Swansea University. The objectives of the course, and of this book, are to introduce the student to the broad arena of environmental modeling, to show how computational models can be used to represent environmental systems, and to illustrate how such models can improve our understanding of the ways in which environmental systems function. Equally important, the book also aims to impart a set of associated analytical and practical skills, which will allow the reader to develop, implement and experiment with a range of computerized environmental models. The emphasis is, therefore, on active engagement in the modeling process rather than on passive learning about a suite of well-established models. A practical approach is adopted throughout, one that tries not to get bogged down in the details of the underlying mathematics and that encourages learning through "hands on" experimentation. To this end, a set of software tools and data sets are provided free-of-charge under the General Public License (GPL) and Gnuplot License so that the reader can work through the various examples and exercises presented in each chapter.

Most of the data sets used in this book relate to an area immediately south of Llyn Efyrnwy, Powys in mid-Wales, UK. Apart from the fact that this is a particularly beautiful part of the world, I chose this site because it is one of a relatively small number of locations in the UK at which the solar irradiance measurements used in Chapter 5 are routinely recorded. I should also confess to deriving a certain amount of innocent amusement thinking about the additional challenge that the pronunciation of this particular Welsh place name will present to many readers. If nothing else, it will help to take the reader's mind off the demands of environmental modeling, every now and then.

I am deeply grateful to the UK Ordnance Survey, and in particular Ed Parson, for making available the digital elevation data used in Chapters 2 and 10. Thanks are also due to the UK MetOffice for permission to use the various meteorological data sets pertaining to Llyn Efyrnwy, and also to the staff at the British Atmospheric Data Centre (BADC), which is operated by the UK's Natural Environmental Research Council (NERC), for providing the excellent service through which I was able to access these data.

This book was put together using a range of "open source" software, including the GNU/Linux operating system. Most of the figures were produced using gnuplot (http://www.gnuplot.info/); the majority of the remainder were created using the PSTricks class in LaTeX. LyX (http://www.lyx.org/) and LaTeX (http://www.latex-project.org/) were used to produce the camera-ready copy, and the Beamer class in LaTeX was used to create the presentation files. I should like to thank the developers of all of these software packages.

I should also acknowledge the many cohorts of undergraduate students at Swansea University who have acted as a test bed for much of the material presented in this book: a sea of blank faces is undoubtedly the most immediate and effective signal that the material being presented is inadequately explained or otherwise confusing, and I hope that the salutary lessons that my students have taught me along the way have resulted in a clearer exposition in this book. On a more positive note, I am deeply gratified by those students who, having been introduced to computer-based environmental modeling for the first time, have honed their newly acquired skills and gone on to greater things. I hope that this is also the case for the readers of this book.

Finally, I should like to acknowledge the support of various friends and colleagues who have offered help and advice, and above all provided much-needed injections of humor at numerous points during the production of this book: to Mat Disney, Philip Lewis ("Lewis") and Tristan Quaife at University College London, to Tim Fearnside, Sietse Los, Adrian Luckman, Peter North and Rory Walsh at Swansea, and to Paul Mather at the University of Nottingham, *diolch yn fawr iawn* (thank you very much). Paul Mather deserves special mention for kindly reading through the drafts of this book and for providing many detailed comments and helpful suggestions. As it has become conventional to say at this point, though, any oversights, omissions or errors that remain are mine alone. I understand that in the world of "closed source" computer software such things are often described as "features"; I hope, however, that "features" of this type are few and far between herein. *Pob lwc!* (Good luck!)

**Mike Barnsley**
School of the Environment and Society
Swansea University, UK

# Chapter 1

# Models and Modeling

**Topics**

- Why model?

- The modeling process

- A typology of models

- Systems analysis and systems dynamics

## 1.1   WHY MODEL?

A model is a simplified representation of a more complex phenomenon, process or system; an environmental model is one that pertains to a specific aspect of either the natural or the built environment. Environmental models have been developed to represent, among other things, elements of Earth's climate system, hydrological processes, ecosystems and biogeochemical cycles. The principal purposes of these models are threefold: to increase knowledge, and hence to reduce uncertainty, of the phenomenon, process or system that the model purports to represent (i.e., to improve understanding); to provide a tool with which to estimate the state of the phenomenon, process or system at times and locations other than those for which observations are presently available (i.e., to facilitate prediction); and to provide a framework within which "what if" questions can be asked about possible changes to the state and operation of the phenomenon, process or system under specified conditions (i.e., to perform simulations).

The production of an environmental model involves a process of abstraction: in the sense that most environmental models deal with abstract concepts and ideas (i.e., mathematical formulae and computational code) rather than physical objects

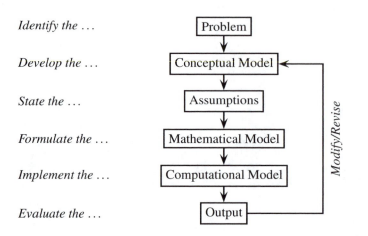

Figure 1.1: Schematic representation of the modeling process.

and events (i.e., conceptualization); in terms of identifying and extracting the most important elements of the phenomenon, process or system and discarding the least significant ones (i.e., selection); and in the sense of summarizing the essence of the phenomenon, process or system (i.e., encapsulation). Building a well-designed model therefore forces one to examine carefully, analytically and in detail the component elements of an environmental system, the processes and structures that govern the relationships and interactions between them, and the spatial and temporal scales over which they operate.

## 1.2 THE MODELING PROCESS

In principle, the process of designing, building and using an environmental model can be divided into a series of discrete stages. These stages are shown schematically in Figure 1.1 and are described in detail below. In practice, the boundaries between the different stages are not always well defined and progression from one stage to the next is seldom as straightforward or as linear as Figure 1.1 implies. Nevertheless, this diagram provides a useful framework within which to introduce the basic concepts.

### 1.2.1 Identifying the Nature and Scope of the Problem

The first step is to identify the specific science question, or problem, that is to be addressed and then to establish both whether and how a model will help to answer this question (Wainwright and Mulligan 2004). The problem should be sufficiently well defined and focused so that it is amenable to solution using the knowledge, skills and resources at hand (these factors influence the tractability of the problem), but it should also be sufficiently generic so that it is of more than just parochial interest (this implies that a compromise is negotiated between the specificity and the generality of the model). If the problem is poorly defined at the outset, the model-building process

Table 1.1: Four main phases of systems analysis (after Huggett 1980).

| Phase | Actions |
| --- | --- |
| Lexical | Define the system boundaries (closure). Choose the system components, i.e., state variables (entitation). Estimate the values (i.e., the state) of the state variables (quantitation). |
| Parsing | Define verbally, statistically or analytically the relationships between the state variables. |
| Modeling | Model construction. Model operationalization (i.e., running the model). |
| Analysis | Model validation and verification (i.e., compare the results of the model with observations of the target system). |

is likely to be more difficult, more time-consuming and more complex. Worse still, the resultant model may not be appropriate to the task for which it was originally intended.

A related consideration is the scope of the model, in terms of those elements of the science question that the model is, and is not, intended to address. The scope of the model may have to be limited in various ways to produce a tractable solution. For instance, the model may need to be designed so that it represents a selected part of the target environmental system, a particular spatial domain, a specified period of time, or perhaps a combination of all three.

### 1.2.2 Developing the Conceptual Model

After specifying the science question, the next step is to develop a conceptual model of the problem. The term *conceptual model* is used here to refer to a model that is expressed verbally or in written or diagrammatic form (i.e., concepts), as distinct from one that is represented in terms of mathematical formulae (i.e., a mathematical model) or one that is constructed from physical materials (i.e., a physical model).

The development of a conceptual model necessarily involves a comprehensive analysis of the target phenomenon, process or system with the aim of identifying its component parts, their respective inputs and outputs, the relationships between them, and the processes and structures that govern their interaction. This stage in the model-building process is therefore closely related to the lexical and parsing phases of systems analysis, a branch of science concerned with the study of complex systems, including their composition, structure, function and operation (Huggett 1980, Table 1.1). In each case, it is assumed that the "real world" can be divided into a number of more or less discrete systems, which can be further sub-divided into their component parts and processes, identified by careful analysis and detailed observation (Hardisty *et al.* 1993).

Table 1.2: Important definitions in environmental modeling.

| Element | Definition | Example |
|---------|-----------|---------|
| Constant | Quantity whose value does not vary in the target system. | Speed of light. |
| Parameter | Quantity whose value is constant in the case considered, but may vary in different cases. | Total solar radiation at the top of Earth's atmosphere. |
| Variable | Quantity whose value may change freely in response to the functioning of the system. | Amount of precipitation. |
| Relation | Functional connection or correspondence between two or more system elements. | Rainfall, run-off and soil erosion. |
| Relationship | State of being related. | — |
| Process | Operation or event, operating over time (temporal process) or space (spatial process) or both, which changes a quantity in the target system. | Evapotranspiration. |
| Scale | Relative dimension, in space and time, over which processes operate and measurements are made. | Local, regional, global; diurnal, seasonal, annual. |
| Structure | Manner in which component parts of a system are organized. | — |
| System | Set of related elements (e.g., constants, parameters and variables), the relations between them, the functions or processes that govern these relations and the structure by which they are organized. | Forest ecosystem, drainage basin, global carbon cycle, Earth's climate. |

The component elements of an environmental system typically include inputs, outputs, constants, parameters, variables (also known as stocks, stores, pools and reservoirs), processes (flows), relations (links or connectors) and structures (Edwards and Hamson 1989); see Table 1.2 for definitions. The boundaries, or limits, of the target environmental system must also be specified. In this context, environmental systems are sometimes classified in terms of their degree of openness: open systems, also known as forced systems, have exogenous (or forcing) variables; closed systems, also known as unforced systems, have no exogenous variables (i.e., all of the variables are endogenous to the system) (Hardisty *et al.* 1993).

The process of developing a conceptual model is often aided by using diagrams to represent the component parts of the target system and the connections between them (e.g., Figure 1.2). These diagrams vary from the simple to the complex, from the

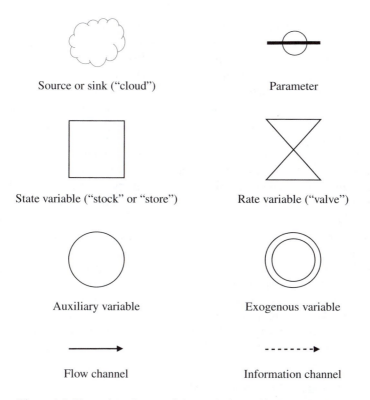

Source or sink ("cloud")

Parameter

State variable ("stock" or "store")

Rate variable ("valve")

Auxiliary variable

Exogenous variable

Flow channel

Information channel

Figure 1.2: Examples of some of the symbols used in Forrester diagrams.

schematic to the formalized. The symbols employed for this purpose differ somewhat between studies, although Forrester diagrams (Forrester 1973) and their derivatives are used quite widely, particularly in the field of ecosystem dynamics (Ford 1999, Deaton and Winebrake 2000).

### 1.2.3 Stating the Assumptions

Every environmental model is founded on a set of assumptions. These assumptions may be made so that a complex environmental system can be simplified sufficiently to produce a working model (e.g., when deciding which elements of the target system should be included in the model and which should be omitted) or they may reflect the limits to current knowledge of the target environmental system (e.g., concerning the nature and form of the relations between its component parts). The validity and the scope of any such assumptions ultimately determine the value of the resultant model (Edwards and Hamson 1989, Wainwright and Mulligan 2004).

Some of the assumptions that are made when a model is first created may later be found to be incorrect, in which case it is often possible to revise the assumptions in subsequent versions of the model. Some may be known to be wrong at the outset, but nevertheless they may be retained because they have a relatively insignificant effect on the output of the model and are necessary for reasons of simplicity or efficiency (Wainwright and Mulligan 2004). What is important is that each of the assumptions inherent in a model is recognized, understood and stated explicitly (Edwards and Hamson 1989, Wainwright and Mulligan 2004). The primary reasons for this are twofold: first, it clarifies the nature, purpose and limitations of the model, not least in the mind of the modeler; second, it helps potential users of the model to understand its scope (i.e., the range of conditions over which the model is known, or thought, to be valid), to challenge the assumptions on which it is based and, hence, to develop improved versions of the model in the future. For both the modeler and the user, it is particularly important to consider the consequences, whether they are intended or not, of the assumptions made in the model and to identify any assumptions that may have been made implicitly (i.e., without recognizing that this is the case) (Edwards and Hamson 1989).

Although it is not always mentioned in this context, the modeler should make clear the spatial and temporal scales over which the relations and processes that are being modeled operate. Many environmental processes operate, or are dominant, over a restricted range of spatial and temporal scales, and different processes operate at different scales; this phenomenon is sometimes referred to as "domains of scale" in the field of landscape ecology (Wiens 1989). It is inadvisable, therefore, to apply a model outside the range of scales for which it is designed; that is to say, most models are scale dependent.

### 1.2.4   Formulating the Mathematical Model

The next stage is to represent the conceptual model in mathematical terms; that is, using mathematical tools and concepts, such as variables, functions and equations. This process can be described as one of formulating the mathematical model, since it involves translating the conceptual model into mathematical formulae (Edwards and Hamson 1989). It is frequently the most challenging stage in the development of a model. Sometimes this is because the solution to the problem demands the use of advanced mathematical techniques, although each of the models considered in this book requires only basic skills in algebra and trigonometry. More often it is because there is more than one way in which the system can be represented mathematically and it is not immediately apparent which approach is best. Therefore, deriving a suitable mathematical formulation of a model is often a trial-and-error process, but it is also a skill that improves with practice.

The range and the diversity of mathematical models that have been developed to study environmental systems are considerable, and various different schemes have been proposed to group them by type. While these classification schemes differ at the level of detail, most are founded on a common set of principles that include the following considerations: the extent to which the model is derived from theory or from observations (i.e., empirical models versus theoretically informed models); the degree to which random events and effects play a major role in the target system and, hence, in the model (i.e., deterministic versus stochastic, or probabilistic, models); the level of knowledge or understanding of the target system that the model purports to represent (i.e., black box versus white box models); whether the model is operated predominantly in forward or inverse mode (i.e., forward-mode versus inverse-mode models); whether the model deals with environmental processes that are static or dynamic with respect to time and space (i.e., static versus dynamic models); whether the model deals with environmental processes that can be considered to operate in a discrete or continuous manner with respect to time and space (i.e., discrete versus continuous models); and whether the model parameters are lumped or distributed (i.e., lumped-parameter versus distributed-parameter models). Each of these considerations is examined in greater detail in the sections that follow.

### Empirical and Theoretical Models

An empirical model is one that is primarily or solely based on observations, as distinct from one that is derived from theory. In models of this type, the relationships between the component elements of an environmental system are established by examining measurements of the variables concerned. The form of each of these relationships is then defined by a mathematical function (Edwards and Hamson 1989). The decision as to which one of a set of candidate mathematical functions (e.g., Figure 1.3) should be used typically involves a compromise between how well the candidate functions fit the data and the relative simplicity of their mathematical form; the decision, however, is not informed by theory (Edwards and Hamson 1989). Regression analysis is widely used in this context to fit the function to the data.

While empirical models are frequently valuable in terms of making predictions in the cases for which they are developed, they typically lack generality (i.e., they are specific to a particular circumstance or set of data). It is often difficult, therefore, to employ them at other spatial locations or different points in time.

### Deterministic and Stochastic (Probabilistic) Models

Loosely defined, a deterministic model is one in which the outputs (i.e., the results) are uniquely and consistently determined by the inputs (i.e., the values used to drive the model) (Edwards and Hamson 1989). Deterministic models therefore function in the same way, and produce exactly the same output, each time they are run using a particular set of input values. This definition can be recast a little more precisely as follows: the state (i.e., the value) of a variable in a deterministic model is uniquely and consistently determined by the initial conditions of the model (i.e., the values of the constants and parameters input to the model) and, subsequently, the previous states of the variable itself. In contrast to empirical models, the implication is that deterministic models are based on assumptions, theory or knowledge of the nature and form of the relations between the variables in the target system. Deterministic models tend, therefore, to offer a much greater degree of generality than empirical models. Consequently, if they are properly configured and suitably implemented, deterministic models can usually be applied to different spatial locations and points in time from those for which they were originally developed and tested.

In contrast, stochastic (or probabilistic) models are ones in which random events and effects play an important role (Edwards and Hamson 1989). The state of the

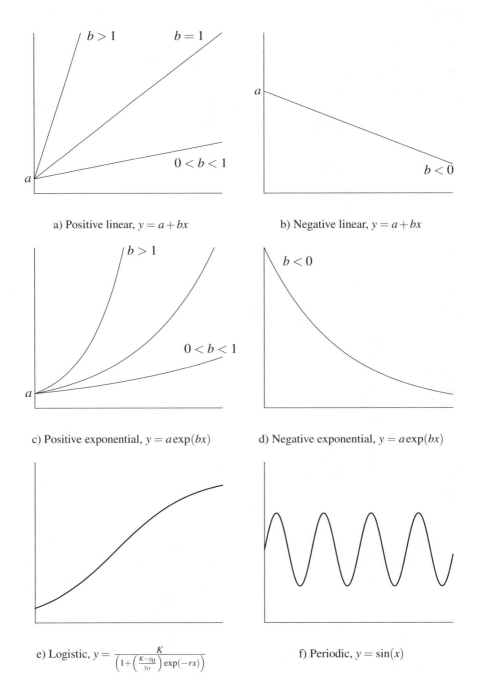

a) Positive linear, $y = a + bx$

b) Negative linear, $y = a + bx$

c) Positive exponential, $y = a\exp(bx)$

d) Negative exponential, $y = a\exp(bx)$

e) Logistic, $y = \dfrac{K}{\left(1 + \left(\frac{K - y_0}{y_o}\right)\exp(-rx)\right)}$

f) Periodic, $y = \sin(x)$

Figure 1.3: Forms of various mathematical functions.

model variables is therefore described by probability distributions, rather than single values. As a result, the output from such a model will vary from run to run even when the input values used are the same. This class of model is therefore suitable in those circumstances where apparently random fluctuations in the system processes, and hence in the system variables, render deterministic models inappropriate (Kirkby *et al.* 1993). The random fluctuations that are observed in the system, and hence in the model, may be due to environmental processes and events that are truly random in nature or they may be pseudo-random; that is, where knowledge of a potentially deterministic process is inadequate or incomplete, such that it has to be treated as though it is random.

## *Black Box and White Box Models*

Mathematical models can also be classified according to the degree to which the composition, structure and operation of the target system is known and, hence, is represented in the model. The two extreme cases in this respect are usually referred to as white box (or clear box) models and black box models. In white box models, the internal workings of the target system are known, completely understood and clearly stated. This *a priori* information may be based on detailed observation of the target system or it may be derived from theory (Kirkby *et al.* 1993). By contrast, in black box models the target system is treated as a sealed unit, with no attempt made to understand the variables of which it is composed or the relations between them. Whereas white box models are based on knowledge or theory of the target system, black box models are usually defined empirically (Kirkby *et al.* 1993). In practice, most models fall somewhere between these two extremes; as such, they might be described as gray box models.

## *Forward and Inverse Modeling*

Mathematical models are typically specified in terms of functional relations between two or more variables of the target system. For example, all other things being equal, the growth and abundance of a particular plant species in a given environment is functionally related to the ambient light conditions, the air temperature, the amount of precipitation and the availability of nutrients in the soil. A functional relation, such as this, can be expressed mathematically as follows:

$$growth = f(light, temperature, precipitation, nutrients) \qquad (1.1)$$

Equation 1.1 indicates that the variable *growth* is a function $f$ of four other system variables: *light*, *temperature*, *precipitation*, and *nutrients*. Thus, the value of *growth* is related to, and is in some way dependent on, the values of these four variables. If the form of the function $f$ is known, or can be derived from theory or else be obtained by induction from a set of observations (i.e., empirically), it can be expressed as a rule, which allows the value of the dependent variable (i.e., *growth*) to be determined for any value of the independent variables (i.e., *light*, *temperature*, *precipitation*, and *nutrients*). For instance, if $y = f(x)$ describes a functional relation between two system variables $x$ and $y$, and the function $f$ is defined in terms of the rule $f(x) = x^2$, then $y = f(3) = 3^2 = 9$ when $x = 3$. This will be referred to as forward modeling, for reasons that will become clear once inverse modeling is introduced, below.

Sometimes the modeler may be more interested in the inverse functional relation between two or more variables in the target environmental system. For instance, the modeler may wish to use knowledge of the relation between temperature and plant growth to reconstruct past climatic conditions (e.g., temperature) from an analysis of the abundance of a given plant species (e.g., inferred from pollen counts in cores taken from lake sediments). Inverse functional relations, such as this, can be expressed mathematically as follows:

$$x = f^{-1}(y) \tag{1.2}$$

Equation 1.2 states that there is a function $f^{-1}$, which is the inverse of function $f$, that relates the value of variable $y$ to that of variable $x$ (Biggs 1989, Piff 1992). Note that this does not imply that $x$ is dependent on $y$ (e.g., ambient temperature is not controlled in a direct way by the abundance of a particular plant species), merely that they are functionally related in some way. The rule for $f^{-1}$ is obtained by reversing the rule for $f$ (Grossman 1995). For example, if $y = f(x)$ and $f(x) = x^2$, then $x = f^{-1}(y)$ where $f^{-1}(y) = \sqrt{y}$. This is referred to as the inverse model. For simple empirical models, obtaining the rule for the inverse function is relatively straightforward: the values of variable $y$ are regressed on the corresponding values of variable $x$, rather than *vice versa*. For more complex deterministic and stochastic models, however, the problem is much more challenging and involves the use of analytical and numerical solutions that are beyond the scope of this book.

One area in which the application of inverse models is increasingly common is the field of Earth observation, also known as terrestrial remote sensing. In part, this area of science is concerned with the reflectivity, at different wavelengths, of Earth's surface materials (e.g., vegetation, soil, rocks, snow, ice and water). The reflectivity of these materials is known to be a function of their respective physical, chemical and biological properties. Various models have been developed to represent these functional relations. These models can be used to predict the amount of solar radiation (i.e., sunlight) reflected by a terrestrial surface, based on information about its physical, chemical and biological properties. The inverse functional relation, however, is typically of greater interest for practical purposes; that is, the ability to estimate the properties of Earth's surface materials from measurements of spectral reflectance made by sensors mounted onboard aircraft or Earth-orbiting satellites. This requires the use of sophisticated analytical and numerical techniques to invert the forward model against the measured data (Goel 1989, Kuusk 1995, Verstraete *et al.* 1996).

*Static Models, Dynamic Models, Equilibrium, Stability and Feedback*

Mathematical models can be divided into two further categories: static and dynamic (Ford 1999). The former deal with systems that do not change, or at least are thought not to change, appreciably with respect to time. This type of model focuses on the

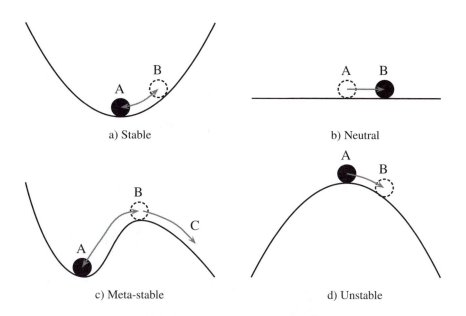

Figure 1.4: Graphical representation of different system states. The system state is represented by the position of a ball on the topographical surface.

processes, or forces, that keep the system in a state of equilibrium. Dynamic models, in contrast, deal with systems that change over time, which is much more common in environmental systems. Dynamic mathematical models are typically constructed from difference equations or differential equations (see Chapter 8).

The need to understand the behavior of an environmental system with respect to time leads the modeler to consider the important issue of system stability. A number of generic system states can be envisaged in this context; these states are known as stable, neutral, meta-stable and unstable (Huggett 1980, Figure 1.4). In a stable system, the system returns to its original state (position A, Figure 1.4a) having been perturbed (i.e., moved to position B, Figure 1.4a) by an external force or process. In a neutral system (Figure 1.4b), the action of an external force or process causes no change in the state of the system variables. In a meta-stable system, the system is, initially, in a weak stable state (position A, Figure 1.4c), but the effect of an external force or process may be to move it to an unstable transitional state (position B, Figure 1.4c) from which it may either return to its original state or change to a different one (position C, Figure 1.4c). Finally, an unstable system is one in which the system state changes, perhaps irrevocably, as a result of an external force or process (Figure 1.4d).

The relative stability of an environmental system is partly controlled by feedback mechanisms. Feedback is the process by which a fraction of the output from a given system, or part thereof (i.e., a sub-system), is returned (i.e., is fed back) as input to the same system or sub-system. This process is represented schematically in Figure 1.5. Feedback is an important feature of many environmental systems, not least because

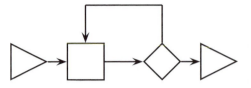

Figure 1.5: Schematic representation of a feedback relation.

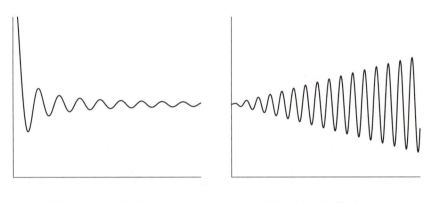

a) Negative feedback                                    b) Positive feedback

Figure 1.6: Negative and positive feedback relations.

it makes their operation more complex and their behavior more difficult to predict. Feedback is said to be either negative or positive (Figure 1.6). Negative feedback is produced by deviation-damping processes that counteract the effect of a perturbation to a system and, hence, tend to maintain the stability of the system: a system that displays a propensity for this type of behavior is said to exhibit homeostasis. In contrast, positive feedback is the result of processes that amplify an initial perturbation to a system and that tend to keep the system changing toward a new state: this tendency is sometimes referred to as homeorhesis. A notable example of a positive feedback loop in Earth's climate system is one that involves sea ice and albedo (i.e., the average reflectivity of a surface). In this feedback loop, warmer atmospheric conditions (the initial perturbation) result in increased melting of the Arctic and Antarctic ice sheets. This reveals more of the relatively dark (i.e., lower albedo) polar oceans. These, in turn, absorb more solar radiation, which causes the oceans to warm further, so that still more of the polar ice sheets melt, and so on.

Before proceeding further, it is worth noting that many environmental systems not only exhibit feedback mechanisms but also branching or splitting events (Figure 1.7). These occur where two or more inputs feed into a single process, or where two or more outcomes are possible depending on certain conditions (Huggett 1980). As is the case with feedback mechanisms, branching events make the behavior of the system more complex and, hence, more difficult to predict.

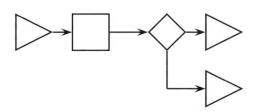

Figure 1.7: Schematic representation of a branching or splitting relation.

Feedback mechanisms and branching events are examined in greater detail in the context of the Daisyworld model, a simple biospheric feedback model, which is analyzed in Chapter 9.

### Continuous and Discrete Models

Mathematical models can also be divided into two further categories: discrete and continuous. The distinction between the two relates to the way in which each treats time (or spatial location or both) as a variable. In discrete models, time proceeds in a series of finite steps that are usually, although not necessarily, of equal length; each step relates to a specific interval, such as an hour, a day, a month or a year. The state of the system is updated at the end of each time step, taking into account changes in the system variables resulting from the processes operating on the system over that time step. In continuous models, by contrast, the system state is updated instantaneously in response to continuous changes in the forcing factors. Discrete and continuous models also differ in terms of the mathematical equations on which they are founded: the former are expressed in terms of difference equations; the latter employ differential equations. Discrete and continuous mathematical models are explored in detail, in the context of studies of population growth, in Chapter 8.

### Distributed-Parameter and Lumped-Parameter Models

Environmental systems often exhibit considerable spatial heterogeneity in the value of their variables and in the processes that control them. Some environmental models account for this spatial variation explicitly in their mathematical formulation. This is achieved by dividing the geographical domain of the model into a number of discrete spatial units, such as tessellating polygons (particularly square cells, which are then known as rasters), triangulated irregular networks (TINs) or irregularly shaped spatial objects (Wainwright and Mulligan 2004). Models that assign different values of the system parameters and variables to each of these spatial units are known as distributed-parameter models. This approach is not always feasible, however. Sometimes this is because of computational constraints; on other occasions it may be due to a lack of data on the spatial variability of the model parameters and variables. In such circumstances, it may be necessary to assign a single "lumped" value across the whole of the model's spatial domain; models such as this are often referred to as lumped-parameter models.

*Analytical and Numerical Solutions to Mathematical Models*

It is possible, in many cases, to construct a mathematical model of an environmental system by analytical means; that is, by formulating the model so that it is expressed concisely in terms of mathematical equations, functions, variables and constants. This approach is sometimes referred to as an analytical or closed-form solution. Not all environmental models are amenable to this approach, however; some must be solved numerically. The numerical approach relies heavily on the power of modern computers to solve the model by performing a sequence of operations rapidly over and over again, each time applying the operations to the result of the previous iteration (British Computer Society 1998). Sometimes an estimate, or guess, is made regarding the correct solution, to which the iterative procedure is initially applied. Provided that the mathematical model has been formulated and implemented correctly, the expectation is that the procedure will move ever closer to the correct solution with each successive iteration. Depending on the nature of the model, and in some cases the accuracy of the initial estimate, the iterative procedure will eventually reach the correct solution or else a very close approximation to it. The best known example of this approach is Newton's iterative method for calculating $\sqrt{x}$ (Harris and Stocker 1998). Examples of the analytical and numerical approaches to modeling are examined in detail in Chapter 7.

### 1.2.5    Implementing the Computational Model

*General Considerations*

Once an appropriate formulation has been derived for the mathematical model, the next step is to convert the equations and formulae into some form of software that can be run on a computer; that is, to implement the computational model. There is an enormous range of options available to the modeler in this context, including spreadsheet packages, specialized modeling software, high-level computer programming and scripting languages, and integrated modeling environments (Table 1.3). Each has its advantages and disadvantages. Ultimately, the decision regarding which software the modeler chooses to use is likely to be conditioned by issues of availability, cost, personal preference and prior experience, in addition to the ease with which the software is learned and its suitability for the task at hand.

It has been argued that it is easier to learn how to use spreadsheet packages than formal programming languages (Hardisty *et al.* 1993). Spreadsheets are certainly ubiquitous and, hence, are familiar to many computer users. Most spreadsheet packages offer the added advantage of built-in facilities for data visualization, statistical analysis and function-fitting. Spreadsheets have been used quite widely, therefore, in environmental modeling; for example, in conservation biology and landscape ecology (Donovan and Welden 2001, 2002). Nevertheless, spreadsheets have intrinsic limitations that present problems when producing more sophisticated environmental models, especially those requiring iteration to reach the solution. Most importantly, perhaps, spreadsheets do not encourage a structured approach to model building, and the potential for sharing and reusing code is limited.

Table 1.3: Selection of tools suitable for implementing computer-based environmental models.

| Category | Name | Reference |
|---|---|---|
| Spreadsheets | Calc | http://www.openoffice.org/ |
| | Excel | http://office.microsoft.com/ |
| | Gnumeric | http://www.gnome.org/projects/ |
| | Kspread | http://www.koffice.org/ |
| | Quattro Pro | http://www.corel.com/ |
| Specialized modeling environments | MODELMAKER | http://www.modelkinetix.com/ |
| | POWERSIM | http://www.powersim.com/ |
| | SIMILE | http://simulistics.com/ |
| | STELLA | http://www.iseesystems.com/ |
| | VENSIM | http://www.vensim.com/ |
| High-level computer programming languages | BASIC | http://www.freebasic.net/ |
| | C | Kernighan and Ritchie (1988) |
| | C++ | Oualline (1995) |
| | FORTRAN | Hahn (1994) |
| | JAVA | Niemeyer and Knudsen (2005) |
| | PASCAL | Buchanan (1989) |
| | VB | Willis and Newsome (2005) |
| Scripting languages | awk | Aho *et al.* (1988) |
| | JavaScript | Flanagan (2001) |
| | Perl | Wall and Schwartz (1993) |
| | Python | Lutz (1996) |
| | PHP | Atkinson (2000) |
| | Rexx | Cowlishaw (1990) |
| | Ruby | Thomas and Hunt (2000) |
| | Tcl/Tk | Welch *et al.* (2003) |
| Integrated modeling environments | IDL | http://www.ittvis.com/ |
| | MATLAB | http://www.mathworks.com/ |
| | OCTAVE | http://www.gnu.org/software/octave/ |
| | R | http://www.r-project.org/ |
| | S | http://www.insightful.com/ |

Specialized modeling software also offers many important benefits, including the ability to construct models diagrammatically using graphical interfaces, as opposed to typing lines of computer code, and the provision of an extensive range of built-in analytical and numerical functions and data visualization routines. This type of software tends to be made available on a commercial basis, however; that is, at some financial cost to the modeler. It may also be difficult to share models developed in one software package with users who employ a different modeling suite. As a result, there is a danger that the modeler becomes "locked in" to a product offered by a particular vendor. Finally, the modeling software may restrict the user to a particular set of modeling techniques, depending on the functions and the range of expressions offered by the software.

High-level computer programming and scripting languages represent yet another option for the modeler. In many senses this is the most flexible option, albeit one that requires the modeler to learn the commands, grammar and syntax of the chosen language. Despite suggestions to the contrary, however, this is not a major challenge and, importantly, the reward is a highly transferable skill that can be applied in a wide range of contexts. The choice of which programming language to use is largely a matter of personal preference. Although there are differences between programming languages, most can be used effectively to implement a computer-based environmental model. Moreover, many programming languages share a set of common features so that once the user has learned one language it is relatively easy to master another.

Irrespective of the specific software that is used for the purpose, the process of implementing a computational model involves a translation between the vocabulary, grammar and syntax of one language (i.e., the functions and equations of mathematics) and those of another (i.e., the instructions of a spreadsheet or high-level computer programming language). Just as in the translation between natural languages, such as English and French, there is rarely a single "correct" solution to this process, although some solutions might be regarded as being intrinsically better than others. The flexibility and versatility of computer software means that the implementation can often be realized in a number of ways. In some instances, the differences may amount to little more than a matter of programming style, broadly analogous to the variations in style of written prose among literary authors; on other occasions, however, the differences may have a significant impact in terms of the efficiency of the resulting code, including the speed with which the result is computed and the demands that the program places on various aspects of the computer's resources, such as memory usage and hard-disk access. Thus, some implementations may be considered to be more elegant or otherwise preferable to others.

*Computer Software Used in This Book*

A simple, but powerful, high-level scripting language, known as awk, is used in this book. The acronym awk is derived from the surnames of the individuals who initially developed the awk language and utility: Aho, Weinberger and Kernighan (Aho *et al.* 1988). There are several versions of awk. The version used here is the one produced by the GNU Project (http://www.gnu.org/), which is known as gawk

(Dougherty 1996, Robbins 2001). gawk has two main advantages in this context: it is an interpreted language, meaning that the computer interprets each line of the program (or script) as it runs, so that the modeler does not have to compile the code before it is executed by the computer; and it is data-driven, in as much as it provides a simple mechanism for reading data from files structured in terms of fields (columns) and records (lines or rows). The second of these two features greatly simplifies the process of incorporating data to models. More generally, gawk provides a powerful and flexible tool for data manipulation and a convenient framework within which to develop a range of environmental simulation models. It is also available free-of-charge for use on a wide range of computer platforms and operating systems.

While the gawk programming language is used to manipulate data stored in files and to implement computational models, a separate utility is required to visualize the inputs to, and outputs from, models. A software package known as gnuplot is used for this purpose. gnuplot is primarily, but not exclusively, an interactive, command-driven, function and data plotting program (Williams and Kelly 1998). It can be used to generate two-dimensional (2D) scatterplots and line diagrams in either rectangular or polar coordinates, as well as plots of points, lines, vectors and surfaces in three dimensions (3D). It is able to fit user-defined functions to a data set and to output the results to a wide range of graphical file formats, as well as to the computer screen and printer. Apart from its analytical and plotting capabilities, gnuplot has two other important advantages: it is available free-of-charge under the terms of its license (Appendix C) and versions of the software are available for use on a wide variety of computer platforms and operating systems.

### Model Parameterization

Before proceeding further, it is worth noting that many computational models require what is sometimes known as parameterization (also referred to as parametrization or parameter estimation); that is, the values of the parameters and the initial values of the variables used in the model need to be determined so that the model produces sensible results (i.e., the model output matches observations of the target system under the conditions being modeled). This process is also known as tuning or calibrating the model. Parameterization can be performed graphically (i.e., visually), statistically (e.g., using least-squares estimation, often known as regression) or analytically (e.g., by solving a set of simultaneous linear equations).

### 1.2.6 Evaluating the Model

*Verification and Validation*

Once the computational model has been implemented, the next step is to run the model to check, first, that it works and, second, that it produces acceptable results. If the result of the first of these tests is unsatisfactory, the implementation should be checked and, where appropriate, revised. The second test is addressed by a process known as model verification and validation (or V&V). The difference between *verification* and *validation* is a subtle one: verification involves an assessment that the computational model satisfies the criteria set out in the model specification (has the modeler built the model correctly?); validation is concerned with assessing whether the computational model is suitable for its intended purpose (has the modeler built the correct model?).

The obvious way to validate a model is to examine how it performs over a range of conditions compared to observations of the target system. This usually involves an evaluation of the goodness-of-fit between the modeled and observed values for a given set of conditions, typically using an index of association or correlation, such as the coefficient of determination, $R^2$, or the chi-squared value, $\chi^2$. The closer the fit, the more accurate the model.

There are several reasons why the values output from a model may differ from observations of the target system made under the same set of conditions. The reasons include incorrect assumptions made about the target system in the specification of the conceptual model, errors or inappropriate methods employed in the construction of the mathematical model, mistakes or incorrect techniques applied in the implementation of the computational model, inexact arithmetic (i.e., rounding errors) performed in the computational model, uncertainty in the data used to parameterize the model, and errors in the data used to test the model (i.e., measurement error). If the output from a model differs significantly from observations of the target system, the sources of error need to be investigated. This investigation may then lead to the model being revised and some or all of the model-development cycle (Figure 1.1) being revisited.

In some instances it may only be possible to measure (or to infer) the values of some of the variables used in the model. This situation might be described as a partial validation. In the extreme case, it may be necessary to validate a model by comparing the results that it produces with those of another model designed to represent the same system. If one compares a simple model to a more complex one, the conditions of the more complex model should be set to replicate those of the simpler one. This form of validation is generally less satisfactory, in that it only highlights differences between the two models, unless the model used as the standard has been previously validated rigorously (e.g., against observations of the target system). Sometimes, however, it is the only practical solution.

*Accuracy, Error and Precision*

It is important, in the context of model verification and validation, to be aware of the distinction between the terms accuracy, error and precision. Broadly speaking, the term accuracy refers to the fidelity with which a model represents the processes and relations of the target environmental system, but it is also used in a somewhat narrower sense to indicate the degree to which the model output conforms to the actual (or true) values for the corresponding system. Interpreted in the latter way, accuracy is the complement of error (i.e., 95% accuracy implies 5% error).

Error is often assessed by comparing the output from a model against a set of independent observations (measurements) made on the variables incorporated in the model, and is usually expressed in terms of the average (mean) difference between the observed and modeled values for those variables (Hardisty *et al.* 1993). The root mean square error (RMSE), which is given as follows,

$$\text{RMSE} = \sqrt{\frac{1}{N} \sum_{i=1}^{N} \left(x_{\text{observed}} - x_{\text{model}}\right)^2} \tag{1.3}$$

is a commonly used measure of error. Note that the value so obtained is only an estimate of the error of the model because it is based on a sample set of observations, typically acquired over a limited range of the total set of possible conditions that the model variables can take.

Precision has two related meanings: the first indicates the degree of agreement among a group of related observations or model outputs; the second refers to the units of the least significant digit of an observation or model output. Note that it is possible for the output of a model to be precise but inaccurate, or to be accurate but imprecise. The difference between error and precision is perhaps best illustrated by the following example: if the value of some property predicted by our model is $-0.4515$, while the expected value is $1905.25$, then the modeled value is very precise (it is provided to four decimal places) but very inaccurate (there is a large error).

*Sensitivity Analysis*

A further component of model evaluation is the investigation of a model's sensitivity to the values with which it is initialized. If a small change in the value of an input parameter produces a large change in the model's output, the model is said to be sensitive to that parameter and the parameter is said to have a high influence on the model (Ford 1999). Conversely, if a large change in a parameter produces a small change in the model's output, the model is said to be insensitive to the parameter and the parameter is said to have a low influence on the model.

The process of establishing the sensitivity of a model to its parameters is known as sensitivity analysis (Saltelli *et al.* 2000, Saltelli *et al.* 2004). Knowledge gained by performing sensitivity analysis on a model can help to elucidate the way in which the modeled system functions and to identify those parameters of the model whose values need to be specified most accurately. The ultimate aim of sensitivity analysis is to focus attention on the critical parts of the model and the environmental system that it purports to represent.

In practice, sensitivity analysis is usually performed by perturbing the values of the model parameters by known amounts, measuring the effect that these variations have on the model outputs. The simplest method is to vary the value of one parameter at a time, while the values of the other parameters are held constant. This approach, known as one-at-a-time (OAT) or univariate sensitivity analysis, is used to quantify the influence that each parameter exerts on the model. This method has a significant limitation, however, in that it does not account for the sensitivity of the model to two or more parameters, which may have limited influence when considered in isolation, that interact to produce major changes in the model output in combination. This problem requires the use of more sophisticated multivariate sensitivity analysis techniques, such as Monte Carlo simulation with simple random or Latin Hypercube sampling (Saltelli *et al.* 2000, Saltelli *et al.* 2004).

Figure 1.8: Photograph of Llyn Efyrnwy.

## 1.3   Llyn Efyrnwy

Many of the data sets and some of the models examined in this book relate to the area
around Llyn Efyrnwy in the Berwyn Mountains (*Mynydd y Berwyn*), Powys, Wales,
UK (52°46′58″ N, 3°30′46″ W; Ordnance Survey (OS) grid reference SJ008178; Figure 1.8).   Llyn Efyrnwy is a man-made reservoir, which was created in 1888 by the
construction of a large masonry dam, supplying water to the city of Liverpool approximately 70 miles away.   Covering an area of $8.24\,km^2$, Llyn Efyrnwy is the largest
lake in Wales. Much of the surrounding area is a dedicated wildlife reserve, which is
owned by Severn Trent Water (http://www.stwater.co.uk/) and managed by the Royal
Society for the Protection of Birds (RSPB) (http://www.rspb.org.uk/). The nature reserve covers an area of roughly 6475 hectares (16,000 acres). Various species of bird
can be seen around the lake and in the sessile oak woodlands, heather moorlands and
coniferous plantations beyond (Table 1.4). The meteorological data sets that are used
in Chapters 2 through 5 and Chapter 10 relate to the MetOffice station located close
to Llyn Efyrnwy (52°44′55″ N, 3°28′16″ W; OS grid reference SJ 008178; elevation
235 m above Ordnance Datum Newlyn (ODN)).

## 1.4   Structure and Objectives of the Book

In the chapters that follow, the reader is taken through the various stages of model
development outlined above.  One of the challenges in this context is that the reader

Table 1.4: Some of the bird species found in the area surrounding Llyn Efyrnwy.

| Habitat | Common name | Latin name |
|---|---|---|
| Lake | Common Sandpiper | *Actitis hypoleucos* |
| | Goosander | *Mergus merganser* |
| | Great Crested Grebe | *Podiceps cristatus* |
| | Peregrine | *Falco peregrinus* |
| Oak woodlands | Nuthatch | *Sitta europaea* |
| | Pied Flycatcher | *Ficedula hypoleuca* |
| | Redstart | *Phoenicurus phoenicurus* |
| | Siskin | *Carduelis spinus* |
| | Wood Warbler | *Phylloscopus sibilatrix* |
| Heather moorlands | Black Grouse | *Tetrao tetrix* |
| | Brambling | *Fringilla montifringilla* |
| | Buzzard | *Buteo buteo* |
| | Hen Harrier | *Circus cyaneus* |
| | Merlin or Pigeon Hawk | *Falco columbarius* |
| | Red Grouse | *Lagopus lagopus* |
| Coniferous plantations | Coal Tit | *Parus ater* |
| | Common or Red Crossbill | *Loxia curvirostra* |
| | Goldcrest | *Regulus regulus* |
| | Goshawk | *Accipiter gentilis* |
| | Nightjar | *Caprimulgus europaeus* |
| | Raven | *Corvus corax* |

must be introduced to three distinct themes: the mathematics underpinning individual models, the computer software or programming language in which the mathematical model is implemented, and the computer software that is used to visualize the inputs to, and outputs from, the computational model. One approach is to cover each of these topics separately, introducing them in sequence. This implies, however, that the reader must learn about a range of environmental models before finding out how these models can be implemented in computer code, or else learn how to write computer programs before being introduced to the models themselves. Either way, the reader is likely to become discouraged before he or she has fully implemented and tested a single environmental model. This book therefore adopts a different approach; one in which, as far as possible, elements of all three themes (the mathematical foundations of a model, the aspects of the computer programming language that are required to implement it, and the components of the data visualization software that are needed to explore its output) are introduced in combination. The material in each chapter therefore progresses incrementally from topics that require fairly basic data visualization and data manipulation techniques through to ones that demand somewhat more sophisticated mathematical modeling and computer programming procedures.

Chapter 2 provides an introduction to the computer software (gnuplot) used to visualize a range of environmental data sets, including the output from environmental models, examined throughout this book. The ability to present data graphically is central to the model-building process and, more generally, to understanding the operation and the behavior of environmental systems. Data visualization allows the modeler to explore the form and the strength of relationships between the system variables prior to building a model and, subsequently, to examine its output. This chapter therefore provides a tutorial on how to use gnuplot to handle a number of different types of data and to create a range of different plot styles. Other features and facilities of gnuplot are introduced, as they are required, in later chapters.

Chapter 3 provides an introduction to gawk, the scripting language that is used in this book to process a range of environmental data sets and to implement several environmental models. Rather than diving straight into the intricacies of model implementation, however, this chapter aims to familiarize the reader with the basics of gawk by showing how it can be used to process a small example data set. The data set contains measurements of precipitation made at Llyn Efyrnwy every 12 hours throughout 1998. The data are given in the format provided by the MetOffice. gawk is used to reformat the data so that they can be visualized more readily in gnuplot.

In Chapter 4, further elements of the gawk scripting language are introduced in the context of measuring and modeling wind speed, wind energy and wind power at Llyn Efyrnwy. Wind speed measurements made at hourly intervals are processed using gawk to calculate the annual mean wind speed and the relative frequency distribution of different wind speeds at this site. gnuplot is used to fit a probability density function (PDF) model of wind speed to the measured data. This model is used to determine the likelihood with which wind speeds capable of driving a small wind energy conversion system (WECS) are observed at Llyn Efyrnwy. gawk is then used to implement a model of the potential for electricity generation by wind power at Llyn Efyrnwy, based on established mathematical formulae.

The transition from measurements to models continues in Chapter 5, in which the amount of solar radiation reaching Earth's surface, known as solar irradiance, is examined. A mathematical model is constructed from equations and formulae published in the scientific literature. This model is implemented in gawk. The resulting computational model is used to predict the amount of solar radiation incident at Llyn Efyrnwy at different times of the year. The model is validated by comparing its output to hourly measurements of total solar irradiance made at the local meteorological station.

Chapter 6 continues the solar radiation theme by modeling its interaction with plant canopies on Earth's surface. Rather than returning to the extensive literature in this area for existing mathematical models, however, this chapter takes a different approach, developing a simple model from scratch. The intention is to demonstrate the process of model development, from specification of the conceptual model and its assumptions, through formulation of the mathematical model and its implementation in gawk code, to testing the model against observations of the target system. Several iterations of the model development cycle (Figure 1.1) are required before the model replicates adequately the observations.

The final model developed in Chapter 6 is reformulated a number of times in Chapter 7 to demonstrate the difference between analytical and numerical solutions to environmental models. In the process, various additional features of the gawk scripting language, such as arrays, iterative methods and control-flow constructs, are introduced.

Chapter 8 covers a range of discrete and continuous models of population growth. It provides a basic introduction to the mathematics of difference equations and differential equations, in addition to constrained (or density-dependent) and unconstrained (or density-independent) models of population growth. This chapter also shows how chaotic (i.e., unpredictable) behavior can be produced, in certain circumstances, by deterministic systems. The initial models are developed further so that they take into account competition for resources among individuals of a single species (intra-specific competition) and between individuals of two or more species (inter-specific competition), in addition to the effect of predator-prey relationships. This requires the introduction of some basic techniques for numerical integration, namely the methods of Euler and Runge-Kutta, and further aspects of control-flow structures in the gawk scripting language.

Many of the elements introduced in the first eight chapters are brought together in Chapter 9, which examines a model feedback mechanism between the biota (the living organisms) and the abiotic environment on an imaginary planet, known as Daisyworld. For instance, the Daisyworld model explores the growth of two species of daisy over time (Chapter 8), as a function of planetary temperature. The planetary temperature is, in turn, partly controlled by changes in the amount of solar radiation that is incident on the planet's surface (Chapter 5). The two species of daisy differ solely with respect to color: one is dark, the other is light, compared to the soil substrate in which they grow. Thus, the model is also concerned with the interaction between solar radiation and plant canopies, or, more specifically, the fractions of incident radiation that are reflected or absorbed by the daisies (Chapters 6 and 7). While the focus of this chapter is the operation and implications of the Daisyworld model, a number of technical aspects are also covered. These aspects include the application of user-defined functions in gawk to produce modular, and hence more manageable, computer code and a consideration of sensitivity analysis as a tool for understanding the operation of the model and for exploring its computational implementation.

In the final chapter, Chapter 10, the model of incident solar radiation at Earth's surface that was introduced in Chapter 5 is extended to account for the effects of sloping terrain. The revised model is applied to data on terrain gradient and aspect derived from a digital elevation model (DEM) covering an area immediately to the south of Llyn Efyrnwy. This necessitates an introduction to handling 2D arrays in gawk. In the second part of the chapter, the information on terrain slope is used to predict the local drainage direction (LDD) network of the study area. This network is compared to the "blue line" features (i.e., rivers and streams) extracted from the corresponding OS digital topographic map. The LDD network and "blue line" features are visualized together using gnuplot's vector plotting capabilities.

A series of additional exercises are set throughout the book, which the reader is encouraged to try. The recommended solutions to these exercises are presented in Appendix E. It is worth noting, however, that there are often a number of different ways of solving a specific problem, so that the reader may find alternative solutions to the ones suggested in Appendix E.

## 1.5  Resources on the CD-ROM

The CD-ROM that accompanies this book contains copies of the computer software (i.e., gawk and gnuplot), data sets, gawk programs and gnuplot scripts used in the following chapters. Copies of gawk and gnuplot can be found in the utils directory on the CD-ROM. Instructions on how to install this software on desktop computers running a version of either the GNU/Linux or the Microsoft Windows operating systems are given in Appendix A, which also provides details on where to find and how to download copies of the latest versions of these packages. Note that both gawk and gnuplot are provided free-of-charge and can therefore be installed on as many computers as required within the fairly broad constraints of their respective licenses (Appendix B and Appendix C, respectively).

As previously noted, the gawk scripting language is used throughout this book to process a range of environmental data sets and to develop computer implementations of various environmental models. Both the data sets and the gawk programs are stored in ASCII (American Standard Code for Information Interchange) text files. These files can therefore be created and edited using, among other things, basic text editor software; however, it is recommended that standard word-processing packages not be used for this purpose. The resources associated with each chapter (i.e., data sets, gawk programs and gnuplot scripts) are stored in separate directories on the CD-ROM, labeled chapter2, chapter3, chapter4 and so on. Updates and errata will be made available via http://stress.swan.ac.uk/~mbarnsle/envmod/.

Files containing material suitable for lecture presentations are also included on the CD-ROM. These files can be found in the present sub-directory and are provided in Portable Document Format (PDF), which can be displayed from a laptop or desktop computer, via a data projector, using suitable software, such as Adobe Acrobat Reader. This software can be downloaded free-of-charge from the following web site: http://www.adobe.com/products/acrobat/readstep2.html. Once a file has been loaded into Acrobat Reader, the presentation can be set to full screen mode by typing Ctrl + L .

## 1.6 Typographical Conventions

The following typographical conventions are used throughout the book. When the reader is required to type an instruction on the command line (i.e., in a GNU/Linux shell or at the Microsoft Windows command prompt), the instruction is presented in a rectangular box as shown below.

```
gawk -f myprog.awk mydata.dat                                          1
```

The number on the right-hand side of the box is used to refer to different instructions within each chapter. Sometimes the command line will be too long to fit on a single line of the book. In this case, the ⇨ symbol is used to indicate that the command line continues on subsequent lines, as shown below.

```
gawk -f myprog.awk -v variable1=3.1415 -v variable2=0.125 ⇨           2
    ⇨-v variable3=23.67 mydata.dat > result.dat
```

Note that this command should be entered on one line and that the blank space before the first of the ⇨ symbols is significant and should be respected. The command line shown above redirects the output from the program into a data file (result.dat in this instance). The contents of a data file are presented in a box with rounded corners, such as that illustrated below.

```
1  199801010900      1.80     -0.30
2  199801012100      7.70      1.60
3  199801020900      6.80      3.90
4  199801022100      6.50      3.80
5  199801030900      9.70      4.80
```

The individual lines of data (i.e., records) in the file may or may not be numbered. Where they are numbered, the line numbers appear of the right-hand side of the page.

Instructions to be typed in gnuplot are shown in a rectangular box with a gray background, as illustrated below.

```
reset                                                                  1
set style data lines                                                   2
plot 'mydata.dat' using 1:2                                            3
```

The line numbers on the right-hand side of this box are used to refer to individual gnuplot instructions. Finally, the gawk code for the computational models developed in this book is presented in the form of listings between horizontal lines, such as that shown in Program 1.1. Once again, for the sake of convenience, line numbers

Program 1.1: introduction.awk

```
(NR >1  && $10 != -999){                                             1
  wind_speed=$10;                                                    2
  sum_speed=sum_speed+wind_speed;                                    3
  num_obs=num_obs+1;                                                 4
}                                                                    5
                                                                     6
END{                                                                 7
  mean_wind_speed=0.515*sum_speed/num_obs;                           8
  printf("Mean wind speed=%3.1f m/s\n", mean_wind_speed);            9
}                                                                   10
```

are appended to the right-hand side of the listing; these numbers are not part of the computer code and should not be included when entering the program.

# Chapter 2

# Visualizing Environmental Data

**Topics**

- Scientific data visualization

**Methods and techniques**

- Generating 2D plots in gnuplot

- Handling time-series data in gnuplot

- Visualizing 3D data sets in gnuplot

- Exporting graphics from gnuplot

- Saving gnuplot commands in script files

## 2.1  INTRODUCTION

The ability to present data graphically is central to the model-building process and, more generally, to understanding the operation and the behavior of environmental systems. When designing a model, for instance, graphical representations of data can help the modeler identify the major trends and anomalies that are characteristic of a particular environmental system. They can also reveal the degree of influence that specific components of a system exert on the state of the system as a whole. This information can, in turn, be used to decide which parameters and variables should be built into a model (i.e., those that have a major influence on the system state) and which ones may safely be omitted (i.e., those that have a negligible impact). Simi-

Table 2.1: Examples of "open source" software for scientific data visualization.

| Software | Description and URL |
|----------|---------------------|
| DX | IBM Open Visualization Data Explorer<br>URL: http://www.opendx.org/ |
| GMT | Generic Mapping Tools<br>URL: http://gmt.soest.hawaii.edu/ |
| gnuplot | A scientific plotting package<br>URL: http://www.gnuplot.info/ |
| Grace | WYSIWYG 2D plotting tool<br>URL: http://plasma-gate.weizmann.ac.il/Grace |
| Gri | Interpreted language for scientific graphics<br>URL: http://gri.sourceforge.net/ |
| XGobi | Data visualization system for multidimensional data<br>URL: http://www.research.att.com/areas/stat/xgobi |

larly, graphical representations of data can help the modeler understand the ways in which individual elements of the system interact and, hence, elucidate the environmental processes that govern their interaction. Later, during the model construction and testing phases, the validity and the accuracy of a model can be evaluated by visually comparing the data that it generates with *in situ* measurements of the corresponding environmental system. Where this procedure highlights errors in the output from the model, particularly systematic ones, it may initiate a further cycle of analysis, explanation and model enhancement. The ultimate objective of this process is to reduce uncertainty in the output from the model and, through this, to understand better the environmental system that the model purports to represent.

Environmental data sets typically contain very large numbers of values. They are also frequently multidimensional, in the sense that the data often represent measurements made on more than one variable (e.g., incident solar radiation, precipitation, temperature and wind speed) at a particular location and point in time. When the data describe measurements made at more than one spatial location and at different points in time, the dimensionality of the data increase further still. These characteristics impose special demands on the computer software used to present environmental data graphically, some of which (e.g., the ability to handle 3D surfaces, contours and vectors) are not well supported by standard business graphics software, such as that associated with many spreadsheet packages. Consequently, a range of specialist software has been developed to visualize scientific data, including those relating to environmental systems (Table 2.1).

The software of choice here is gnuplot. As is the case with many "open source" software packages, gnuplot is available free-of-charge for use on a variety of computer platforms, including GNU/Linux, MacOS X and Microsoft Windows (Williams and Kelly 1998, DiBona *et al.* 1999, Raymond 1999). gnuplot is used throughout this book to generate everything from simple *x-y* scatter plots to time-series diagrams, contour maps and vector-flow plots, and to visualize 3D surfaces. The purpose of

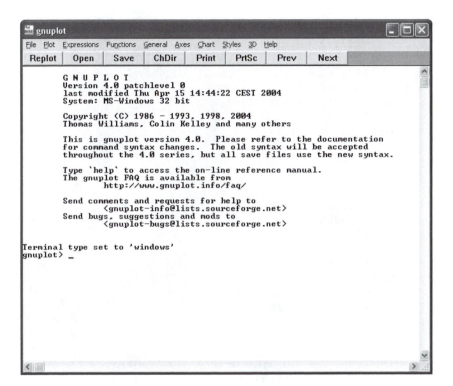

Figure 2.1: GUI for the version of gnuplot for Microsoft Windows.

this chapter is, therefore, to introduce this software through a structured tutorial. This is intended to demonstrate some of gnuplot's graphical and analytical capabilities. It also illustrates a number of scientific data visualization techniques that can be applied to environmental data sets (Tufte 2001). The reader is strongly encouraged to run the examples presented in this chapter on his or her own computer, and to experiment more freely with this software. Further aspects of gnuplot's plotting capabilities are examined in subsequent chapters.

Two versions of gnuplot are included on the CD-ROM that accompanies this book (henceforth referred to simply as "the CD-ROM"): one is for use with GNU/Linux; the other is for use with Microsoft Windows. They are virtually identical in terms of functionality, although the version for Microsoft Windows has a built-in graphical user interface (GUI) that allows the user to interact with the software by means of point-and-click techniques (Figure 2.1). In this book, however, the instructions used to generate plots are entered via gnuplot's command-line interface; that is, they are typed-in beside the gnuplot> prompt. One reason for doing this is to standardize the presentation of the GNU/Linux and Microsoft Windows versions of the software. More importantly, though, the command-line approach confers significant advantages in the long run, including the ability to maintain a record of how individual data sets were plotted and the potential to automate the visualization of several related data sets. Moreover, the learning curve associated with the command-line method is about

```
  1 19980101T0900     1.80      -0.30                        1
  2 19980101T2100     7.70       1.60                        2
  3 19980102T0900     6.80       3.90                        3
  4 19980102T2100     6.50       3.80                        4
  5 19980103T0900     9.70       4.80                        5
  6 19980103T2100     6.50       0.20                        6
  7 19980104T0900     4.20       2.30                        7
  8 19980104T2100     4.80       0.20                        8
  9 19980105T0900     3.20       0.60                        9
 10 19980105T2100     2.30      -0.20                       10

706 19981227T0900     5.50       1.80                      706
707 19981227T2100     5.70       2.70                      707
708 19981228T0900     3.90       0.30                      708
709 19981228T2100     3.80       0.30                      709
710 19981229T0900     3.00      -0.30                      710
711 19981229T2100     6.20       3.00                      711
712 19981230T0900     9.00       6.10                      712
713 19981230T2100     7.40       4.10                      713
714 19981231T0900     5.40       3.70                      714
715 19981231T2100     7.00       4.80                      715
```

Figure 2.2: Extract from the data on air temperature at Llyn Efyrnwy throughout 1998 (le98temp.dat; first and last 10 lines only).

the same as that of the point-and-click approach, largely because a small set of core commands is used to generate many different types of plot. Detailed instructions on how to install and run gnuplot are given in Appendix A.

## 2.2   CREATING 2D PLOTS

### 2.2.1   Creating a Simple *x-y* Plot

The file le98temp.dat, which can be found on the CD-ROM, contains information on air temperatures at Llyn Efyrnwy recorded by an automatic weather station (AWS) operated by the UK MetOffice. The AWS records, among other things, the minimum and maximum air temperature at this site every 12 hours. The file le98temp.dat is derived from the AWS data set for 1998 (Figure 2.2). It consists of four columns of data: the first contains the sequential identifying number of the measurement, the second indicates the date and time at which that measurement was made, and the third and fourth refer to the maximum and minimum air temperature (°C), respectively, recorded over the preceding 12 hours. Note that the date and time information, presented in the second column, is given in the format *YYYYMMDD*T*hhmm*, where *YYYY* denotes the calendar year (e.g., 1998), *MM* is the month of the year (i.e., 01 to 12, for January through to December), *DD* is the day of the month (i.e., 01 to 31), *hh* is the number of complete hours that have passed since midnight (i.e., 00 to 11)

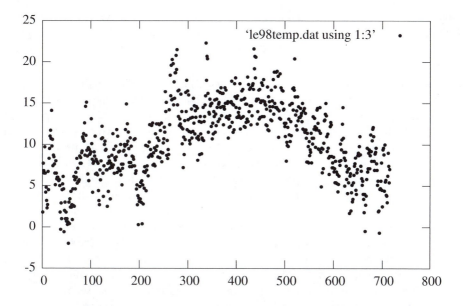

Figure 2.3: Simple *x-y* plot of the maximum air temperature measured at Llyn Efyrnwy every 12 hours throughout 1998.

and *mm* is the number of complete minutes that have elapsed since the start of the hour (i.e., 00 to 59). The letter T, known as the time designator, is used to separate the date and time components. This format conforms to the "basic" version of the international standard for date and time notation (International Organization for Standardization (ISO) 8610:2004; Appendix D.1).

Suppose that one wished to examine how the maximum air temperature at Llyn Efyrnwy varied throughout 1998. A simple way to do this would be to present the data in the form of a 2D scatter-plot in which maximum air temperature (*y*-axis) is plotted against measurement number (*x*-axis). This can be achieved in gnuplot by typing the following instruction next to the gnuplot> command-line prompt and then pressing the ⟵ or Enter key to submit the instruction (i.e., to run the command).

```
plot 'le98temp.dat' using 1:3                                        1
```

This instruction produces a plot resembling the one presented in Figure 2.3, which should appear in a separate pop-up window on the computer screen.

In the example given above, the plot command generates a simple *x-y* plot of the data contained in the named file. Note that the file name must be surrounded by single quotation marks. If the file is not located in the working directory, its full or relative path name must be given (e.g., /mnt/cdrom/chapter2/le98temp.dat under GNU/Linux or d:\chapter2\le98temp.dat in Microsoft Windows; Appendix A). The rest of the plot command (i.e., using 1:3) informs gnuplot which columns of the data file should be used to generate the plot. In this example, the data in columns 1 (measurement number) and 3 (maximum air temperature) of the file are plotted along the *x*-axis and *y*-axis, respectively. The general syntax is using x:y. Thus, the same data could be plotted with the measurement number on the *y*-axis and the maximum air temperature on the *x*-axis, if so desired, by altering the last part of the command line so that it reads using 3:1.

It is possible to obtain the same result using point-and-click methods in the version of gnuplot for Microsoft Windows, although the command-line approach is preferred here. As was noted earlier, the command-line approach confers significant benefits in the long run. For example, it is possible to store a sequence of gnuplot commands in an ASCII text file, sometimes known as a "script" or a "macro", which can be run in a "batch" whenever the user requires (see Section 2.7). There are at least two reasons why this is good practice. First, it ensures that a permanent record is kept of the commands employed to generate a plot and the order in which they should be applied. This can be used, for instance, to reproduce a plot quickly should it be accidentally deleted, or if disk space (or bandwidth) are limited and it is not possible to store (or transmit) the resulting graphics file. Second, it introduces the possibility of automating the production of standard plots from a number of related data sets. Thus, the same commands can be applied, subject perhaps to only minor modification, to a collection of data files. As an example, one might wish to generate separate plots of the maximum air temperatures recorded at Llyn Efyrnwy over several years, where the data for each year are contained in separate files.

It is acknowledged, however, that some readers may have little or no experience with command-line interfaces because the point-and-click approach is now standard in many elements of human–computer interaction. Apart from the need to remember the syntax of specific commands, one aspect of the command-line approach that new users generally find challenging is the requirement, on occasion, to type in relatively long commands. This process can be tiresome, especially if a simple error is made when entering a command, so that the whole command line must be typed in again. The impact of this problem can be reduced by storing the commands in script files, as outlined above, but gnuplot also helps in this respect by maintaining a list of the most recently issued commands. This list is known as the command-line history. It is possible to step back through the command-line history by pressing the ⬆ key in the gnuplot command-line window; similarly, the ⬇ key can be used to step forward through the list. Thus, a previously entered command can be issued again without modification by pressing the ⬅ or Enter key when the appropriate command appears on the screen. It is also possible to edit previously submitted commands, using the ⬅ and ➡ keys to move left or right, respectively, along the command line and by deleting or inserting text as required.

**Exercise 2.1**: Use the ⬆ key to recall the last plot command. Edit this command line so that it reads plot 'le98temp.dat' using 3:1 and then press the ⬅ or Enter key to submit the revised command. How does the resulting plot differ from the output presented in Figure 2.3? Use the ⬆ key again to recall and run the initial plot command once more.

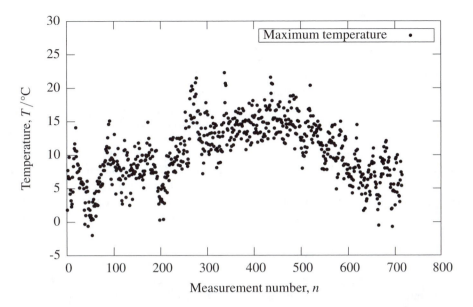

Figure 2.4: Simple *x-y* plot of the maximum air temperature measured at Llyn Efyrnwy every 12 hours throughout 1998, with labeled axes and a boxed key.

### 2.2.2 Labeling the Axes of a Plot

The plot presented in Figure 2.3 is rather bald. It portrays the seasonal variation in maximum air temperature at Llyn Efyrnwy, with the lowest values toward the start and end of the measurement sequence (i.e., January and December) and the highest values toward the middle (i.e., July and August), which is typical of this mid-latitude northern hemisphere site. Neither of the axes has been labeled, so that the contents of the plot can only be guessed at by the casual reader. Moreover, the automatically generated key, which identifies the source of the data, employs the arguments of the plot command to label the data series, instead of using something more instructive. These deficiencies can be overcome by entering a few extra instructions, listed below, via the gnuplot command line, which produce the plot shown in Figure 2.4.

```
set xlabel "Measurement number, n"                              2
set ylabel "Temperature, T/degree Celsius"                      3
set key top right box                                           4
set yrange [-5:30]                                              5
plot 'le98temp.dat' using 1:3 title "Maximum temperature"       6
```

The set command used on lines 2 through 5, above, assigns values to various parameters that control the appearance of plots in gnuplot. In this particular example, lines 2 and 3 specify the labels that should be placed beside the *x*- and *y*-axes, respectively. Note that the text used to define these labels must be enclosed by double quotation marks, so that gnuplot knows where the labels begin and end. Line 4 indicates that a key should be placed in the top right-hand corner of the plot, and that

Table 2.2: Selected options of the `plot` command in gnuplot.

| Option | Abbreviation | Interpretation |
| --- | --- | --- |
| using | u | Use the data in these columns to create the plot. |
| with | w | Plot the data in the following style (see Table 2.3). |
| title | t | Give this data series the following title in the key. |

this should be surrounded by a box. Line 5 sets the range of values that should be displayed on the *y*-axis, here using a lower limit of $-5\,°C$ and an upper limit of $30\,°C$. These changes do not take effect, however, until the data have been plotted once more (line 6). The `plot` command issued on line 6 also gives an explicit title to the data series (`title "Maximum temperature"`); this information is used to produce the key.

### 2.2.3   Plotting Multiple Data Series

As has already been noted, the file `le98temp.dat` contains information on both the maximum and minimum air temperatures recorded at Llyn Efyrnwy. Typically, one might wish to plot both series in the same figure, perhaps to gauge the diurnal temperature range at this site. This can be achieved by modifying the previous `plot` command (line 6) so that it reads as follows:

```
plot 'le98temp.dat' u 1:3 t "Maximum temperature", \     7
     'le98temp.dat' u 1:4 t "Minimum temperature"         8
```

Note that the \ ("backslash") character on the right-hand side of line 7, above, which is known as the line continuation symbol, is used to split a single long gnuplot command over two or more lines so that it is easier to enter and read. Note that the line continuation symbol must be the last character on the line. In this example, the line continuation symbol is used to continue the `plot` command from line 7 to line 8. Thus, lines 7 and 8 represent a single gnuplot command. When entered manually on the gnuplot command line, the line continuation symbol causes the command prompt to change from gnuplot> on the first line to > on the continuation lines. Also, note that the keywords `using` and `title` have been abbreviated to u and t, respectively, on lines 7 and 8, to reduce the amount of typing required. gnuplot allows many of its basic keywords to be abbreviated in this way (see, for example, Table 2.2 and Table 2.3). The output of the revised `plot` command is presented in Figure 2.5.

The syntax of the command on lines 7 and 8 is relatively straightforward. The `plot` keyword is followed by a comma-separated list of the file (or files) containing the data series that are to be plotted, together with instructions on how each series should be presented. In this example, the first series is contained in column 3 of the file `le98temp.dat` and is plotted against the data in column 1 of that file (i.e., maximum temperature on the *y*-axis versus measurement number on the *x*-axis). The second series is stored in column 4 of the same file and is also plotted against the data in column 1 of that file (i.e., minimum temperature on the *y*-axis versus measurement number on the *x*-axis). Different titles are used to label each data series in the key.

Table 2.3: Selected data style options of the `plot` command in gnuplot.

| Option | Abbreviation | Interpretation |
|---|---|---|
| `lines` | l | Connect consecutive data points $(x, y)$ with straight line segments. Requires two columns of data. |
| `points` | p | Plot a symbol at each data point $(x, y)$. Requires two columns of data. |
| `linespoints` | lp | Plot a symbol at each data point $(x, y)$ and connect consecutive data points with straight line segments. Requires two columns of data. |
| `impulses` | i | Draw a vertical line from the $x$-axis ($y = 0$) to each data point $(x, y)$. Requires two columns of data. |
| `dots` | d | Plot a small dot at each data point $(x, y)$. Requires two columns of data. |
| `boxes` | - | Draw a vertical bar (box) centered on the $x$-value from the $x$-axis ($y = 0$) to each data point $(x, y)$. Requires two columns of data. |
| `xerrorbars` | xe | Plot a dot at each data point $(x, y)$ and a horizontal error bar from $(x - \Delta x, y)$ to $(x + \Delta x, y)$, or from $(x_{\min}, y)$ to $(x_{\max}, y)$. Requires three $(x, y, \Delta x)$ or four $(x, y, x_{\min}, x_{\max})$ columns of data. |
| `yerrorbars` | ye | As above, but with vertical error bars. Requires three $(x, y, \Delta y)$ or four $(x, y, y_{\min}, y_{\max})$ columns of data. |
| `xyerrorbars` | xye | As above, but with both horizontal and vertical error bars. Requires four $(x, y, \Delta x, \Delta y)$ or six $(x, y, x_{\min}, x_{\max}, y_{\min}, y_{\max})$ columns of data. |
| `xerrorlines` | xerrorl | As `xerrorbars`, but with consecutive data points $(x, y)$ connected by straight line segments. Requires three $(x, y, \Delta x)$ or four $(x, y, x_{\min}, x_{\max})$ columns of data. |
| `filledcurves` | filledc | Plots either the current curve closed and filled, or the region between the current curve and a given axis, horizontal or vertical line, or a point, filled with the current drawing color. |
| `vectors` | v | Draws vectors from $(x, y)$ to $(x + \Delta x, y + \Delta y)$. Requires four columns of data $(x, y, \Delta x, \Delta y)$. |

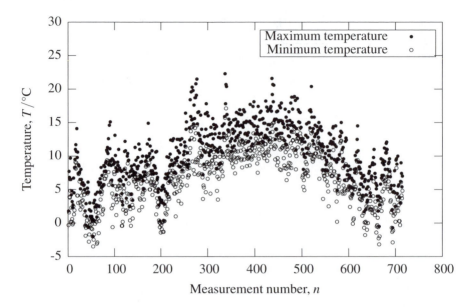

Figure 2.5: Simple *x-y* plot of the maximum and minimum air temperatures measured at Llyn
Efyrnwy every 12 hours throughout 1998.

### 2.2.4   Plotting with Different Data Styles

In Figure 2.5, individual measurements from both data series are represented by point
symbols, although gnuplot offers a large number of other data styles (Table 2.3). The
default data style can be changed using the `set style data` command, as outlined
below:

```
set style data lines                                                   9
set xrange [200:300]                                                  10
replot                                                                11
```

Thus, line 9 sets the default data style to be `lines`, so that consecutive data points in
each series are connected by straight line segments. The command on line 10 is used
to restrict the range of values displayed along the *x*-axis to measurement numbers 200
to 300, which correspond to data acquired between March 12, 1998 and July 1, 1998.
This instruction demonstrates how gnuplot can be used to visualize a chosen subset
of a data file, allowing the user to examine those data in greater detail. These changes
do not take effect, however, until the data have been plotted once more, which is
performed on line 11. The `replot` command used here submits the preceding `plot`
command (lines 7 and 8 in this example) again, taking into account any changes that
have been made to the axes labels, plotting style, and such, since that command was
last issued. The result of these commands is shown in Figure 2.6.

   The style of each data series can be controlled separately using the keyword `with`,
which abbreviates to `w` (Table 2.2). For instance, the following gnuplot command uses
a combination of straight lines and point symbols (`linespoints` or `lp`) to represent

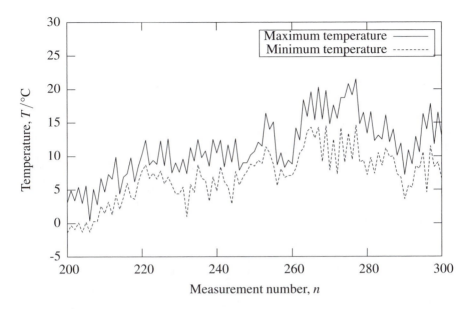

Figure 2.6: Plot of the maximum and minimum air temperatures measured at Llyn Efyrnwy
every 12 hours between March 12, 1998 and July 1, 1998, illustrating the `lines`
data style.

the data from the first series (i.e., maximum temperature) and vertical bars (`boxes`) to
represent the data from the second series (i.e., minimum temperature; see Figure 2.7).

```
plot 'le98temp.dat' u 1:3 t "Maximum temperature" w lp,\        12
     'le98temp.dat' u 1:4 t "Minimum temperature" w boxes      13
```

The second of these two data styles is probably not the most appropriate choice in
these circumstances, but it helps to illustrate the different styles that are available.

## 2.3   PLOTTING TIME-SERIES DATA

Although the file `le98temp.dat` contains time-series data, the values have not been
treated explicitly as such thus far. Instead, the preceding examples have plotted the
maximum and minimum air temperature data against the sequential measurement
number. This section therefore investigates how gnuplot can be used to visualize
explicit time-series data. Only a few extra commands are involved. These commands
indicate which axis should be employed to represent the time dimension, how the date
and time information is formatted (so that the input data can be correctly interpreted),
and how the tic-marks along the time axis should be labeled. An example of each of
these commands is shown on lines 14 to 20, below:

```
set xdata time                      14
set timefmt "%Y%m%dT%H%M"           15
set format x "%d/%m"                16
```

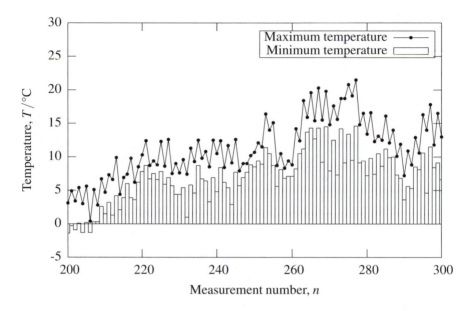

Figure 2.7: Plot of the maximum and minimum air temperatures measured at Llyn Efyrnwy
         every 12 hours between March 12, 1998 and July 1, 1998, illustrating the
         linespoints and boxes data styles.

```
set  xlabel  "Day/Month  (1998)"                                          17
set  xrange  ["19980401T0000":"19980701T0000"]                           18
plot  'le98temp.dat'  u  2:3  t  "Maximum  temperature"  w  lp, \        19
      'le98temp.dat'  u  2:4  t  "Minimum  temperature"  w  lp          20
```

Line 14, for instance, instructs gnuplot to treat the data plotted on the $x$-axis as
date and time values. Line 15 specifies the format in which these values are stored
(Table 2.4). This is given by the format string, "%Y%m%dT%H%M", which indicates that
the date and time values comprise a four-digit number denoting the year (%Y), a two-
digit number representing the month (%m), a two-digit number for the day of month
(%d), the letter T, which is used to separate the date and time components, a two-digit
number for the hour of day (%H) and a two-digit number for the minutes (%M). So,
for example, 9 am on January 5, 1998 is represented as 19980105T0900. This format
corresponds to the "basic" version of the ISO 8610:2004 notation for times and dates
(Appendix D.1) and, hence, the structure of the data in column two of temp98le.dat.

Line 16 instructs gnuplot to use a combination of the day of month (%d) and the
month of year (%m), separated by a forward-slash (or "solidus") symbol (/), to label
the tic-marks along the $x$-axis. Line 17 alters the text string used to label this axis.
Line 18 specifies the range of values to plot along the $x$-axis, in this case starting at
"19980401T0000" (i.e., midnight on March 1, 1998) and ending at "19980701T0000"
(i.e., midnight on July 1, 1998). Finally, the plot command, which is split over two
lines (19 and 20) using the line continuation symbol (\), displays the data using the
linespoints (lp) style for both time series (Figure 2.8).

Table 2.4: Selected time and date format specifiers in gnuplot.

| Specifier | Interpretation |
|---|---|
| %a | Abbreviated name of the day of the week (Sun, Mon, … , Sat) |
| %A | Full name of the day of the week (Sunday, Monday, … , Saturday) |
| %b | Abbreviated name of the month (Jan, Feb, … , Dec) |
| %B | Full name of the month (January, February, … , December) |
| %d | Day of the month (1–31) |
| %H | Hour (00–23) |
| %j | Day of the year (1–366) |
| %m | Number of the month (01–12) |
| %M | Minute (00–59) |
| %S | Second (00–59) |
| %y | Year (two digits; 00–99) |
| %Y | Year (four digits; e.g., 2006) |

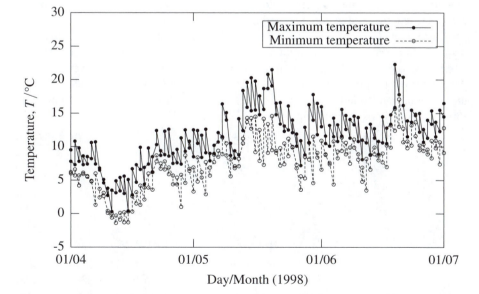

Figure 2.8: Time-series plot of the maximum and minimum air temperatures measured at Llyn Efyrnwy every 12 hours between March 12, 1998 and July 1, 1998.

```
 1  19980101T0900      0.00                                              1
 2  19980101T2100     14.60                                              2
 3  19980102T0900      5.20                                              3
 4  19980102T2100      1.60                                              4
 5  19980103T0900     35.80                                              5
 6  19980103T2100      5.60                                              6
 7  19980104T0900      6.60                                              7
 8  19980104T2100     20.80                                              8
 9  19980105T0900      4.40                                              9
10  19980105T2100      3.00                                             10
11  19980106T0900      4.00                                             11
13  19980107T0900      2.40                                             12
```

Figure 2.9: Extract from the data on precipitation at Llyn Efyrnwy during 1998 (le98rain.dat; first 12 lines only).

### 2.3.1  Plotting Multiple Time-Series

It is possible to display two or more time-series, drawn from different data files, in a single plot. Suppose, for example, that one wished to present information on both air temperature and precipitation at Llyn Efyrnwy. The latter is contained in a file called le98rain.dat (Figure 2.9), which can be found on the CD-ROM. The general structure of this file is similar to that of le98temp.dat. More specifically, it consists of three columns of data denoting the sequential measurement number, the date and time at which the measurements were made (every 12 hours) and the total precipitation (in mm) that accumulated over the preceding 12 hours.

Since temperature and precipitation are measured in different units (°C and mm, respectively), and may also exhibit a very different range of values, it is not always possible to plot them effectively on the same scale on the $y$-axis. To overcome this problem, gnuplot provides a second $y$-axis, known as the $y2$-axis, which is drawn up the right-hand side of the plot (Figure 2.10). The following gnuplot commands illustrate how this facility is employed.

```
set y2label "Precipitation, P/mm"                                       21
set ytics nomirror                                                      22
set y2tics nomirror                                                     23
set xlabel "Month (1998)"                                               24
set format x "%b"                                                       25
set xrange ["19980101T0000":"19990101T0000"]                            26
plot 'le98temp.dat' u 2:3 t "Maximum temperature" w lp, \               27
     'le98rain.dat' u 2:3 axes x1y2 t "Precipitation" w i               28
```

Line 21 sets the label to be drawn along the $y2$-axis. Line 22 instructs gnuplot that the tic-marks drawn on the $y$-axis (the left-hand side of the plot) should not be mirrored on the $y2$-axis (the right-hand side of the plot), and *vice versa* on line 23. Thus, a separate plotting scale is used for each of the two $y$-axes. Line 24 resets the label to be drawn along the $x$-axis. Line 25 instructs gnuplot to use the three-letter abbreviation for the name of the month (i.e., Jan, Feb, ... , Dec) to label the tic-marks

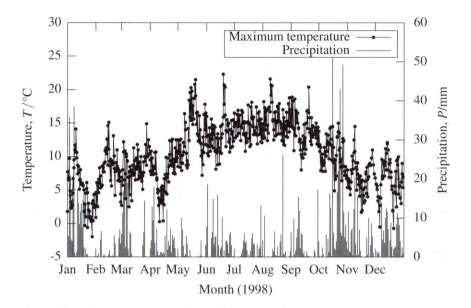

Figure 2.10: A plot of two time-series (maximum air temperature and precipitation at Llyn Efyrnwy) contained in separate data files.

along the *x*-axis (see Table 2.4). Line 26 specifies the range of values to plot along the *x*-axis, in this case starting at `"19980101T0000"` (i.e., midnight on January 1, 1998) and ending at `"19990101T0000"` (i.e., midnight on January 1, 1999). Lines 27 and 28 plot the two data series. Note that the data are taken from separate files (`le98temp.dat` and `le98rain.dat`) and that the temperature data are plotted with the `linespoints` (`w lp`) style, while the precipitation data are plotted with the `impulses` (`w i`) style. The latter draws vertical bars rising from the *x*-axis to each data point. gnuplot is given an explicit instruction (line 28) to plot the second data series on the first *x*-axis (`x1`; i.e., the one drawn along the base of the plot) and the second *y*-axis (`y2`; i.e., the one drawn up the right-hand side of the plot), `axes x1y2`. Unless specified in this way, gnuplot assumes that all data are plotted using the `x1y1` (bottom and left-hand side) axes. For completeness sake it is worth noting that the *x*2-axis refers to the axis along the top of the plot.

### 2.3.2 Further Control over Plotting Styles

The preceding sections provide a brief introduction to a selection of gnuplot's 2D plotting capabilities, including its ability to handle time-series data. Although the possibilities are not explored further here, it is worth noting that gnuplot also offers considerable control over both line and point styles, including their size (for points), width (for lines) and color (for both lines and points). Some of the command-line options that control these properties are listed in Table 2.5. The nature of the line and point styles available depends on the graphical device, known as a "terminal" in gnuplot parlance, to which the plot is sent. Thus, the default point and line styles

Table 2.5: Selected command-line options to control the appearance of data series in gnuplot.

| Option | Abbreviation | Interpretation |
|--------|--------------|----------------|
| pointtype | pt | Symbol type used to denote data points. |
| pointsize | ps | Relative size of point symbol. |
| linetype | lt | Line type used to connect consecutive data points. |
| linewidth | lw | Relative line width. |

differ between the versions of gnuplot for Microsoft Windows and GNU/Linux. This issue is revisited later in this chapter (see Table 2.6 in Section 2.5). For the moment it is sufficient to note that the available styles can be inspected using the test command.

```
test                                                                    29
```

Figure 2.11, for example, was generated using the test command and gnuplot's epslatex "terminal". The epslatex terminal produces a special type of Encapsulated PostScript (EPS) file, and is used to generate most of the figures presented in this book. In the epslatex terminal, line type three (lt 3) corresponds to a fine dashed line and point type seven (pt 7) corresponds to a filled black circle (Figure 2.11). A significant advantage of the epslatex terminal is that it allows various mathematical symbols (e.g., $\alpha$, $\beta$, $\gamma$, $W{\cdot}m^{-2}$ and °C), as well as different typographical fonts, to be used in the resulting plot. Similar capabilities are also available in the jpeg, pdf, png, postscript, svg and x11 (i.e., GNU/Linux) terminal types (see Table 2.6 in Section 2.5), but not in the default terminal for Microsoft Windows (windows). In the latter case, the required symbols must be emulated; for instance, $W{\cdot}m^{-2}$ can be represented as W.m^{-2}, °C as degree Celsius, $\rho$ as rho, and so on. The reader is referred to the file visual.gp on the CD-ROM to inspect the commands used to generate the figures presented in this chapter.

## 2.4    PLOTTING IN THREE DIMENSIONS

It was noted earlier that one of the characteristic features of environmental data sets is that they tend to be multidimensional, often highly so; that is, they frequently consist of co-located measurements made simultaneously (or, at least, quasi-simultaneously) on several different properties (e.g., temperature, precipitation and wind speed) of a particular environmental system. It is possible, of course, to explore the inter-relationships between these properties by plotting them in pairs using, for example, simple *x-y* scatter diagrams. This is not a particularly efficient or effective approach, though, because it may require the interpretation of tens or even hundreds of such plots, depending on the number of independent variables (i.e., dimensions) involved. Consequently, various alternative methods have been developed, including Chernoff faces (Chernoff 1973), star glyphs (Fienberg 1979) and parallel coordinates (Inselberg 1985), each of which attempts to represent multidimensional data in a 2D plot.

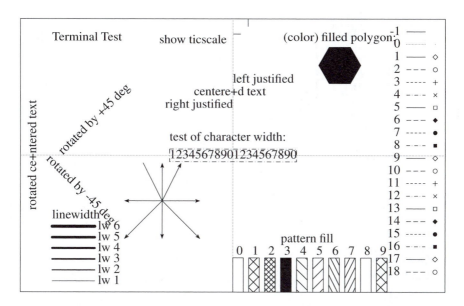

Figure 2.11: Output from gnuplot's `test` command for the `epslatex` terminal type.

Here, however, attention is focused on the visualization of 3D data sets, especially those in which two of the three dimensions describe the spatial location of the measurement (e.g., the Easting and Northing, or latitude and longitude, derived from a map) and the third relates to the value of an environmental property recorded at that location. Some of the techniques that are examined can also be applied to aspatial data sets in which the dimensions are defined by co-located or contemporaneous measurements of three different environmental properties.

### 2.4.1 Description of the Digital Elevation Data Set

The data used in this section are contained in the file `efyrnwy.dem` (Figure 2.12), which can be found on the CD-ROM. These data are derived from a single tile of the UK Ordnance Survey Land-Form PROFILE® digital elevation model (DEM), and are reproduced here by kind permission of Ordnance Survey. The original data comprise a grid of terrain elevation values posted at 10 m spatial intervals, covering a 5 km × 5 km area southwest of Llyn Efyrnwy between Eastings 295000 and 300000 and Northings 315000 and 320000 in the UK National Grid coordinate system. These data have been re-sampled onto an array of 51 × 51 elevation values, with a 100 m spatial posting, to reduce the volume of data that has to be handled here. The resulting file, `efyrnwy.dem`, consists of three columns of data that describe, respectively, the Easting, Northing and elevation of the re-sampled data points (Figure 2.12). The Eastings and Northings are measured in meters relative to a fixed reference point in the UK National Grid, known as the false origin; the elevation values are measured in meters relative to the datum at Newlyn, Cornwall, known as the Ordnance Datum Newlyn (ODN).

```
295000   315000    407                                                    1
295000   315100    410                                                    2
295000   315200    419                                                    3
295000   315300    422                                                    4
295000   315400    421                                                    5
295000   315500    415                                                    6
295000   315600    398                                                    7
295000   315700    385                                                    8
295000   315800    367                                                    9
295000   315900    390                                                    10

300000   319100    347                                                    2592
300000   319200    348                                                    2593
300000   319300    363                                                    2594
300000   319400    381                                                    2595
300000   319500    391                                                    2596
300000   319600    393                                                    2597
300000   319700    382                                                    2598
300000   319800    364                                                    2599
300000   319900    351                                                    2600
300000   320000    353                                                    2601
```

Figure 2.12: First 10 lines and last 10 lines of the Llyn Efyrnwy DEM data file, `efyrnwy.dem`
(© Crown Copyright License Number NC/03/11298).

## 2.4.2   Visualizing 3D Data in gnuplot

gnuplot provides a command, `splot`, which can be used to plot 3D data, such as those contained in `efyrnwy.dem`. The syntax of the `splot` command is very similar to that of its 2D counterpart, `plot`. The `splot` keyword is followed by the name of the data file, the columns of data in this file that should be plotted on the $x$-, $y$- and $z$-axes, respectively, and any instructions relating to the way in which the data should be presented. For example, Figure 2.13 was generated as follows:

```
reset                                                                    30
unset key                                                                31
set style data points                                                    32
set xlabel "Easting/m"                                                   33
set ylabel "Northing/m"                                                  34
set zlabel "Elevation/m"                                                 35
set xtics 295000, 1000, 300000                                          36
set ytics 315000, 1000, 320000                                          37
set ztics 250, 100, 550                                                  38
splot 'efyrnwy.dem' u 1:2:3                                              39
```

The `reset` command (line 30) causes all of the parameters that were modified by the `set` commands issued in the preceding section (axis labels, plotting ranges and so on) to be restored to their default values. Line 31 instructs gnuplot not to add a key to subsequent plots. More generally, the `unset` command is used to "turn off"

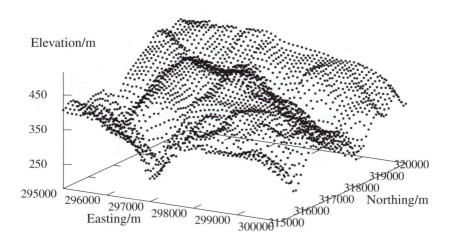

Figure 2.13: Visualization of the Llyn Efyrnwy DEM. Data reproduced by kind permission of Ordnance Survey (© Crown Copyright License Number NC/03/11298).

or reset specific features of a plot. Line 32 selects point symbols as the default data style. Lines 33 through 35 specify the labels that are to be placed beside the *x*-, *y*- and *z*-axes, respectively. Lines 36 through 38 control the position of the tic-marks that are to be placed along the axes. The three values associated with each of these commands specify the position of the first tic-mark, the interval between the tic-marks, and the position of the final tic-mark, respectively, on the given axis. Thus, line 36 indicates that the tic-marks on the *x*-axis should commence at a value of 295000 and be plotted at 1000 unit intervals, ending at 300000. Finally, line 39 generates a 3D plot of the data in the file `efyrnwy.dem`, plotting the data in columns 1, 2 and 3 on the *x*-, *y*- and *z*-axes, respectively (Figure 2.13).

Strictly speaking, Figure 2.13 is not a 3D plot but a visualization of a 3D data set projected onto a 2D plane (i.e., a flat piece of paper), sometimes known as a 2.5D (two-and-a-half dimensional) plot. To avoid having to use this sort of convoluted linguistics, however, the term 3D plot will be used as shorthand to describe figures of this type throughout the rest of the book.

### 2.4.3 Altering the View Direction

While Figure 2.13 provides an indication of the topography near Llyn Efyrnwy, it is impossible to obtain a complete appreciation of the nature of the terrain by observing it from just one direction. The problem is that data points located in the foreground partially obscure those in the background. Moreover, it is often difficult to distinguish between the two because we do not have a proper sense of perspective. This problem can be overcome, to a certain extent, by rotating the plot so that the terrain is viewed from a number of directions (Figure 2.14). In gnuplot, this can be controlled using the `set view` command, as follows:

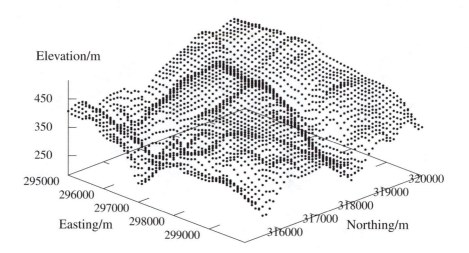

Figure 2.14: Viewing the Llyn Efyrnwy DEM from a different direction. Data reproduced
by kind permission of Ordnance Survey (© Crown Copyright License Number
NC/03/11298).

```
set view 45, 45, 1, 1                                                    40
replot                                                                   41
```

Line 40 sets the view direction to be 45° with respect to the $y$-axis (effectively tilting
the plot toward the viewer) and 45° with respect to the $z$-axis (rotating the plot around
that axis). By comparison, the default values used to produce Figure 2.13 are 60° and
30°, respectively. The final two arguments on line 40 control, in turn, the scaling of
the plot in both the $x$- and $y$-axes, and in the $z$-axis. A value greater than 1 stretches
that axis; a value less than 1 compresses it. In this case, the plot has not been re-
scaled along any axis. Line 41 plots the data again, taking into account the revised
view direction (Figure 2.14).

> **Exercise 2.2**: Plot the Llyn Efyrnwy DEM data set, viewing it from several
> directions. Experiment with the scaling of the $x$-, $y$- and $z$-axes.

### 2.4.4   Generating a 3D Surface Plot

It is often more appropriate to visualize 3D data as a quasi-continuous surface rather
than a cloud of data points. One solution is to connect adjacent data points using
straight line segments so that they form a wire-frame model (Figure 2.15). Various
methods can be used for this purpose depending on the distribution of the original
data and the intended application of the derived surface model (Burrough and Mc-
Donnell 1998). If the data points are distributed irregularly across the measurement
space, two approaches are generally possible. The first involves the construction of a

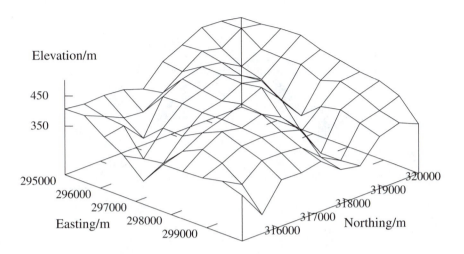

Figure 2.15: Wire-frame surface (10 × 10 element grid) generated from the Llyn Efyrnwy
DEM. Data reproduced by kind permission of Ordnance Survey (© Crown Copyright License Number NC/03/11298).

triangulated irregular network (TIN), in which the data points are connected so that
they form a network of triangular surface facets of irregular size and shape. This approach is frequently applied to digital elevation data, often using a method known as
Delaunay Triangulation (Watson 1992, Bonham-Carter 1994). The second approach
involves interpolating the original data onto a regular rectangular grid. Several methods can be used to achieve this, including trend-surface analysis, inverse-distance
weighting and kriging. The relative merits of TIN and grid-based approaches are discussed extensively by Burrough and McDonnell (1998); both methods are commonly
implemented in current GIS.

Simple wire-frame surface models can be generated in gnuplot using the dgrid3d
command. This performs an interpolation of the original $z$-axis values onto a regular
rectangular grid using the inverse distance weighting (IDW) method. The interpolated
$z$-value for a particular intersection point on the resulting grid is computed from the
weighted average of the original $z$-values in the local neighborhood, with the weights
being inversely proportional to the distance between the grid point and the data points
according to Equation 2.1:

$$\text{weight} = \frac{1}{\Delta x^w + \Delta y^w} \tag{2.1}$$

where $\Delta x$ and $\Delta y$ are the distances along the $x$- and $y$-axes between the original data
point and the interpolated grid point, and $w$ is a weighting factor (Williams and
Kelly 1998). Thus, the closer the original data point is to an interpolated grid point,
the greater the influence it has on the interpolated $z$ value. Moreover, as the value of
the weighting factor, $w$, is increased, data points further from the grid point have a
decreasing effect on the interpolated $z$ value.

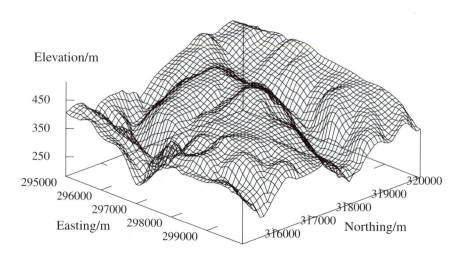

Figure 2.16: Wire-frame surface (51 × 51 element grid) generated from the Llyn Efyrnwy
DEM. Data reproduced by kind permission of Ordnance Survey (© Crown Copyright License Number NC/03/11298).

The `dgrid3d` command is used to construct a wire-frame model of the terrain around Llyn Efyrnwy even though the digital elevation data are provided in the form of a regular grid of values. If the `dgrid3d` command is not used, the resulting straight line segments connect the data points in the sequence in which they occur in the data file, in the manner of a child's join-the-dots diagram, instead of their immediate spatial neighbors in the grid. Thus, Figure 2.15 is generated as follows:

```
set style data lines                                                          42
set dgrid3d 10, 10, 16                                                        43
replot                                                                        44
```

Line 42 sets the default data style to be `lines`, ensuring that points on the interpolated grid are connected by straight line segments. Line 43 instructs gnuplot to treat the data as scattered point values and to generate from these a gridded data set. The first two arguments to this command determine the number of rows and columns in the output grid. In this case, the grid is specified in terms of 10 equally spaced samples along both the $x$- and $y$-axes (i.e., a $10 \times 10$ grid). The third argument is the weighting factor, $w$, in Equation 2.1. Finally, line 44 plots the digital elevation data taking these commands into account.

The density of the interpolated grid can be varied by altering the arguments to the `dgrid3d` command. For instance, Figure 2.16 was produced as follows:

```
set dgrid3d 51, 51, 16                                                        45
replot                                                                        46
```

The use of a $51 \times 51$ element grid makes sense in this case because it matches the number of points in the original data set. The resulting plot reveals much more in-

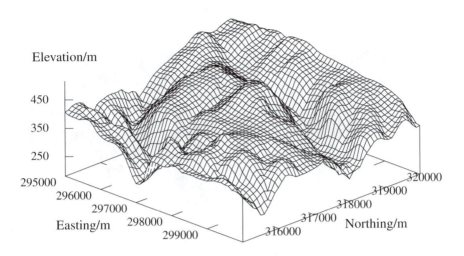

Figure 2.17: Wire-frame surface (51 × 51 element grid) generated from the Llyn Efyrnwy DEM, with hidden line removal. Data reproduced by kind permission of Ordnance Survey (© Crown Copyright License Number NC/03/11298).

formation about the surface topography in this area, including the main ridges and valleys, as well as more subtle geomorphological features. In other cases, selecting the correct density for the grid may be a matter of trial and error. If the grid is too dense, it may take a long time to generate the wire-frame surface even on a fast computer. Interpretation of the resulting patterns may also be difficult. If the grid is too sparse, important detail may be lost, as in Figure 2.15.

It is worth noting that wire-frame surface plots can be rotated and viewed from different directions in gnuplot either by using the set view command, described in Section 2.4.3, or by moving the mouse to the left or right within the plot window while holding down the left-hand mouse button. Similarly, diagonal movements of the mouse within the plot window will rotate the surface plot around a different axis, while upward and downward movements of the mouse will increase or decrease the vertical ($z$-axis) scaling of the plot, respectively.

### 2.4.5   Hidden-Line Removal

Although the dense grid used to generate Figure 2.16 means that a greater amount of detail is evident in the resulting wire-frame model, visual interpretation is made difficult by the mass of intersecting lines in the plot. This problem can be alleviated by removing those lines that would normally be hidden from view if the surface was solid (i.e., so that vertical protrusions obscure the features behind them; Figure 2.17). This process is known as hidden-line removal and the relevant gnuplot commands are

```
set hidden3d                                                        47
replot                                                              48
```

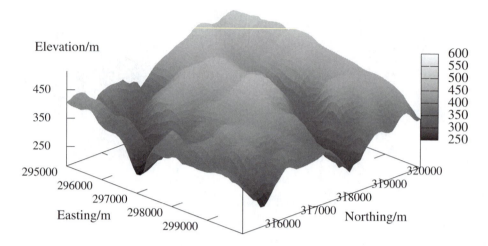

Figure 2.18: Grayscale rendered surface model of the terrain elevation close to Llyn Efyrnwy.
         Data reproduced by kind permission of Ordnance Survey (© Crown Copyright
         License Number NC/03/11298).

Thus, line 47 enables hidden-line removal and line 48 plots the data accordingly. Note
that it is easier to identify the main geomorphological features of the terrain around
Llyn Efyrnwy (e.g., ridges, valleys and breaks of slope), and that parts of the $x$- and
$y$-axes are hidden, in the resulting plot.

### 2.4.6   Producing Solid Surface Models

The visualization of 3D data can be enhanced further still by generating a solid
(cf. wire-frame) surface model, in which each facet of the surface is rendered with a
different graytone or color depending on its elevation, gradient, aspect or other environmental property (Figure 2.18). The benefit of this approach is that the human
interpreter is provided with two visual cues with which to comprehend the data, the
general shape of the surface and the gray tone or color of each surface facet. This can
be achieved in gnuplot by issuing the following commands.

```
set style line 9                                                          49
set pm3d at s hidden3d 9                                                   50
unset hidden3d                                                            51
splot 'efyrnwy.dem' u 1:2:3 with pm3d                                      52
```

Thus, line 49 determines the line style used to draw the wire-frame grid (a solid
black line in this example). Line 50 instructs gnuplot to produce a grayscale or
color visualization of the elevation data and to render this onto the wire-frame surface (pm3d at surface, which abbreviates to pm3d at s). The rendering can also be
placed at the base of the plot (pm3d at b), on top of the plot (pm3d at t), or some
combination of these (e.g., pm3d at bst). Here, the pm3d style is asked to perform

hidden-line removal on the wire-frame grid and the surface model (`hidden3d`; line 50), although it uses a different algorithm from the one employed in the preceding examples. Consequently, line 51 "unsets" the standard hidden-line removal algorithm. Lastly, line 52 plots the data once more, this time using the `pm3d` style.

**Exercise 2.3**: Using the mouse, rotate the solid 3D surface plot of the Llyn Efyrnwy DEM about its *x*-, *y*- and *z*-axes. Experiment with the `set pm3d` command, altering the position of the rendered surface with respect to the plot (e.g., `pm3d at b`). Remember to `replot` the data each time.

### 2.4.7   Contouring 3D Surface Plots

Another way of visualizing 3D data is to produce a contour plot. This is perhaps most commonly applied to environmental data sets for which the *x*- and *y*-axes refer to the spatial location of the measurements made on the *z*-axis variable. An obvious example is the topographic map, where the contours denote lines of equal terrain elevation (Watson 1992). Other environmental examples include plots of barometric air pressure (isobars), precipitation (isohyets), temperature (isotherms), water depth (isobaths) and salinity (isohalines). Contour plots can also be used to visualize data sets in which the contours portray variation in the value of the *z*-axis property as a function of two other aspatial properties, which define the *x*- and *y*-axes.

Contours can be added to a wire-frame surface model in gnuplot as follows:

```
unset pm3d                                                        53
set hidden3d                                                      54
set contour base                                                  55
set cntrparam levels discrete 250, 300, 350, 400, 450, 500       56
splot 'efyrnwy.dem' u 1:2:3                                       57
```

Line 53 instructs gnuplot to cease plotting the data as a solid surface model. As a result, subsequent 3D plots are produced as standard wire-frame models. Line 54 invokes the standard algorithm for hidden line removal. Line 55 indicates that the contours should be drawn on the base of the plot: the other options are to draw them on the wire-frame surface itself (`set contour surface`) or on both the base and the surface of the plot (`set contour both`). Rather than accepting the default values, line 56 instructs gnuplot to modify the contour parameters (`cntrparam`) so that the contour lines are drawn at pre-defined intervals in elevation; in this example they are plotted at 50 m intervals from 250 m to 500 m. Finally, line 57 plots the data as a wire-frame model and a set of contours (Figure 2.19).

Sometimes it is helpful to remove the 3D surface altogether, leaving only the contour plot. The simplest way to do this in gnuplot is to enter the following commands:

```
unset surface                                                    58
set view map                                                     59
set size square                                                  60
replot                                                           61
```

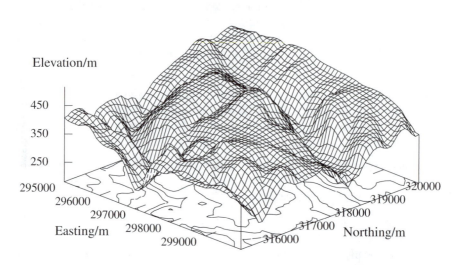

Figure 2.19: Visualization of the Llyn Efyrnwy DEM, with a contour map added to the base of the plot. Reproduced by kind permission of Ordnance Survey (© Crown Copyright License Number NC/03/11298).

where line 58 removes the 3D surface and line 59 sets the view direction so that one is effectively looking vertically down onto the plot. This command is equivalent to set view 0,0,1,1. It gives the plot the appearance of a planimetric map even though it is generated using the splot command. Line 60 causes gnuplot to produce a plot with an aspect ratio (i.e., the ratio of the length of the *y*-axis to the length of the *x*-axis) of 1. This produces a square plot, appropriate to the presentation of a conventional planimetric map, unless the window size is altered manually using the mouse. Line 61 plots the data taking these changes into account. The result is not shown here.

It is possible to gain greater control over the appearance of the resulting contour plot, and hence to produce a better quality figure, by saving the data that define the positions of the contour lines in a separate data file. These can then be read back into gnuplot and drawn using the plot command, as shown below.

```
set terminal push                                                    62
set terminal table                                                   63
set output 'contours.dat'                                            64
splot 'efyrnwy.dem' using 1:2:3                                       65
set output                                                           66
set terminal pop                                                     67
set grid                                                             68
plot 'contours.dat' index 0 u 1:2, \                                 69
     'contours.dat' index 1 u 1:2, \                                 70
     'contours.dat' index 2 u 1:2, \                                 71
     'contours.dat' index 3 u 1:2, \                                 72
     'contours.dat' index 4 u 1:2, \                                 73
     'contours.dat' index 5 u 1:2                                    74
```

The `set terminal push` command (line 62) stores information concerning the current terminal type and its settings, which can subsequently be recalled using the `set terminal pop` command (line 67). This pair of commands provides a convenient way of saving the default terminal type (e.g., the one employed when gnuplot starts up), before the terminal type is changed for whatever reason, and of restoring it later. Significantly, this approach is independent of the operating system on which gnuplot is run. Thus, the same commands can be employed without modification on different computers; for example, on one running GNU/Linux and another running Microsoft Windows. These commands are therefore said to be "portable" between operating systems.

In the example given above, after storing information on the default terminal type (line 62), gnuplot is instructed to use the `table` terminal type until further notice (line 63). Instead of producing graphical output, this terminal type generates several columns of data in ASCII text format (Figure 2.20). Line 64 ensures that these data are output to a file called `contours.dat` located in the working directory, as opposed to displaying them on the screen. If a file by this name already exists in that directory, its contents are over-written; otherwise, a new file is created. Line 65 then plots the contours, storing the *x,y,z* values that denote their locations in the named data file. Note that the data are organized so that contour lines are listed in descending order of elevation. Moreover, gnuplot separates the data for contour lines at the same elevation (e.g., 500 m) by a single blank line, and data from contours at different elevations (e.g., 300 m and 250 m) by two blank lines (e.g., lines 851 and 852 in Figure 2.20). Line 66 closes the output file.

Line 67 restores the default terminal type so that subsequent plots are directed to the computer screen rather than to a data file. Line 68 draws a grid across the resulting plot. Lines 69 through 74 plot the data contained in the newly created file, `contours.dat`. Note that this is a single `plot` command, which is split over several lines using the line continuation symbol (\). This command plots the data contained in the first two columns of the data file (i.e., `u 1:2`). The `index` keyword provides control over which section of the data file to plot. Thus, `index 0` tells gnuplot to read in the data from the start of the file up to the first pair of consecutive blank lines (lines 58 and 59 in Figure 2.20), `index 1` refers to the subsequent block of data up to the next pair of blank lines, and so on. Six contour levels were specified (250 m, 300 m, 350 m, 400 m, 450 m and 500 m), so `contours.dat` contains six blocks of data (blocks 0 to 5). Note that lines in the data file that start with a hash symbol (#) are treated as comments and are ignored by gnuplot (they are not plotted).

The result of the gnuplot commands described above is presented in Figure 2.21. One feature evident in this figure is the valley that runs roughly WNW–ESE, roughly along Northing 318000. This valley contains the Afon Conwy, a large stream that drains into the Afon Efyrnwy below Llyn Efyrnwy.

gnuplot offers further control over the production of contour plots, including a range of contour interpolation routines, such as piecewise linear, cubic splines and B-splines. These are controlled via options to the `set cntrparam` command. For more professional-quality output, however, it is generally necessary to use specialist software for digital mapping (Jones 1997, Burrough and McDonnell 1998).

```
                                                                              1
#Surface 0 of 1 surfaces                                                      2
                                                                              3
# Contour 0, label:        500                                                4
  295000    319420    500                                                     5
  295100    319456    500                                                     6
  295200    319500    500                                                     7
  295300    319500    500                                                     8
  295400    319420    500                                                     9
  295500    319400    500                                                    10


  296380    319900    500                                                    55
  296400    319975    500                                                    56
  296500    320000    500                                                    57
                                                                            58
                                                                            59
# Contour 1, label:        450                                              60
  298400    319400    450                                                    61
                                                                            62
  298500    319700    450                                                    63
                                                                            64
  296794    317600    450                                                    65
  296800    317590    450                                                    66
  296843    317500    450                                                    67
  296816    317400    450                                                    68
  296800    317362    450                                                    69
  296729    317300    450                                                    70


  299800    315317    300                                                   840
  299700    315369    300                                                   841
  299650    315400    300                                                   842
  299700    315450    300                                                   843
  299729    315500    300                                                   844
  299777    315600    300                                                   845
  299800    315630    300                                                   846
  299833    315700    300                                                   847
  299854    315800    300                                                   848
  299900    315843    300                                                   849
  300000    315850    300                                                   850
                                                                           851
                                                                           852
# Contour 5, label:        250                                             853
  297100    315000    250                                                   854
  297000    315067    250                                                   855
  296990    315000    250                                                   856
```

Figure 2.20: Extracts from Llyn Efyrnwy contour line data file (`contours.dat`; lines 1–10, 55–70 and 840–856).

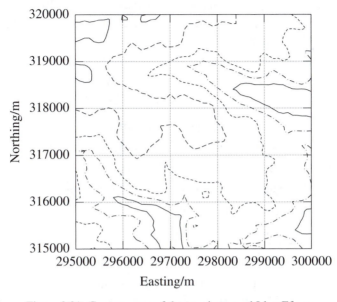

Easting/m

Figure 2.21: Contour map of the terrain around Llyn Efyrnwy.

**Exercise 2.4**: Using gnuplot, produce a contour plot similar to that shown in Figure 2.21 based on the Llyn Efyrnwy DEM. The resulting plot should display elevation contours at 25 m intervals from 300 m to 500 m.

## 2.5 PRINTING PLOTS

The method of printing a plot in the version of gnuplot for Microsoft Windows is to click the right-hand button of the mouse over the title bar of the plot window, select "Options" and then "Print ... ". This sequence should reveal a standard printer dialog box with options to select the appropriate printer, paper size and orientation. Clicking on the OK button will send the current plot to the selected printer.

The method for printing plots in the version of gnuplot for GNU/Linux is slightly different and is driven from the command-line interface as follows:

```
set output '|lpr'          75
set terminal push          76
set terminal postscript    77
replot                     78
set output                 79
set terminal pop           80
```

Line 75 sets the destination to which the plot should be output. In this example, the plot is "piped" (|) through to the printer (lpr). Information about the current (default) terminal type is stored (line 76) before the terminal type is changed to postscript,

Table 2.6: Selected output file formats supported by gnuplot.

| Format | Description |
|---|---|
| aqua | For display on Mac OS X |
| emf | Enhanced Metafile Format |
| pdf | Adobe Portable Document Format (PDF) |
| png | Portable Network Graphics file format |
| postscript | PostScript and Encapsulated PostScript file format |
| svg | Scalable Vector Graphics file format |
| table | Dump an ASCII table of $x\ y\ [z]$ values to output |
| windows | For use with Microsoft Windows |
| x11 | For use with X servers (e.g., under GNU/Linux and UNIX) |

on the assumption that the printer is PostScript-compatible (line 77). Line 78 plots the current figure, sending the resulting PostScript commands directly to the printer instead of the screen. Lines 79 and 80, respectively, close the pipe to the printer and send subsequent plots to the previously stored default terminal (i.e., to the screen).

## 2.6  EXPORTING GRAPHICS FILES

As well as printing directly from gnuplot, there is often a need to incorporate plots into reports and other types of document. This book, for example, has been produced using LyX (http://www.lyx.org), a document preparation system based on the LaTeX typesetting package (Goossens *et al.* 1994, Lamport 1994). Among other things, this software allows EPS figures, such as the ones presented in this chapter, to be included within a document (Merz 1996, Goossens *et al.* 1997). Standard word-processing packages, such as OpenOffice (http://www.openoffice.org/), Corel WordPerfect and Microsoft Word, have similar functionality. Each is able to import figures stored in a number of different formats supported by gnuplot, including vector formats, such as Enhanced Metafile Format (EMF) and Scalable Vector Graphics (SVG), and bitmap images, such as Portable Network Graphics (PNG) (Table 2.6). In general, vector formats are preferable because the size of the plot can be altered subsequently in the word-processing package without loss of resolution.

Output from the version of gnuplot for Microsoft Windows can be saved as an EMF file as follows:

```
set terminal push                              81
set terminal emf "Times Roman" 12              82
set output 'myplot.emf'                        83
replot                                         84
set output                                     85
set terminal pop                               86
```

Line 82 instructs gnuplot to create the next plot in the form of an EMF file, using 12 pt Times Roman font for the text. Line 83 ensures that the plot is output to a file called `myplot.emf` in the working directory. Line 84 re-issues the previous `plot` or `splot` command, sending the results to the named file. Line 85 closes this file and line 86 restores the previous terminal type. The resulting graphics file, `myplot.emf`, can be imported directly into most standard word-processing or graphics packages. The exact method differs from one package to another, but the relevant functions can often be found under the "Insert" option on the main menu bar.

In the version of gnuplot for Microsoft Windows it is also possible to copy the current plot to the clipboard and from there paste it directly into a document (e.g., a Word file). This is done by clicking the right-hand button of the mouse over the title bar of the plot window, selecting "Options" and then "Copy to Clipboard". In most Microsoft Windows packages the contents of the clipboard can then be pasted into the current document by typing `Ctrl` + `v` .

## 2.7   COMMAND-LINE SCRIPTS

In Section 2.2.1, one of the justifications given for using command-line entry, rather than point-and-click methods, was the ability to store the gnuplot commands used to create a plot (or plots) in a simple ASCII text file. This file can then be employed to process similar data sets using the same set of commands. Files of this type are sometimes known as scripts or macros. In the context of gnuplot, a script is simply an ASCII text file that contains a series of valid gnuplot commands. For example, the commands listed in lines 1 through 86 throughout this chapter could be saved in an ASCII text file called `visual.plt` using a standard text editor. The commands contained in this file can then be invoked ("run") by typing

```
load 'visual.plt'                                                    87
```

at the `gnuplot>` command-line prompt, assuming that `visual.plt` is located in the working directory. If the script file is located somewhere else on the computer, the full or relative path name of the file must be given (e.g., `/mnt/cdrom/chapter2/visual.plt` for GNU/Linux or `d:\chapter2\visual.plt` for Microsoft Windows). Thus, the `load` command (line 87) reads in and executes each line of the named script file as though it had been typed interactively on the command line (Williams and Kelly 1998).

## 2.8   SUMMARY

This chapter provides a general introduction to the graphical and analytical functions of gnuplot, which is given as an example of the type of computer software that can be used to visualize a range of environmental data sets. The coverage is by no means exhaustive either in terms of what can be achieved with gnuplot or, indeed, the full scope of scientific data visualization. gnuplot is, however, used throughout the rest of the book to examine key data sets and to plot the output from various environmental models. In doing so, other features of the software are explored. In the meantime, it is worth noting that gnuplot comes with extensive built-in documentation, which

can be accessed by typing `help`, or `help <command>` for help on a specific command
(e.g., `help plot`).

---

**Exercise 2.5**: Use a text editor to create and save a file that contains the
gnuplot commands required to generate the plot shown in Figure 2.10. The
script file should output the resulting plot as an EMF or PNG file. Use the
gnuplot command line to load and run the script file. Paste the graphics
file that it generates into a word-processor document to check that it has
been created properly. N.B. If you use word-processing software instead
of a text editor to create and save the gnuplot script file, you must ensure
that the file is saved in ASCII text format. This can often be achieved by
using the word-processor's "Save as" facility.

---

## SUPPORTING RESOURCES

The following resources are provided in the `chapter2` sub-directory on the CD-ROM:

| | |
|---|---|
| `le98temp.dat` | Air temperature data set for Llyn Efyrnwy, 1998 |
| `le98rain.dat` | Precipitation data set for Llyn Efyrnwy, 1998 |
| `efyrnwy.dem` | Digital elevation data set for Llyn Efyrnwy (© Crown Copyright License Number NC/03/11298) |
| `visual.plt` | gnuplot commands listed in this chapter |
| `visual.gp` | gnuplot commands used to generate the EPS figures presented in this chapter |

# Chapter 3

# Processing Environmental Data

**Topics**

- Data pre-processing and post-processing

- Records and fields

**Methods and techniques**

- Creating and running a simple gawk program

- Selecting and manipulating specific fields and records of data

- Command-line switches

- Formatting the output from a gawk program

- Output redirection

- Handling "missing data"

## 3.1 INTRODUCTION

The ability to manipulate data is an important part of the environmental modeling process, and not only within the model itself. Where a model requires input values, for instance, the data must be presented in a form that the model can handle. This may differ, however, from the format in which the data are originally provided. In such circumstances the data must be reformatted so that they are compatible with the model; this task is sometimes referred to as data pre-processing. Similarly, the data output from a model may have to be post-processed so that they can be compared to, or combined with, other data sets. These might include *in situ* measurements of the target environmental system or the output from another model. Moreover, it may be necessary to post-process the output from a model so that it can be visualized effectively or presented in summary form as part of a report.

The data sets examined in the preceding chapter represent a case in point. The files used to explore the temporal variations of air temperature (le98temp.dat) and precipitation (le98rain.dat) at Llyn Efyrnwy are derived from the AWS data sets recorded by the UK MetOffice. The original data were pre-processed by the author, however, to simplify the introduction to gnuplot. Specifically, several columns of data were removed to reduce the size of the data sets, and the date and time information was reformatted to conform to the ISO 8610:2004 standard. Figure 3.1, for example, shows part of the original precipitation data set (rain981e.dat) used to generate the file le98rain.dat, which was examined in the preceding chapter (Figure 2.10).

Many different tools can be employed to process environmental data sets in the various ways outlined above. These include spreadsheets and databases, as well as bespoke software written by the modeler in a computer programming language of his or her choice. Each has its advantages and disadvantages, some of which are discussed in Chapter 1. A simple, but powerful, computer programming language, known as awk, is used in this book. The acronym awk is derived from the surnames of the individuals who developed the language, specifically Aho, Weinberger and Kernighan (Aho *et al.* 1988). There are several versions of awk; the one used here is the GNU Project (http://www.gnu.org/) implementation, which is known as gawk (Dougherty 1996, Robbins 2001). gawk has a number of important advantages in this context, some of which are outlined in Table 3.1. Significantly, gawk not only provides a powerful tool for data manipulation, but also a convenient framework in which to develop a range of environmental simulation models. These properties are demonstrated in subsequent chapters. By way of introduction, though, the current chapter explores how gawk can be used to reformat a particular data set so that it can be visualized in gnuplot.

## 3.2   STRUCTURE OF THE LLYN EFYRNWY PRECIPITATION DATA

The exercises covered in this chapter make use of the file rain981e.dat, which is included on the CD-ROM. This file contains information on the precipitation that accumulated at Llyn Efyrnwy over the preceding 12 hours recorded twice daily (at 09:00 and 21:00 GMT) throughout 1998 by an AWS (Figure 3.1). Unlike the files used in the preceding chapter, the data are given in the original format supplied by the UK MetOffice, as downloaded from the BADC web site (http://badc.nerc.ac.uk/).

The data in rain981e.dat can be thought of as being divided into rows and columns, similar to the structure of a spreadsheet. The rows and columns are known formally as records and fields, respectively. There are 717 records in rain981e.dat, each of which contains 10 fields of data (Figure 3.1). In this example, the fields are separated by white space (i.e., blank spaces or tab stops) and every record is terminated by a newline character, which is not displayed but ensures that each record is placed on a separate line. In other files, commas are sometimes used to separate fields; these are known as comma-separated value (CSV) files.

| ID | IDTYPE | MET_DOM | YEAR | MONTH | DAY | END_HOUR | COUNT | AMT | DUR |
|---|---|---|---|---|---|---|---|---|---|
| 425000 | RAIN | NCM | 1998 | 1 | 1 | 900 | 12 | 0 | -999 |
| 425000 | RAIN | NCM | 1998 | 1 | 1 | 2100 | 12 | 146 | -999 |
| 425000 | RAIN | NCM | 1998 | 1 | 2 | 900 | 12 | 52 | -999 |
| 425000 | RAIN | NCM | 1998 | 1 | 2 | 2100 | 12 | 16 | -999 |
| 425000 | RAIN | NCM | 1998 | 1 | 3 | 900 | 12 | 358 | -999 |
| 425000 | RAIN | NCM | 1998 | 1 | 3 | 2100 | 12 | 56 | -999 |
| 425000 | RAIN | NCM | 1998 | 1 | 4 | 900 | 12 | 66 | -999 |
| 425000 | RAIN | NCM | 1998 | 1 | 4 | 2100 | 12 | 208 | -999 |
| 425000 | RAIN | NCM | 1998 | 1 | 5 | 900 | 12 | 44 | -999 |
| 425000 | RAIN | NCM | 1998 | 1 | 5 | 2100 | 12 | 30 | -999 |
| 425000 | RAIN | NCM | 1998 | 1 | 6 | 900 | 12 | 40 | -999 |
| 425000 | RAIN | NCM | 1998 | 1 | 6 | 2100 | 12 | -999 | -999 |
| 425000 | RAIN | NCM | 1998 | 1 | 7 | 900 | 12 | 24 | -999 |
| 425000 | RAIN | NCM | 1998 | 1 | 7 | 900 | 24 | 148 | -999 |
| 425000 | RAIN | NCM | 1998 | 1 | 7 | 2100 | 12 | 66 | -999 |
| 425000 | RAIN | NCM | 1998 | 1 | 8 | 900 | 12 | 136 | -999 |
| 425000 | RAIN | NCM | 1998 | 1 | 8 | 2100 | 12 | 384 | -999 |
| 425000 | RAIN | NCM | 1998 | 1 | 9 | 900 | 12 | 10 | -999 |
| 425000 | RAIN | NCM | 1998 | 1 | 9 | 2100 | 12 | 2 | -999 |
| 425000 | RAIN | NCM | 1998 | 1 | 10 | 900 | 12 | 0 | -999 |
| 425000 | RAIN | NCM | 1998 | 1 | 10 | 2100 | 12 | 0 | -999 |
| 425000 | RAIN | NCM | 1998 | 1 | 11 | 900 | 12 | 0 | -999 |
| 425000 | RAIN | NCM | 1998 | 1 | 11 | 2100 | 12 | 12 | -999 |
| 425000 | RAIN | NCM | 1998 | 1 | 12 | 900 | 12 | 2 | -999 |

Figure 3.1: Partial contents of the file rain981e.dat (first 25 records).

Table 3.1: Selected properties of the gawk programming language.

| | |
|---|---|
| Data driven | Reading data from files, selecting the required elements from these data, manipulating the chosen elements and writing the results to files are very much simpler than in many other computer programming languages. |
| Powerful | Extensive built-in control-flow constructs and functions. Can be extended through user-defined functions. |
| C-like syntax | Syntax similar to the C programming language (Kernighan and Ritchie 1988), which is widely used in industry, commerce and academia. Consequently, it is relatively easy to migrate to C having learned gawk. |
| Interpreted | Interprets each line of the program at run-time. Removes the additional distractions and complexities associated with languages in which the code must be compiled (i.e., converted into a binary format that can be executed by the computer). |
| Platform independent | Available for a host of different computer platforms, including GNU/Linux, UNIX, MacOS X and Microsoft Windows/DOS. |
| Free | Covered by the General Public License (GPL; see Appendix B), sometimes referred to as the *Copyleft* agreement. Can be freely copied and redistributed. Legal to place a copy of gawk on any number of computers without infringing copyright. |

The first record of rain98le.dat consists of brief textual descriptions of the data contained in the corresponding fields of the remaining records; these are, in effect, the field names (Table 3.2). For example, the first field contains the number used by the MetOffice to identify (ID) the rain gauge at this meteorological station, the second field specifies the type of measurement made at this site (RAIN), and the third field indicates that the data are reported in the National Climate Message (NCM) format (see http://badc.nerc.ac.uk/data/surface/ukmo_guide.html for further details).

It is important to note that the precipitation values given in the ninth field of Figure 3.1 (AMT) are reported in tenths of a millimeter (0.1 mm). Thus, a value of 18 indicates that 1.8 mm of precipitation accumulated over the preceding 12 hours. Furthermore, the value −999 is used to indicate "missing data". This may refer to data that have been lost for some reason (e.g., owing to instrument failure) or that are not collected at this site.

## 3.3   CREATING AND RUNNING A SIMPLE gawk PROGRAM

The first thing that one might wish to do with the file rain98le.dat is to examine its contents on the computer screen. Various command-line tools and software packages have been developed for this purpose, including text editors and spreadsheets. Here, though, gawk is used to achieve the same result, primarily because it serves

Table 3.2: Interpretation of the data fields in the file `rain981e.dat`.

| Field | Field name | Data contained in this field |
| --- | --- | --- |
| 1 | ID | Meteorological station ID number |
| 2 | IDTYPE | Meteorological station type |
| 3 | MET_DOM | Meteorological domain |
| 4 | YEAR | Year |
| 5 | MON | Month (1 to 12) |
| 6 | DAY | Day of the month (1 to 31) |
| 7 | END_HOUR | Hour of observation (09:00 or 21:00 GMT) |
| 8 | COUNT | Period (hours) over which the observations were recorded |
| 9 | AMT | Precipitation amount (in units of 0.1 mm) |
| 10 | DUR | Precipitation duration (minutes) |

as a convenient introduction to the structure of a basic **gawk** program. Thus, the following instruction must be entered on the command line to inspect the contents of `rain981e.dat` using **gawk**, assuming that the file `rain981e.dat` is located in the working directory:

```
gawk '{print}' rain981e.dat                                              1
```

If the file is located elsewhere on the computer, its full or relative pathname must be used (e.g., `/mnt/cdrom/chapter3/rain981e.dat` in GNU/Linux or `d:\chapter3\rain981e.dat` in Microsoft Windows).

In the example given above, the keyword `gawk` invokes the **gawk** utility (Robbins 2001). This interprets the **gawk** program specified by the user. Here, the program `{print}` is entered in full via the command line. The **gawk** utility applies the "rules" specified in the program to the data contained in the file named on the command line `rain981e.dat`. Thus, the general syntax is `gawk 'program' datafile`. The program consists of a pair of curly braces, `{}`, and a single **gawk** statement, `print`. The curly braces denote the beginning and end of this particular program, and the `print` statement instructs the computer to print out each record of data exactly as it appears in the named file. The output is sent to the computer screen unless otherwise specified. This process, and the resulting output, is illustrated schematically in Figure 3.2. Note that a pair of single quotation marks is placed around the **gawk** program to prevent it from being interpreted as an operating system command; this procedure is known as *quoting*. The quotation marks are not part of the program and are only required when the program is typed in full on the command line (cf. supplied from a program file; see Section 3.5).

| ID     | IDTYPE | MET_DOM | YEAR | MON | DAY | END_HOUR | COUNT | AMT | DUR  |
|--------|--------|---------|------|-----|-----|----------|-------|-----|------|
| 425000 | RAIN   | NCM     | 1998 | 1   | 1   | 900      | 12    | 0   | -999 |
| 425000 | RAIN   | NCM     | 1998 | 1   | 1   | 2100     | 12    | 146 | -999 |
| 425000 | RAIN   | NCM     | 1998 | 1   | 2   | 900      | 12    | 52  | -999 |
| 425000 | RAIN   | NCM     | 1998 | 1   | 2   | 2100     | 12    | 16  | -999 |
| ⋮      | ⋮      | ⋮       | ⋮    | ⋮   | ⋮   | ⋮        | ⋮     | ⋮   | ⋮    |

Input data file: rain98le.dat

{print} Program → GAWK Utility

| ID     | IDTYPE | MET_DOM | YEAR | MON | DAY | END_HOUR | COUNT | AMT | DUR  |
|--------|--------|---------|------|-----|-----|----------|-------|-----|------|
| 425000 | RAIN   | NCM     | 1998 | 1   | 1   | 900      | 12    | 0   | -999 |
| 425000 | RAIN   | NCM     | 1998 | 1   | 1   | 2100     | 12    | 146 | -999 |
| 425000 | RAIN   | NCM     | 1998 | 1   | 2   | 900      | 12    | 52  | -999 |
| 425000 | RAIN   | NCM     | 1998 | 1   | 2   | 2100     | 12    | 16  | -999 |
| ⋮      | ⋮      | ⋮       | ⋮    | ⋮   | ⋮   | ⋮        | ⋮     | ⋮   | ⋮    |

Output: Computer screen

Figure 3.2: Operation of a gawk program. The arrows show the flow of data through the gawk utility and the supply of instructions from the gawk program, which controls the way the data are processed.

## 3.4   USING gawk TO PROCESS SELECTED FIELDS

If gawk could only echo the contents of a data file to the computer screen, it would be a very limited tool indeed. It is, in contrast, a remarkably powerful and flexible utility, one that can be used to select particular items of data from a file and to process them according to the user's instructions (Robbins 2001). gawk facilitates this by breaking up a data file into its constituent records and by splitting each record into its component fields; moreover, it performs this task automatically. By default a record equates to a single line of data in a file and a field corresponds to a single item of data in a record, where individual fields are separated by white space. This behavior can be altered, however, as required. For instance, gawk can also be instructed how to handle CSV files by setting the field separator to be a comma, although this possibility is not explored here.

Suppose, for example, that one wished to extract from the file rain98le.dat those data pertaining to the 12-hourly accumulation of precipitation, expressed in units of millimeters (cf. tenths of a millimeter), together with the year, month, day and time at which the observations were made; that is, to recreate the file le98rain.dat used in the preceding chapter. This process involves selecting the appropriate fields from each record of the input data file, and in the case of the precipitation values converting them into millimeters by dividing by 10.0. This can be achieved using gawk by typing the instruction shown on the following command line:

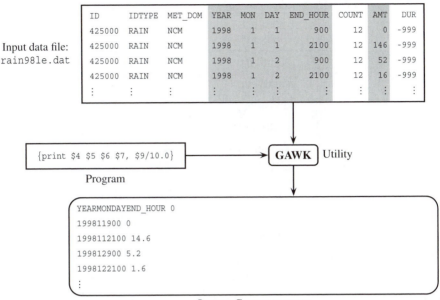

| | ID | IDTYPE | MET_DOM | YEAR | MON | DAY | END_HOUR | COUNT | AMT | DUR |
|---|---|---|---|---|---|---|---|---|---|---|
| Input data file:<br>rain98le.dat | 425000 | RAIN | NCM | 1998 | 1 | 1 | 900 | 12 | 0 | -999 |
| | 425000 | RAIN | NCM | 1998 | 1 | 1 | 2100 | 12 | 146 | -999 |
| | 425000 | RAIN | NCM | 1998 | 1 | 2 | 900 | 12 | 52 | -999 |
| | 425000 | RAIN | NCM | 1998 | 1 | 2 | 2100 | 12 | 16 | -999 |
| | ⋮ | ⋮ | ⋮ | ⋮ | ⋮ | ⋮ | ⋮ | ⋮ | ⋮ | ⋮ |

{print $4 $5 $6 $7, $9/10.0}  ⟶  **GAWK** Utility

Program

```
YEARMONDAYEND_HOUR 0
199811900 0
1998112100 14.6
199812900 5.2
1998122100 1.6
⋮
```

Output: Computer screen

Figure 3.3: Operation of a gawk program designed to manipulate and print selected fields of an input data file. The areas of the input data file highlighted in gray indicate the fields selected for processing by the gawk program.

```
gawk '{print $4 $5 $6 $7, $9/10.0}' rain98le.dat                    2
```

The operation and output of this program is shown schematically in Figure 3.3.

Although the revised command line is slightly longer than the one employed in the preceding example, the general syntax is the same; only the program differs. More specifically, the print statement is now followed by six entities, known as arguments, which instruct gawk exactly what to print and in which order. If no arguments are supplied, gawk prints the entire record. A more detailed explanation of the revised program is given below.

gawk uses the dollar sign ($) followed by a number to refer to a specific field within a record (Robbins 2001). So, for example, $1 refers to the first field, $2 refers to the second, and so on. Thus, the revised program instructs gawk to print out the values in the fourth ($4), fifth ($5), sixth ($6), seventh ($7) and ninth ($9) fields of each record in the input data file, rain98le.dat. The data in the other fields are discarded. The selected fields contain data on the year, month, day, time and the accumulated precipitation, respectively. The values in the ninth field ($9) are divided by 10.0 before they are printed (i.e., $9/10.0) to convert them into millimeters. Note that the comma symbol (,) used in this program instructs gawk to leave a space between the corresponding arguments in the output; where the comma is omitted the arguments are joined together (concatenated). Here, the comma symbol is used to insert a space between the date and time information and the precipitation values.

Table 3.3: Selected mathematical operators available in gawk.

| Symbol | Operator | Example |
|---|---|---|
| + | addition | `$2+$3` |
| - | subtraction | `$2-$3` |
| / | division | `$3/$2` |
| * | multiplication | `$3*6.5` |
| ** | exponentiation | `$2**2` |
| ^ | exponentiation | `$2^2` |
| sqrt | square root | `sqrt($2)` |
| sin | sine | `sin(0.5)` |
| cos | cosine | `cos($3)` |
| log | natural logarithm | `log($4)` |
| exp | exponential of $x$ | `exp($2/$3)` |
| rand | random number generator | `rand()` |

It may be helpful to think of the gawk utility as a filter, which can be configured to allow some or all of the input data file to pass through and, in this case, to be displayed on the computer screen (Figure 3.3). It also allows the input data to be modified as it is filtered (e.g., dividing the contents of the ninth field by 10.0). In this context, gawk provides a wide range of mathematical operators that can be used to manipulate data (Table 3.3). These operators are explored in subsequent chapters.

**Exercise 3.1**: Modify the program specified in the command line given on page 65 so that it reads '{print $4, $5, $6, $7, $9/10.0}'. Run the revised program. How does the output differ from that shown in Figure 3.3? What effect do the additional comma symbols have? What happens if the commas are removed?

## 3.5  STORING THE gawk PROGRAM IN A FILE

In the examples studied thus far, the gawk programs are entered via the command line. This method is convenient if the program is short and it is unlikely that it will be needed again (sometimes described as "throw away" code). Indeed, the facility to run simple one-line programs in this way is one of the main advantages of gawk. For longer programs, however, and where there is a need to apply the same set of instructions to several data files, this approach can be time-consuming, inefficient and error-prone. The preferred alternative, therefore, is to store the program in a separate file.

Program 3.1: selcols.awk

```
{print $4 $5 $6 $7, $9/10.0}                                        1
```

Program 3.1 shows the contents of a file, selcols.awk, such as might be created using a standard text editor. The program is identical to the one employed in the preceding example (see page 65): the only difference is that it is stored in a file, rather than entered directly via the command line. The program stored in selcols.awk can be applied to the data contained in rain98le.dat as follows:

```
gawk -f selcols.awk rain98le.dat                                   3
```

assuming that both files are located in the working directory; if not, either the full or the relative pathname of each file must be provided. Note that the -f symbol, which is known as a switch, instructs the gawk utility to read the program from a named file, in this case selcols.awk. The rest of the command line indicates that the rule contained in selcols.awk should be applied to the data in rain98le.dat. Also note that the quotation marks, which were employed to demarcate the gawk program in each of the preceding examples, are no longer required because the program is stored in a file. The output produced by the revised command line is identical to that presented in Figure 3.3.

## 3.6   USING gawk TO PROCESS SELECTED RECORDS

Close inspection of Figure 3.3 reveals something unintended in the first record (line) of the output, which reads YEARMONDAYEND_HOUR 0. This result is produced because the program instructs the gawk utility to apply the print $4 $5 $6 $7, $9/10.0; statement to every record in the input data file, even though it is not really appropriate to the first of these because it contains the text strings that specify the field names. In the circumstances, gawk has done its best to apply the print statement to these data, including "dividing" one of the text strings (i.e., AMT) by 10.0. The result in this case is reported as 0, but this is clearly not what is intended.

One way to overcome this problem is to edit the input data file by deleting the first record. This is unsatisfactory, though, because the original (unmodified) version of the file may be needed at some point in the future. Equally, keeping two or more subtly different versions of the same file (i.e., modified and unmodified) consumes additional disk space on the computer and, more importantly, is a recipe for confusion in the future. A much better approach is to alter the gawk program so that it skips the first record of the input data file (i.e., the text containing the field names), but processes subsequent records in the prescribed way. Program 3.2, selcols2.awk, demonstrates one way in which this can be achieved. This program can be run from the command line as follows:

```
gawk -f selcols2.awk rain98le.dat                                  4
```

Program 3.2: selcols2.awk

```
(NR>1){print $4 $5 $6 $7, $9/10.0}                                    1
```

A new feature, (NR>1), known as a pattern, is introduced in Program 3.2. The gawk utility reads each record of the data file in turn and checks whether the current record matches this pattern. If it does, gawk performs the corresponding action, which in this case consists of printing the specified fields, before proceeding to read the next record of data from the named file. If the current record does not match the pattern, however, the action is not performed, and gawk moves on to the next record. This sequence is repeated until every record in the data file has been read and processed.

The pattern used in this program employs one of gawk's special features, the built-in variable NR (Robbins 2001). The term *variable* refers to a named entity that can be used to store a specific value (i.e., a number or a text string) within a program (British Computer Society 1998). Variables are so-called because the value that each holds can change (i.e., vary) in response to the actions performed by the program. The term *built-in* indicates that this particular variable is integral to (i.e., is automatically provided by) the gawk programming language (cf. user-defined variables, which are created by the computer programmer as he or she requires; see Chapter 4).

The built-in variable NR is used by gawk to keep track of the total number of records that have been read from the input data file thus far (Robbins 2001). Thus, NR has the value 1 when the program reads the first record from the input data file, it has the value 2 for the second record, and so on. The pattern (NR > 1) checks to see whether the value of NR is greater than 1. This pattern matches the second and subsequent records of the data file, but not the first. As a result, the corresponding action, {print $4 $5 $6 $7, $9/10.0}, is performed on all of the records in the input data file with the exception of the first. This process is illustrated schematically in Figure 3.4.

The preceding example illustrates the general syntax of a gawk program, namely (*pattern*){*action*}. The pairing of a pattern and an action is sometimes known as a rule (Robbins 2001) because it describes what action should be taken if the current record matches the specified pattern. A simple analogy from everyday life would be "if it is raining (*pattern*), take an umbrella {*action*}".

It is worth noting that the pattern (NR>1) involves a comparison between two numbers (i.e., the value of the built-in variable NR for the current data record and the numerical constant 1). The outcome of this comparison is either true or false. If the value of NR is greater than 1, the pattern matches the current record (i.e., the pattern evaluates as true) and, hence, the corresponding action is performed; otherwise, if the value of NR is less than or equal to 1, the pattern does not match the current record (i.e., the pattern evaluates as false) and the action is skipped. gawk provides other numerical comparison operators, which are listed in Table 3.4. These operators can be used individually or in combination to construct sophisticated patterns that allow specific parts of a data set to be selected for processing, as is demonstrated later in this chapter.

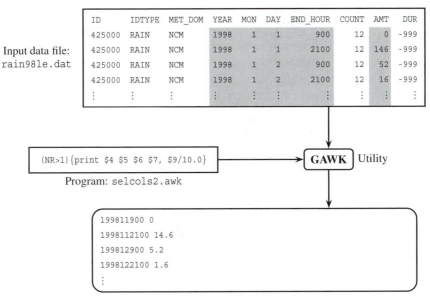

Figure 3.4: Operation of a gawk program designed to print selected records and fields of an input data file. The areas of the input data file highlighted in gray correspond to the records and fields selected by the gawk program.

Table 3.4: Numerical comparison operators in gawk.

| Operator | Example | Evaluates as ... |
|----------|---------|------------------|
| == | x==y | True if $x$ and $y$ are equal; otherwise false. |
| != | x!=y | True if $x$ and $y$ are *not* equal; otherwise false. |
| > | x>y | True if $x$ is greater than $y$; otherwise false. |
| < | x<y | True if $x$ is less than $y$; otherwise false. |
| >= | x>=y | True if $x$ is greater than or equal to $y$; otherwise false. |
| <= | x<=y | True if $x$ is less than or equal to $y$; otherwise false. |

Table 3.5: Selected `printf` format control specifiers.

| Format control | Interpretation | Examples |
|---|---|---|
| `%i` | Integer | `50`, `-347267` |
| `%e` | Scientific (exponential) notation | `1.950e+03` |
| `%f` | Floating-point number | `3.1415927`, `-29.621` |
| `%g` | Scientific or floating-point notation, whichever uses fewer characters | `1.950e+03` or `1950.00` |
| `%s` | Character string | `AMT`, `DUR` |
| `\t` | Tab space character | |
| `\n` | Newline character | |

Program 3.3: selcols3.awk

```
(NR >1) { printf ( "%i%i%i%i %f\n", $4, $5, $6, $7, $9/10.0) }          1
```

## 3.7   CONTROLLING THE FORMAT OF THE OUTPUT

The `print` statement, used in each of the preceding programs, provides basic control over the way in which the output is presented. A related command, `printf`, offers much greater flexibility, allowing one to specify exactly how the output should be formatted (Robbins 2001). In general terms, the syntax of the `printf` statement is as follows:

```
printf("format-string", arg1, arg2, ..., argN)
```

In addition to the `printf` keyword, the `printf` statement contains a comma-separated list of arguments (arg1, arg2, ..., argN) and a format string (`"format-string"`): the former specifies what is to be printed and in which order; the latter comprises a string of characters that specifies exactly how the output should be formatted, by means of format specifiers (Table 3.5). Each argument has a corresponding format specifier.

Program 3.3, which is functionally identical to Program 3.2, illustrates how the `printf` statement is used. There are five arguments in this case: the first four are the contents of fields 4 (year), 5 (month), 6 (day) and 7 (hour) of the input data file, respectively; the fifth is the value in the ninth field of the input data file divided by the numerical constant 10.0 (i.e., accumulated precipitation in millimeters (mm)). The corresponding format string (`"%i%i%i%i %f\n"`) indicates that each of the first four arguments should be printed as an integer value (`%i`); that is, a number without a fractional part or, more formally, a member of the infinite set of positive and negative natural numbers ($\ldots -3, -2, -1, 0, 1, 2, 3 \ldots$). By contrast, the format string indicates that the fifth argument should be printed as a floating-point number (`%f`);

Table 3.6: Selected `printf` format modifiers. Note that ␣ marks a blank space in the output.

| Modifier | Interpretation | Examples | | |
|---|---|---|---|---|
| | | Value | Modifier | Result |
| *width* | Minimum width in characters | 62 | %9i | ␣␣␣␣␣␣␣62 |
| | | 62 | %2i | 62 |
| | | 3.1415927 | %1i | 3 |
| | | 3.1415927 | %9f | ␣3.141593 |
| | | "string" | %9s | ␣␣␣string |
| | | "string" | %2s | string |
| *.prec* | Precision | 3.1415927 | %9.2f | ␣␣␣␣␣3.14 |
| | | | %9.4f | ␣␣␣3.1416 |
| — | Left-justify | 3.1415927 | %-9.2f | 3.14␣␣␣␣␣ |
| *leading zero* | Pad output with leading zeros | 1 | %02s | 01 |
| | | 1 | %03s | 001 |

that is, a number with a fractional part or, more formally, a member of the infinite set of positive and negative real numbers, such as 3.1415927. The format string also indicates that the fifth argument should be separated from the other four arguments by a single blank space, and that a newline character (\n) should be inserted at the end of each line in the output. The effect of the `printf` format specifiers, therefore, is to control how particular values are presented in the output. For instance, applying the format specifier `%i` to the numerical constant 3.1415927 causes it to be printed out as the integer value 3, by truncating the fractional part. Similarly, the format specifier `%f` converts the integer constant 1 to a floating-point number, which is printed out as 1.000000.

> **Exercise 3.2**: Run Program 3.3, `selcols3.awk`, on the data contained in `rain98le.dat` and confirm that the output is the same as that shown in Figure 3.4. Edit the program by removing the newline character (\n) from the format string. What effect does this have when you re-run the program on the same data set? What is the role of the newline character?

It is possible to achieve still greater control over the format of the output from the `printf` statement by using format modifiers (Table 3.6). Among other things, these can be employed to allocate different field widths in the output for each argument, to specify the precision (i.e., the number of decimal places) with which floating-point numbers are reported, to left-justify or right-justify within a field, and to pad-out the values in particular fields with leading zeros.

Program 3.4: selcols4.awk

```
(NR >1) {                                                                      1
    printf ("%4i%02i%02iT%04i  %5.1f\n", $4, $5, $6, $7, $9/10.0) ;            2
}                                                                              3
```

Program 3.4 shows how format modifiers are used with the `printf` statement. In this example, the value of the first argument (i.e., the year number) is printed as an integer and is allocated a space at least four characters wide in the output (`%4i`; e.g., `1998`). The second and third arguments (i.e., the month and the day) are also printed as integers, each being allocated a space at least two characters wide in the output. If the value of either of these two arguments consists of just one character (i.e., a number in the range `0` to `9`) it is padded out by a leading zero (`%02i`); for example, the value `1` is output as `01`. Similarly, the fourth argument, also printed as an integer, is allocated a space at least four characters wide, padded with leading zeros where necessary (`%04i`; e.g., the value `900` is output as `0900`). The format string also specifies that the third and fourth arguments should be separated by the letter "T", so that the output conforms to the ISO 8610:2004 standard for the representation of date and time information in numerical format (see Appendix D). Lastly, the fifth argument is printed out as a floating-point number, is allocated a space of at least five characters wide in the output, is reported to one decimal place (1 d.p.; `%5.1f`) and is separated from the other arguments by a single blank space.

In addition to the changes that have been made to the format string, Program 3.4 (`selcols4.awk`) is also laid out slightly differently from Program 3.3 (`selcols3.awk`). More specifically, the action part of Program 3.4 is spread over three lines of code, and the second line is indented by one tab stop, whereas it is presented on a single line in Program 3.3. This makes no difference to the functioning of the code, but it helps to improve its readability. Note that line 2 of Program 3.4 is terminated by a semi-colon (`;`). Strictly speaking, this is not required by the **gawk** utility, but it serves two useful purposes: first, it confirms that the line of code is meant to end at this point; second, the convention of terminating a statement with a semi-colon eases the transition to coding in other programming languages, such as C, where it is obligatory. Matters of programming style such as these are developed further in subsequent chapters, but it is important to understand that they are guidelines rather than rules and that, ultimately, consistency is as important as programming style (Oualline 1997).

Program 3.4 can be applied to the data in `rain98le.dat` by typing the following instruction on the command line:

```
gawk  -f selcols4.awk rain98le.dat                                             6
```

The first 12 lines of the resulting output are shown in Figure 3.5. Notice how the time and date values have been padded with leading zeroes where required; thus, 9 am on January 1, 1998 is reported as `19980101T0900` (cf. `199811T900`). As a result, the two fields of data output are properly formatted and aligned. Note, also, the spurious negative precipitation value reported at 9 pm on January 6, 1998 (i.e., the final record

```
19980101T0900     0.0
19980101T2100    14.6
19980102T0900     5.2
19980102T2100     1.6
19980103T0900    35.8
19980103T2100     5.6
19980104T0900     6.6
19980104T2100    20.8
19980105T0900     4.4
19980105T2100     3.0
19980106T0900     4.0
19980106T2100   -99.9
```

Figure 3.5: First 12 lines of output produced by Program 3.4 applied to the data in
rain98le.dat. Note the spurious negative precipitation reported at 9 pm on January 6, 1998 (last record shown; see text for details).

shown in Figure 3.5). This has arisen because, for whatever reason, a precipitation
value was not recorded by the AWS on this occasion. Instead, the "missing data"
value (−999) was inserted in this record of the input data file (rain98le.dat). The
gawk program, selcols4.awk, has faithfully divided this number by 10.0 to produce
a value of −99.9 in the output, although this is clearly not what was intended. The
following sections explore two possible solutions to this problem: one of which uses
gnuplot; the other, gawk. First, though, a method for saving the output from a gawk
program to a file is demonstrated.

## 3.8  REDIRECTING THE OUTPUT TO A FILE

In each of the examples examined thus far, the results of data processing are directed
to the computer screen for immediate inspection. It is sometimes necessary, though,
to save the results for future reference so that they can be combined with other data
sets, incorporated within a technical report, or simply output to a printer. This can
be achieved by sending the output to a file, rather than to the screen. This process is
known as redirection and is demonstrated below.

```
gawk -f selcols4.awk rain98le.dat > rain98le.out                  7
```

The request for redirection is indicated by the > symbol on the command line, which
sends the results of the gawk program selcols4.awk (which is, in turn, applied to
the input file rain98le.dat) to the output data file rain98le.out. Note that the redirection symbol (>) forms part of the command line under GNU/Linux and Microsoft
Windows, and should not be confused with the "greater than" logical operator (>),
which is part of the gawk programming language. It may also help to think of the
redirection symbol as an arrow that points in the direction in which the data flows,
i.e., from the gawk program (on the left) to the output data file (on the right).

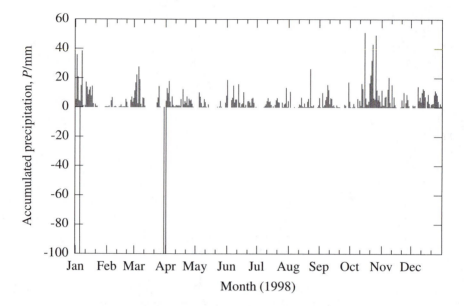

Figure 3.6: Hourly precipitation accumulation at Llyn Efyrnwy throughout 1998 illustrating the "missing data" problem.

## 3.9   VISUALIZING THE OUTPUT DATA

The data file, `rain98le.out`, produced in the preceding section can be visualized in gnuplot using the following commands, which should be familiar from Chapter 2:

```
reset                                                        1
unset key                                                    2
set xdata time                                               3
set timefmt "%Y%m%dT%H%M"                                    4
set format x "%b"                                            5
set xlabel "Month (1998)"                                    6
set ylabel "Accumulated precipitation, P/mm"                 7
set xrange ["19980101T0000":"19990101T0000"]                8
plot 'rain98le.out' u 1:2 w i                                9
```

These commands produce the plot shown in Figure 3.6. The problem with Figure 3.6 is immediately apparent: "missing data" values are plotted as spurious negative precipitation readings. These points badly distort the plot and, in any case, are highly misleading. A cosmetic solution to this problem is to limit the range of data values displayed on the *y*-axis to numbers greater than zero (i.e., by typing `set yrange [0:*]` in gnuplot) because it is not possible to have negative precipitation values. This merely hides the problem, though; it does not solve it. A much better approach is to use the "missing data" facility in gnuplot, which is demonstrated below.

```
set datafile missing "-99.9"                                10
replot                                                       11
```

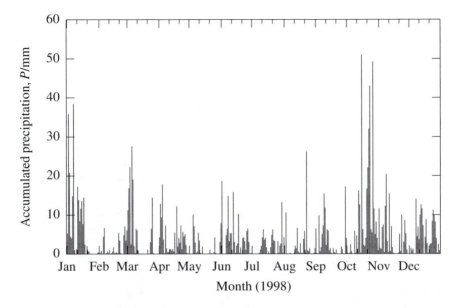

Figure 3.7: Hourly precipitation accumulation at Llyn Efyrnwy throughout 1998 with the "missing data" values removed by gnuplot.

Line 10 tells gnuplot which character string is used to denote "missing data" in the data file. In the original precipitation data set (rain98le.dat) the missing data value is −999, but in the output data file (rain98le.out) it is −99.9 because the raw precipitation values are divided by 10.0 to convert them into millimeters. Thus, line 10 tells gnuplot to treat any data item that has the value −99.9 on the y-axis as missing data. As a result, these data records are not plotted. Line 11 plots the data once again, this time taking the missing data value into account. The result is shown in Figure 3.7, which indicates clearly that Llyn Efyrnwy received rainfall every month during 1998, although it peaked in January, March and, in particular, November.

## 3.10   LOGICAL OR BOOLEAN OPERATORS IN gawk

Another approach to the "missing data" problem is to modify the gawk program selcols4.awk so that records containing the missing data value are skipped, in much the same way that the first record is already skipped because it contains the names of the data fields rather than the data themselves. This approach is exemplified in Program 3.5, selcols5.awk, which is run from the command line as follows:

```
gawk -f selcols5.awk rain98le.dat > le98rain.dat
```
8

Program 3.4 (selcols4.awk) and Program 3.5 (selcols5.awk) differ solely in terms of the pattern specified on line 1 of the code. In selcols5.awk the pattern has been extended so that it matches those records for which the record number is greater than one (NR>1; i.e., the second and subsequent records, but not the first) and

Program 3.5: selcols5.awk

```
(NR >1  && $9 != -999) {                                                    1
  printf ("%4i%02i%02iT%04i  %5.1f\n",  $4,  $5,  $6,  $7,  $9/10.0) ;2
}                                                                           3
```

Table 3.7: Logical or Boolean operators in gawk.

| Operator | Meaning | Example | Evaluates as ... |
|----------|---------|---------|------------------|
| && | and | a && b | True if both a *and* b are true; otherwise, false. |
| \|\| | or | a \|\| b | True if a *or* b *or* both are true; otherwise, false. |
| ! | not | !a | True if a is false. False if a is true. |

for which the value in the ninth field is not equal to -999 ($9!=-999; i.e., the missing data value). The double ampersand symbol (&&) employed in this pattern is known as a logical, or Boolean, operator (Table 3.7) and is interpreted as "and". The symbol != is one of the comparison operators introduced earlier in this chapter (Table 3.4) and means "is not equal to". It is used here to check whether the ninth field of the current record (i.e., the raw precipitation value) contains the missing data value (-999). In short, the revised pattern skips the first record of the input data file, which contains the titles for each of the fields in the remaining records, and any other record in that file for which the accumulated precipitation value is missing.

> **Exercise 3.3**: Using a text editor, write a gawk program to process the file temp98le.dat, which is included on the CD-ROM, so that the output is similar to the contents of the file called le98temp.dat, which was studied in the preceding chapter. Run the program on the command line, redirecting the output to a new file, le98temp.out. Visualize the results in gnuplot, and confirm that the plot is similar to that shown in Figure 2.8.

## 3.11   SUMMARY

This chapter introduces the gawk programming language and uses it to pre-process an environmental data set. The principal advantage that gawk offers in this context is the comparative ease with which data can be read from a file (or files), and with which specific parts of the data can be selected and processed. As an example, a data set containing information on the accumulated precipitation at Llyn Efyrnwy every 12 hours throughout 1998 is converted into a format suitable for visualization in gnuplot. Further aspects of the gawk programming language are introduced in the chapters that follow, allowing a wider range of environmental data processing and modeling exercises to be performed.

## SUPPORTING RESOURCES

The following resources are provided in the `chapter3` sub-directory of the CD-ROM:

| | |
|---|---|
| `rain98le.dat` | Precipitation data for Llyn Efyrnwy, 1998, in MetOffice format |
| `selcols.awk` | gawk program to print all records and fields in a data file |
| `selcols2.awk` | gawk program to select specific records and fields of data from a file |
| `selcols3.awk` | gawk program to select specific records and fields from a file and print them out in a given format |
| `selcols4.awk` | gawk program to select specific records and fields from a file and print them out in a given format using format modifiers |
| `selcols5.awk` | gawk program to select specific records and fields from a file, skipping records with "missing data" values |
| `preciptn.bat` | Command line instructions used in this chapter. |
| `preciptn.plt` | gnuplot commands used to generate the figures presented in this chapter |
| `preciptn.gp` | gnuplot commands used to generate figures presented in this chapter as EPS files |

# Chapter 4

# Wind Speed and Wind Power

**Topics**

- Wind power as a source of natural renewable energy

- Estimating wind energy and wind power

**Methods and techniques**

- Variables and comments in gawk

- Conditional statements in gawk — the if else construct

- Actions performed *after* reading data from a file — gawk's END block

- Writing, plotting and fitting functions in gnuplot

## 4.1 INTRODUCTION

This chapter examines the potential for producing electricity from wind power at Llyn Efyrnwy using a combination of a wind turbine and an electricity generator, known as a wind energy conversion system (WECS) (Freris 1990, Manwell *et al.* 2002, Twidell and Weir 2006). Wind speed data recorded by an AWS over the course of a year are analyzed to determine the mean wind speed at Llyn Efyrnwy and the extent to which the wind speed varies around this value. A mathematical model is fitted to these data to estimate the frequency with which wind speeds capable of driving a small WECS are experienced at Llyn Efyrnwy. A separate mathematical model is used to estimate the power that the wind produces and, hence, the amount of electricity that could be generated by a small WECS based at this site. Further aspects of gnuplot and gawk are also introduced.

Figure 4.1: Part of the Taff Ely wind farm, South Wales, UK.

Wind power is an increasingly attractive source of natural renewable energy that may help to reduce the current reliance on fossil fuels, such as coal, oil and gas, and hence limit the emission of carbon dioxide ($CO_2$) and other "greenhouse" gases into the atmosphere (Golding 1977, Gipe 1995). The emission of greenhouse gases is the subject of an international treaty, commonly referred to as the Kyoto Protocol, which was adopted by the United Nations Framework Convention on Climate Change (UNFCCC) held in Kyoto, Japan, on December 11, 1997. By March 15, 1999, 84 countries had signed up to the Kyoto Protocol, and in so doing agreed to place limits on the emission of various greenhouse gases, such as $CO_2$, methane ($CH_4$), nitrous oxide ($N_2O$), hydrofluorocarbons (HFCs), perfluorocarbons (PFCs) and sulfur hexafluoride ($SF_6$), into the atmosphere. Signatories to the treaty are also required to develop technologies for carbon sequestration (i.e., the removal of carbon from the atmosphere), to promote energy efficiency and to increase the use of renewable forms of energy, including wind, wave and solar power (http://unfccc.int/resource/convkp. html). The current list of signatories to the Kyoto Protocol (189 as of June 12, 2006) can be found at http://unfccc.int/resource/conv/ratlist.pdf.

Although wind power is a potentially significant source of renewable energy, it is not a panacea. There are many important issues that need to be addressed, not least of which are the intermittent nature of the wind, the general lack of data on the wind "resource" (especially its variability with respect to space and time), the impact that wind "farms" have on landscape quality and visual amenity (a subjective and often highly emotive topic), the large distances that sometimes separate the most suitable sites for wind farms from the major centers of energy demand, and the effects that wind turbines have on both people (e.g., through noise pollution) and wildlife (e.g., via collisions with the turbine rotors) (Twidell and Weir 2006).

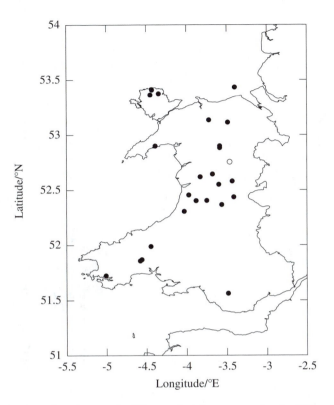

Figure 4.2: Location of onshore and offshore wind farms operating in Wales, UK in 2006 (closed circles). The position of Llyn Efyrnwy is marked by the open circle.

In the UK, most of the wind farms currently in operation are located in the north and west of the country, including Wales (Figure 4.1). The spatial distribution of the 23 wind farms operating in Wales in 2006 is shown in Figure 4.2, while summary statistics about these sites are given in Table 4.1. More wind farms are planned in the near future, including a large off-shore installation in Swansea Bay, South Wales, at a site known as Scarweather Sands. In comparison to these major developments, however, this chapter has a modest objective: to evaluate the energy resource that might be delivered by a small WECS based at Llyn Efyrnwy.

Gipe (1999) identifies three categories of small WECS based on the radius ($r$) of the turbine rotors: micro ($r \leq 1.25\,\text{m}$), mini ($1.25\,\text{m} < r \leq 2.75\,\text{m}$) and household ($r > 2.75\,\text{m}$). For the purpose of this exercise, it is assumed that the WECS installed at Llyn Efyrnwy is required to meet part of the electricity demand from a small number of residential dwellings and, hence, that the radius of the turbine rotors is 3 m. In this context, it is helpful to note that the average annual consumption of electricity per UK household is estimated to be 4600 kilowatt hours (4600 kW·h; http://www.dti.gov.uk/ energy/statistics/publications/energy-consumption/page17658.html).

The potential for generating electricity from wind power at a particular site, such as Llyn Efyrnwy, is controlled by a number of factors. These factors include the

Table 4.1: Location and characteristics of wind farms operating in Wales, UK in 2006.

| Site name | Operational since | Number of turbines | Turbine rating (MW) | Capacity (MW) | Latitude(°N) | | Longitude(°W) | |
|---|---|---|---|---|---|---|---|---|
| Mynydd Clogau | Apr 2006 | 17 | 0.85 | 14.45 | 52°34'49" | 52.5803° | 3°25'59" | 3.4331° |
| Hafoty Ucha 3 extension | Jan 2006 | 1 | 0.85 | 0.85 | 52°54'01" | 52.9003° | 4°23'18" | 4.3883° |
| Tir Mostyn and Foel Goch | Sep 2005 | 25 | 0.85 | 21.25 | 53°06'52" | 53.1144° | 3°29'17" | 3.4881° |
| Cefn Croes | Apr 2005 | 39 | 1.50 | 58.50 | 52°24'18" | 52.4050° | 3°45'03" | 3.7508° |
| Castle Pill Farms | Sep 2004 | 1 | 0.50 | 0.50 | 51°43'27" | 51.7242° | 5°00'26" | 5.0072° |
| Llangwyryfon | Feb 2004 | 11 | 0.85 | 9.35 | 52°18'26" | 52.3072° | 4°01'43" | 4.0286° |
| North Hoyle (Offshore) | Dec 2003 | 30 | 2.00 | 60.00 | 53°26'00" | 53.4333° | 3°24'00" | 3.4000° |
| Moel Maelogen | Dec 2002 | 2 | 1.30 | 2.60 | 53°08'08" | 53.1356° | 3°43'24" | 3.7233° |
| Haffoty Ucha 2 extension | Dec 2002 | 2 | 0.85 | 1.70 | 52°54'01" | 52.9003° | 3°35'18" | 3.5883° |
| Blaen Bowi | Jul 2002 | 3 | 1.30 | 3.90 | 51°59'17" | 51.9881° | 4°26'43" | 4.4453° |
| Cemmaes | Mar 2002 | 18 | 0.85 | 15.30 | 52°38'38" | 52.6439° | 3°40'45" | 3.6792° |
| Parc Cynog | Feb 2001 | 5 | 0.72 | 3.60 | 51°51'28" | 51.8578° | 4°34'40" | 4.5778° |
| Haffoty Ucha 1 | Sep 1998 | 1 | 0.60 | 0.60 | 52°53'01" | 52.8836° | 3°35'18" | 3.5883° |
| Mynydd Gorddu | Apr 1998 | 19 | 0.50 | 10.20 | 52°27'18" | 52.4550° | 3°58'20" | 3.9722° |
| Llyn Alaw | Oct 1997 | 34 | 0.60 | 20.40 | 53°21'48" | 53.3633° | 4°26'59" | 4.4497° |
| Centre for Alternative Technology | Apr 1997 | 1 | 0.60 | 0.60 | 52°37'09" | 52.6192° | 3°49'54" | 3.8317° |
| Rheidol | Jan 1997 | 8 | 0.30 | 2.40 | 52°24'12" | 52.4033° | 3°52'56" | 3.8822° |
| Carno "A" and "B" | Oct 1996 | 56 | 0.60 | 33.60 | 52°33'02" | 52.5506° | 3°36'01" | 3.6003° |
| Trysglwyn | Jul 1996 | 14 | 0.40 | 5.60 | 53°22'28" | 53.3744° | 4°20'43" | 4.3453° |
| Dyffryn Brodyn | Dec 1994 | 11 | 0.50 | 5.50 | 51°52'09" | 51.8692° | 4°33'25" | 4.5569° |
| Bryn Titli | Jul 1994 | 22 | 0.45 | 9.90 | 52°22'03" | 52.3675° | 3°33'51" | 3.5642° |
| Taff Ely | Aug 1993 | 20 | 0.45 | 9.00 | 51°33'46" | 51.5628° | 3°29'09" | 3.4858° |
| Llandinam "P&L" | Dec 1992 | 103 | 0.30 | 30.90 | 52°26'11" | 52.4364° | 3°24'49" | 3.4136° |
| Rhyd-y-Groes | Dec 1992 | 24 | 0.30 | 7.20 | 53°24'31" | 53.4086° | 4°25'47" | 4.4297° |
| Mawla (Moel Maelogen) | – | 1 | 1.30 | 1.30 | 53°08'08" | 53.1350° | 3°43'24" | 3.723° |

Source: http://www.bwea.com/. Latitude and longitude values are given in both degrees, minutes and seconds (DMS) and decimal degrees (DD).

mean wind speed close to ground level ($\overline{u_0}$), typically expressed in meters per second ($m \cdot s^{-1}$), the variation of wind speed over time (i.e., the frequency with which the wind speed varies and the range of wind speeds encountered), and the mean air density ($\rho$), expressed in kilograms per cubic meter ($kg \cdot m^{-3}$), which is in turn dependent on altitude and air temperature. As a general rule, the higher the mean wind speed, the greater the wind power and, hence, the more electricity is generated; however, the actual amount of electricity generated is also controlled by variation of the wind speed about the mean value. Thus, most wind turbines have a "cut in" wind speed at which they start to produce power (Manwell *et al.* 2002, Twidell and Weir 2006). If the wind speed drops below this value, during periods of calm or light breezes for example, no electricity is generated. Similarly, there is often a "cut out" wind speed above which it may be necessary to stop the turbine for safety reasons. In practice, though, many WECS continue to operate at very high wind speeds, albeit at a reduced level of efficiency, with a high power output (Twidell and Weir 2006). On balance, therefore, a site with relatively stable wind conditions is generally preferable to one at which the wind speed is variable, provided that the mean wind speed is sufficiently high (Gipe 1999). Consequently, the two main steps in evaluating the potential power output from a small WECS at Llyn Efyrnwy involve (i) determining the annual mean wind speed at this site and (ii) quantifying the variation of wind speed over time.

## 4.2 DESCRIPTION OF THE WIND SPEED DATA

The exercises covered in this chapter make use of the file wind98le.dat, which can be found on the CD-ROM. This file contains data on the mean wind speed and wind direction recorded at hourly intervals throughout 1998 by an AWS based at Llyn Efyrnwy (Table 4.2 and Figure 4.3). The hourly mean wind speed is reported to the nearest Knot, where one Knot is approximately equal to $0.515\,m \cdot s^{-1}$. The hourly mean wind direction indicates the direction from which the wind blew, in arc degrees (°) relative to true (geographical) north, to the nearest 10°. The file also contains information on the maximum gust speed, the maximum gust direction and the time of the maximum gust, although these data are not used here and, hence, the fields concerned are omitted from Figure 4.3. The data are supplied by the UK MetOffice in ASCII text format via the BADC web site (http://badc.nerc.ac.uk/).

Hourly mean values of wind speed are not ideal in terms of predicting the likely performance of a WECS because they mask fluctuations in wind speed that take place over shorter timescales: more frequent observations are, however, seldom available (Twidell and Weir 2006). Moreover, the data in wind98le.dat describe measurements made at a single height above the ground (10 m; the international standard), so that there is no information on how the wind speed varies with elevation, which would help to determine the optimum configuration of a WECS at Llyn Efyrnwy. Despite these limitations, the data can be used to give a preliminary indication of the potential for electricity generation from wind power at this site, which is often the best that can be achieved in the absence of a dedicated survey (Gipe 1995).

| ID | IDTYPE | MET_DOM | YEAR | MON | DAY | END_HOUR | COUNT | MDIR | MSPEED |
|---|---|---|---|---|---|---|---|---|---|
| 794801 | WIND | ESAWWIND | 1998 | 1 | 1 | 0 | 1 | 210 | 9 |
| 794801 | WIND | ESAWWIND | 1998 | 1 | 1 | 100 | 1 | 210 | 8 |
| 794801 | WIND | ESAWWIND | 1998 | 1 | 1 | 200 | 1 | 220 | 9 |
| 794801 | WIND | ESAWWIND | 1998 | 1 | 1 | 300 | 1 | 240 | 2 |
| 794801 | WIND | ESAWWIND | 1998 | 1 | 1 | 400 | 1 | 240 | 1 |
| 794801 | WIND | ESAWWIND | 1998 | 1 | 1 | 500 | 1 | 230 | 6 |
| 794801 | WIND | ESAWWIND | 1998 | 1 | 1 | 600 | 1 | 230 | 3 |
| 794801 | WIND | ESAWWIND | 1998 | 1 | 1 | 700 | 1 | 200 | 3 |
| 794801 | WIND | ESAWWIND | 1998 | 1 | 1 | 800 | 1 | 210 | 11 |
| 794801 | WIND | ESAWWIND | 1998 | 1 | 1 | 900 | 1 | 200 | 11 |
| 794801 | WIND | ESAWWIND | 1998 | 1 | 1 | 1000 | 1 | 200 | 15 |
| 794801 | WIND | ESAWWIND | 1998 | 1 | 1 | 1100 | 1 | 190 | 14 |
| 794801 | WIND | ESAWWIND | 1998 | 1 | 1 | 1200 | 1 | 200 | 21 |
| 794801 | WIND | ESAWWIND | 1998 | 1 | 1 | 1300 | 1 | 200 | 20 |
| 794801 | WIND | ESAWWIND | 1998 | 1 | 1 | 1400 | 1 | 200 | 22 |
| 794801 | WIND | ESAWWIND | 1998 | 1 | 1 | 1500 | 1 | 190 | 22 |
| 794801 | WIND | ESAWWIND | 1998 | 1 | 1 | 1600 | 1 | 180 | 22 |
| 794801 | WIND | ESAWWIND | 1998 | 1 | 1 | 1700 | 1 | 190 | 20 |
| 794801 | WIND | ESAWWIND | 1998 | 1 | 1 | 1800 | 1 | 200 | 23 |
| 794801 | WIND | ESAWWIND | 1998 | 1 | 1 | 1900 | 1 | 230 | 20 |
| 794801 | WIND | ESAWWIND | 1998 | 1 | 1 | 2000 | 1 | 230 | 22 |
| 794801 | WIND | ESAWWIND | 1998 | 1 | 1 | 2100 | 1 | 240 | 21 |
| 794801 | WIND | ESAWWIND | 1998 | 1 | 1 | 2200 | 1 | 240 | 20 |
| 794801 | WIND | ESAWWIND | 1998 | 1 | 1 | 2300 | 1 | 240 | 21 |

Figure 4.3: Extract from the file wind981e.dat (first 25 records and first 10 fields only).

Table 4.2: Explanation of the data fields in `wind98le.dat`.

| Field | Name | Contents |
|-------|------|----------|
| 1 | ID | Meteorological station ID number |
| 2 | IDTYPE | Meteorological station type |
| 3 | MET_DOM | Meteorological domain |
| 4 | YEAR | Year |
| 5 | MON | Month (1–12) |
| 6 | DAY | Day of the month (1–31) |
| 7 | END_HOUR | Hour of observation (00:00–23:00 GMT) |
| 8 | COUNT | Number of hours over which measurements are made |
| 9 | MDIR | Mean direction from which the wind blows (°) |
| 10 | MSPEED | Mean wind speed (Knots; $1\,\text{Knot} \approx 0.515\,\text{m·s}^{-1}$) |
| 11 | GUST_DIR | Maximum gust direction (°) |
| 12 | GUST_SPEED | Maximum gust speed (Knot) |
| 13 | GUST_TIME | Time of maximum gust (00:00–23:00 GMT) |

## 4.3 CALCULATING THE ANNUAL MEAN WIND SPEED

The mean wind speed, $\bar{u}$, over a particular period of time can be expressed as

$$\bar{u} = \frac{1}{N} \sum_{i=1}^{N} u_i \tag{4.1}$$

where $u_i$ is the $i^{\text{th}}$ observation in a series of $N$ wind speed measurements made at regular intervals of time. Equation 4.1 is translated into gawk code in Program 4.1, which also converts the mean wind speed value from Knots to meters per second (the units conventionally employed in wind energy studies). The structure and operation of this program are explained below.

The pattern on line 1 of Program 4.1 matches those records of the input data file for which the record number is greater than one (NR>1) and (&&) for which the value in the tenth field is not equal to -999 ($10!=-999). The first part of the pattern selects the second and subsequent records of the input data file, but not the first; the second part of the pattern selects only those records that contain a valid hourly mean wind speed measurement, discarding the records in which this value is missing (note that the value -999 is used to indicate missing data in this file). Thus, the two parts of the pattern combine to skip the first record of the file because this contains the field names and any other records in which the wind speed measurement is missing. The corresponding actions, which are grouped within a pair of curly braces on lines 1 and 5, are performed if, and only if, the current record matches both parts of this pattern.

Program 4.1: meanwspd.awk

```
(NR >1  &&  $10 != -999) {                                               1
   wind_speed = $10;                                                     2
   sum_speed = sum_speed + wind_speed;                                   3
   num_obs = num_obs +1;                                                 4
}                                                                        5
                                                                         6
END {                                                                    7
   mean_wind_speed = 0.515* sum_speed / num_obs;                         8
   printf ("Mean  wind  speed=%3.1f  m/s\n", mean_wind_speed);           9
}                                                                       10
```

Line 2 of Program 4.1 introduces a new entity, known as a user-defined variable (cf. a built-in variable, such as NR; see Chapter 3). As the name suggests, this type of variable is created by the programmer to store a specific value for subsequent use within a program. The value stored by a variable can be a number, a single character or a text string. Each variable has a name, chosen by the programmer, by which it is identified. As far as possible, the name of a variable should indicate the nature of the data that the variable holds. The name can comprise any sequence of letters (a–z, A–Z), digits (0–9) and underscores (_), but it may not begin with a digit or contain any blank spaces (Robbins 2001). Underscores are useful in this context because they allow several words to be linked together into a single, hopefully more meaningful, name. Thus, the name of the variable on line 2 is wind_speed. It is also important to note that variable names are case sensitive, so that gawk will treat wind_speed and Wind_Speed as entirely separate variables, which may store very different values. Moreover, user-defined variables are initially assigned the value zero (0) or the empty string ("") in gawk. This process, which is performed automatically by the gawk utility, is sometimes known as initialization.

Line 2 of Program 4.1 takes the value from the tenth field ($10) of the current record of the input data file and stores it in the variable wind_speed, replacing whatever value this variable previously held. This process is known as assignment, and the = symbol is referred to as the assignment operator. Here, the variable wind_speed is assigned the value of the tenth field in the current record. Note that a common mistake is to confuse the assignment operator (a single "equals" sign, =) with one of the comparison operators (a double "equals" sign, ==). The former means "make the variable named on the left-hand side of this operator equal to the value given on the right-hand side"; the latter asks the question "is the value on the left-hand side of this operator equal to that on the right hand side". So, for example, the statement wind_speed=4.5 assigns the variable wind_speed the value 4.5 (i.e., it changes the value of the variable), whereas wind_speed==4.5 checks to see whether the value stored in wind_speed is equal to 4.5 (but the value of the variable remains unchanged).

A second variable, sum_speed, is introduced on line 3 of Program 4.1. This is used to store the sum of the wind speed values read from the input data file. Thus, line 3 takes the wind speed value for the current record, which it stored in wind_speed

Table 4.3: Operation of lines 1–5 of Program 4.1 applied to the data in `wind98le.dat` (Figure 4.3), showing how the values of the variables `wind_speed`, `sum_speed` and `num_obs` change as each record of data is processed.

| NR | $10 | Pattern match? | wind_speed | sum_speed | num_obs |
|---|---|---|---|---|---|
| 1 | MSPEED | No | 0 | 0 | 0 |
| 2 | 9 | Yes | 9 | 9 | 1 |
| 3 | 8 | Yes | 8 | 17 | 2 |
| 4 | 9 | Yes | 9 | 26 | 3 |
| 5 | 2 | Yes | 2 | 28 | 4 |
| ⋮ | ⋮ | ⋮ | ⋮ | ⋮ | ⋮ |
| 1599 | 8 | Yes | 8 | 17 581 | 1598 |
| 1600 | -999 | No | 8 | 17 581 | 1598 |
| 1601 | 8 | Yes | 8 | 17 589 | 1599 |
| ⋮ | ⋮ | ⋮ | ⋮ | ⋮ | ⋮ |
| 8207 | 10 | Yes | 10 | 72 866 | 8201 |
| 8208 | 10 | Yes | 10 | 72 876 | 8202 |
| 8209 | 8 | Yes | 8 | 72 884 | 8203 |

on line 2, and adds this to the sum of the wind speed values examined thus far (`sum_speed+wind_speed`). The result of this operation is stored in `sum_speed`, replacing the value previously held by this variable. The syntax of line 3 may seem a little strange at first, but its effect is to add up (sum) the hourly mean wind speed measurements as the data are read in record-by-record from the file, storing the running total in `sum_speed`.

Line 4 of Program 4.1 introduces a third variable called `num_obs`. As the name may suggest, this variable is used to count the number of records that contain valid observations of hourly mean wind speed (i.e., that match the pattern specified on line 1 of the program). The value of `num_obs` is increased ("incremented") by 1 for each such record (`num_obs=num_obs+1`). Note that Table 4.3 illustrates the way in which the values of the three variables `wind_speed`, `sum_speed` and `num_obs` are updated as each record of the input data file `wind98le.dat` is read and processed.

The mean wind speed over the entire period for which observations are available is calculated by dividing `sum_speed` by `num_obs` once all of the records from the input data file have been processed. gawk provides a special feature, known as the END block (or END rule), that can be used for this purpose. Statements contained in the END block are performed after the main pattern-action block is completed (Figure 4.4). In Program 4.1, the END block is given on lines 7 through 10; it is identified by the keyword END (line 7), which must be typed entirely in uppercase, and a matching pair of opening and closing curly braces, {} (lines 7 and 10, respectively). The curly

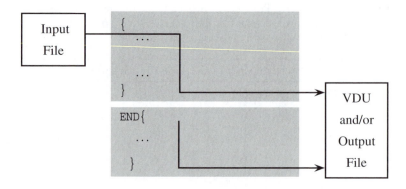

Figure 4.4: gawk's main pattern-action and END blocks.

braces serve to group together all of the statements (i.e., actions) in the END block. In this example, the END block contains two statements (lines 8 and 9 of Program 4.1). Line 8 calculates the mean wind speed in Knots (sum_speed/num_obs), converts this value into units of meters per second by multiplying it by the constant factor 0.515, and stores the result in another new variable, mean_wind_speed. Line 9 employs a printf statement to print some explanatory text followed by the mean wind speed in meters per second, expressed as a three-digit floating-point number reported to one decimal place (%3.1f).

Program 4.1 can be applied to the data in wind98le.dat by typing the following instruction on the command line:

```
gawk -f meanwspd.awk wind98le.dat                                        1
```

This produces the following output:

```
Mean wind speed=4.6 m/s
```

Thus, the mean wind speed at Llyn Efyrnwy throughout 1998 was approximately $4.6\,\mathrm{m \cdot s^{-1}}$. Assuming that this value is representative of the longer-term mean, it is too low to consider siting a commercial wind farm there, but it may be sufficient to power a small WECS to meet at least some of the electricity demand from a limited number of households.

## 4.4   DETERMINING THE MAXIMUM WIND SPEED

The long-term mean wind speed at a particular site is one factor that determines the potential for electricity generation from wind power; variation of the wind speed about the mean value is another. For example, a site that experiences very high winds for relatively short periods of time and calm conditions otherwise is typically less suited to wind energy generation than a location at which the wind speed is more consistent, even if the mean wind speed is slightly lower. The problem of wind speed variability is twofold: first, when the wind speed falls below the cut-in threshold of

Program 4.2: meanmaxw.awk

```
(NR>1 && $10!=-999){                                              1
  wind_speed=$10;                                                 2
  sum_speed+=wind_speed;                                          3
  ++num_obs;                                                      4
  if(wind_speed>max_speed){                                       5
    max_speed=wind_speed;                                         6
  }                                                               7
}                                                                 8
                                                                  9
END{                                                              10
  mean_wind_speed=0.515*sum_speed/num_obs;                        11
  printf("Mean wind speed=%3.1f m/s\n", mean_wind_speed);         12
  max_speed*=0.515;                                               13
  printf("Max. wind speed=%4.1f m/s\n", max_speed);               14
}                                                                 15
```

the WECS, no electricity is generated; second, it is not always possible to exploit the full potential of very high winds. Thus, power may have to be "dumped" when the wind speed exceeds the cut-out threshold of the WECS, and electricity generated during high winds that is surplus to immediate requirements has to be stored in some way (e.g., using batteries) to meet the demand in periods of calm or light winds.

The range of wind speeds encountered at a specific site gives an initial indication of the wind speed variability. The minimum wind speed is, of course, likely to be $0 \, \text{m·s}^{-1}$ (i.e., calm conditions). The maximum wind speed, on the other hand, varies from site to site and must therefore be determined from meteorological observations. For example, the maximum wind speed at Llyn Efyrnwy can be established from the measurements contained in the file wind981e.dat. One way to find this value is to search through the data set manually, although this method is tedious and error-prone. A more effective approach is to modify Program 4.1 so that it determines the maximum, as well as the mean, wind speed. This task involves comparing the wind speed value for each record of the input data file with the maximum wind speed encountered among the preceding records. If the current wind speed is greater than the previous maximum value, then this becomes the new maximum. This procedure is implemented in Program 4.2 (meanmaxw.awk).

The major difference between Programs 4.1 and 4.2 is that several lines of code have been added to the main pattern-action and END blocks in Program 4.2 (lines 5–7 and 13–14, respectively). These are used to establish and print the maximum wind speed recorded in the input data file. More specifically, lines 5 through 7 introduce another of **gawk**'s programming constructs, the if statement, which is sometimes referred to as a control-flow or branching structure (Figure 4.5); thus, the if statement controls the actions that are performed depending on the outcome of a specified condition. The general syntax of an if statement is if(*condition*){actions}, which is interpreted as "perform the following actions if, and only if, the condition is found

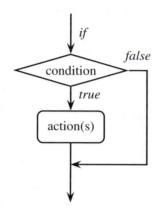

Figure 4.5: gawk's `if` construct.

to be true" (Robbins 2001). In Program 4.2, for example, line 5 checks whether the current wind speed (`wind_speed`) is greater than (`>`) the maximum wind speed encountered thus far (`max_speed`): this is the condition. If this condition is true, `max_speed` is assigned the value of `wind_speed` (i.e., the current wind speed becomes the new maximum), otherwise the value of `max_speed` remains unaltered: this is the action (line 6). Note that a pair of curly braces, `{}`, is used to group the action (or actions) contained within the `if` statement. Also, the action statement is indented by a single tab stop to highlight the fact that it forms part of the `if` statement.

There are several other differences between Programs 4.1 and 4.2, most of which involve small, but important, changes to the syntax of the commands. For instance, line 3 of Program 4.1, which reads `sum_speed=sum_speed+wind_speed`, is replaced by `sum_speed+=wind_speed` on line 3 of Program 4.2. The two statements are functionally equivalent (i.e., they produce the same result), but the latter makes use of the "plus equals" operator (`+=`), which is more concise. Thus, the `+=` operator sums the values given on either side, storing the result in the variable named on the left-hand side (Robbins 2001). So, for example, if the variables x and y initially store the numbers 5 and 3, respectively, the statement `x+=y` modifies x such that, subsequently, its new value is 8; the value of y remains unchanged. Table 4.4 presents the full set of gawk's concise arithmetic assignment operators (Robbins 2001).

A similar modification is made on line 4 of Program 4.2, which keeps track of the number of valid wind speed measurements. The equivalent line of code in Program 4.1 is `num_obs=num_obs+1`, which increases ("increments") the value of `num_obs` by one every time a record matches the pattern specified on line 1. Adding one to the value of a variable is such a common operation in computer programming that gawk, in common with many other programming languages, provides a more concise way of expressing it using the "plus plus" symbol, `++`, which is known as the increment operator. Thus, `++num_obs` means "add one to the value of `num_obs` and store the result in `num_obs`". There is a corresponding decrement operator, `--` ("minus minus"), which subtracts one from the value of a variable. So, for example, if x and y initially store the values 5 and 3, respectively, the statements `++x` and `--y` change their values to 6

Table 4.4: gawk's arithmetic assignment operators.

| Operator | Syntax | Interpretation |
| --- | --- | --- |
| += | x+=y | Adds the values of x and y, assigns the result to x. |
| -= | x-=y | Subtracts the value of y from x, assigns the result to x. |
| *= | x*=y | Multiplies the values of x and y, assigns the result to x. |
| /= | x/=y | Divides the value of x by y, assigns the result to x. |
| %= | x%=y | Finds the remainder of x modulo y, assigns the result to x (e.g., if x=10 and y=3 then x%y=1). |
| **= | x**=y | Raises x to the power of y (i.e., $x^y$), assigns the result to x. |
| ^= | x^=y | Equivalent to the **= operator. |

Table 4.5: gawk's increment and decrement operators.

| Operator | Syntax | Example | Interpretation |
| --- | --- | --- | --- |
| | x++ | y=x++ | Assign y the value of x, then increment x by 1 |
| ++ | ++x | y=++x | Increment x by 1, then assign y this value |
| | x-- | y=x-- | Assign y the value of x, then decrement x by 1 |
| -- | --x | y=--x | Decrement x by 1, then assign y this value |

and 2 subsequently. Table 4.5 presents the full set of gawk's increment and decrement operators (Robbins 2001).

Finally, line 13 of Program 4.2 uses the "times equals" operator, *=, to convert the maximum wind speed value from Knots into meters per second. This operator takes the value of the variable given on its left-hand side and multiplies it by the value on the right-hand side; the result is stored in the variable listed on the left-hand side of the operator. So, for example, if x and y initially have values of 5 and 3, respectively, the statement y*=x assigns y the value 15, while the value of x remains unchanged.

Program 4.2 can be applied to the data in wind98le.dat by typing the following instruction on the command line:

```
gawk -f meanmaxw.awk wind98le.dat
```
2

This produces the following output:

```
Mean wind speed=4.6 m/s
Max. wind speed=19.1 m/s
```

Thus, the maximum mean hourly wind speed at Llyn Efyrnwy during 1998 was $19.1 \, \mathrm{m \cdot s^{-1}}$.

## 4.5   EXPLORING WIND SPEED VARIABILITY

A more comprehensive picture of wind speed variability can be gained by calculating the absolute frequency with which winds of different speeds are experienced at Llyn Efyrnwy. To do so, the wind speed data must be grouped into a number of discrete classes with an interval of, for instance, $1\,\text{m·s}^{-1}$ (e.g., $0.0\,\text{m·s}^{-1} \leq u_0 < 1.0\,\text{m·s}^{-1}$, $1.0\,\text{m·s}^{-1} \leq u_0 < 2.0\,\text{m·s}^{-1}$, and so on). The task then becomes one of counting the number of occasions on which the measured wind speed lies within each of these intervals. Relative frequency values can also be obtained from these data by dividing the raw counts by the total number of observations in the data set. If the number of wind speed measurements in the data set is sufficient, and the observations are representative of the long-term conditions at the site, it is possible to construct a wind speed probability distribution, which can be used to estimate the likelihood with which a given wind speed will occur. This information is important when matching the operational characteristics of a WECS (e.g., its cut-in and cut-out thresholds, and the wind speed at which it is rated for maximum power output) to the local wind regime (Twidell and Weir 2006). Methods for calculating the absolute frequency, relative frequency and probability distributions of wind speed are examined below.

### 4.5.1   Determining the Absolute Frequency Distribution

There are various ways in which the absolute frequency distribution of different wind speeds can be determined from the data contained in `wind98le.dat`, one of which is presented in Program 4.3 (`windfreq.awk`). The code employed in this program sacrifices brevity for simplicity. Thus, the programming constructs are the same as those used in Program 4.2, but the number of lines of code involved is much greater. There are much more concise ways of coding the same problem, but these involve the use of more sophisticated **gawk** programming constructs, notably arrays, which are introduced in subsequent chapters.

The pattern on line 1 of Program 4.3 is identical to that used in Program 4.2, and it serves the same purpose: to skip the first record of the input data file and any other records for which the hourly mean wind speed measurement is missing. Lines 2 and 4, which begin with the hash symbol (#), are known as comments. They are ignored by the **gawk** utility and are included here solely to indicate to the reader the purpose of the subsequent lines of code. The judicious use of comments, such as these, is an example of good programming practice. Used sparingly, comments help the reader to understand what a program is meant to do and how it is intended to function. Line 3 takes the value from the tenth field of the current record of the input data file (i.e., the hourly mean wind speed in Knots), multiplies this value by 0.515 (to convert it into $\text{m·s}^{-1}$) and stores the result in the variable `wind_speed`.

Lines 5 through 24 employ a series of `if` statements to count the number of times that different wind speeds occur. Line 5, for example, checks whether the current hourly mean wind speed is less than $1\,\text{m·s}^{-1}$ (`wind_speed<1`). If this is the case, the value of the variable `speed_1` is increased by 1 (i.e., `++speed_1`). Thus, `speed_1` is used to keep track of the number of times that the wind speed lies in the range

Program 4.3: windfreq.awk (lines 1–46)

```
(NR >1 && $10!=-999){                                                1
  # Wind speed in meters per second                                  2
  wind_speed=0.515*$10;                                              3
  # Count frequency of different wind speeds                         4
  if(wind_speed<1){++speed_1}                                        5
  if(wind_speed>=1 && wind_speed<2){++speed_2}                       6
  if(wind_speed>=2 && wind_speed<3){++speed_3}                       7
  if(wind_speed>=3 && wind_speed<4){++speed_4}                       8
  if(wind_speed>=4 && wind_speed<5){++speed_5}                       9
  if(wind_speed>=5 && wind_speed<6){++speed_6}                      10
  if(wind_speed>=6 && wind_speed<7){++speed_7}                      11
  if(wind_speed>=7 && wind_speed<8){++speed_8}                      12
  if(wind_speed>=8 && wind_speed<9){++speed_9}                      13
  if(wind_speed>=9 && wind_speed<10){++speed_10}                    14
  if(wind_speed>=10 && wind_speed<11){++speed_11}                   15
  if(wind_speed>=11 && wind_speed<12){++speed_12}                   16
  if(wind_speed>=12 && wind_speed<13){++speed_13}                   17
  if(wind_speed>=13 && wind_speed<14){++speed_14}                   18
  if(wind_speed>=14 && wind_speed<15){++speed_15}                   19
  if(wind_speed>=15 && wind_speed<16){++speed_16}                   20
  if(wind_speed>=16 && wind_speed<17){++speed_17}                   21
  if(wind_speed>=17 && wind_speed<18){++speed_18}                   22
  if(wind_speed>=18 && wind_speed<19){++speed_19}                   23
  if(wind_speed>=19){++speed_20}                                    24
}                                                                   25
                                                                   26
END{                                                               27
  print "# Wind speed distribution, Llyn Efyrnwy (1998)";          28
  print "# Speed (m/s)   Frequency";                               29
  printf("%13.1f %10i\n",  0.5,   speed_1);                        30
  printf("%13.1f %10i\n",  1.5,   speed_2);                        31
  printf("%13.1f %10i\n",  2.5,   speed_3);                        32
  printf("%13.1f %10i\n",  3.5,   speed_4);                        33
  printf("%13.1f %10i\n",  4.5,   speed_5);                        34
  printf("%13.1f %10i\n",  5.5,   speed_6);                        35
  printf("%13.1f %10i\n",  6.5,   speed_7);                        36
  printf("%13.1f %10i\n",  7.5,   speed_8);                        37
  printf("%13.1f %10i\n",  8.5,   speed_9);                        38
  printf("%13.1f %10i\n",  9.5,   speed_10);                       39
  printf("%13.1f %10i\n", 10.5,   speed_11);                       40
  printf("%13.1f %10i\n", 11.5,   speed_12);                       41
  printf("%13.1f %10i\n", 12.5,   speed_13);                       42
  printf("%13.1f %10i\n", 13.5,   speed_14);                       43
  printf("%13.1f %10i\n", 14.5,   speed_15);                       44
  printf("%13.1f %10i\n", 15.5,   speed_16);                       45
  printf("%13.1f %10i\n", 16.5,   speed_17);                       46
```

Program 4.3 (continued): windfreq.awk (lines 47–50)

```
printf ("%13.1f %10i\n", 17.5, speed_18);          47
printf ("%13.1f %10i\n", 18.5, speed_19);          48
printf ("%13.1f %10i\n", 19.5, speed_20);          49
}                                                  50
```

$0\,\mathrm{m\cdot s}^{-1} \leq u_0 < 1\,\mathrm{m\cdot s}^{-1}$. Line 6 performs a similar function for wind speeds in the range $1\,\mathrm{m\cdot s}^{-1} \leq u_0 < 2\,\mathrm{m\cdot s}^{-1}$ and uses the variable speed_2 to store the result. This procedure is replicated for each of the other wind speed classes on lines 7 through 24. In total, 20 if statements and variables are used to record the frequency values for winds ranging in speed from $0\,\mathrm{m\cdot s}^{-1}$ to $20\,\mathrm{m\cdot s}^{-1}$ (recall that the maximum wind speed is $19.1\,\mathrm{m\cdot s}^{-1}$; see Section 4.4) in steps of $1\,\mathrm{m\cdot s}^{-1}$.

The END block (lines 27 to 50) prints out two lines of explanatory text plus the frequency data for each of the wind speed classes. Note that the text output by lines 28 and 29 of Program 4.3 commences with a hash symbol (#). This is intended to make the output data easier to use within gnuplot, which treats anything on a line after a hash symbol as a comment (i.e., it ignores lines such as these when plotting data from a file). Lines 30 through 49 use the printf statement to output a pair of values separated by a single blank space. Line 30, for example, prints out the wind speed at the mid-point of the first wind speed class (i.e., $0.5\,\mathrm{m\cdot s}^{-1}$) and the number of records in wind98le.dat which belong to this class (i.e., that exhibit a wind speed in the range of $0\,\mathrm{m\cdot s}^{-1} \leq u_0 < 1\,\mathrm{m\cdot s}^{-1}$). The remaining lines (31 through 49) print out the equivalent values for the other wind speed classes. Note that the format string ("%13.1f %10i\n") used on lines 30 to 49 aligns the columns of data in the output.

Program 4.3 can be applied to the data in wind98le.dat by typing the following instruction on the command line:

```
gawk -f windfreq.awk wind98le.dat > windfreq.out          3
```

The contents of the output data file, windfreq.out, are presented in Figure 4.6. These data can be visualized in gnuplot (Figure 4.7) by typing the following commands:

```
reset                                              1
unset key                                          2
set yrange [0:*]                                   3
set xrange [0:20]                                  4
set xlabel "Mean wind speed, u_0/m.s^{-1}"         5
set ylabel "Absolute frequency"                    6
plot 'windfreq.out' u 1:2 w boxes lw 2             7
```

Line 1 resets gnuplot to its default behavior (see Chapter 2, page 44), and line 2 indicates that the plot should not contain a key. Line 3 sets the range of values to be displayed on the y-axis, constraining the minimum plotted value to zero, but allowing gnuplot to determine the maximum value. Line 4 performs a similar task for the x-axis (i.e., $0\,\mathrm{m\cdot s}^{-1} \leq u_0 < 20\,\mathrm{m\cdot s}^{-1}$). Lines 5 and 6 define the labels to be placed along the x- and y-axes, respectively. Finally, line 7 plots the data in the output file,

```
# Wind speed distribution, Llyn Efyrnwy (1998)
# Speed (m/s)    Frequency
         0.5          753
         1.5          859
         2.5         1216
         3.5         1173
         4.5         1030
         5.5          807
         6.5          679
         7.5          473
         8.5          365
         9.5          268
        10.5          181
        11.5          154
        12.5           95
        13.5           52
        14.5           52
        15.5           30
        16.5           12
        17.5            1
        18.5            2
        19.5            1
```

Figure 4.6: Absolute frequency distribution of the hourly mean wind speed at Llyn Efyrnwy.

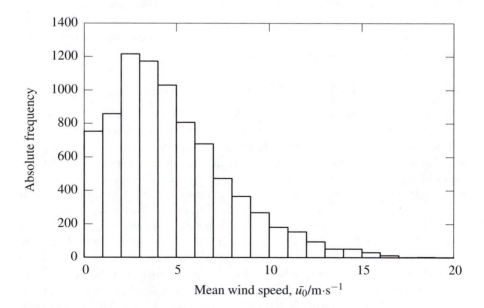

Figure 4.7: Visualization of the absolute frequency distribution of the hourly mean wind speed at Llyn Efyrnwy in 1998.

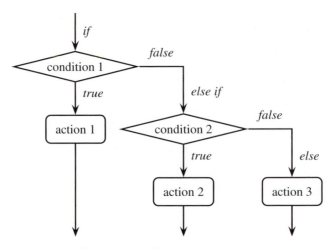

Figure 4.8: gawk's if else construct.

windfreq.out, using the values in field 1 for the *x*-series and those in field 2 for the *y*-series. The data are plotted using the boxes style, which draws a box centered on the *x*-axis value, extending from $y = 0$ to the *y*-axis value. The width of each box is calculated automatically so that it touches the adjacent boxes (Williams and Kelly 1998). Here, the line width of the boxes is set to twice the normal value (lw 2) so that the boxes are more pronounced.

It is clear from Figure 4.6 and Figure 4.7 that the modal wind speed at Llyn Efyrnwy lies in the range $2 \, \text{m·s}^{-1}$ to $3 \, \text{m·s}^{-1}$ and that the wind-speed frequency distribution is negatively skewed. The significance of this is that many WECS start to generate electricity when the wind speed exceeds $3 \, \text{m·s}^{-1}$, which suggests that there will be periods during which no power will be generated by a WECS at Llyn Efyrnwy.

### 4.5.2   Determining the Relative Frequency Distribution

Relative frequency values are typically more useful than absolute frequency values because they allow comparisons to be made between wind speed observations recorded at different sites, as well as between measurements made at the same site on different occasions. Relative frequency values are calculated by dividing the raw counts by the total number of observations in the data set. This procedure is demonstrated in Program 4.4, which also incorporates a number of minor modifications with respect to Program 4.3 intended to improve the efficiency of the code.

An extra variable, num_obs, is employed within the main pattern-action block of Program 4.4 to keep track of the total number of valid wind speed measurements (line 25). The value stored in this variable is used in the END block to convert the absolute counts for the different wind speed classes into relative frequency values (lines 31 to 50), which are printed as floating-point numbers to four decimal places (%12.4f).

The code on lines 5 through 24 has also been amended to make use of a sequence of gawk's if else statements (Figure 4.8), instead of the if statements employed in Program 4.3. To understand why this change has been made, consider the case of a

Program 4.4: windfrq2.awk (lines 1–47)

```
(NR>1 && $10!=-999){                                                      1
  # Wind speed in meters per second                                       2
  wind_speed=0.515*$10;                                                   3
  # Count frequency of different wind speeds                              4
  if(wind_speed<1){++speed_1}                                             5
  else if(wind_speed>=1 && wind_speed<2){++speed_2}                       6
  else if(wind_speed>=2 && wind_speed<3){++speed_3}                       7
  else if(wind_speed>=3 && wind_speed<4){++speed_4}                       8
  else if(wind_speed>=4 && wind_speed<5){++speed_5}                       9
  else if(wind_speed>=5 && wind_speed<6){++speed_6}                      10
  else if(wind_speed>=6 && wind_speed<7){++speed_7}                      11
  else if(wind_speed>=7 && wind_speed<8){++speed_8}                      12
  else if(wind_speed>=8 && wind_speed<9){++speed_9}                      13
  else if(wind_speed>=9 && wind_speed<10){++speed_10}                    14
  else if(wind_speed>=10 && wind_speed<11){++speed_11}                   15
  else if(wind_speed>=11 && wind_speed<12){++speed_12}                   16
  else if(wind_speed>=12 && wind_speed<13){++speed_13}                   17
  else if(wind_speed>=13 && wind_speed<14){++speed_14}                   18
  else if(wind_speed>=14 && wind_speed<15){++speed_15}                   19
  else if(wind_speed>=15 && wind_speed<16){++speed_16}                   20
  else if(wind_speed>=16 && wind_speed<17){++speed_17}                   21
  else if(wind_speed>=17 && wind_speed<18){++speed_18}                   22
  else if(wind_speed>=18 && wind_speed<19){++speed_19}                   23
  else {++speed_20}                                                      24
  ++num_obs;                                                             25
}                                                                        26
                                                                         27
END{                                                                     28
  print "# Wind speed distribution, Llyn Efyrnwy (1998)";                29
  print "# Speed (m/s)   Rel. Freq.";                                    30
  printf("%13.1f %12.4f\n",  0.5,   speed_1/num_obs);                    31
  printf("%13.1f %12.4f\n",  1.5,   speed_2/num_obs);                    32
  printf("%13.1f %12.4f\n",  2.5,   speed_3/num_obs);                    33
  printf("%13.1f %12.4f\n",  3.5,   speed_4/num_obs);                    34
  printf("%13.1f %12.4f\n",  4.5,   speed_5/num_obs);                    35
  printf("%13.1f %12.4f\n",  5.5,   speed_6/num_obs);                    36
  printf("%13.1f %12.4f\n",  6.5,   speed_7/num_obs);                    37
  printf("%13.1f %12.4f\n",  7.5,   speed_8/num_obs);                    38
  printf("%13.1f %12.4f\n",  8.5,   speed_9/num_obs);                    39
  printf("%13.1f %12.4f\n",  9.5,   speed_10/num_obs);                   40
  printf("%13.1f %12.4f\n", 10.5,   speed_11/num_obs);                   41
  printf("%13.1f %12.4f\n", 11.5,   speed_12/num_obs);                   42
  printf("%13.1f %12.4f\n", 12.5,   speed_13/num_obs);                   43
  printf("%13.1f %12.4f\n", 13.5,   speed_14/num_obs);                   44
  printf("%13.1f %12.4f\n", 14.5,   speed_15/num_obs);                   45
  printf("%13.1f %12.4f\n", 15.5,   speed_16/num_obs);                   46
  printf("%13.1f %12.4f\n", 16.5,   speed_17/num_obs);                   47
```

Program 4.4 (continued): windfrq2.awk (lines 48–51)

```
  printf ("%13.1f %12.4f\n", 17.5, speed_18/num_obs);        48
  printf ("%13.1f %12.4f\n", 18.5, speed_19/num_obs);        49
  printf ("%13.1f %12.4f\n", 19.5, speed_20/num_obs);        50
}                                                             51
```

record for which the measured wind speed is $0.5 \text{ m·s}^{-1}$. In Program 4.3, the record is initially checked against the condition for the `if` statement on line 5. Since the wind speed is less than $1 \text{ m·s}^{-1}$, the condition evaluates as true, and the corresponding action is performed (i.e., the value of `speed_1` is incremented by one). The record is then checked against each of the other `if` statements even though, by definition, it cannot fall within any of the other wind speed classes. Thus, the wind speed associated with each record is checked 20 times (i.e., against each of the `if` statements).

Program 4.4 is a little more intelligent in this respect. The `if else` statements operate in a nested manner (Figure 4.8). For example, if the condition on line 5 (`wind_speed<1`) is true for the current record, the corresponding action is performed (`++speed_1`), after which the program skips lines 6 through 24 and, instead, proceeds to line 25. If the condition on line 5 is not met, then (`else`) the condition on line 6 (`wind_speed>=1 && wind_speed<2`) is evaluated. If this condition evaluates as true for the current record, the corresponding action (`++speed_2`) is performed, after which the program proceeds to line 25, and so on. In this way, the program continues to drop down through the `if else` statements until one of the conditions evaluates as true, at which point the corresponding action is performed. If none of the conditions is found to be true for the current record, the final `else` (line 24) acts as a "catch all". Since most of the wind speed values reported in `wind98le.dat` are quite low, this procedure should increase the efficiency, and hence the speed of execution, of the code.

Program 4.4 can be applied to the data in `wind98le.dat` by typing the following instruction on the command line:

```
gawk -f windfrq2.awk wind98le.dat > windfrq2.out        4
```

The contents of the output data file, `windfrq2.out`, are presented in Figure 4.9. These data can be visualized by typing the following commands in gnuplot:

```
set ylabel "Relative frequency"              8
plot 'windfrq2.out' u 1:2 w boxes lw 2       9
```

The resulting plot is presented in Figure 4.10.

**Exercise 4.1**: Write a gawk program to calculate the relative frequency with which the wind blows from the following directions at Llyn Efyrnwy: 315°–15°, 15°–45°, 45°–75°, ... , 275°–315°. Run this program on the data contained in the file `wind98le.dat` and plot the results in gnuplot.

```
# Wind speed distribution, Llyn Efyrnwy (1998)
# Speed (m/s)    Rel. Freq.
         0.5        0.0918
         1.5        0.1047
         2.5        0.1482
         3.5        0.1430
         4.5        0.1256
         5.5        0.0984
         6.5        0.0828
         7.5        0.0577
         8.5        0.0445
         9.5        0.0327
        10.5        0.0221
        11.5        0.0188
        12.5        0.0116
        13.5        0.0063
        14.5        0.0063
        15.5        0.0037
        16.5        0.0015
        17.5        0.0001
        18.5        0.0002
        19.5        0.0001
```

Figure 4.9: Relative frequency distribution of the hourly mean wind speed at Llyn Efyrnwy throughout 1998.

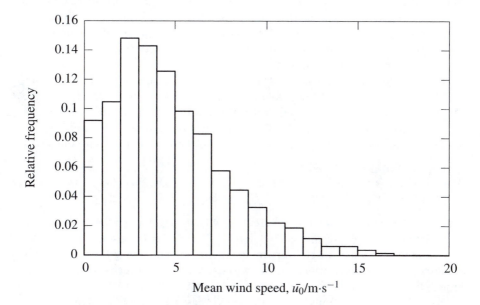

Figure 4.10: Visualization of the relative frequency distribution of hourly mean wind speed at Llyn Efyrnwy throughout 1998.

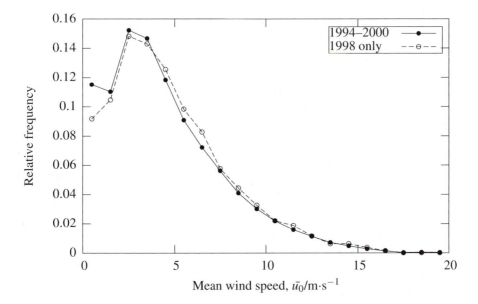

Figure 4.11: Relative frequency distributions of hourly mean wind speed at Llyn Efyrnwy for
1998 and the period 1994 to 2000, inclusive.

### 4.5.3  Probability Distributions and Probability Density Functions

The file `wind98le.dat` contains a sample of over 8000 wind speed measurements.
Even so, it is possible that the data come from a year in which the observed wind
speeds differ significantly from the long-term norm; that is, the 1998 data may be
unrepresentative of the typical conditions at this site. This possibility can be evaluated
by analyzing a longer time-series of data.

Comprehensive measurements of the hourly mean wind speed at Llyn Efyrnwy
are available for the period from 1994 to 2000, inclusive, from the BADC web site.
The relative frequency distribution of these values is given in the file `wind7yr.dat`,
which is included on the CD-ROM. The data from 1998 only and from 1994 through
to 2000 can be compared by plotting them in gnuplot using the following commands:

```
set key top right Left box                                              10
plot 'wind7yr.dat' u 1:2 t "1994-2000" w lp, \                         11
     'windfrq2.out' u 1:2 t "1998 only" w lp                           12
```

The resulting output is shown in Figure 4.11, which suggests that the 1998 data are
broadly typical of the prevailing conditions at this site. It is reasonable, therefore,
to treat Figure 4.10 as a discrete approximation to the underlying (i.e., continuous)
wind speed probability distribution; that is to say, the values on the $y$-axis indicate the
probability with which a set of discrete wind speed classes (i.e., $0 \leq u_0 < 1\,\mathrm{m \cdot s^{-1}}$,
$1 \leq u_0 < 2\,\mathrm{m \cdot s^{-1}}, \ldots, 19\,\mathrm{m \cdot s^{-1}} \leq u_0 < 20\,\mathrm{m \cdot s^{-1}}$) are experienced at Llyn Efyrnwy.

It is possible to represent the wind speed probability distribution at Llyn Efyrnwy
more accurately using a continuous mathematical function, known as a probability

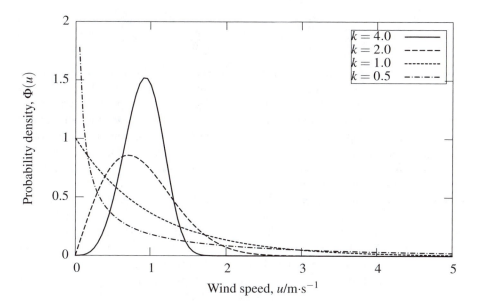

Figure 4.12: Weibull PDFs for various values of the shape parameter ($k$) and a fixed value of the scale parameter ($c = 1$).

density function (PDF). In studies of wind speed it is common to use either the Weibull distribution or the Rayleigh distribution for this purpose (Bowden *et al.* 1983, Manwell *et al.* 2002, Twidell and Weir 2006). The Weibull distribution is defined in terms of two parameters, $c$ and $k$:

$$\Phi(u) = \frac{k}{c} \left(\frac{u}{c}\right)^{k-1} \exp\left[-\left(\frac{u}{c}\right)^{k}\right] \tag{4.2}$$

where $u$ is the wind speed, $\Phi(u)$ is the probability density of the wind speed $u$, $c$ is a scale parameter and $k$ is a shape parameter. The Rayleigh distribution is specified in terms of a single parameter, $c$, and is a special case of the Weibull PDF for $k = 2$:

$$\Phi(u) = \frac{2u}{c^2} \exp\left[-\left(\frac{u}{c}\right)^{2}\right] \tag{4.3}$$

Figure 4.12 shows a number of Weibull PDFs produced using a fixed value of the scale parameter, $c$ (1.0), but different values of the shape parameter, $k$ (0.5, 1.0, 2.0 and 4.0). This figure was generated using the following gnuplot commands:

```
set  xrange  [0:5]                                                     13
set  yrange  [0:2]                                                     14
set  xlabel  "Wind speed, u_0/m.s^{-1}"                                15
set  ylabel  "Probability density, Phi(u)"                            16
set  dummy u                                                           17
phi(u,c,k)=(k/c)*((u/c)**(k-1))*exp((-1*(u/c)**k))                     18
plot  phi(u,1.0,4.0)  t  "k=4.0"  lw 2,  \                             19
```

```
      phi(u,1.0,2.0)  t  "k=2.0"  lw  2,  \                    20
      phi(u,1.0,1.0)  t  "k=1.0"  lw  2,  \                    21
      phi(u,1.0,0.5)  t  "k=0.5"  lw  2                        22
```

Lines 13 through 16 specify the ranges of values and the labels to be plotted on the
*x*- and *y*-axes. Line 17 instructs gnuplot to employ a "dummy" variable, u, to denote
the *x*-axis. This facility makes it possible to refer to the variable plotted on the *x*-axis
(i.e., wind speed) by its conventional name or symbol (i.e., *u*), which is often more
convenient and memorable than the generic symbol *x* (Williams and Kelly 1998).

Line 18 defines the function that is to be plotted, which is an implementation
of Equation 4.2 in gnuplot code. This is known as a user-defined function because
it is defined by the user, as distinct from a built-in function, such as sin(), which
is provided by default in gnuplot. Thus, the mathematical representation given in
Equation 4.2 is translated into gnuplot code. For the most part this is fairly self-
explanatory, although it is worth noting that the symbol ** means "raise to the power
of" in gnuplot; for instance x**3 means raise x to the power of 3 ($x^3$). Moreover,
parentheses are used more extensively in the gnuplot code than in Equation 4.2 to
make absolutely clear the way in which the function is evaluated. For the sake of
convenience, gnuplot allows each user-defined function to be given a unique name
(phi in this example), so that it can be referred to subsequently without having to
type the equation in full each time it is used (e.g., lines 19 through 22).

To the right of the function name on line 18, enclosed in parentheses, is a comma-
separated list of values, which are known as the arguments to the function. In this
example, the function has three arguments, which means that it expects to receive
three input values: the wind speed, *u*, the scale parameter, *c*, and the shape parameter,
*k*. The values of these arguments are passed in that order to the function, where they
are stored locally in the gnuplot variables u, c and k. This process is analogous to the
assignment of variables in gawk (see Chapter 3).

The phi function is employed (or "called") on lines 19 through 22 to plot four
separate Weibull PDFs, all of which share the same value of the scale parameter
($c = 1.0$), but each of which has a different value of the shape parameter ($k = 4.0$,
2.0, 1.0 and 0.5, respectively). Note that the value of *u* is not specified here; instead it
is allowed to vary freely and is plotted on the *x*-axis. Finally, recall that the backslash
(\) character is the line continuation symbol, such that lines 19 through 22 represent
a single gnuplot command.

### 4.5.4   Function Fitting in gnuplot

In practice, rather than specifying the parameter values explicitly in the manner outlined above, the values of *c* and *k* are usually established by fitting the Weibull PDF
to an observed wind speed distribution. This offers two important benefits. First, it
allows the wind speed distribution at a site to be summarized and reported in terms
of just two numbers: the estimated values of *c* and *k*. Second, the estimated values
of *c* and *k* can be used to predict the probability with which different wind speeds are
likely to occur: for instance calm conditions, when no electricity is generated, or very
high winds, which might cause damage to the wind turbine.

```
initial set of free parameter values

c               = 10
k               = 1

After 12 iterations the fit converged.
final sum of squares of residuals : 0.000761955
rel. change during last iteration : -1.04466e-08

degrees of freedom (ndf) : 3
rms of residuals         (stdfit) = sqrt(WSSR/ndf)        : 0.0159369
variance of residuals (reduced chisquare) = WSSR/ndf : 0.000253985

Final set of parameters            Asymptotic Standard Error
=======================            ==========================

c               = 5.32883         +/- 0.3177         (5.961%)
k               = 1.55037         +/- 0.09683        (6.246%)
```

Figure 4.13: Part of the output from the gnuplot function-fitting procedure.

It is, of course, possible to write a short gawk program to fit a mathematical function, such as the Weibull PDF, to a set of observations. Here, however, the curve-fitting capabilities of gnuplot are employed. The precise details of the function-fitting algorithm used by gnuplot need not detain us here, except to note that it makes use of least-squares adjustment (Williams and Kelly 1998). The gnuplot commands needed to achieve this are as follows:

```
set fit logfile "weibull.fit"                               23
c=10                                                        24
k=1                                                         25
fit phi(u,c,k) 'windfrq2.out' u 1:2 via c,k                26
```

Line 23 specifies the name of the file into which the output from the function-fitting procedure is placed; in this example, it is `weibull.fit`. If a file with this name exists in the working directory, the results of the function-fitting procedure are appended to that file; otherwise, a new file is created with this name. Lines 24 and 25 assign initial (estimated) values to the two parameters, $c$ and $k$; this process is directly analogous to the assignment of values to variables in gawk (see Chapter 3). Providing initial estimates of $c$ and $k$ in this way helps gnuplot's function-fitting algorithm to find the correction solution. Note that it is sometimes necessary to experiment with these initial values on a trial-and-error basis until a good fit is achieved. Finally, line 26 fits the function `phi(u,c,k)` to the data contained in fields 1 and 2 of the file `windfrq2.out` by adjusting the values `c` and `k` (i.e., via `c,k`).

The result of the function-fitting procedure is shown in Figure 4.13. This gives the estimated value of the two parameters as $c = 5.32883$ and $k = 1.55037$. Note that the exact values may differ very slightly between computers (e.g., between 32-bit and 64-bit processors). The output also provides a measure of the uncertainty associated with the estimated values of $c$ ($\pm 0.3177$) and $k$ ($\pm 0.09683$), which can be interpreted as confidence limits about the estimated values (Williams and Kelly 1998).

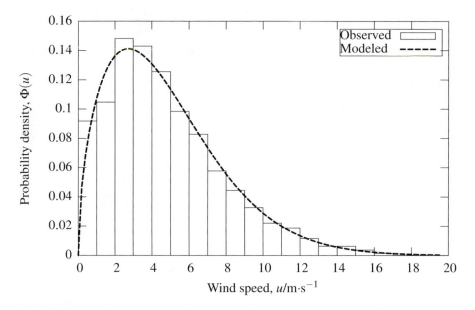

Figure 4.14: Relative frequency distribution of hourly mean wind speed at Llyn Efyrnwy in 1998 and the Weibull PDF fitted to these data ($k \approx 1.536$; $c \approx 5.295$).

The result of the function-fitting procedure can be visualized as follows:

```
set xrange [0:20]                                                    27
set yrange [0:*]                                                     28
plot 'windfrq2.out' u 1:2 t "Observed" w boxes, \                   29
    phi(u,c,k) t "Modeled" w l lt 2 lw 3                            30
```

Lines 27 and 28 set the range of values to be displayed along the *x*- and *y*-axes, respectively. Lines 29 and 30 plot the relative frequency distribution of the observed wind speeds using the boxes style (to indicate the discrete nature of the data) and the Weibull PDF, which has been fitted to these data, using the line style (because it is a continuous function). Note that the values of c and k for Llyn Efyrnwy have already been established by the function-fitting procedure, and are passed by reference to the function phi: in other words, they do not have to be entered explicitly here. The value of u varies freely and is plotted along the *x*-axis. The result is shown in Figure 4.14.

### 4.5.5   Probability of the Wind Speed Exceeding a Given Value

Equation 4.2 can be modified to express the probability of the wind speed, *u*, exceeding a given value, $u'$ (Manwell *et al.* 2002, Twidell and Weir 2006):

$$\Phi(u > u') = \exp\left[-\left(\frac{u'}{c}\right)^k\right] \tag{4.4}$$

This equation can be represented in gnuplot by another function, phi_prime:

```
phi_prime(u_prime,c,k)=exp((-1.0*(u_prime/c)**k))
```
*31*

where u_prime equates to $u'$ in Equation 4.4. Thus, the probability of the wind speed exceeding 1 m·s$^{-1}$ is evaluated by typing the following command in gnuplot:

```
print phi_prime(1.0,c,k)
```
*32*

Note that the values of c and k have already been established by the function-fitting procedure (line 26), and are passed by reference to the user-defined function phi_prime. The print command prints the output from this function to the screen as follows:

```
0.928000829058524
```

So, the probability of the hourly mean wind speed at Llyn Efyrnwy exceeding 1 m·s$^{-1}$ is approximately 0.928. Put another way, this suggests that hourly mean wind speeds greater than 1 m·s$^{-1}$ are experienced, on average, about 92.8% of the time.

In the example given above, the directive phi_prime(1.0,c,k) instructs gnuplot to evaluate the user-defined function phi_prime(u_prime,c,k) for u_prime=1.0 and for the values of c and k previously estimated by the function-fitting procedure. The same result could also have been obtained by giving all three parameter values explicitly, that is:

```
print phi_prime(1.0,5.32883,1.55037)
```
*33*

Any value of u_prime can be used in this context. For instance, the probability that the wind speed exceeds 2.5 m·s$^{-1}$ at Llyn Efyrnwy is evaluated as follows in gnuplot:

```
print phi_prime(2.5,c,k)
```
*34*

This command produces the following result:

```
0.733946698348097
```

> **Exercise 4.2**: Assuming that the cut-in and cut-out wind speeds for the WECS at Llyn Efyrnwy are 3 m·s$^{-1}$ and 20 m·s$^{-1}$, respectively, calculate the probability with which wind speeds capable of driving the WECS are experienced at this site.

## 4.6   WIND ENERGY AND POWER

### 4.6.1   Theoretical Basis

Once the wind speed distribution at Llyn Efyrnwy has been established, the next step is to estimate the power that this produces. It transpires that the energy content of the wind is proportional to the cube of the wind speed, $u_0^3$. This is converted into power by a wind turbine according to the following equation:

$$P = A\frac{\rho u_0^3}{2} \tag{4.5}$$

Table 4.6: Factors affecting the power output of a WECS.

| Factor | Symbol | Units | Typical value |
|---|---|---|---|
| Power output | $P$ | W | – |
| Cross-sectional area of turbine rotors | $A$ | $m^2$ | – |
| Air density | $\rho$ | $kg{\cdot}m^{-3}$ | 1.225 |
| Wind speed | $u$ | $m{\cdot}s^{-1}$ | – |
| Power coefficient | $C_p$ | – | 0.4 |
| Generator efficiency | $N_g$ | – | 0.75 |
| Mechanical efficiency | $N_b$ | – | 0.9 |

where $P$ is the power of the wind measured in Watts (W), $\rho$ is the density of dry air ($\rho = 1.225\,kg{\cdot}m^{-3}$ at sea level under average atmospheric pressure conditions and at $15\,°C$) and $A$ is the cross-sectional area swept by the turbine rotors (Twidell and Weir 2006). The latter, of course, is given by $A = \pi r^2$ where $r$ is the radius in meters of the rotor blades. Careful inspection of Equation 4.5 reveals that if the cross-sectional area of the turbine rotors is doubled, the wind power produced increases by a factor of two, whereas if the wind speed doubles, the power produced by the turbine increases by a factor of eight (Twidell and Weir 2006).

It is impossible to extract all of the power from a free-flowing air mass using a wind turbine, partly because the wind loses momentum as it encounters the rotor blades and partly because it must have some residual kinetic energy to move beyond the turbine (Twidell and Weir 2006). The fraction of wind power extracted varies from one turbine to another and is expressed in terms of the power coefficient, $C_P$, also known as the coefficient of performance (Table 4.6). The theoretical maximum value of $C_P$, known as the Betz limit, is 0.5926 (Betz 1926, Manwell *et al.* 2002). The value of $C_P$ for a given turbine is generally much smaller than the Betz limit and varies with wind speed; for a well-designed turbine, $C_P$ might peak around 0.4. There is also some loss of power via the mechanical components of the wind turbine (e.g., the gearbox and bearings) and the electricity generator. These losses are often expressed in terms of efficiency values where, for example, 1.0 indicates no loss and 0.5 indicates 50 % loss of the potentially available power. The mechanical efficiency, $N_b$, of turbines can differ considerably, and may be as high as 0.95. The efficiency of electricity generators, $N_g$, also varies, with 0.8 being a common maximum value (Manwell *et al.* 2002). Taking all of these factors together, the power output of a WECS can be expressed as follows:

$$P = A\frac{\rho u_0^3}{2}C_P N_b N_g \qquad (4.6)$$

The relationship between wind speed ($m{\cdot}s^{-1}$) and wind power (kW) represented by Equation 4.6 can be visualized in gnuplot using the following commands, the output from which is presented in Figure 4.15:

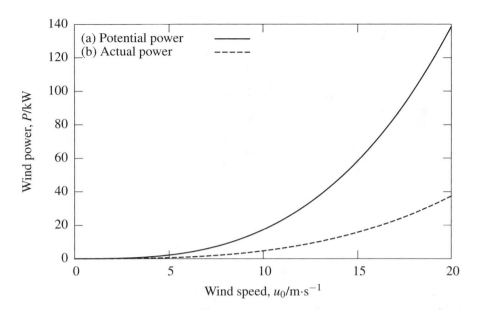

Figure 4.15: Theoretical relationship between wind power ($P$; kW) and wind speed ($u_0$; m·s$^{-1}$), based on assumed values of the cross-sectional area swept by the turbine rotors ($r = 3$ m, $A \approx 28.27$ m$^2$), air density ($\rho = 1.225$ kg·m$^{-3}$), power coefficient ($C_P = 0.4$), generator efficiency ($N_g = 0.75$) and mechanical efficiency ($N_b = 0.9$). (a) Potential power (Equation 4.5) and (b) actual power (Equation 4.6).

```
set xtics auto                                                        35
set xrange [0:20]                                                     36
rho=1.225                # Air density kg/m^3 at sea level            37
radius=3.0               # Rotor radius in meters                     38
area=pi*(radius**2)      # Area swept by turbine rotors               39
Cp=0.4                   # Power coefficient                          40
Ng=0.75                  # Generator efficiency                       41
Nb=0.9                   # Mechanical efficiency                      42
Power(u)=area*(rho*(u**3)/2)/1000                                     43
set xlabel "Wind speed, u_0/m.s^{-1}"                                 44
set ylabel "Wind power, P/kW"                                         45
set key top left Left nobox                                           46
plot Power(u) t "(a) Potential power" lw 2, \                         47
     Power(u)*Cp*Ng*Nb t "(b) Actual power" lw 2                      48
```

A major assumption underlying Figure 4.15 is that the value of $C_P$ is constant at all wind speeds, whereas this is not normally the case. Nevertheless, it serves to highlight the power generation capacity of a small WECS.

A new user-defined function, Power, is declared on line 41. This function has a single argument, u (i.e., the wind speed). The other values required by Power are provided by a number of user-defined variables, which are initialized on lines 35 to 40. The value of $\pi$ is given by the built-in variable pi (line 37) provided by gnuplot.

Program 4.5: wpower.awk

```
(NR>1 && $10!=-999){                                              1
  radius=3.0;              # Length of rotor blades               2
  pi=3.1415927;           # Pi                                    3
  area=pi*(radius**2);    # Cross-sectional area of rotors        4
  rho=1.225;              # Air density at sea level              5
  Cp=0.4;                 # Power coefficient                     6
  Nb=0.9;                 # Mechanical efficiency                 7
  Ng=0.75;               # Generator efficiency                  8
  knots2ms=0.515;         # Convert from Kts to m/s               9
  cut_in=3.0;             # Cut-in wind speed (m/s)              10
  cut_out=20.0;           # Cut-out wind speed (m/s)             11
  wind_speed=$10*knots2ms;                                       12
  if(wind_speed>=cut_in && wind_speed<cut_out){                 13
    power=area*(rho*(wind_speed**3)/2)*Cp*Nb*Ng;               14
    printf("%4i%02i%02iT%04i %8.4f\n", \                         15
      $4, $5, $6, $7, power/1000.0);                            16
  }                                                              17
}                                                                18
```

Note that the multiplication of the various factors on lines 41 and 46, which is implicit in Equations 4.5 and 4.6, must be stated explicitly in gnuplot code using the * operator. Also note that Power outputs values in kW, rather than W, by dividing by 1000 (line 41). The resulting plot compares the potential wind power given by Equation 4.5 with the actual power output of the WECS given by Equation 4.6, taking into account the losses due to $C_P$, $N_b$ and $N_g$. Figure 4.15 clearly shows that a relatively small fraction of the potential power of the wind is captured by the WECS.

It is also common to express the power output from a WECS in terms of power per unit area, or power density (W·m$^{-2}$), by dividing the raw power values by the cross-sectional area swept by the rotor blades (A). This allows a direct comparison to be made between the efficiency of wind turbines of different size (Freris 1990, Manwell *et al.* 2002).

> **Exercise 4.3**: Use gnuplot to create a plot of wind power density versus wind speed based on the values of $A$, $C_P$, $N_b$ and $N_g$ given in Table 4.6.

### 4.6.2    Application to Llyn Efyrnwy Data

Equation 4.6 can also be applied to the data in wind98lv.dat, using Program 4.5 (wpower.awk), by typing the following instruction on the command line:

```
gawk -f wpower.awk wind98le.dat > wind98le.pwr                    5
```

Line 1 of Program 4.5 should be familiar from previous programs: it is a pattern that is used to skip the first record of the named data file, which contains textual information as opposed to numerical data, and any record for which the hourly mean wind speed measurement is missing. Lines 2 to 9 initialize a number of the variables used in Equation 4.6. This is not the most efficient way of achieving this, because these actions are repeated for each record that is read from the named data file, but it will suffice for now. Lines 10 and 11 specify the values of the two variables (`cut_in` and `cut_out`) that are used to define the wind speeds at which electricity generation using the WECS at Llyn Efyrnwy cuts in ($3 \, \text{m·s}^{-1}$) and out ($20 \, \text{m·s}^{-1}$), respectively. Line 12 converts the hourly mean wind speed measurement for the current data record from Knots to meters per second, and stores the result temporarily in the variable `wind_speed`. Line 13 checks to see whether this value is greater than or equal to the cut-in wind speed and less than the cut-out wind speed. If this condition holds true for the current record, the actions on lines 14 through 16 are performed.

Line 14 calculates the power output arising from the hourly mean wind speed for the current record, taking into account the values assigned to the other variables. This is, in effect, a direct translation of Equation 4.6 into gawk code, and is very similar to the implementation of this equation in gnuplot demonstrated in the preceding section. Note that, as in gnuplot, the operator `**` means "raise to the power of" in gawk. Thus, `wind_speed**3` means raise the hourly mean wind speed measurement for the current record to the power of 3 ($u^3$).

Lines 15 and 16 use a `printf` statement to print out the date and time of the wind speed measurement and the mean power that this produces given the assumed properties of the WECS specified on lines 2 through 11. Recall that the `printf` format string indicates that the first four arguments — fields 4 (`$4`, year), 5 (`$5`, month), 6 (`$6`, day) and 7 (`$7`, time) of the current record — should be treated as integer values (`%i`) and the fifth — the value of the variable `power` divided by 1000 (i.e., power in kW·h) — as a floating-point number (`%f`). In addition, these lines of code instruct the gawk utility to allow a minimum width of four characters in the printout for the first and fourth arguments (`%4d`), two for the second and third (`%2d`) and eight for the fifth (`%8f`). The second and third arguments are also padded with leading zeroes if the values that they represent are less than two characters wide (`%02d`). Thus, the value 1 is printed out as 01. The fifth argument is printed out with four digits after the decimal place (`%8.4f`). Finally, the printout proceeds to the next line on the screen or output file (`\n`) after printing all of the arguments for the current record of data. The resulting output, `wind981v.pwr`, is shown in Figure 4.16. Note that no electricity is generated at 3 am, 4 am, 6 am or 7 am on January 1, 1998 because the wind speed is too low.

### 4.6.3  Visualizing the Output

The data file, `wind981v.pwr`, output by Program 4.5 can be visualized in gnuplot by entering the following instructions, all of which have been encountered previously:

```
unset key                                                    49
set xdata time                                               50
set timefmt "%Y%m%dT%H%M"                                    51
```

```
19980101T0000    0.4656
19980101T0100    0.3270
19980101T0200    0.4656
19980101T0500    0.1380
19980101T0800    0.8501
19980101T0900    0.8501
19980101T1000    2.1555
19980101T1100    1.7525
19980101T1200    5.9148
19980101T1300    5.1094
19980101T1400    6.8007
19980101T1500    6.8007
```

Figure 4.16: Partial contents of the file `wind98lv.pwr` (first 12 records).

```
set  xrange  ["19980101T0000":"19981231T2359"]                52
set  yrange  [0:*]                                            53
set  format  x  "%b"                                          54
set  xlabel  "Month (1998)"                                   55
set  ylabel  "Power, P/kW.h"                                  56
plot  'wind98le.pwr'  u  1:2  w  i                            57
```

The result is presented in Figure 4.17. It is clear from this figure that the power output from a WECS based at Llyn Efyrnwy is likely to be variable, with output exceeding 10 kW·h on occasion, but with prolonged periods when it is much less than this, and sometimes when no power is produced at all. Despite this limitation, electricity generation is possible throughout the year and storage batteries might be used to smooth out some, if not all, of the variations in power output to meet the needs of the intended users.

It is possible to show that the average hourly power output produced over the whole year in this case study is approximately 1.2 kW. The average annual consumption of energy per household in the UK is currently about 4600 kW·h, which roughly equates to 0.5 kW per hour. Thus, all other things being equal, a small WECS based at Llyn Efyrnwy, having the characteristics outlined previously in this chapter, might be expected to meet or supplement the electricity demand of up to two households. In reality the situation is more complex, but this analysis serves as an initial guide.

## 4.7  SUMMARY

This chapter examines the potential for electricity generation from wind power at Llyn Efyrnwy, based on a WECS with a turbine 3 m in radius. A number of simplifying assumptions are made about energy demand and storage, as well as the performance of the hypothetical WECS. Based on these assumptions, gawk and gnuplot are used to manipulate a set of wind speed data recorded by an AWS to evaluate the magnitude and temporal variability of the wind "resource" at this site and to simulate the likely power output of the WECS. Although it was not expressed as such, these two

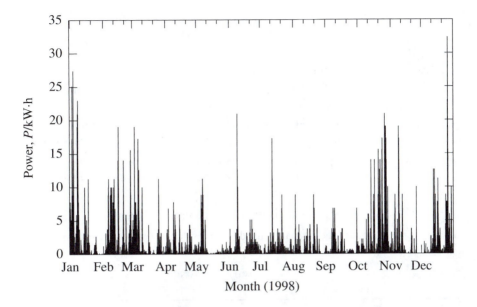

Figure 4.17: Time series of the simulated power output from a small WECS at Llyn Efyrnwy.

tasks involve the specification of a simple model of wind power generation (Equation 4.6), its implementation in computer code, and its application to environmental data recorded at Llyn Efyrnwy. This general approach is explored more extensively in the following chapter.

> **Exercise 4.4**: Write a gawk program to calculate the average hourly power output produced over the entire year by the WECS at Llyn Efyrnwy using the data contained in the file `wind98le.pwr`.

## SUPPORTING RESOURCES

The following resources are provided in the `chapter4` sub-directory of the CD-ROM:

| | |
|---|---|
| `wind98le.dat` | Hourly mean wind speed and wind direction data set for Llyn Efyrnwy, 1998, in MetOffice format |
| `meanwspd.awk` | gawk program to calculate the annual mean wind speed from the data in `wind98le.dat` |
| `meanmaxw.awk` | gawk program to calculate the annual mean and maximum wind speeds from the data in `wind98le.dat` |

| `windfreq.awk` | **gawk** program to determine the absolute frequency with which winds of different speeds occur at Llyn Efyrnwy based on the data in `wind98le.dat` |
| `windfrq2.awk` | **gawk** program to determine the relative frequency with which winds of different speeds occur at Llyn Efyrnwy based on the data in `wind98le.dat` |
| `wind7yr.dat` | Relative frequency distribution of hourly mean wind speeds measured at Llyn Efyrnwy between 1994 and 2000, inclusive |
| `wpower.awk` | **gawk** program to calculate the wind power (kW·h) at Llyn Efyrnwy throughout 1998, based on the data contained in `wind98le.dat` |
| `wind.bat` | Command line instructions used in this chapter |
| `wind.plt` | **gnuplot** commands used to generate the figures presented in this chapter |
| `wind.gp` | **gnuplot** commands used to generate figures presented in this chapter as EPS files |

# Chapter 5

# Solar Radiation at Earth's Surface

**Topics**

- Environmental significance of solar radiation incident on Earth's surface

- Global, direct and diffuse solar irradiance

- Radiation laws and the electromagnetic spectrum

- Measuring and modeling solar irradiance

**Methods and techniques**

- Actions performed *before* reading data from a file — the BEGIN block

- Passing parameter values to gawk programs via the command line

- Trigonometric functions in gawk

- More control-flow constructs in gawk — the for loop

## 5.1 INTRODUCTION

Energy is emitted from the sun in the form of electromagnetic radiation. This energy drives Earth's climate and its environmental systems (Beer *et al.* 2000, IPCC 2001). Thus, for example, Earth's atmosphere, oceans and land surface are heated by the absorption of a fraction of the shortwave solar radiation that is incident upon them. This radiation is subsequently emitted at longer wavelengths causing the atmosphere to be heated further, depending on the balance between the rates at which it absorbs and emits longwave radiation (McGuffie and Henderson-Sellers 1997). The resulting spatial differences in atmospheric heating generate momentum in the air mass, which among other things produces winds at Earth's surface (Twidell and Weir 2006).

Incident solar radiation, in conjunction with air temperature, also plays a key role in the growth and development of vegetation through the process of photosynthesis (Monteith and Unsworth 1995). The latter involves the assimilation of carbon into a range of complex organic molecules, including carbohydrates such as sugars, starch and cellulose (Jones 1983, Campbell and Norman 1998); these organic molecules are subsequently converted into plant tissue, commonly referred to as biomass. The assimilated carbon is drawn down (or "sequestered") from the atmosphere in the form of $CO_2$. The production of vegetation biomass therefore represents an important "sink" for this major greenhouse gas (IPCC 2001). Consequently, some countries plan to establish forests to sequester atmospheric $CO_2$ as part of their commitment to the Kyoto Protocol. This represents one element of so-called emission reduction credit (ERC) trading, which allows major new sources of greenhouse gases to be offset by emission reductions elsewhere.

Incident solar radiation is also an important source of renewable energy in its own right. In this context, it can be "harvested" to heat water for domestic or industrial use, to warm air for space heating or for drying crops, and to generate electricity using photovoltaic cells (Twidell and Weir 2006).

The amount of solar radiation incident on Earth's surface varies both spatially and with the time of day and year. The spatial variation is a function of differences in latitude and various properties of the local terrain, such as its altitude, gradient and aspect. The temporal variation is governed by the angle of the sun above the horizon, the turbidity of the atmosphere and the degree of cloud cover.

This chapter explores the temporal variation of incident solar radiation at Llyn Efyrnwy using both *in situ* measurements and a computational model. The former provides a further example of how gawk and gnuplot can be employed to analyze environmental data sets. The latter illustrates how a computational model can be constructed from published mathematical formulae and used to simulate the temporal trends in incident solar radiation observed at Llyn Efyrnwy. Importantly, the model provides a general tool that can, in principle, be applied to other sites, including those for which *in situ* measurements are unavailable.

## 5.2   DESCRIPTION OF THE SOLAR IRRADIANCE DATA

The exercises covered in this chapter make use of the data file radt981e.dat, which is included on the CD-ROM. This file contains measurements of the total broadband solar radiation incident at Llyn Efyrnwy, also known as the global solar irradiance. These data were recorded at hourly intervals throughout 1998 by an AWS operated by the MetOffice (see Figure 5.1 and Table 5.1). The solar irradiance values are given in Watt hours per square meter ($W \cdot h \cdot m^{-2}$). This is a non-standard unit of energy, but one that is nevertheless widely used; it describes the solar power in Watts (W) expended over a period of one hour (h) per square meter of Earth's surface ($m^{-2}$). The data were obtained from the BADC web site in ASCII text format.

| ID | IDTYPE | MET_DOM | YEAR | MON | DAY | END_HOUR | COUNT | GLOBAL | DIFFUSE | DIRECT |
|---|---|---|---|---|---|---|---|---|---|---|
| 7948 | DCNN | ESAWRADT | 1998 | 1 | 1 | 0 | 1 | 0 | -999 | -999 |
| 7948 | DCNN | ESAWRADT | 1998 | 1 | 1 | 100 | 1 | 0 | -999 | -999 |
| 7948 | DCNN | ESAWRADT | 1998 | 1 | 1 | 200 | 1 | 0 | -999 | -999 |
| 7948 | DCNN | ESAWRADT | 1998 | 1 | 1 | 300 | 1 | 0 | -999 | -999 |
| 7948 | DCNN | ESAWRADT | 1998 | 1 | 1 | 400 | 1 | 0 | -999 | -999 |
| 7948 | DCNN | ESAWRADT | 1998 | 1 | 1 | 500 | 1 | 0 | -999 | -999 |
| 7948 | DCNN | ESAWRADT | 1998 | 1 | 1 | 600 | 1 | 0 | -999 | -999 |
| 7948 | DCNN | ESAWRADT | 1998 | 1 | 1 | 700 | 1 | 0 | -999 | -999 |
| 7948 | DCNN | ESAWRADT | 1998 | 1 | 1 | 800 | 1 | -1 | -999 | -999 |
| 7948 | DCNN | ESAWRADT | 1998 | 1 | 1 | 900 | 1 | 13 | -999 | -999 |
| 7948 | DCNN | ESAWRADT | 1998 | 1 | 1 | 1000 | 1 | 44 | -999 | -999 |
| 7948 | DCNN | ESAWRADT | 1998 | 1 | 1 | 1100 | 1 | 24 | -999 | -999 |
| 7948 | DCNN | ESAWRADT | 1998 | 1 | 1 | 1200 | 1 | 27 | -999 | -999 |
| 7948 | DCNN | ESAWRADT | 1998 | 1 | 1 | 1300 | 1 | 33 | -999 | -999 |
| 7948 | DCNN | ESAWRADT | 1998 | 1 | 1 | 1400 | 1 | 13 | -999 | -999 |
| 7948 | DCNN | ESAWRADT | 1998 | 1 | 1 | 1500 | 1 | 8 | -999 | -999 |
| 7948 | DCNN | ESAWRADT | 1998 | 1 | 1 | 1600 | 1 | 5 | -999 | -999 |
| 7948 | DCNN | ESAWRADT | 1998 | 1 | 1 | 1700 | 1 | 2 | -999 | -999 |
| 7948 | DCNN | ESAWRADT | 1998 | 1 | 1 | 1800 | 1 | 2 | -999 | -999 |
| 7948 | DCNN | ESAWRADT | 1998 | 1 | 1 | 1900 | 1 | 0 | -999 | -999 |
| 7948 | DCNN | ESAWRADT | 1998 | 1 | 1 | 2000 | 1 | 0 | -999 | -999 |
| 7948 | DCNN | ESAWRADT | 1998 | 1 | 1 | 2100 | 1 | 0 | -999 | -999 |
| 7948 | DCNN | ESAWRADT | 1998 | 1 | 1 | 2200 | 1 | 0 | -999 | -999 |
| 7948 | DCNN | ESAWRADT | 1998 | 1 | 1 | 2300 | 1 | 0 | -999 | -999 |

Figure 5.1: Partial contents of the file radt981e.dat (first 25 records).

Table 5.1: Explanation of the data fields in `radt98lv.dat`.

| Field | Name | Contents |
|---|---|---|
| 1 | ID | Meteorological station ID number |
| 2 | IDTYPE | Meteorological station type |
| 3 | MET_DOM | Meteorological domain |
| 4 | YEAR | Year |
| 5 | MON | Month (1–12) |
| 6 | DAY | Day of the month (1–31) |
| 7 | END_HOUR | Hour of observation (00:00–23:00 GMT) |
| 8 | COUNT | Number of hours over which measurements are made |
| 9 | GLOBAL | Global solar irradiance (W·h·m$^{-2}$) |
| 10 | DIFFUSE | Diffuse solar irradiance (W·h·m$^{-2}$) |
| 11 | DIRECT | Direct solar irradiance (W·h·m$^{-2}$) |

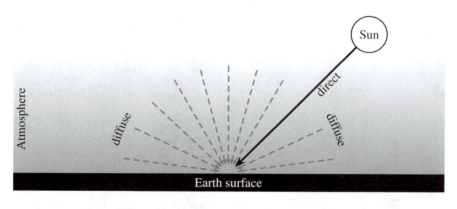

Figure 5.2: Direct and diffuse solar irradiance.

Table 5.1 describes the contents of the 11 fields of data in `radt98le.dat`. Note that fields 9 through 11 contain three separate components of the solar irradiance: global, diffuse and direct (Figure 5.2). Direct irradiance is that which arrives at a point on Earth's surface from the direction of the sun; it is dominated by the direct solar beam (Monteith and Unsworth 1995). Diffuse irradiance is that which arrives from all other directions in the hemisphere above the surface, having been scattered out of the direct solar beam by clouds, gases and particulate matter within the atmosphere, or by reflection from adjacent areas of terrain (Monteith and Unsworth 1995). Global irradiance is the sum of the direct and diffuse components. In this context, therefore, global irradiance refers to the total amount of solar radiation incident at a particular location on Earth's surface, and not to the amount received by the planet as a whole.

Program 5.1: selradt.awk

```
(NR >1 && $9 >=0) {                                                    1
   printf ("%4i%02i%02iT%04i %4i\n", $4, $5, $6, $7, $9);              2
}                                                                      3
```

```
19980101T0000    0
19980101T0100    0
19980101T0200    0
19980101T0300    0
19980101T0400    0
```

Figure 5.3: First five lines of the output file, radt98le.out, produced using Program 5.1.

Approximately 70 meteorological stations in the UK report solar irradiance data. Most record hourly measurements of global and diffuse irradiance, from which the direct component can be computed by subtraction. Just three stations in the UK report data on direct irradiance. The AWS at Llyn Efyrnwy provides only global irradiance values (field 9 in radt98le.dat), so that fields 10 (diffuse irradiance) and 11 (direct irradiance) of radt98le.dat are filled with the "missing data" value (-999).

## 5.3   ANALYZING THE OBSERVATIONS

### 5.3.1   Data Extraction and Pre-Processing

Program 5.1 is designed to extract data on the global solar irradiance at Llyn Efyrnwy, as well as the dates and times at which these observations were made, from the file radt98le.dat. The code is very similar to that employed in the preceding chapters. Line 1 consists of a pattern that matches each record in the data file with the exception of the first (NR>1), provided that the value in the ninth field of the record is greater than or equal to zero ($9>=0). Thus, the pattern skips the first record of the data file, which contains the field names, and discards any other record in which the global solar irradiance value is less than zero. Negative values of global solar irradiance are indicative of missing data (-999) and of unwanted artifacts that are produced by the irradiance sensor in the AWS under very low light conditions (e.g., at dawn and dusk). Provided that the current record matches both parts of the pattern, the corresponding action (line 2) is performed. The action statement prints the values in the fourth (year), fifth (month), sixth (day), seventh (time) and ninth (global solar irradiance) fields of the relevant records in the specified format.

Program 5.1 can be run from the command line as follows:

```
gawk -f selradt.awk radt98le.dat > radt98le.out                        1
```

The first few lines of the output file, radt98le.out, are shown in Figure 5.3.

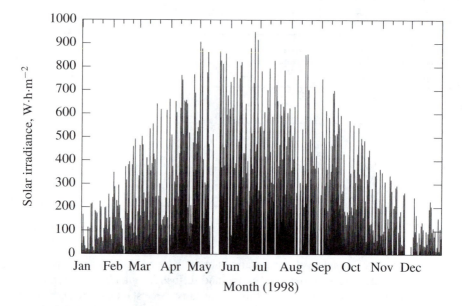

Figure 5.4: Hourly variation in total solar irradiance at Llyn Efyrnwy throughout 1998.

### 5.3.2  Visualizing the Output

The data contained in the file radt98le.out can be visualized by typing the following commands in gnuplot:

```
reset                                                         1
unset key                                                     2
set xdata time                                                3
set timefmt "%Y%m%dT%H%M"                                     4
set format x "%b"                                             5
set xlabel "Month (1998)"                                     6
set ylabel "Solar irradiance, W.h.m^-2"                       7
set xrange ["19980101T0000":"19990101T0000"]                 8
plot 'radt98le.out' u 1:2 w i                                 9
```

These commands should be familiar from the preceding chapters. They produce the plot shown in Figure 5.4, which shows the variation in global solar irradiance at Llyn Efyrnwy throughout 1998. There is a clear seasonal component to this variation, as one might expect, with the largest irradiance values recorded during the northern hemisphere summer months, especially between May and August, when the sun is highest in the sky. By comparison, the daily maximum values are much smaller in the northern hemisphere winter months, most notably between November and February, when the sun is lower in the sky. It should also be apparent that there are two extended periods of missing data: one in May, the other at the end of November and the beginning of December. Superimposed on the seasonal trend, there is also significant variation over the diurnal cycle, as well as from day to day, although these effects are difficult to discern in Figure 5.4.

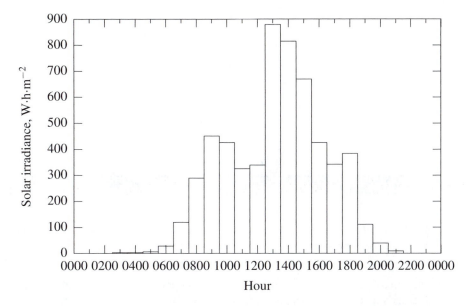

Figure 5.5: Diurnal variation in total solar irradiance at Llyn Efyrnwy on June 21, 1998.

The diurnal variation can be seen more clearly by focusing on measurements made over the course of a single day. For example, gnuplot can be instructed to plot only those data recorded on June 21, 1998, the date of the northern hemisphere summer solstice, by issuing the following commands:

```
set format x "%H%M"                                    10
set xlabel "Hour"                                      11
set xrange ["19980621T0000":"19980621T2359"]          12
plot 'radt98le.out' u 1:2 w boxes                      13
```

These commands produce the output shown in Figure 5.5. Line 10, for example, sets the labels used for each of the tic-marks along the *x*-axis so that these indicate the time, in hours (%H) and minutes (%M), at which the measurements were made. Line 11 updates the *x*-axis label accordingly. Line 12 constrains the range of values to be plotted on the *x*-axis, displaying only those measurements recorded between 00:00 and 23:59 Greenwich Mean Time (GMT) on June 21, 1998. Note that the time and date information specified on this line is placed inside a pair of double quotation marks and is given in the format used in the data file radt98le.out. Finally, line 13 plots the data using the boxes style, taking account of the preceding instructions.

The pattern of variation evident in Figure 5.5 broadly follows the trend that one might expect over the course of a diurnal cycle, with the maximum value of solar irradiance recorded around solar noon, when the sun is at its highest point in the sky, decreasing systematically toward dawn (around 03:00 GMT) and dusk (around 21:00 GMT). The simple diurnal trend is modified, however, by the effect of cloud cover. Clouds affect the solar irradiance signal in two main ways (Figure 5.6). First, they attenuate (i.e., reduce) the direct solar beam by obscuring the solar disk; in general,

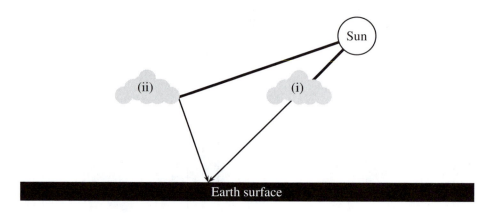

Figure 5.6: Impact of cloud cover on incident solar radiation. (i) Attenuation of the direct solar
beam; (ii) reflection from clouds over adjacent terrain.

the greater the cloud cover, the more the direct irradiance is likely to be attenuated.
Second, clouds located over adjacent areas of terrain reflect solar radiation onto the
target area, increasing the diffuse irradiance that it receives. The trend presented in
Figure 5.5 suggests that cloud cover may have reduced the direct solar irradiance,
particularly between 10:00 and 12:00 GMT.

> **Exercise 5.1**: Make appropriate modifications to Program 5.1 so that it
> prints out the global solar irradiance values recorded by the AWS at Llyn
> Efyrnwy each day at solar noon only (i.e., 12:00 GMT). Run the revised
> program and redirect the output to a new file. Plot the data in the output
> file using gnuplot. Comment on the observed trend.

## 5.4   MODELING SOLAR IRRADIANCE

Globally, the number of sites at which solar irradiance data are routinely collected is
very limited. Even in the UK, a comparatively well-instrumented nation, only 70 or so
meteorological stations regularly record data of this type on behalf of the MetOffice;
this compares to roughly 1200 sites that contribute standard hourly weather observations. The coverage elsewhere in the world is typically much less comprehensive
(Gueymard 2003). As a consequence, it is often the case that *in situ* measurements
of solar irradiance are unavailable for the site of interest. There are two possible
courses of action in these circumstances. The first is to establish a dedicated meteorological station, with the equipment necessary to measure solar irradiance, at the
chosen location; this is not always viable, though, for reasons of accessibility and
cost. The second is to construct a simulation model that can be used to estimate the
solar irradiance at the site. This approach is explored in the remainder of this section.

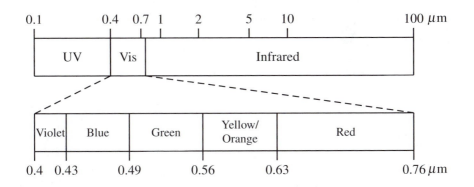

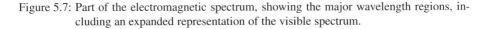

Figure 5.7: Part of the electromagnetic spectrum, showing the major wavelength regions, including an expanded representation of the visible spectrum.

### 5.4.1   Formulating the Mathematical Model

One of the first steps involved in building an environmental model is to identify the mathematical formulae on which it will be based. Suitable formulae will often have been developed in previous studies reported in the scientific literature and these can be used to construct the mathematical model. Sometimes alternative formulations exist for parts of the model, in which case the task becomes one of selecting the most appropriate formulae and of assembling these into a single coherent model. The selection criteria one might employ in this context include keeping the complexity of the mathematics concerned to a minimum, limiting the number of parameters in the model for which input data are required, and constraining the computational load required to run the model. Many of these issues are considered in the subsections that follow, after a brief introduction to the fundamental principles of solar radiation.

*Fundamental Principles of Solar Radiation*

The sun emits energy in the form of electromagnetic radiation over a continuous range of wavelengths known as the electromagnetic spectrum (Figure 5.7). The sun's behavior in this respect is similar to that of a perfect (or "blackbody") radiator at a temperature of approximately 5800 K (Campbell and Norman 1998), where K refers to kelvin, a temperature scale closely related to degrees Celsius (K = °C + 273.15). Blackbody radiators absorb and subsequently emit all of the radiation incident upon them. The amount of energy emitted by a blackbody radiator varies with wavelength according to Planck's equation,

$$M(\lambda) = \frac{2\pi hc^2}{\lambda^5 (e^{\frac{hc}{\lambda kT}} - 1)} \tag{5.1}$$

where $\lambda$ is the wavelength of the radiation, measured in meters, $h$ is Planck's constant ($h = 6.626 \times 10^{-34}$ J·s where J stands for Joule, a unit of energy), $c$ is the speed of light ($c \approx 3 \times 10^8$ m·s$^{-1}$), $k$ is Boltzmann's constant ($k = 1.3807 \times 10^{-23}$ J·K$^{-1}$), $T$ is the

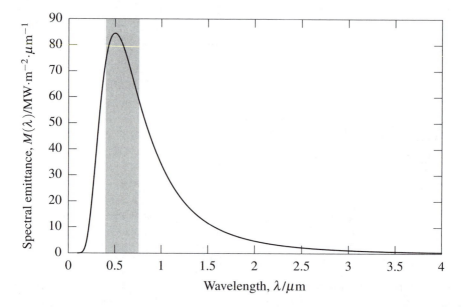

Figure 5.8: Spectral distribution of the radiant energy emitted by a blackbody radiator at 5800 K (e.g., the sun) in Mega-Watts per square meter per micrometer ($MW \cdot m^{-2} \cdot \mu m^{-1}$). The region shaded in gray indicates the visible spectrum.

temperature of the blackbody in kelvin and $M(\lambda)$ is known as the spectral emittance of the blackbody.

Equation 5.1 produces the emittance curve shown in Figure 5.8 for a blackbody radiator at a temperature of 5800 K and for wavelengths up to $4\mu m$, where $\mu m$ denotes a micrometer (i.e., one millionth of a meter). Note that the part of the electromagnetic spectrum that is visible to the human eye lies roughly between $0.4\mu m$ and $0.76\mu m$ (Figure 5.7), which corresponds closely to the region of maximum emission in the solar spectrum (Figure 5.8).

The exact wavelength of maximum emission by a blackbody radiator is given by Wien's displacement law,

$$\lambda_{max} = \frac{2897}{T} \tag{5.2}$$

where $\lambda_{max}$ is the wavelength of peak emittance in micrometers. Assuming that the sun has an average surface temperature of 5800 K, $\lambda_{max} = 2897/5800 \approx 0.5\mu m$, which corresponds to the part of the electromagnetic spectrum that the human eye perceives as green light. The average temperature of Earth's surface, on the other hand, is roughly 287 K ($\sim 13\,°C$). Consequently, its emission spectrum peaks around $10\mu m$, in a part of the spectrum known as the infrared.

The total amount of radiation emitted by a blackbody radiator across all wavelengths, $M$, is given by the Stefan-Boltzmann equation,

$$M = \sigma T^4 \tag{5.3}$$

where $\sigma$ is the Stefan-Boltzmann constant ($\sigma = 5.67 \times 10^{-8}\,\text{W·m}^{-2}\text{·K}^{-4}$). Hence, assuming that the sun behaves like a blackbody radiator at a temperature of 5800 K,

$$
\begin{aligned}
M_{\text{sun}} &\approx \left(5.67 \times 10^{-8}\,\text{W·m}^{-2}\text{·K}^{-4}\right) \times (5800\,\text{K})^4 \\
&\approx 6.4165 \times 10^7\,\text{W·m}^{-2} \\
&\approx 64\,\text{MW·m}^{-2}
\end{aligned}
\tag{5.4}
$$

Note that this value (64 MW·m$^{-2}$) is equivalent to the area under the curve in Figure 5.8. By the same token, assuming that Earth behaves like a blackbody radiator at a temperature of 287 K,

$$
\begin{aligned}
M_{\text{earth}} &\approx \left(5.67 \times 10^{-8}\,\text{W·m}^{-2}\text{·K}^{-4}\right) \times (287\,\text{K})^4 \\
&\approx 385\,\text{W·m}^{-2}
\end{aligned}
\tag{5.5}
$$

This value is many orders of magnitude smaller than $M_{\text{sun}}$, but it is highly significant in terms of its contribution to warming Earth's atmosphere.

### Effect of the Earth–Sun Distance

Because of its finite size and its large distance from the sun, Earth receives only a small fraction of the total radiation emitted by the sun; this fraction is governed by the inverse square law (Schott 1997, Schowengerdt 1997), which states that the intensity of the solar radiation incident on a surface perpendicular to the sun's rays is inversely proportional to the square of its distance from the sun. This can be expressed mathematically as follows:

$$
E_0(\lambda) = M(\lambda) \left(\frac{r_{\text{sun}}}{d}\right)^2
\tag{5.6}
$$

where $E_0(\lambda)$ is the solar radiation incident at the top of Earth's atmosphere, known as the exo-atmospheric solar spectral irradiance, $M(\lambda)$ is the spectral emittance of the sun, $r_{\text{sun}}$ is the radius of the sun ($r \approx 6.96 \times 10^8\,\text{m}$) and $d$ is the mean distance between Earth and the sun ($d \approx 1.4 \times 10^{11}\,\text{m}$). Applying Equation 5.6 to the data contained in Figure 5.8 yields the curve shown in Figure 5.9.

It is possible to calculate the total exo-atmospheric solar irradiance across all wavelengths, $E_0$, by replacing $M(\lambda)$ with $M_{\text{sun}}$ in Equation 5.6 as follows:

$$
\begin{aligned}
E_0 &= M_{\text{sun}} \left(\tfrac{r_{\text{sun}}}{d}\right)^2 \\
&\approx 6.4165 \times 10^7\,\text{W·m}^{-2} \left(\frac{6.96 \times 10^8\,\text{m}}{1.5 \times 10^{11}\,\text{m}}\right)^2 \\
&\approx 1380\,\text{W·m}^{-2}
\end{aligned}
\tag{5.7}
$$

This value describes the area beneath the curve in Figure 5.9 and is commonly known as the solar constant, although this name is somewhat misleading because $E_0$ varies slightly during the course of the solar cycle and more significantly over longer periods of time (Lean 1991, Lee *et al.* 1995, Crommelynck and Dewitte 1997). The term total solar irradiance (TSI) is rather more apposite. Regardless of the name used to refer to

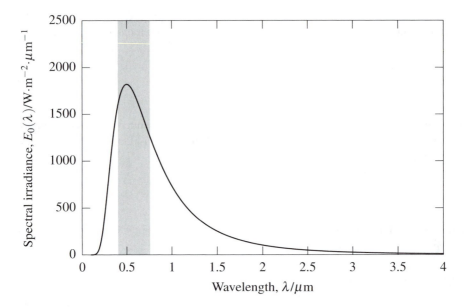

Figure 5.9: Solar spectral irradiance at the top of the atmosphere, assuming that the sun behaves
as a blackbody radiator at 5800 K. The region shaded in gray indicates the extent
of the visible spectrum.

this quantity, its value is fundamental to understanding global environmental change
because it defines the magnitude of the external "forcing" on Earth's climate system.

For various reasons, the actual solar spectrum at the top of Earth's atmosphere is
more complex than that shown in Figure 5.9. Numerous reference spectra have been
produced to represent this complexity (see Figure 5.10). Coverage of the derivation
of these spectra is beyond the scope of this book, except to note that they are based
on data obtained from a range of sources, including sensors carried aboard Earth-
orbiting satellites, space shuttle missions and high-altitude aircraft (Fligge *et al.* 2001,
Gueymard *et al.* 2002).

*Attenuation of Incident Solar Radiation by Earth's Atmosphere*

Only a fraction of the total exo-atmospheric solar irradiance reaches Earth's surface
(Figure 5.10). Some is absorbed by gases and particulate matter, the latter commonly
known as aerosols, on its downward path through the atmosphere; some is scattered
back out to space by aerosols and clouds (Figure 5.11). The combined effect of these
two processes is to attenuate the incident solar radiation as it passes down through
the atmosphere. The degree of attenuation depends on the composition of the atmosphere, in terms of its gaseous constituents and the concentrations of different types
of aerosol, and on the path length that the solar radiation has to travel through the
atmosphere to the ground. The path length is, in turn, dependent on the altitude of the
terrain and the angle of the sun above the horizon, which is a function of the latitude
of the site and the time of day and year.

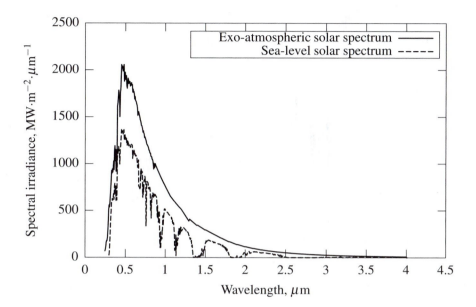

Figure 5.10: Reference solar spectral irradiance at the top of the atmosphere (solid line) and at sea level (dashed line) for a 1.5 air mass (AM1.5) atmospheric path length. (Source of data: http://rredc.nrel.gov/solar/standards/am0/.)

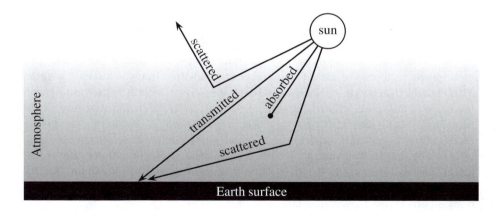

Figure 5.11: Scattering, absorption and transmission of solar radiation on its passage down through Earth's atmosphere.

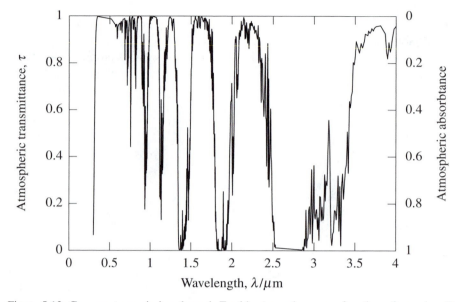

Figure 5.12: Gaseous transmission through Earth's atmosphere as a function of wavelength. (Source of data: Vermote *et al.* 1997.)

The fraction of incident solar radiation that is absorbed by atmospheric gases varies as a function of wavelength (Vermote *et al.* 1997, Figure 5.12), but is relatively stable over space and time. By contrast, the types and amounts of aerosols within the atmosphere, sometimes referred to as the aerosol mass loading, vary considerably from one location to another and over relatively short periods of time. The main natural sources of atmospheric aerosols are dust from volcanic eruptions and desert storms, as well as smoke from forest fires. The principal anthropogenic sources are sulfates produced by the combustion of fossil fuels, such as oil and coal, and smoke resulting from the intentional burning of vegetation biomass.

Accurate representation of the passage of incident solar radiation through Earth's atmosphere requires the use of comprehensive radiative transfer models, such as MODTRAN (Kneizys *et al.* 1996) and 6S (Vermote *et al.* 1997), which are beyond the scope of this book. Instead, the state of the atmosphere is characterized by means of a single coefficient, $\tau$ ($0 \leq \tau \leq 1$), where $\tau = 1$ implies that the incident solar radiation is transmitted through the atmosphere without attenuation (strictly possible only in a perfect vacuum) and $\tau = 0$ denotes complete attenuation of the solar signal (Monteith and Unsworth 1995). In practice, $\tau$ varies with wavelength but, to simplify matters further, a single value of $\tau$ is used here to describe the average atmospheric transmission across the entire solar spectrum.

*Estimating the Direct and Diffuse Components of the Solar Irradiance*

Several methods for calculating the direct and diffuse solar radiation incident on a horizontal element of Earth's surface are reported in the scientific literature. The

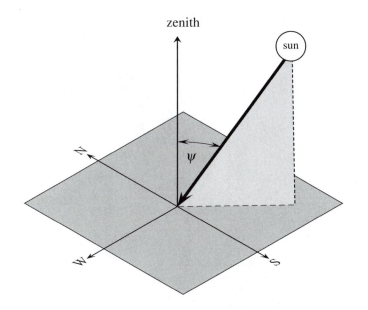

Figure 5.13: Solar zenith angle, $\psi$, with respect to a flat horizontal surface.

mathematical formulations given by Campbell and Norman (1998), adapted from Liu and Jordan (1960), are used here because of their relative simplicity. Specifically,

$$I_{\text{direct}} = E_0 \tau^m \cos \psi \tag{5.8}$$

$$I_{\text{diffuse}} = 0.3 \left(1 - \tau^m\right) E_0 \cos \psi \tag{5.9}$$

$$I_{\text{global}} = I_{\text{direct}} + I_{\text{diffuse}} \tag{5.10}$$

where $I_{\text{direct}}$, $I_{\text{diffuse}}$ and $I_{\text{global}}$ are, respectively, the direct, diffuse and global broadband solar irradiance at ground level, $E_0$ is the total exo-atmospheric solar irradiance, $\tau$ is the atmospheric transmittance, $m$ is the air mass number and $\psi$ is the solar zenith angle (i.e., the angle of the sun relative to a point directly above the observer; Figure 5.13).

For the purpose of this chapter, $E_0$ is assumed to be 1380 W·m$^{-2}$ (Equation 5.7) and $\tau$ is assumed to be 0.7. The latter value is representative of clear-sky conditions. According to Campbell and Norman (1998), the air mass number, $m$, is given by

$$m = \frac{p_{\text{alt}}}{p_{\text{sea}}} \cdot \frac{1}{\cos \psi} \tag{5.11}$$

for $\psi \leq 80°$; where $p_{\text{alt}}$ and $p_{\text{sea}}$ are the atmospheric pressure at the altitude of the study site and at sea level, respectively, and $\psi$ is the solar zenith angle.

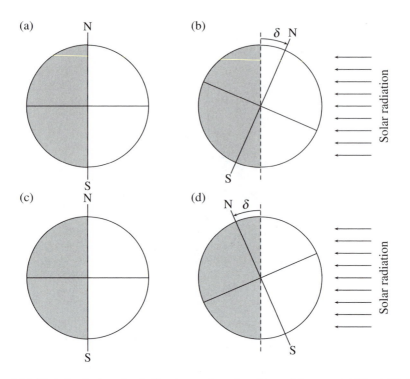

Figure 5.14: Variation in the solar declination angle at four times of the year: (a) March 21 ($\delta =$ 0°), (b) June 21 ($\delta = +23.4°$), (c) September 21 ($\delta = 0°$) and (d) December 21 ($\delta = -23.4°$). Light areas indicate sunlit parts of Earth's surface. (After Twidell and Weir 2006.)

*Calculating the Solar Zenith Angle and the Solar Declination Angle*

To evaluate Equations 5.8 and 5.9, the solar zenith angle, $\psi$, must either be known or be calculated. The solar zenith angle varies with latitude, time of year and time of day as follows:

$$\cos \psi = \sin \phi_L \sin \delta + \cos \phi_L \cos \delta \cos \theta \tag{5.12}$$

where $\phi_L$ is the latitude of the study site ($\phi_L = 52.756°N$ at Llyn Efyrnwy), $\delta$ is the solar declination angle and $\theta$ is the hour angle of the sun (Twidell and Weir 2006).

The solar declination angle describes the angle of the plane of the sun with respect to Earth's equator (Figure 5.14). This angle varies as a function of the time of year between approximately $+23.4°$ on June 21 (i.e., the northern hemisphere summer solstice) and $-23.4°$ on December 21 (i.e., the northern hemisphere winter solstice; Figure 5.15). Once again, numerous methods for calculating $\delta$ are reported in the literature but approximate values, suitable for use here, may be obtained from the following equation:

$$\delta \approx -23.4 \cos \left( \frac{360(\text{DoY} + 10)}{365} \right) \tag{5.13}$$

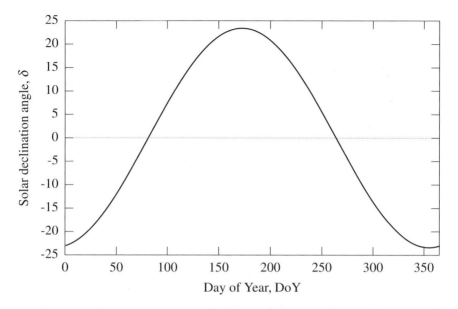

Figure 5.15: Variation in the solar declination angle, $\delta$, as a function of the day of year, DoY.

where DoY is the day of year (Jones 1983), and where DoY = 1 on January 1, DoY = 2 on January 2, and so on. More accurate values of the solar declination angle can be obtained from the standard meteorological tables (List 2000).

The hour angle of the sun, $\theta$, describes the angle through which Earth has rotated with respect to local solar noon (Twidell and Weir 2006). Since Earth rotates 360° every 24 hours, or 15° per hour, approximate values of $\theta$ are given by the following equation:

$$\theta \approx 15(12 - h)\qquad\qquad(5.14)$$

where $h$ is the local solar time in hours, ranging from 0 to 24 (Figure 5.16), such that $\theta = 0°$ at local solar noon.

### 5.4.2 Implementing the Computational Model

Once an appropriate formulation for the mathematical model has been selected, the next step is to convert the equations into gawk code; that is, to implement the computational model. This process involves a translation between the vocabulary, grammar and syntax of one language (i.e., the symbols and equations of mathematics) and those of another (i.e., the programming language gawk). Just as in the translation between natural languages, such as English and French, there is rarely a single "correct" solution, although some solutions might be regarded as being intrinsically better than others. The flexibility of gawk means that the implementation can often be realized in several ways. In some instances, the differences may amount to little more than a matter of programming style, broadly analogous to the variations in the style of

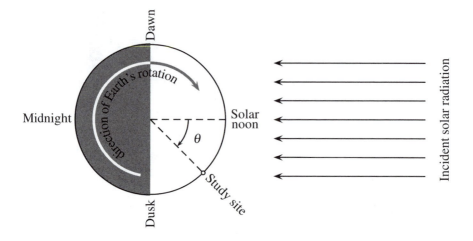

Figure 5.16: Solar hour angle, $\theta$, as though viewing Earth vertically downward from a point directly above the North Pole.

written prose among literary authors. On other occasions, however, the differences may have a significant impact in terms of the efficiency of the resulting code, such as its computational speed or the demands that it places on various aspects of the computer's resources, including memory usage and hard-disk access. Thus, some implementations may be considered to be more elegant or otherwise preferable to others.

Program 5.2, `solarrad.awk`, represents a fairly straightforward implementation of the mathematical equations outlined above, which tries to balance brevity versus clarity of the code.

*The* BEGIN *Block*

The whole of Program 5.2 is set within a BEGIN block, or BEGIN rule, which is a special feature of the gawk programming language. The BEGIN block is used to group together a set of statements that are processed *before* those contained in the main pattern-action block (i.e., before reading data from a file or files; Figure 5.17); it is analogous to the END block, introduced in Chapter 4, which is used to group statements that are processed *after* the main pattern-action block has been completed. In this particular example, however, the program does not contain a main pattern-action block, only a BEGIN block; this is because the aim is to generate estimated values of the solar irradiance at Llyn Efyrnwy, rather than to process data input from a file. This type of computational model is sometimes referred to as a simulation model.

The BEGIN block commences on line 1 of Program 5.2 with the keyword BEGIN, which must be typed entirely in upper case and must be followed immediately, on the same line, by an opening curly brace ({). The BEGIN block ends on line 35 with a closing curly brace (}). The gawk statements between the curly braces (lines 2 through 34) constitute the body of the BEGIN block.

Program 5.2: solarrad.awk

```
BEGIN{                                                              1
    latitude=52.756;              # Latitude (degrees)             2
    E_0=1380.0;                   # Exo-atmos. solar irradiance    3
    tau=0.7;                      # Atmos. transmittance           4
    hour_angle=0;                 # Solar hour angle (degrees)     5
    DOY=1;                        # Day of year                    6
    p_alt=1000;                   # Atmos. pressure (altitude)     7
    p_sea=1013;                   # Atmos. pressure (sea level)    8
                                                                   9
    deg2rad=(2*3.1415927)/360;  # Degrees to radians              10
                                                                  11
    latitude*=deg2rad;            # Latitude in radians           12
    hour_angle*=deg2rad;          # Hour angle in radians         13
                                                                  14
    # Solar declination angle (Eq. 5.13)                          15
    declination = (-23.4*deg2rad)* \                              16
        cos(deg2rad*(360*(DOY+10)/365));                          17
                                                                  18
    # Cosine of the solar zenith angle (Eq. 5.12)                 19
    cos_zenith=sin(latitude)*sin(declination)+ \                  20
        cos(latitude)*cos(declination)*cos(hour_angle);           21
                                                                  22
    # Atmospheric air mass (Eq. 5.11)                             23
    air_mass=(p_alt/p_sea)/cos_zenith;                            24
                                                                  25
    # Direct, diffuse and global (total) solar                    26
    # irradiance (Eq. 5.8, 5.9 and 5.10)                          27
    I_direct=(E_0)*(tau**air_mass)*cos_zenith;                    28
    I_diffuse=0.3*(1-(tau**air_mass))*(E_0)*cos_zenith;           29
    I_global=I_direct+I_diffuse;                                  30
                                                                  31
    # Output results                                              32
    printf("%3i %7.3f %7.3f %7.3f\n", \                           33
        DOY, I_global, I_direct, I_diffuse);                      34
}                                                                 35
```

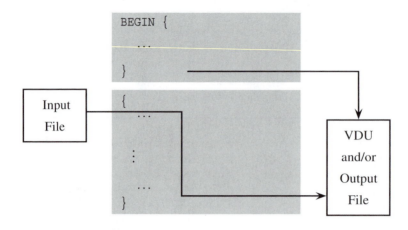

Figure 5.17: gawk's BEGIN and main pattern-action blocks.

*Naming Conventions for Variables*

Lines 2 through 8 of Program 5.2 initialize a number of variables that are used within the program. The names given to these variables are intended, as far as possible, to be indicative of the corresponding terms in the mathematical equations (see Table 5.2). Several of the variable names have been constructed from two words joined together by an underscore symbol (_). In the case of $p_{alt}$, for example, p and alt are conjoined to form p_alt. This convention is quite commonly used, although other formats for variable names are also employed. As a general rule, the use of very short variable names, such as x, y and z, is not recommended because they are not self-explanatory. By the same token, very long variable names are best avoided because they tend to inhibit, rather than improve, the readability of the code. Above all, it is important to be consistent when choosing names for variables, whichever convention is adopted. Also remember, as far as gawk is concerned, p_alt, p_Alt and P_alt are three entirely separate variables that may store completely different values.

    Note that Program 5.2 uses fixed values for several of the key parameters (lines 2 through 8). More specifically, it is assumes that $\phi_L = 52.756°$, $E_0 = 1380\,\text{W·m}^{-2}$, $\tau = 0.7$, $\theta = 0°$ and DoY $= 1$. Thus, Program 5.2 simulates the direct and diffuse broadband solar irradiance received at Llyn Efyrnwy ($\phi_L = 52.756°$) under clear-sky conditions ($\tau = 0.7$) at solar noon ($\theta = 0°$) on the first day of January (DoY $= 1$). Similarly, the atmospheric pressure at the site ($p_{alt}$) and at sea level ($p_{sea}$) is assumed to be 1000 mb and 1013 mb, respectively, where mb denotes pressure in millibars (lines 7 and 8). This is a reasonable assumption in the circumstances, although the code could be modified to use data on atmospheric pressure recorded by the AWS.

*Trigonometric Functions in gawk*

One characteristic of gawk, which is common to other programming languages used for scientific computing, is that the arguments to trigonometric functions (e.g., sin,

Table 5.2: Symbols used in the mathematical model (Equations 5.7 through 5.14) and variables used in the computational model (Program 5.2) of solar irradiance at Llyn Efyrnwy.

| Term | Symbol | Variable | Equations |
|------|--------|----------|-----------|
| Latitude | $\phi_L$ | `latitude` | 5.12 |
| Total exo-atmospheric solar irradiance | $E_0$ | `E_0` | 5.7 |
| Atmospheric transmittance | $\tau$ | `tau` | 5.8, 5.9 |
| Atmospheric pressure at altitude of site | $p_{alt}$ | `p_alt` | 5.11 |
| Atmospheric pressure at sea level | $p_{sea}$ | `p_sea` | 5.11 |
| Solar hour angle | $\theta$ | `hour_angle` | 5.12, 5.14 |
| Day of year | DoY | `DOY` | 5.13 |
| Solar declination angle | $\delta$ | `declination` | 5.12, 5.13 |
| Cosine of solar zenith angle | $\cos(\psi)$ | `cos_zenith` | 5.8, 5.9, 5.12 |
| Air mass number | $m$ | `air_mass` | 5.8, 5.9, 5.11 |
| Direct solar irradiance | $I_{direct}$ | `I_direct` | 5.8 |
| Diffuse solar irradiance | $I_{diffuse}$ | `I_diffuse` | 5.9 |
| Global solar irradiance | $I_{global}$ | `I_global` | 5.10 |

cos and tan) must be expressed in radians. Several of the values that are given as input to the model, however, are expressed in arc degrees (e.g., the latitude of Llyn Efyrnwy and the solar hour angle). Consequently, a multiplication factor is established on line 10 of Program 5.2, which is used to convert between arc degrees and radians. Since there are $2\pi$ radians and $360°$ in a circle, the conversion factor is simply $2\pi/360$; where $\pi \approx 3.1415927$. This value is stored in the variable `deg2rad`. Thus, lines 12 and 13 multiply the original values of the variables `latitude` and `hour_angle` by `deg2rad` to convert them into radians. Recall that `latitude*=deg2rad` is merely a more concise form of the statement `latitude=latitude*deg2rad`.

*Converting Equations into Code*

Equation 5.13, which calculates the solar declination angle on any day of the year, is implemented on lines 16 and 17 of Program 5.2. This single gawk statement is split over two lines of code for presentation purposes using the line continuation symbol (\), which is the last character on line 16. The conversion factor `deg2rad` is used twice in this statement to transform the angles concerned from arc degrees to radians. Also note that the multiplication operator ($\times$), which is implicit in Equation 5.13, must be entered explicitly using the $*$ symbol in the corresponding gawk code; for example, $360(\text{DoY} + 10)$ in the equation becomes `360*(DOY+10)` in the code.

Lines 20 and 21 represent an implementation of Equation 5.12, which is used to determine the cosine of the solar zenith angle. One of the terms in this equation is the solar declination angle ($\delta$ or declination), which explains why that value must be calculated first (i.e., on lines 16 and 17). Once again, this single long gawk statement is split over two lines of the program for presentation purposes; in all other ways lines 20 and 21 represent a direct translation from Equation 5.12 into gawk code.

Line 24 implements Equation 5.11. This line of code therefore calculates the air mass number given the cosine of the solar zenith angle (lines 20 and 21) and the atmospheric pressure at the site and at sea level (lines 7 and 8).

Lines 28 through 30 calculate the direct, diffuse and global broadband solar irradiance at Llyn Efyrnwy based on Equations 5.8 through 5.10. Recall that in gawk the operator ** means "raise to the power of", so that the term $\tau^m$ in Equations 5.8 and 5.9 can be expressed as tau**air_mass (lines 28 and 29). Finally, lines 33 and 34 print out the resulting values for the global, direct and diffuse irradiance and the day of year in the chosen format.

*Running the Computational Model*

Program 5.2 can be run from the command line as follows:

```
gawk -f solarrad.awk > solarrad.out
```
2

This command line produces the following output:

```
1  158.991    81.425    77.566
```

Thus, the model indicates that approximately 159 W·m$^{-2}$ of global solar irradiance is received at Llyn Efyrnwy under clear-sky conditions at solar noon on January 1. This total comprises roughly 81.5 W·m$^{-2}$ of direct solar flux and 77.5 W·m$^{-2}$ of diffuse solar flux.

Strictly speaking, the results output by the model are instantaneous values; that is, the model predicts the solar irradiance at a given instant in time (i.e., solar noon). Consequently, the model output is not directly comparable with the measurements made by the AWS, which are integrated over the preceding hour. For the purpose of this exercise, however, it is assumed that the instantaneous values generated by the model represent the mean irradiance conditions during the preceding hour. This is a reasonable first approximation in most instances. Given this assumption, it should be noted that the predicted value of total solar irradiance (159 W·h·m$^{-2}$) is much higher than the measured value recorded by the AWS at solar noon on January 1, 1998 (27 W·h·m$^{-2}$), possibly because of the prevailing cloud and atmospheric conditions on that day, but it is quite similar to the value recorded by the AWS at the same time on the following day (169 W·m$^{-2}$).

### 5.4.3   Enhancing the Implementation

*Entering Parameter Values via the Command Line*

One of the limitations of Program 5.2 is that the file containing the code must be opened, edited and saved using a text editor to alter the value of any of the variables. To simulate the global solar irradiance at solar noon for every day of the year, for example, the code would have to be edited 365 times, changing the value of DOY (line 6) on each occasion. This is clearly unsatisfactory. A better approach would be to assign values to selected variables when the model is run, via the command line for instance.

gawk allows the value of one or more variables to be specified on the command line using the -v switch (or flag), which has the general syntax -v *variable=value*. For example, the following command line assigns the variable DOY the value 2 at the start of the program:

```
gawk -f solarrad.awk -v DOY=2
```
*3*

To take advantage of this facility, however, line 6 must be removed from Program 5.2; otherwise, it would override the value of DOY entered on the command line.

*Control-Flow Constructs in gawk*

The ability to initialize a variable on the command line, in the manner outlined above, is a very useful feature of **gawk**, one that will be used extensively throughout this book. Nevertheless, it does relatively little to help where the objective is to simulate the variation in solar irradiance throughout the year. Even using the -v flag, the program must still be run 365 times in its current configuration, with a different value of DOY entered via the command line on each occasion. Clearly, this is a very time-consuming process and one that is likely to be error-prone. A more efficient approach would be to get the computer to do the hard work by building this procedure into the program itself, so that the value of DOY is varied automatically between 1 and 365. gawk provides a number of simple mechanisms to perform this type of repetitive task. These mechanisms are known as looping or control-flow constructs.

gawk has three control-flow constructs: the for, while and do while statements. Only the first of these is examined here; the other two are introduced in subsequent chapters. The for statement performs a particular sequence of operations a given number of times. It has the following general syntax (Robbins 2001):

```
for(initialization; condition; increment){
    body
}
```

The following fragment of **gawk** code, which is represented graphically in Figure 5.18, presents a practical example of a for loop:

```
for(DOY=1; DOY<=365; ++DOY){
    body
}
```

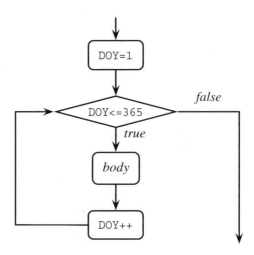

Figure 5.18: A `for` loop designed to perform a set of instructions contained in the body 365 times. The value of the variable DOY increases by 1 each time, from 1 to 365.

This code is intended to perform the set of instructions contained in the body of the loop (which is not specified here) 365 times, using a different value for the variable DOY on each occasion, commencing with 1, culminating with 365, and increasing in steps of 1. Thus, the code fragment starts by initializing the value DOY to 1 (DOY=1). The code fragment then assesses whether the value of DOY is less than or equal to 365 (DOY<=365); this is known as the condition. If the condition DOY<=365 is true, the code in the body is executed. The value of DOY is then increased by 1 (++DOY), so that DOY=2 subsequently. The condition is then re-evaluated using the new value of DOY. If the condition remains true (i.e., the value of DOY is still less than or equal to 365), the code in the body of the loop is executed again taking into account the new value of DOY, and so on. The variable DOY is, in effect, used to count the number of times the loop has been traversed thus far, and a variable used in this way is sometimes known as a counter. The program continues to loop around these statements until the condition is false (i.e., the value of DOY exceeds 365). At this point the program either proceeds to the next line of code after the `for` loop or, if there are no further lines of code, stops.

Program 5.3 (`solarrd2.awk`) presents a modified version of `solarrad.awk` that makes use of the example `for` loop outlined above. The loop starts on line 15 with the keyword `for`, the control statements and an opening curly brace, and ends on line 35 with a closing curly brace. Lines 16 through 34 of this program therefore constitute the body of the loop. These have been indented by an extra tab space to make it clear to the reader where the loop begins and ends. The other difference with respect to Program 5.2 is that the value of DOY is no longer initialized in the first few lines of the program, since this is now performed by the `for` loop (line 15 in Program 5.3). Program 5.3 can be run from the command line as follows:

```
gawk -f solarrd2.awk > solarrd2.out
```
4

Program 5.3: solarrd2.awk

```
BEGIN{                                                              1
   latitude=52.756;              # Latitude (degrees)              2
   E_0=1380.0;                   # Exo-atmos. solar irradiance     3
   tau=0.7;                      # Atmos. transmittance            4
   hour_angle=0;                 # Solar hour angle (degrees)      5
   p_alt=1000;                   # Atmos. pressure (altitude)      6
   p_sea=1013;                   # Atmos. pressure (sea level)     7
                                                                   8
   deg2rad=(2*3.1415927)/360;    # Degrees to radians              9
                                                                  10
   latitude*=deg2rad;            # Latitude in radians            11
   hour_angle*=deg2rad;          # hour angle in radians          12
                                                                  13
   # for-loop to cycle through each day of year                   14
   for(DOY=1;DOY<=365;++DOY){                                     15
      # Solar declination angle (Eq. 5.13)                        16
      declination = (-23.4*deg2rad)* \                            17
         cos(deg2rad*(360*(DOY+10)/365));                         18
                                                                  19
      # Cosine of the solar zenith angle (Eq. 5.12)               20
      cos_zenith=sin(latitude)*sin(declination)+ \                21
         cos(latitude)*cos(declination)*cos(hour_angle);          22
                                                                  23
      # Atmospheric air mass (Eq. 5.11)                           24
      air_mass=(p_alt/p_sea)/cos_zenith;                          25
                                                                  26
      # Direct, diffuse and global (total) solar                  27
      # irradiance (Eq. 5.8, 5.9 and 5.10)                        28
      I_direct=(E_0)*(tau**air_mass)*cos_zenith;                  29
      I_diffuse=0.3*(1-(tau**air_mass))*(E_0)*cos_zenith;         30
      I_global=I_direct+I_diffuse;                                31
                                                                  32
      # Output results                                            33
      printf("%3i %7.3f %7.3f %7.3f\n", \                         34
         DOY, I_global, I_direct, I_diffuse);                     35
   }                                                              36
}                                                                 37
```

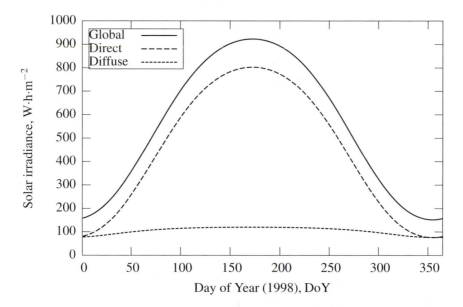

Figure 5.19: Simulated variation in the total, direct and diffuse solar irradiance as a function of the day of year (DoY).

### 5.4.4    Visualizing the Simulated Variation in Solar Irradiance

The data contained in the output file, solarrd2.out, can be visualized in gnuplot (Figure 5.19) by entering the following commands:

```
reset                                                    14
set xrange [0:365]                                       15
set style data lines                                     16
set key top left Left box                                17
set xlabel "Day of Year (1998), DoY"                     18
set ylabel "Solar irradiance, W.h.m^-2"                  19
plot 'solarrd2.out' u 1:2 t "Global" lw 2, \             20
     'solarrd2.out' u 1:3 t "Direct" lw 2, \             21
     'solarrd2.out' u 1:4 t "Diffuse" lw 2               22
```

Line 14 instructs gnuplot to reset all of the plotting parameters (e.g., the axis labels, plotting ranges, data types and so on) to their default values. Line 15 sets the range of values plotted on the *x*-axis, commencing at 0 (i.e., the start of the year) and ending at 365 (i.e., the final day of the year). Line 16 sets the default data style. Line 17 places a key in the top left-hand corner of the plot, enclosed in a box in which the text is left-justified. Lines 18 and 19 define the labels to be placed along the *x*- and *y*-axes, respectively. Lines 20 through 22 plot the data for the day of year (*x*-axis) against the global, direct and diffuse irradiance values (*y*-axis), respectively. These data are contained in the first, second, third and fourth fields of the file solarrd2.out. Note that the final gnuplot command is split over three lines (20 through 22) for presentation purposes using the line continuation symbol (\).

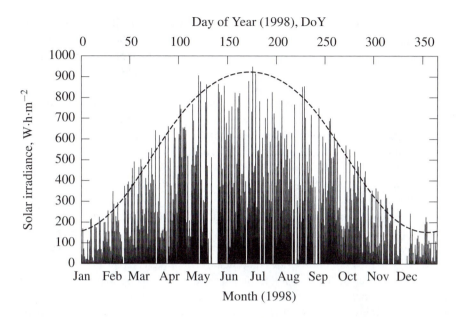

Figure 5.20: Comparison between observed (vertical impulses) and simulated (solid line) total
solar irradiance as a function of the time of year.

The seasonal pattern evident in Figure 5.19 is similar to that seen in Figure 5.5.
Moreover, it appears that the majority of the variation in global solar irradiance is
accounted for by changes in the level of direct solar irradiance, which is a function
of the solar zenith angle (Equation 5.8). The diffuse solar irradiance displays the
same basic trend, but the range of values is much smaller. Indeed, a reasonable first
approximation would be to assume that the diffuse component is constant (roughly
$100\,\mathrm{W\cdot h\cdot m^{-2}}$) throughout the year on clear days (Campbell and Norman 1998).

It is possible to make a direct comparison between the measured and simulated
variation in total solar irradiance by plotting the two data sets together (Figure 5.20).
This can be achieved by entering the following commands in gnuplot:

```
unset key                                                          23
set xdata time                                                     24
set timefmt "%Y%m%dT%H%M"                                          25
set format x "%b"                                                  26
set xlabel "Month (1998)"                                          27
set x2label "Day of Year (1998), DoY"                             28
set xrange ["19980101T0000":"19990101T0000"]                      29
set x2range [0:365]                                                30
set xtics nomirror                                                 31
set x2tics nomirror auto                                           32
plot 'radt98le.out' u 1:2 w i, \                                   33
     'solarrd2.out' u 1:2 w l axes x2y1 lw 2                       34
```

Line 23 removes the key from the figure. Line 24 indicates that the data plotted on the
first *x*-axis (i.e., the measured irradiance values) should be treated as a time series, and

line 25 indicates the format in which the time and date information are given. Line 26 indicates that the labels for the tic-marks on the first $x$-axis should be given by the shortened name of the month (i.e., Jan, Feb, ..., Dec). Lines 27 and 28 specify the labels that should be placed along the first (lower) and second (upper) $x$-axes, while lines 29 and 30 control the range of values to be plotted on each of these axes. Lines 31 and 32 ensure that the tic-marks used on each of the $x$-axes are not mirrored on the other one (i.e., only one set of tic-marks is plotted on each $x$-axis). Finally, lines 33 and 34 plot the measured and simulated total solar irradiance values: the former are read from the file `radt98le.out`, created earlier in this chapter, and are plotted using the `impluses` data style on the standard (`x1y1`) axes; the latter are read from the file `solarrd2.out`, generated using Program 5.3, and are plotted using the `lines` data style on the `x2y1` axes.

The observations presented in Figure 5.20 are recorded at hourly intervals throughout the day, whereas the simulation model estimates the irradiance values at solar noon only. Taking this into account, it appears that the estimated values correspond quite closely to the maximum values recorded by the AWS (i.e., measurements likely to be made at, or close to, solar noon) on some days. It is likely that the atmosphere was particularly clear and cloud-free on these occasions, consistent with the assumptions made in the computational model (i.e., $\tau = 0.7$). The lower solar irradiance values measured by the AWS on the other days may result from either a more turbid atmosphere (i.e., $\tau < 0.7$) or the influence of cloud cover. These possibilities could be tested by examining data on atmospheric visibility and cloud cover recorded by the AWS, but this is left as an exercise for the reader.

---

**Exercise 5.2**: Make appropriate modifications to Program 5.3 so that it simulates the total solar irradiance at Llyn Efyrnwy as a function of the time of day (hour) over a single day of the year. Assume that $E_0 = 1380\,\text{W·m}^{-2}$, $\tau = 0.7$, $\phi_L = 52.756°$, $p_{alt} = 1000\,\text{mb}$ and $p_{sea} = 1013\,\text{mb}$. Run the revised computational model for June 21, 1998 (DoY $= 172$). Use the data that this generates to produce a plot similar to the one shown in Figure 5.20, which presents a comparison between the observed and the simulated hourly variation in total solar irradiance. Comment on the likely reasons for the differences between the observed and simulated responses.

---

## 5.5  SUMMARY

This chapter explores the temporal variation of incident solar radiation received at Llyn Efyrnwy, using *in situ* measurements and a simulation model. The simulation model is constructed from mathematical formulae that are published in the scientific literature. Alternative mathematical formulations exist for various parts of the model considered here; in each case, the simplest of these is selected. Sometimes, however, the appropriate mathematical formulae will not have been developed previously, or else the existing formulations may be considered too complex, may require too many input values, or too much computing power, for the task at hand. In these circumstances one may be required to formulate a new mathematical model *ab initio*. This is exemplified in the following chapter, which illustrates how one might begin with an extremely simple mathematical model, gradually enhancing this as necessary so that its output conforms ever more closely to observations of the environmental process or system that it is intended to represent.

## SUPPORTING RESOURCES

The following resources are provided in the chapter5 sub-directory of the CD-ROM:

| | |
|---|---|
| radt98le.dat | Hourly mean wind speed and wind direction data set for Llyn Efyrnwy, 1998, in MetOffice format |
| selradt.awk | gawk program to extract the measurements of global solar irradiance from radt98le.dat and output them in a format suitable to be plotted using gnuplot |
| solarrad.awk | gawk program to estimate the global, direct and diffuse solar irradiance received at Llyn Efyrnwy at solar noon on a specific day of the year |
| solarrd2.awk | gawk program to estimate the global, direct and diffuse solar irradiance received at Llyn Efyrnwy at solar noon on each day of the year |
| solar.bat | Command line instructions used in this chapter |
| solar.plt | gnuplot commands used to generate the figures presented in this chapter |
| solar.gp | gnuplot commands used to generate figures presented in this chapter as EPS files |

# Chapter 6

# Light Interaction with a Plant Canopy

**Topics**

- Light interaction with plant canopies

- Significance of light interaction in energy budget and plant growth studies, and for Earth observation

- Developing a model of light interaction with a plant canopy

- Stages in the model development cycle: specification, formulation, implementation and evaluation

- Representing the multiple scattering of light within a plant canopy

## 6.1  INTRODUCTION

In the preceding chapter, attention was focused on the amount of solar radiation that reaches Earth's surface, using both measurements and models. This chapter builds on those foundations, examining how incident solar radiation interacts with different materials on Earth's surface. Emphasis is placed on plant canopies, in part because of the important dynamic role that they play in Earth's climate system and, in particular, the global carbon cycle.

Knowledge of the principles and processes underlying radiation interaction with Earth's surface is central to the study of many environmental issues, including those concerned with Earth's energy budget (McGuffie and Henderson-Sellers 1997). In that context, solar radiation incident on Earth's surface is ultimately either absorbed

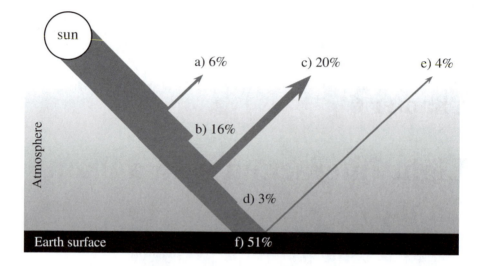

Figure 6.1: Earth's shortwave radiation budget, showing the approximate percentages of the incident solar radiation that are (a) reflected by the atmosphere, (b) absorbed by the atmosphere, (c) reflected by clouds, (d) absorbed by clouds, (e) reflected by the land and ocean surface and (f) absorbed by the land and ocean surface.

by the surface or reflected from it, the latter passing back up through the atmosphere and out to space (Figure 6.1). The balance between absorption and reflection is an important factor in Earth's climate system (McGuffie and Henderson-Sellers 1997), in part because the absorbed fraction is subsequently re-radiated at longer wavelengths, contributing to the warming of Earth's atmosphere (Figure 6.2).

Earth's surface is composed of various types of material, including vegetation, soil, water, snow and ice, which differ in terms of physical and chemical composition. As a result, each reflects and absorbs different fractions of the incident solar radiation. The fractions reflected and absorbed also vary as a function of wavelength. Healthy green vegetation, for example, reflects very little of the incident solar radiation at blue ($\sim 0.46\,\mu$m) and red ($\sim 0.69\,\mu$m) wavelengths because plant pigments, particularly chlorophyll *a* and *b*, absorb strongly at these wavelengths (Figure 6.3). By contrast, vegetation generally reflects a slightly larger fraction of the incident solar radiation at green wavelengths ($\sim 0.55\,\mu$m) and almost half of the incoming solar radiation at near infra-red (NIR) wavelengths ($\sim 0.85\,\mu$m) (Gausman 1977, Tucker and Garratt 1977, Grant 1987, Jacquemoud and Baret 1990). The latter is due to the greatly reduced absorption by, and hence the intense scattering of radiation within and between, plant leaves in the NIR. Thus, the spatial disposition of surface materials, together with their spectral reflectance and absorptance characteristics, plays an important role in determining Earth's energy balance.

A fraction of the incident solar radiation at wavelengths between $0.4\,\mu$m and $0.7\,\mu$m is absorbed by vegetation for photosynthesis (Campbell and Norman 1998). This quantity is often referred to as the fraction of absorbed photosynthetically ac-

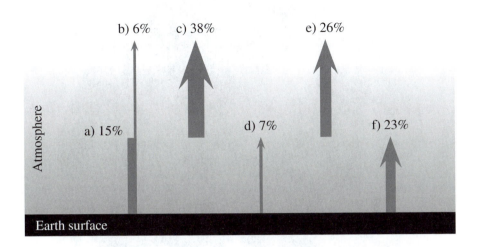

Figure 6.2: Earth's longwave radiation budget, showing the approximate percentages of long-wave radiation emission from the surface that are (a) absorbed by atmospheric gases or (b) escape to space, (c) the net emission by atmospheric gases, (d) the sensible heat flux from the surface, (e) the net emission by clouds and (f) the latent heat flux from the surface.

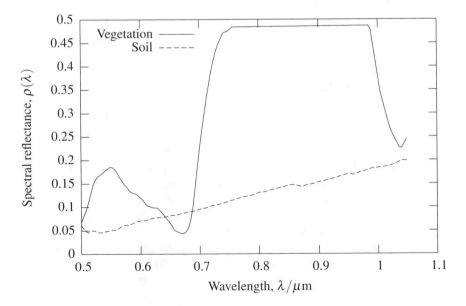

Figure 6.3: Typical spectral reflectance curves for vegetation (solid line) and soil (dashed line) in the range $0.5\,\mu$m (blue/green) to $1.05\,\mu$m (near-infrared).

Figure 6.4: Upward-looking hemispherical photograph taken from within a deciduous tree canopy, giving an indication of the horizontal and vertical distribution of light within the canopy.

tive radiation (fAPAR). In conjunction with temperature and the availability of water and nutrients, the fAPAR governs the growth and development of vegetation canopies (Jones 1983, Monteith and Unsworth 1995). The fAPAR is a function of the amount of above-ground plant material (biomass) and its 3D structure (i.e., the size and shape of the leaves and the other plant parts, their inclination and orientation in terms of zenith and azimuth angles, and their spatial distribution within the canopy). In turn, the fAPAR controls the uptake of $CO_2$ by plants through photosynthesis, as well as the release of water by evapotranspiration. Understanding the interception of solar radiation by vegetation canopies is therefore central to studies of the global carbon cycle and carbon sequestration (Lal *et al.* 1999, Cao *et al.* 2003), of plant growth and crop yield (Russell *et al.* 1989, Bouman 1992, Goudriaan and van Laar 1994) and of hydrometeorology (Ward and Robinson 1999, Beven 2001).

The vertical and horizontal distribution of light within a vegetation canopy is also an important ecological factor, which has significant implications for the strategies adopted by plants in terms of their competition for the available light resource (Figure 6.4). Thus, some species seek competitive advantage by growing tall quickly to gain access to the maximum amount of solar radiation, which is found at the top of the canopy (Whittaker 1975). These species, however, must use much of the energy that they derive from photosynthesis in the production of woody biomass for the stem (or trunk) and branches, which are required to support their foliage. In contrast, other species may become adapted to life at the base of the canopy, where the level of available light is generally greatly reduced, but much less effort has to be put into the production of woody biomass (Whittaker 1975). They may also be able to take advantage of light that penetrates through gaps in the crown of the canopy, which produces "sun flecks" at its base.

Lastly, there is considerable interest in the interaction of solar radiation with plant canopies in the context of satellite remote sensing (Goel 1987, Gao and Lesht 1997, Pinty and Verstraete 1998). This arises from the need to develop inventories of the amount, condition and productivity of vegetation, in both managed and unmanaged landscapes, at regional and global scales (Barnsley 1999). Among other things, this information is required by the dynamic global vegetation models (DGVMs) and land-surface parameterization schemes incorporated within the present generation of global climate models (Woodward *et al.* 1995, Sellers *et al.* 1996, Friend *et al.* 1997, Cox *et al.* 1998, Cramer *et al.* 2001). Data from sensors mounted aboard Earth-orbiting satellites can assist in initializing these models and in constraining their output through the use of data assimilation techniques. These data describe the amount of solar radiation reflected from Earth's surface, however, rather than the properties of interest. Turning them into useful information therefore demands a model that relates specific properties of the land surface to measurements of reflected radiation, and *vice versa*. Once again, the development of such models requires an understanding of how solar radiation interacts with Earth's surface.

## 6.2 DEVELOPING A MODEL OF LIGHT INTERACTION WITH PLANT CANOPIES

A model was constructed in the preceding chapter to simulate the amount of solar radiation incident on Earth's surface. The model is based on mathematical equations that are widely reported in the relevant scientific literature. A similar approach could easily be adopted in the present chapter, where the aim is to represent the interaction of solar radiation with a plant canopy. There are various candidate formulae and models from which to choose in this context; see, for example, Verhoef (1984), Camillo (1987), Goel (1987), Verstraete *et al.* (1990), Gao (1993), Gastellu-Etchegorry *et al.* (1996), Kuusk (1996) and North (1996). There are many other situations, however, in which appropriate mathematical formulae and models have yet to be developed, or where the existing formulations are considered to be too complex, or else demand too much input data or too many computational resources, for the task at hand. In these circumstances, one must formulate the mathematical model from scratch (*ab initio*). To illustrate what this involves, existing models of radiation interaction with plant canopies are eschewed here. Instead, a new model is developed from first principles. The primary objective in doing so is to highlight the main stages involved in model development and, more specifically, to demonstrate that the model-building process often requires several iterations around the cycle of model specification, formulation, implementation and evaluation. Thus, a very simple model is constructed at first. The sophistication of the model is subsequently increased so that its output conforms ever more closely to reality. The reader might like to note that the final version of the model reported in this chapter is similar to the "Adding method" model developed by Cooper *et al.* (1982).

### 6.2.1   Specifying the Conceptual Model

The first step in building a model is to incorporate knowledge of the environmental processes and parameters involved, together with any assumptions, approximations and simplifications that need to be made, in a conceptual model. The reasons for doing so are twofold. First, it clarifies the nature and purpose of the model, not least in the mind of the modeler. Second, it provides information to help potential users of the model understand its scope and limitations, to challenge its underlying assumptions, approximations and simplifications, and to develop improved versions of the model in the future.

Here, the conceptual model is summarized by the following three statements:

1. Different materials on Earth's surface reflect different fractions of the incident solar radiation (i.e., reflectance varies according to the type of surface material).

2. The fraction of incident solar radiation reflected by a given surface material is dependent on the wavelength of the radiation (i.e., reflectance also varies spectrally).

3. A given area of Earth's surface may be covered by a mixture of materials.

In short, the fraction of incident solar radiation that is reflected from a given area of Earth's surface is a function of both the wavelength of the radiation and the relative proportions of ground covered by different surface materials.

Four simplifying assumptions are also made, namely that (i) the sun is located directly overhead, (ii) the surface area is flat and level, (iii) radiation travels either vertically downward from the sun to Earth's surface or vertically upward from Earth's surface into space, and (iv) the atmosphere has no appreciable effect on either the downwelling or the upwelling radiation. It should be evident that these are quite drastic simplifications. Nevertheless, they help to focus attention on the core elements of the problem, ignoring for now the more complex effects produced by variations in the angle of the sun, the 3D structure of the surface materials, and the physical and chemical composition of the atmosphere. These can, of course, be added into the model at a later stage, if so required.

### 6.2.2   Formulating the Mathematical Model

The second stage is to represent the conceptual model in mathematical terms, using functions and equations. This is sometimes known as formulating the mathematical model because it involves translating the conceptual model into mathematical formulae (Edwards and Hamson 1989). It is also frequently the most challenging stage in

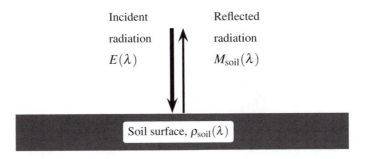

Figure 6.5: Reflection from a soil surface. See text for details.

the development of a model. Sometimes this is because the solution to the problem demands the use of advanced mathematical techniques (although each of the models developed here requires only basic skills in algebra). More often it is because there are several different ways in which the system can be represented mathematically and it is not immediately apparent which approach is best. Consequently, deriving a suitable mathematical formulation is often a trial-and-error process, but it is also a skill that improves with practice. Finally, it is worth noting that some models cannot be expressed mathematically, at least not in concise analytical terms. These must therefore be solved numerically. This issue will not be considered further here, but it is addressed extensively in Chapter 7.

The ability to "see" a model in mathematical terms is often aided by the use of diagrams to represent the various component elements of the model system and the relationships between them. This is frequently achieved using Forrester diagrams (Forrester 1961, 1969, 1973), which represent the structure of a system in terms of sources, sinks, reservoirs (also known as pools or stocks), flows (or processes), converters and inter-relationships (also known as links or connectors). These are widely used to study system dynamics in ecological modeling (Ford 1999, Deaton and Winebrake 2000). Here, though, a less formal approach is adopted, starting with a simple schematic representation of the reflection of solar radiation from a bare soil surface (Figure 6.5).

*Reflection from a Soil Surface*

Figure 6.5 shows two fluxes (or streams) of radiation. These are (i) the incident flux, which travels vertically downward from the sun to the soil surface and (ii) the reflected flux, which travels vertically upward from the soil surface to space. Note that the arrows are intended to represent aggregate fluxes across the whole of the soil surface, not just the locations to and from which they point. Strictly speaking, therefore, they should be referred to as flux densities because they describe radiant fluxes over a given surface area (Monteith and Unsworth 1995). The incident flux density is usually known as irradiance ($E$; see also Chapter 5), while the reflected flux density is referred to as radiant exitance ($M$). Both are measured in units of Watts per square meter ($W \cdot m^{-2}$).

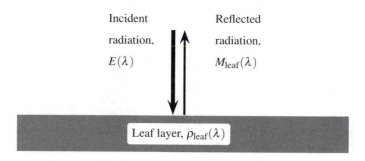

Figure 6.6: Reflection from a layer of leaves.

Irradiance and radiant exitance typically vary as a function of the wavelength of the radiation, which is denoted by $\lambda$. Thus, the terms spectral irradiance, $E(\lambda)$, and spectral radiant exitance, $M(\lambda)$, are used to refer to values of these quantities at a specific wavelength or measured over a narrow range of wavelengths, known as a waveband. Both are measured in units of Watts per square meter per micrometer ($\mathrm{W \cdot m^{-2} \cdot \mu m^{-1}}$). Recall from the preceding chapter that a micrometer, $\mu m$, is a unit of wavelength denoting one millionth of a meter.

The fraction of incident solar radiation that is reflected from a soil surface is, by definition, governed by the reflectance of the soil. This is, in turn, a function of its physical and chemical composition. The reflectance of soil also varies according to wavelength, which can be expressed as follows:

$$\rho_{\mathrm{soil}}(\lambda) = \frac{M_{\mathrm{soil}}(\lambda)}{E(\lambda)} \tag{6.1}$$

where $\rho$ denotes reflectance. It should be evident from this that the spectral reflectance of soil, $\rho_{\mathrm{soil}}(\lambda)$, must lie in the range 0 to 1 ($0 \le \rho_{\mathrm{soil}}(\lambda) \le 1$), because it is physically impossible for the amount of radiation reflected from the soil surface at a given wavelength to be negative (which would imply that the soil absorbs more radiation than is incident upon it) or to exceed unity (which would imply that it reflects more radiation than it receives). Equation 6.1 can be rearranged to yield the amount of radiation reflected from the soil given values of $E(\lambda)$ and $\rho_{\mathrm{soil}}(\lambda)$, as follows:

$$M_{\mathrm{soil}}(\lambda) = E(\lambda) \times \rho_{\mathrm{soil}}(\lambda) \tag{6.2}$$

*Reflection from a Leaf Layer*

Now consider another area of ground that is covered entirely by a layer of healthy green leaves through which no soil is visible from above (Figure 6.6). Assume for now that the leaves are sufficiently thick and closely spaced, without overlapping, so that none of the incident solar radiation penetrates down to the soil substrate in which the vegetation grows. In these circumstances, the fraction of incident solar radiation that is reflected from the leaf layer is governed by the spectral reflectance

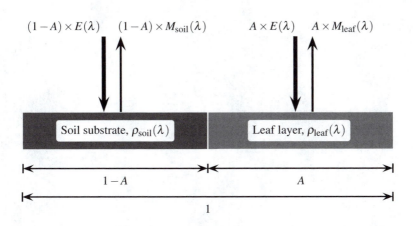

$(1-A) \times E(\lambda)$     $(1-A) \times M_{\text{soil}}(\lambda)$       $A \times E(\lambda)$    $A \times M_{\text{leaf}}(\lambda)$

Soil substrate, $\rho_{\text{soil}}(\lambda)$       Leaf layer, $\rho_{\text{leaf}}(\lambda)$

$1-A$       $A$

$1$

Figure 6.7: Reflection from a mixed soil and leaf surface.

of the leaves, $\rho_{\text{leaf}}(\lambda)$. This is, in turn, a function of their physical and chemical composition. The spectral reflectance of the leaves can therefore be expressed as

$$\rho_{\text{leaf}}(\lambda) = \frac{M_{\text{leaf}}(\lambda)}{E(\lambda)} \tag{6.3}$$

and hence

$$M_{\text{leaf}}(\lambda) = E(\lambda) \times \rho_{\text{leaf}}(\lambda) \tag{6.4}$$

where the value of $\rho_{\text{leaf}}(\lambda)$ lies in the range 0 to 1 ($0 \leq \rho_{\text{leaf}}(\lambda) \leq 1$).

*Reflection from a Mixed Soil and Leaf Surface*

Now consider the case where part of the surface area is covered by a dense layer of leaves, while the remainder is exposed soil (Figure 6.7). Assume that the fractional area of ground covered by leaves is $A$, where $0 \leq A \leq 1$, so that the fractional area of exposed soil is $1 - A$. This is a reasonable first approximation to a simple row crop at a relatively early stage of growth (Figure 6.8).

If solar radiation at a given wavelength, $E(\lambda)$, is incident uniformly across the entire area, the fraction of the radiation that is reflected from the surface as a whole can be written as follows:

$$\rho_{\text{surface}}(\lambda) = \frac{M_{\text{surface}}(\lambda)}{E(\lambda)} \tag{6.5}$$

where

$$M_{\text{surface}}(\lambda) = ((1-A) \times M_{\text{soil}}(\lambda)) + (A \times M_{\text{leaf}}(\lambda)) \tag{6.6}$$

Figure 6.8: Photograph of a sugar beet crop, showing the mixture of leaves (dark areas) and soil (light areas) visible when the canopy is viewed vertically downward (image size approximately $1.4\,\text{m} \times 0.8\,\text{m}$).

since the soil covers $1 - A$, and the leaves cover $A$, of the total surface area. Moreover, substituting Equations 6.2 and 6.4 into Equation 6.6 gives

$$M_{\text{surface}}(\lambda) = (A \times E(\lambda) \times \rho_{\text{leaf}}(\lambda)) + ((1 - A) \times E(\lambda) \times \rho_{\text{soil}}(\lambda)) \tag{6.7}$$

Equation 6.7 states that the amount of solar radiation incident on the leaf layer is proportional to the area of ground that its covers, $A$. It then reflects a fraction, $\rho_{\text{leaf}}(\lambda)$, of this radiation back out to space. The total amount of radiation reflected from the leaf layer is therefore $A \times E(\lambda) \times \rho_{\text{leaf}}(\lambda)$. Likewise, the soil receives some fraction of the total incident solar radiation, $E(\lambda)$, proportional to the area that it covers, $(1 - A)$, and reflects a fraction of this, $\rho_{\text{soil}}(\lambda)$, back out to space, which equates to $(1 - A) \times E(\lambda) \times \rho_{\text{soil}}(\lambda)$. The total amount of radiation reflected from the surface as a whole is therefore the sum of these two components.

Equation 6.7 can be rearranged to determine the average spectral reflectance for the surface as a whole by dividing by $E(\lambda)$.

$$\rho_{\text{surface}}(\lambda) = \frac{M_{\text{surface}}(\lambda)}{E(\lambda)} = A\rho_{\text{leaf}}(\lambda) + (1 - A)\rho_{\text{soil}}(\lambda) \tag{6.8}$$

So, for example, if $A = 0.5$ (i.e., one half of the surface area is covered by leaves), $\rho_{\text{leaf}}(\lambda) = 0.475$ (i.e., the leaves reflect just under one half of the solar radiation that is incident on them) and $\rho_{\text{soil}}(\lambda) = 0.125$ (i.e., the soil reflects one eighth of the solar radiation incident on it), the spectral reflectance for the surface as a whole is given by

$$
\begin{aligned}
\rho_{\text{surface}}(\lambda) &= (0.5 \times 0.475) + ((1 - 0.5) \times 0.125) \\
&= 0.2375 + 0.0625 \\
&= 0.3
\end{aligned}
\tag{6.9}
$$

The values of $\rho_{\text{leaf}}(\lambda)$ and $\rho_{\text{soil}}(\lambda)$ used in this example are typical of those at NIR wavelengths ($\sim 0.85 \mu m$); that is, wavelengths slightly longer than the human eye can see (Figure 6.3). By comparison, at red wavelengths ($\sim 0.65 \mu m$), which are visible to the human eye, typical values of $\rho_{\text{leaf}}(\lambda)$ and $\rho_{\text{soil}}(\lambda)$ are 0.04 and 0.08, respectively. These produce a value of $\rho_{\text{surface}}(\lambda) = 0.06$ for the same fractional areas of leaves and soil.

### Checking the Mathematical Model

It is good practice to get into the habit of checking a model, in the manner outlined above, at a number of different stages during its development, to confirm that it is performing as expected. The mathematical model given in Equation 6.8, for example, can be tested by presenting it with a range of input values. These should, at the very least, consist of extreme or "special" case values. For instance, when $A = 0$ (i.e., there are no leaves present), $\rho_{\text{surface}}(\lambda)$ should equal $\rho_{\text{soil}}(\lambda)$. Similarly, when $A = 1$ (i.e., no soil is visible), $\rho_{\text{surface}}(\lambda)$ should equal $\rho_{\text{leaf}}(\lambda)$.

Clearly, few environmental models are as simple as the one considered here, where the formulation can be checked quickly and simply by mental arithmetic. More complex models generally demand much greater rigor and more extensive testing to establish that they operate as intended. Indeed, it may only be possible to do so after they have been implemented in computer code, perhaps by performing a large number of model "runs" (simulations) using various sets of input values.

### 6.2.3 Implementing the Computational Model

The third step is to implement the mathematical model (Equation 6.8) in computer code; that is, to create the computational model. There are various ways that this can be realized in gawk, one of which is given in Program 6.1 (`mixture.awk`). This implementation is intended to simulate the reflectance of a mixed soil and leaf surface for a range of values of $A$ (i.e., the fractional area of ground covered by leaves), varying between 0 and 1 in increments of 0.1.

### Program Structure

The first 13 lines of code in Program 6.1 comprise a series of comments, denoted by the hash (#) symbol at the start of each line. These are used to indicate the purpose of the code, to show how the program should be invoked via the command line, and to identify the major variables concerned. Note that the terms $A$, $\rho_{\text{leaf}}(\lambda)$, $\rho_{\text{soil}}(\lambda)$ and $\rho_{\text{surface}}(\lambda)$ from the mathematical model (Equation 6.8) are represented by the

Program 6.1: mixture.awk

```
# Simple model of solar radiation interaction with a mixed    1
# soil and vegetation surface. The program calculates the     2
# average spectral reflectance of a surface covered by the    3
# specified areal fractions of soil and vegetation (leaves).  4
#                                                             5
# Usage: gawk -f mixture.awk -v rho_leaf=value \              6
#    -v rho_soil=value [ > output_file ]                      7
#                                                             8
# Variables:                                                  9
# area_leaf    Fractional area covered by leaves             10
# rho_leaf     Leaf spectral reflectance                     11
# rho_soil     Soil spectral reflectance                     12
# rho_surface Average spectral reflectance of surface        13
                                                             14
BEGIN{                                                       15
  for(area_leaf=0;area_leaf<=1;area_leaf+=0.1){              16
    rho_surface=(area_leaf*rho_leaf)+ \                      17
      ((1-area_leaf)*rho_soil);                              18
    print area_leaf, rho_surface;                            19
  }                                                          20
}                                                            21
```

variables area_leaf, rho_leaf, rho_soil and rho_surface, respectively, in the code. More generally, a combination of carefully chosen variable names, instructive comments and clear layout tends to produce code that is easier to read, understand and maintain; it is, in effect, self-documenting (British Computer Society 1998).

The rest of the code is contained within the BEGIN block (lines 15 to 21) because the intention is to use the computational model to generate data (i.e., to perform a simulation), rather than to process data stored in a file. A for loop (line 16) is used to vary the fractional area of ground covered by leaves (area_leaf) between 0 (bare soil) and 1 (complete vegetation cover) in steps of 0.1. For each iteration around the loop, the reflectance of a surface with that particular mixture of leaves and soil is calculated (rho_surface; lines 17 and 18). Recall that the backslash symbol (\) at the end of line 17 is the line continuation symbol. This allows a single long gawk statement to be split over two or more lines of code. Finally, the program prints out the fractional area covered by leaves and the reflectance of the mixed soil and vegetation surface (line 19).

It is important to note that values for the spectral reflectance of the leaves and the soil are not specified ("hard wired") in this program. These values must therefore be provided via the command line using the -v switch (e.g., -v rho_leaf=0.475). The intention in doing so is to keep the program as flexible as possible, so that it can be used to simulate the reflectance of the mixed leaf and soil surface at any given wavelength. If appropriate values of these variables are not provided on the command line, gawk will set them to zero (0), with possibly unintended consequences.

### 6.2.4 Running the Model

This implementation of the model can be run from the command line by typing

```
gawk -f mixture.awk -v rho_leaf=0.475 -v rho_soil=0.125 > ↻
    ↻mixture.nir
```
*1*

to simulate the variation in spectral reflectance at NIR wavelengths as a function of the fractional area covered by leaves, and

```
gawk -f mixture.awk -v rho_leaf=0.04 -v rho_soil=0.08 > ↻
    ↻mixture.red
```
*2*

to perform an equivalent simulation at red wavelengths. Notice that a separate -v flag is used for each variable whose value is initialized on the command line.

### 6.2.5 Evaluating the Output from the Computational Model

The next step is to examine graphically the results that the model produces. This can be achieved by typing the following commands in gnuplot:

```
set xlabel "Fraction of ground covered by leaves, A"        1
set ylabel "Spectral reflectance, rho_surface(lambda)"      2
set key top left                                            3
set yrange [0:0.7]                                          4
set data style lp                                           5
plot 'mixture.red' t "Red (0.65 micrometers)", \            6
     'mixture.nir' t "NIR (0.85 micrometers)"               7
```

These commands assume that the data files mixture.red and mixture.nir, generated by Program 6.1, are located in the working directory; otherwise, their full or relative pathnames must be given. The results are presented in Figure 6.9.

It is evident from Figure 6.9 that the model formulation presented in Equation 6.8, and implemented in Program 6.1, defines a straight-line relationship between the fraction of ground covered by leaves ($A$; area_leaf) and the spectral reflectance for the surface as a whole ($\rho_{surface}(\lambda)$; rho_surface). Indeed, it might be described as a linear mixture model (Ichoku and Karnieli 1996). Detailed observations of real plant canopies, however, suggest that the actual trends are more like those shown in Figure 6.10. Specifically, spectral reflectance exhibits a curvilinear relationship with increasing vegetation amount at both red and NIR wavelengths: for the former, the relationship is indirect (negative); for the latter, it is direct (positive). Both exhibit asymptotic trends, with the asymptote being reached at lower levels of vegetation cover in the red than in the NIR. Note that spectral reflectance is plotted as a function of leaf area index (LAI) in Figure 6.10. The difference between this and fractional ground cover is explored more fully in the next section.

The differences between Figures 6.9 and 6.10 suggest that the approach adopted thus far has been too simplistic. Consequently, the next section revisits the conceptual model and considers how it might be made to correspond more closely to reality, updating the mathematical and computational models accordingly.

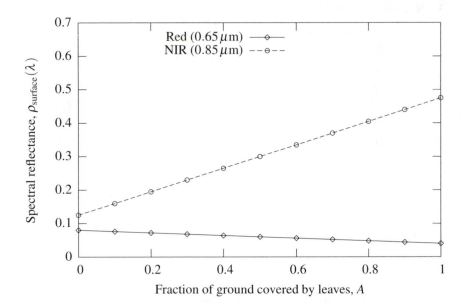

Figure 6.9: Output from a simple light-interaction model for a surface comprising a mixture of soil and leaves (Program 6.1). Red: $\rho_{\text{leaf}}(\lambda) = 0.04$, $\rho_{\text{soil}}(\lambda) = 0.08$. NIR: $\rho_{\text{leaf}}(\lambda) = 0.475$, $\rho_{\text{soil}}(\lambda) = 0.125$.

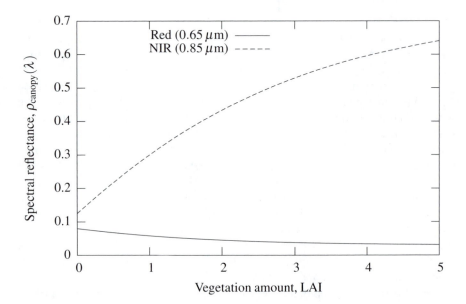

Figure 6.10: Expected variation in the spectral reflectance of a simple plant canopy at red and NIR wavelengths as a function of vegetation amount, LAI.

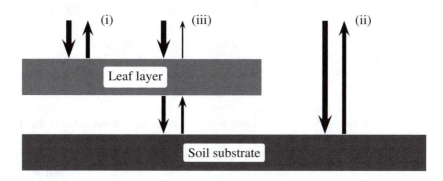

Figure 6.11: Two-layer model of light interaction with a plant canopy.

**Exercise 6.1**: Develop a revised version of the simple light-interaction model that simulates the spectral reflectance of an area of ground covered by a mixture of three surface materials (e.g., vegetation, soil and snow or ice). Begin by specifying the conceptual model, stating any assumptions that must be made. Next, formulate the mathematical model, basing this on Equation 6.8. Implement the model in gawk, using Program 6.1 as a template. Finally, run the revised model and examine the output.

## 6.3 A TWO-LAYER LIGHT INTERACTION MODEL

### 6.3.1 Improving the Conceptual Model

The model that has been developed thus far simulates the spectral reflectance for an area of ground comprising a mixture of leaves and soil. It portrays the surface as a single layer, or slab, in which the leaves and soil lie side by side (Figure 6.7). Clearly, this is not a very realistic representation of the physical structure of most plant canopies, in which the leaves are usually located some distance above the soil substrate. In terms of the conceptual model, this would be better represented by two discrete layers; that is, by a separate leaf-layer suspended above the soil substrate (Figure 6.11).

In the revised conceptual model, the leaf layer covers some fraction ($A$) of the total surface area when viewed vertically downward from above, while the remainder ($1 - A$) represents one or more gaps through which the soil substrate can been seen. In Figure 6.11 the leaf layer is represented as a single continuous slab, but it might equally well be a discontinuous fragmented layer covering the same fractional area of ground. Below the leaf layer, the soil substrate extends across the entire surface area. As a result of this modification, incoming solar radiation can now traverse three different pathways through the plant canopy before being reflected back out to space (Figure 6.11). Thus, it may (i) be reflected directly from the leaf layer, (ii) pass down through a gap in the leaf layer, reflect from the soil substrate and escape back out to space through the same gap in the leaf layer, or (iii) pass down through the leaf layer, reflect from the soil substrate and pass back up through the leaf layer.

The revised conceptual model introduces two new phenomena: one explicitly, namely transmission; the other implicitly, namely absorption. Transmission refers to the passage of radiation through an object, such as a leaf, while transmittance is the fraction of solar radiation incident upon an object that is transmitted through it (Glickman 2000). Absorption refers to radiation that is retained by a substance (i.e., radiation that is neither reflected nor transmitted), while absorptance is the fraction of the incident radiation that is absorbed (Glickman 2000). Both transmittance and absorptance vary as a function of wavelength in most instances. It is common, therefore, to refer to spectral transmittance and spectral absorptance when dealing with radiation at a particular wavelength or in a narrow spectral waveband.

### 6.3.2   Reformulating the Mathematical Model

The symbols $A$, $\rho_{\text{leaf}}(\lambda)$ and $\rho_{\text{soil}}(\lambda)$ are employed with the same meaning as before, while $\tau_{\text{leaf}}(\lambda)$ is used to refer to the spectral transmittance of the leaf layer, $\alpha_{\text{leaf}}(\lambda)$ the spectral absorptance of the leaf layer, and $\alpha_{\text{soil}}(\lambda)$ the spectral absorptance of the soil substrate, noting that

$$\rho_{\text{leaf}}(\lambda) + \tau_{\text{leaf}}(\lambda) + \alpha_{\text{leaf}}(\lambda) = 1 \tag{6.10}$$

Thus, solar radiation incident upon the leaf layer is partitioned three ways: some is reflected from it, some is transmitted through it, and some is absorbed by it. Following the law of conservation of energy, the reflected, transmitted and absorbed fractions must account for the total radiation striking the leaf layer (i.e., they must sum to 1). Similarly,

$$\rho_{\text{soil}}(\lambda) + \alpha_{\text{soil}}(\lambda) = 1 \tag{6.11}$$

which implies that the soil substrate is assumed to be completely opaque, that is $\tau_{\text{soil}}(\lambda) = 0$.

The two-layer model can therefore be formulated in terms of the three pathways shown in Figure 6.11, namely

(i)     $A\rho_{\text{leaf}}(\lambda)$

(ii)    $(1-A)\rho_{\text{soil}}(\lambda)$

(iii)   $A\tau_{\text{leaf}}(\lambda)\rho_{\text{soil}}(\lambda)\tau_{\text{leaf}}(\lambda)$

and where the spectral reflectance from the vegetation canopy as a whole is the sum of these three terms, i.e.,

$$\begin{aligned}\rho_{\text{canopy}}(\lambda) &= A\rho_{\text{leaf}}(\lambda) + (1-A)\rho_{\text{soil}}(\lambda) \\ &\quad + A\tau_{\text{leaf}}(\lambda)\rho_{\text{soil}}(\lambda)\tau_{\text{leaf}}(\lambda)\end{aligned} \tag{6.12}$$

While the profusion of Greek symbols in this formulation might make it look rather daunting, bear in mind that it involves nothing more demanding than simple addition and multiplication.

It may be helpful to think of Equation 6.12 in terms of fractions and percentages. Thus, the fraction of incident solar radiation intercepted by some part of the leaf layer is $A$. For example, if the leaf layer completely covers the ground below ($A = 1$), the fraction intercepted is 1 (all of the incident solar radiation) because there are no gaps through which the solar radiation can pass uninterrupted down to the soil substrate. By contrast, if the leaf layer covers half of the area of ground below ($A = 0.5$), the fraction intercepted is 0.5. Put another way, 50% of the solar radiation will be intercepted by the leaf layer, while the remainder will pass down through a gap toward the soil substrate. Similarly, $\rho_{leaf}(\lambda)$ describes the fraction of solar radiation striking the upper surface of the leaf layer that is reflected back out to space. So, for example, if $\rho_{leaf}(\lambda) = 0.475$ at NIR wavelengths, 47.5% of the solar radiation striking the leaf layer will be reflected upward from it; the remainder will be absorbed by it or transmitted through it. Moreover, it is possible to combine these two values to determine the fraction of incident radiation that is reflected upward from the leaf layer. This relates to pathway (i) in Figure 6.11 and is given by $A\rho_{leaf}(\lambda)$. Thus, if $A = 0.5$ and $\rho_{leaf}(\lambda) = 0.475$ at NIR wavelengths, the result is $0.5 \times 0.475 = 0.2375$. In other words, 23.75% of the incident NIR radiation is reflected directly from the leaf layer. Note that the equivalent figure at red wavelengths is 2%, assuming that $\rho_{leaf}(\lambda) = 0.04$.

For incident solar radiation to traverse pathway (ii) in Figure 6.11 it must pass down through a gap in the leaf layer $(1 - A)$, be reflected from the soil substrate $(\rho_{soil}(\lambda))$ and pass back up through the same gap in the leaf layer. Note that in this model the fraction of radiation reflected from the soil substrate that passes back up through the gap in the leaf layer is 1 (i.e., all of it) because the model assumes that radiation travels vertically downward and upward through the canopy. So, if the radiation has passed down through a gap in the leaf layer it must, by definition, pass back up through the same gap on its return path. The fraction of incident solar radiation that traverses this pathway through the canopy is therefore $(1 - A)\rho_{soil}(\lambda)$. Assuming that $A = 0.5$ and $\rho_{soil}(\lambda) = 0.125$ at NIR wavelengths, $(1 - A)\rho_{soil}(\lambda) = 0.5 \times 0.125 = 0.0625$. Thus, 6.25% of the incident NIR radiation takes this route through the model canopy. The equivalent figure at red wavelengths is 4%, assuming $A = 0.5$ and $\rho_{leaf}(\lambda) = 0.08$.

Finally, for incident solar radiation to traverse pathway (iii) in Figure 6.11 it must be intercepted by and transmitted through the leaf layer on its downward path, be reflected upward from the soil substrate, and then be transmitted back up through the leaf layer. Again, the fraction of this radiation that is intercepted by the leaf layer on its upward path from the soil substrate is 1 because it passed down through the leaf layer to reach the soil substrate in the first place. Given the assumptions made in the model, it must retrace the same route on its way back up. Nevertheless, the second interception leads to further attenuation of the upwelling radiation because only a fraction of this is transmitted through the leaf layer. Therefore, the fraction of NIR radiation that traverses this particular pathway is $A\tau_{leaf}(\lambda)\rho_{soil}(\lambda)\tau_{leaf}(\lambda) \approx$

Program 6.2: twolayer.awk

```
# Simple model of light interaction with a two-layer      1
# plant canopy (i.e., a leaf layer suspended above a      2
# soil substrate).                                         3
#                                                          4
# Version 1.                                               5
#                                                          6
# Usage: gawk -f twolayer.awk -v rho_leaf=value \          7
#              -v tau_leaf=value -v rho_soil=value \       8
#              [ > output_file ]                           9
#                                                         10
# Variables:                                              11
# area_leaf  Fractional area covered by leaves            12
# rho_leaf   Leaf spectral reflectance                    13
# tau_leaf   Leaf spectral transmittance                  14
# rho_soil   Soil spectral reflectance                    15
# rho_canopy Canopy spectral reflectance                  16
# gap        Fractional area of gaps in leaf layer        17
                                                          18
BEGIN{                                                    19
   for(area_leaf=0;area_leaf<=1;area_leaf+=0.1){          20
      gap=1-area_leaf;                                     21
      rho_canopy=area_leaf*rho_leaf + (gap*rho_soil) + \   22
         (area_leaf*tau_leaf*rho_soil*tau_leaf);           23
      print area_leaf, rho_canopy;                         24
   }                                                       25
}                                                          26
```

0.0141, assuming that $A = 0.5$, $\rho_{\text{leaf}}(\lambda) = 0.475$, $\tau_{\text{leaf}}(\lambda) = 0.475$ and $\rho_{\text{soil}}(\lambda) = 0.125$. Note that $\tau_{\text{leaf}}(\lambda) = \rho_{\text{leaf}}(\lambda)$ is a reasonable approximation in most instances. Thus, approximately 1.41% of the incident NIR radiation takes this route through the model canopy. The equivalent figure at red wavelengths is approximately 0.006%, assuming that $\tau_{\text{leaf}}(\lambda) = \rho_{\text{leaf}}(\lambda) = 0.04$ and $\rho_{\text{soil}}(\lambda) = 0.08$.

### 6.3.3  Implementing the Two-Layer Model in gawk

Implementing the two-layer model given in Equation 6.12 involves making a few minor modifications to Program 6.1. These are presented in Program 6.2. They consist of (i) the introduction of a new variable, gap, the value of which is set to 1-area_leaf (line 21), which is intended solely to improve the readability of the code, (ii) some changes to the model calculation and the output of results (lines 22 and 23) that are required to represent Equation 6.12, and (iii) a small change to the variables whose values are printed out (line 24). Once again, recall that the backslash (\) at the end of line 22 is the line continuation symbol.

### 6.3.4 Running the Two-Layer Model

The two-layer model can be run from the command line by typing

```
gawk -f twolayer.awk -v rho_leaf=0.475 -v tau_leaf=0.475 -↷    3
   ↷v rho_soil=0.125 > 2layer.nir
```

assuming that $\rho_{leaf}(\lambda) = 0.475$ and $\rho_{soil}(\lambda) = 0.125$, and that $\tau_{leaf}(\lambda) = \rho_{leaf}(\lambda)$ at NIR wavelengths. Note, though, that Program 6.2 allows one to evaluate the effect of differences between $\tau_{leaf}(\lambda)$ and $\rho_{leaf}(\lambda)$, if so desired. The equivalent simulation can be performed for red wavelengths by typing the following instruction on the command line:

```
gawk -f twolayer.awk -v rho_leaf=0.04 -v tau_leaf=0.04 -v ↷    4
   ↷rho_soil=0.08 > 2layer.red
```

assuming that $\rho_{leaf}(\lambda) = 0.04$, $\rho_{soil}(\lambda) = 0.08$ and that $\tau_{leaf}(\lambda) = \rho_{leaf}(\lambda)$.

### 6.3.5 Evaluating the Two-Layer Model

It is possible to generate a plot of the output from the two-layer model by typing the following commands in gnuplot (Figure 6.12):

```
set ylabel "Spectral reflectance, rho_canopy(lambda)"        8
plot '2layer.red' t "Red (0.65 micrometers)", \              9
   '2layer.nir' t "NIR (0.85 micrometers)"                  10
```

These commands assume that the files 2layer.red and 2layer.nir, created using Program 6.2, are located in the working directory; otherwise, their full or relative pathnames must be provided.

It is clear from Figure 6.12 that the results produced using the two-layer light interaction model do not correspond well to observations of real vegetation canopies (Figure 6.10). Indeed, they do not differ significantly from those of the one-layer model. The reasons for this are examined in the following two sections, in which several further modifications are made to the model so that its output conforms more closely to expectations.

## 6.4 ACCOUNTING FOR MULTIPLE SCATTERING

### 6.4.1 Enhancing the Conceptual and Mathematical Models

It is instructive to note that the first and second terms in Equation 6.12, $A\rho_{leaf}(\lambda)$ and $(1-A)\rho_{soil}(\lambda)$, relate to solar radiation that has interacted once only with the plant canopy, either with the leaf layer or with the soil substrate, but not both. These terms therefore describe what are known as single scattering events. In contrast, the third term in Equation 6.12, $A\tau_{leaf}(\lambda)\rho_{soil}(\lambda)\tau_{leaf}(\lambda)$, relates to radiation that has interacted three times with elements of the vegetation canopy. More specifically, it describes radiation that is first transmitted down through the leaf layer, then reflected from the soil substrate, and finally transmitted back up through the leaf layer. This is

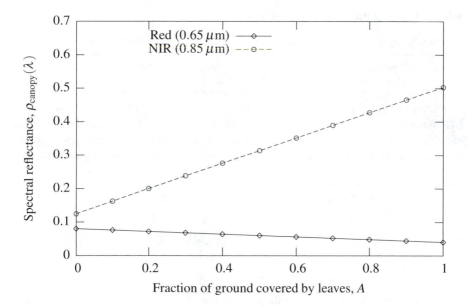

Figure 6.12: Spectral reflectance of a plant canopy as a function of the fractional cover of leaves based on the two-layer light interaction model given in Program 6.2.

commonly known as multiple scattering, since the radiation is scattered (i.e., reflected or transmitted) more than once on its path through the canopy. It is also common to refer to the "order" of scattering. Thus, first-order scattering refers to radiation that has been scattered once, second-order scattering refers to radiation that has been scattered twice, and so on.

Bearing this in mind, it is possible that the performance of the two-layer model could be improved by incorporating higher-order scattering effects. Figure 6.13, for example, illustrates the path that incident solar radiation might take through the two-layer canopy as a result of fifth-order scattering. This can be represented as follows:

$$A \times \tau_{\text{leaf}}(\lambda) \times \rho_{\text{soil}}(\lambda) \times \rho_{\text{leaf}}(\lambda) \times \rho_{\text{soil}}(\lambda) \times \tau_{\text{leaf}}(\lambda) \tag{6.13}$$

Note that, here, $\rho_{\text{leaf}}(\lambda)$ denotes reflectance downward from the underside (or abaxial surface) of the leaf layer, as well as upward from the upper (or adaxial) surface. This assumes that the reflectance of the abaxial and adaxial surfaces of the leaf layer are identical; while this is not always the case for individual leaves in real vegetation canopies, it is a reasonable approximation in this instance.

Adding the fifth-order scattering term (Equation 6.13) into the two-layer light interaction model gives

$$\begin{aligned} \rho_{\text{canopy}}(\lambda) \quad = \quad & A\rho_{\text{leaf}}(\lambda) + (1-A)\rho_{\text{soil}}(\lambda) \\ & + A\tau_{\text{leaf}}(\lambda)\rho_{\text{soil}}(\lambda)\tau_{\text{leaf}}(\lambda) \\ & \boxed{+ A\tau_{\text{leaf}}(\lambda)\rho_{\text{soil}}(\lambda)\rho_{\text{leaf}}(\lambda)\rho_{\text{soil}}(\lambda)\tau_{\text{leaf}}(\lambda)} \end{aligned} \tag{6.14}$$

where the fifth-order term is highlighted in the box.

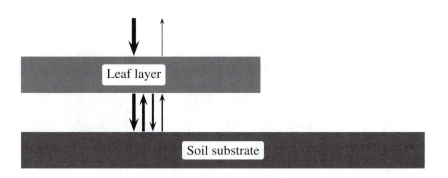

Figure 6.13: Fifth-order multiple scattering in a two-layer model of light interaction. Note that the arrows, which indicate the radiation pathway through the model canopy, have been offset slightly with respect to one another for the sake of clarity.

### 6.4.2 Implementing the Revised Two-Layer Model

An implementation of the revised two-layer model (Equation 6.14) is presented in Program 6.3. It differs only marginally from Program 6.2. Specifically, the spectral reflectance of the canopy (rho_canopy) is now calculated over four lines of code (22 through 25), with the fifth-order scattering term given on lines 24 and 25.

### 6.4.3 Running the Revised Two-Layer Model

The revised model can be run from the command line by typing

```
gawk -f twolayr2.awk -v rho_leaf=0.475 -v tau_leaf=0.475 -↩     5
    ↪v rho_soil=0.125 > 2layer2.nir
```

assuming that $\rho_{\text{leaf}}(\lambda) = 0.475$, $\rho_{\text{soil}}(\lambda) = 0.125$ and that $\tau_{\text{leaf}}(\lambda) = \rho_{\text{leaf}}(\lambda)$ at NIR wavelengths. It is also possible to perform an equivalent simulation for red wavelengths by typing

```
gawk -f twolayr2.awk -v rho_leaf=0.04 -v tau_leaf=0.04 -v ↩     6
    ↪rho_soil=0.08 > 2layer2.red
```

assuming that $\rho_{\text{leaf}}(\lambda) = 0.04$, $\rho_{\text{soil}}(\lambda) = 0.08$ and that $\tau_{\text{leaf}}(\lambda) = \rho_{\text{leaf}}(\lambda)$.

### 6.4.4 Evaluating the Revised Two-Layer Model

A plot of the output produced by the revised two-layer model (Program 6.3) can be generated in gnuplot using the following commands:

```
plot '2layer2.red' t "Red (0.65 micrometers)", \      11
     '2layer2.nir' t "NIR (0.85 micrometers)"         12
```

assuming that the data files 2layer2.red and 2layer2.nir are located in the working directory; otherwise, their full or relative pathnames must be given (Figure 6.14).

Program 6.3: twolayr2.awk

```
# Simple model of light interaction with a two-layer        1
# plant canopy (i.e., a leaf layer suspended above a        2
# soil substrate), incorporating fifth-order scattering.    3
#                                                            4
# Version 2                                                  5
#                                                            6
# Usage: gawk -f twolayr2.awk -v rho_leaf=value \            7
#                -v tau_leaf=value -v rho_soil=value \       8
#                [ > output_file ]                           9
#                                                            10
# Variables:                                                 11
# area_leaf   Fractional area covered by leaves              12
# rho_leaf    Leaf spectral reflectance                      13
# tau_leaf    Leaf spectral transmittance                    14
# rho_soil    Soil spectral reflectance                      15
# rho_canopy  Canopy spectral reflectance                    16
# gap         Fractional area of gaps in leaf layer          17
#                                                            18
BEGIN{                                                       19
    for(area_leaf=0;area_leaf<=1;area_leaf+=0.1){            20
        gap=1-area_leaf;                                     21
        rho_canopy=area_leaf*rho_leaf + (gap*rho_soil) + \   22
            (area_leaf*tau_leaf*rho_soil*tau_leaf) + \       23
            (area_leaf*tau_leaf*rho_soil*rho_leaf* \         24
            rho_soil*tau_leaf);                              25
        print area_leaf, rho_canopy;                         26
    }                                                        27
}                                                            28
```

Disappointingly, the results still do not correspond well to the responses expected for real plant canopies (Figure 6.10); the inclusion of fifth-order scattering in the model has not had a significant impact in this instance. The reason for this is clear in hindsight; multiple scattering is represented in the model as the arithmetic product of a number of fractional values. For example, if $\rho_{leaf}(\lambda) = 0.475$, $\tau_{leaf}(\lambda) = 0.475$ and $\rho_{soil}(\lambda) = 0.125$, as is typical at NIR wavelengths, and $A = 0.5$, the fraction of the incident solar radiation that traverses the pathway through the model canopy shown in Figure 6.13 is $0.5 \times 0.475 \times 0.125 \times 0.475 \times 0.125 \times 0.475 \approx 0.00084$, i.e., less than 0.1% of the total incident flux. By comparison, the fraction that is scattered once only, by the leaf layer or the soil substrate, is $A\rho_{leaf}(\lambda) + (1-A)\rho_{soil}(\lambda) = 0.3$. The contribution of single scattering to the total canopy reflectance at NIR wavelengths is therefore approximately 357 times larger than that of fifth-order scattering (i.e., 0.3/0.00084). The equivalent value at red wavelengths is greater than 300,000. For the two-layer model, therefore, the inclusion of fifth-order multiple scattering does not significantly affect the total spectral reflectance of the model canopy.

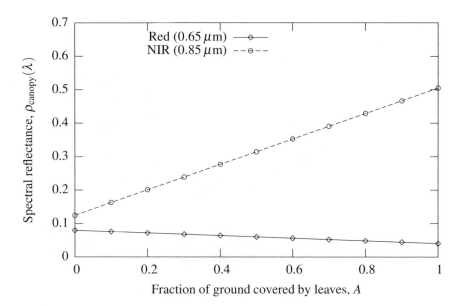

Figure 6.14: Spectral reflectance of a plant canopy as a function of the fractional cover of leaves based on the two-layer model including fifth-order scattering.

## 6.5  MULTIPLE LEAF-LAYER MODELS

### 6.5.1  Enhancing the Conceptual Model

Since the inclusion of fifth-order multiple scattering has not, on its own, significantly increased the realism of the two-layer model, other enhancements must be sought. One possibility is to increase the number of leaf layers in the model to generate, for example, a three-layer model that contains two leaf-layers and a soil substrate. This is a slightly more realistic representation of the 3D structure of real vegetation canopies. It also has four important ramifications for the model. First, it increases the number of pathways that radiation can trace through the model canopy (Figure 6.15). Second, it raises the possibility of specifying separately the fractional area of ground covered by leaves in each layer. This allows variations in the density of plant material to be modeled as a function of depth into the canopy, if so desired. Third, it enables variations in the spatial clumping of plant material to be modeled by representing different degrees of overlap between the leaf layers. Fourth, a different measure must be used to represent the total amount of vegetation in the canopy as a whole. The LAI, which is the total single-sided area of leaves per unit area of ground, is widely employed in this context. If, for example, both leaf layers in Figure 6.15 completely cover the ground below ($A_{\text{leaf layer 1}} = 1$ and $A_{\text{leaf layer 2}} = 1$), then LAI $= 2 \times 1 = 2$. If, instead, each leaf layer covers only half of the ground below ($A_{\text{leaf layer 1}} = 0.5$ and $A_{\text{leaf layer 2}} = 0.5$), then LAI $= 2 \times 0.5 = 1$. Thus, $0 \leq \text{LAI} \leq N$ in this model, where $N$ is the number of leaf layers.

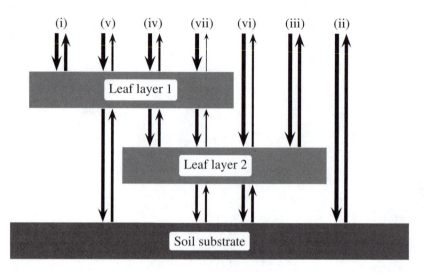

Figure 6.15: Schematic representation of the various pathways (i–vii) that incident solar radiation can take through a three-layer model of a plant canopy.

The remainder of this section examines the three-layer light interaction model shown in Figure 6.15. To simplify things slightly, it is assumed that the fraction of ground covered by each of the leaf layers is identical. Otherwise, the conceptual basis and assumptions of the model remain the same as for the two-layer case.

### 6.5.2   Formulating the Three-Layer Model

The mathematical formulation of the three-layer model is simply an extension of the two-layer case and is given by Equation 6.15.

$$
\begin{aligned}
\rho_{\text{canopy}}(\lambda) \quad = \quad & A\rho_{\text{leaf}}(\lambda) \\
& + (1-A)(1-A)\rho_{\text{soil}}(\lambda) \\
& + (1-A)A\rho_{\text{leaf}}(\lambda) \\
& + A\tau_{\text{leaf}}(\lambda)A\rho_{\text{leaf}}(\lambda)\tau_{\text{leaf}}(\lambda) \\
& + A\tau_{\text{leaf}}(\lambda)(1-A)\rho_{\text{soil}}(\lambda)\tau_{\text{leaf}}(\lambda) \\
& + (1-A)A\tau_{\text{leaf}}(\lambda)\rho_{\text{soil}}(\lambda)\tau_{\text{leaf}}(\lambda) \\
& + A\tau_{\text{leaf}}(\lambda)A\tau_{\text{leaf}}(\lambda)\rho_{\text{soil}}(\lambda)\tau_{\text{leaf}}(\lambda)\tau_{\text{leaf}}(\lambda)
\end{aligned}
\tag{6.15}
$$

Note that the seven terms listed in Equation 6.15 correspond, in order, to the pathways numbered (i) to (vii) in Figure 6.15. These terms and pathways are explained more fully in Table 6.1. Note that the fifth and sixth terms, pathways (v) and (vi), are arithmetically equivalent. Also remember that if radiation passes down through a gap in a leaf layer on its way into the canopy it must pass back up through that gap on its return path because it is assumed that radiation travels vertically upward and

Table 6.1: Explanation of the pathways that radiation can traverse through the three-layer model of a plant canopy presented in Figure 6.15 and the corresponding terms in Equation 6.15.

| Path | Term | Explanation |
|------|------|-------------|
| (i) | $A\rho_{\text{leaf}}(\lambda)$ | Intercept and reflect up from top leaf-layer. |
| (ii) | $(1-A)(1-A)\rho_{\text{soil}}(\lambda)$ | Pass down through gaps in both leaf layers, reflect up from soil substrate and pass back up through the same gaps. |
| (iii) | $(1-A)A\rho_{\text{leaf}}(\lambda)$ | Pass down through gap in top leaf-layer, reflect up from lower leaf-layer, and pass back up through gap in top leaf-layer. |
| (iv) | $A\tau_{\text{leaf}}(\lambda)A\rho_{\text{leaf}}(\lambda)\tau_{\text{leaf}}(\lambda)$ | Transmit down through top leaf-layer, reflect up from lower leaf-layer, transmit back up through top leaf-layer. |
| (v) | $A\tau_{\text{leaf}}(\lambda)(1-A)\rho_{\text{soil}}(\lambda)\tau_{\text{leaf}}(\lambda)$ | Transmit down through top leaf-layer, pass down through gap in lower leaf layer, reflect up from soil substrate, pass back up through gap in lower leaf-layer and transmit up through top leaf-layer. |
| (vi) | $(1-A)A\tau_{\text{leaf}}(\lambda)\rho_{\text{soil}}(\lambda)\tau_{\text{leaf}}(\lambda)$ | Pass down through gap in top leaf-layer, transmit down through lower leaf-layer, reflect up from soil, transmit up through lower leaf-layer and pass back up through gap in top leaf-layer. |
| (vii) | $A\tau_{\text{leaf}}(\lambda)A\tau_{\text{leaf}}(\lambda)\rho_{\text{soil}}(\lambda)\tau_{\text{leaf}}(\lambda)\tau_{\text{leaf}}(\lambda)$ | Transmit down through both leaf layers, reflect up from soil substrate, and transmit up through both leaf layers. |

downward through the canopy. A similar logic applies to radiation that is intercepted by, and transmitted through, one or both of the leaf layers. For this reason, the parameters $A$ and $(1 - A)$ only influence the radiation on its downward path into the model canopy.

Spatial overlap between the two leaf-layers is handled implicitly in Equation 6.15. Thus, for any value of $A > 0$, some of the incident solar radiation interacts with both leaf layers (terms (iv) and (vii) in Equation 6.15), which implies that the two leaf-layers overlap to some extent, even at very low levels of vegetation cover. Moreover, the fraction of radiation that intercepts both leaf layers (and by implication the degree of overlap between them) increases as the value of $A$ grows.

One of the limitations of this model is that there is no way in which radiation can be multiply scattered an even number of times, even though this is possible in real vegetation canopies. So, for example, the first, second and third terms in Equation 6.15 describe single scattering events; the fourth, fifth and sixth terms pertain to third-order multiple scattering; while the final term relates to fifth-order multiple scattering. Although Equation 6.15 encompasses all possible first-order and third-order scattering terms, the same cannot be said for the fifth-order terms. The latter can also occur as a result of radiation being reflected upward and downward several times between the leaf layers, or between the bottom leaf-layer and the soil substrate. If all of the pathways that result in fifth-order scattering are included, three extra terms (shown in boxes below) must be added to Equation 6.15 to give Equation 6.16.

$$
\begin{aligned}
\rho_{\text{canopy}}(\lambda) \;=\; & A\rho_{\text{leaf}}(\lambda) \\
& + (1 - A)(1 - A)\rho_{\text{soil}}(\lambda) \\
& + (1 - A)A\rho_{\text{leaf}}(\lambda) \\
& + A\tau_{\text{leaf}}(\lambda)A\rho_{\text{leaf}}(\lambda)\tau_{\text{leaf}}(\lambda) \\
& \boxed{+ A\tau_{\text{leaf}}(\lambda)A\rho_{\text{leaf}}(\lambda)\rho_{\text{leaf}}(\lambda)\rho_{\text{leaf}}(\lambda)\tau_{\text{leaf}}(\lambda)} \\
& + A\tau_{\text{leaf}}(\lambda)(1 - A)\rho(\lambda)_{soil}\tau_{\text{leaf}}(\lambda) \\
& \boxed{+ A\tau_{\text{leaf}}(\lambda)(1 - A)\rho(\lambda)_{soil}\rho_{\text{leaf}}(\lambda)\rho(\lambda)_{soil}\tau_{\text{leaf}}(\lambda)} \\
& + (1 - A)A\tau_{\text{leaf}}(\lambda)\rho(\lambda)_{soil}\tau_{\text{leaf}}(\lambda) \\
& \boxed{+ (1 - A)A\tau_{\text{leaf}}(\lambda)\rho(\lambda)_{soil}\rho_{\text{leaf}}(\lambda)\rho(\lambda)_{soil}\tau_{\text{leaf}}(\lambda)} \\
& + A\tau_{\text{leaf}}(\lambda)A\tau_{\text{leaf}}(\lambda)\rho_{\text{soil}}(\lambda)\tau_{\text{leaf}}(\lambda)\tau_{\text{leaf}}(\lambda)
\end{aligned}
$$

$$(6.16)$$

Note that the second and third of these new terms are arithmetically equivalent.

### 6.5.3  Implementing the Three-Layer Model

Program 6.4 is an implementation of the three-layer model represented in Figure 6.15, taking into account up to fifth-order scattering, based on Equation 6.16. The code differs only slightly from that of the two-layer model (Program 6.3). Specifically, a new variable, LAI, is used to store the total area of leaves in the canopy (line 21),

Program 6.4: 3layers.awk

```
# Program to calculate the spectral reflectance of a        1
# vegetation canopy modeled as two plane-parallel layers    2
# of leaves above a soil substrate.                         3
#                                                           4
# Usage: gawk -f 3layers.awk -v rho_leaf=value \           5
#               -v tau_leaf=value -v rho_soil=value \      6
#               [ > output_file ]                          7
#                                                           8
# Variables:                                                9
# area_leaf     Fractional area of leaves in each leaf layer  10
# LAI           Leaf Area Index                             11
# rho_leaf      Leaf spectral reflectance                   12
# tau_leaf      Leaf spectral transmittance                 13
# rho_soil      Soil spectral reflectance                   14
# rho_canopy    Canopy spectral reflectance                 15
# gap           Fractional area of gaps in each leaf layer  16
#                                                           17
BEGIN {                                                     18
  for(area_leaf=0;area_leaf<=1;area_leaf+=0.1){            19
    # Initialize variables                                 20
    LAI=2*area_leaf;                                       21
    gap=(1-area_leaf);                                     22
    # Calculate and print total canopy reflectance         23
    rho_canopy=area_leaf*rho_leaf + (gap*gap*rho_soil) + \  24
      (gap*area_leaf*rho_leaf) + \                         25
      area_leaf*tau_leaf*area_leaf*rho_leaf*tau_leaf + \   26
      area_leaf*tau_leaf*area_leaf*rho_leaf*rho_leaf*\     27
        rho_leaf*tau_leaf + \                              28
      2*(area_leaf*tau_leaf*gap*rho_soil*tau_leaf) + \     29
      2*(gap*area_leaf*tau_leaf*rho_soil*rho_leaf*\        30
        rho_soil*tau_leaf) + \                             31
      area_leaf*tau_leaf*area_leaf*tau_leaf*rho_soil* \    32
        tau_leaf*tau_leaf;                                 33
    print LAI, rho_canopy;                                 34
  }                                                         35
}                                                           36
```

while the fraction of incident solar radiation traversing each of the pathways through the model canopy is calculated on lines 24 through 33.

### 6.5.4   Running the Three-Layer Model

Program 6.4 can be run from the command line by typing

```
gawk -f 3layers.awk -v rho_leaf=0.475 -v tau_leaf=0.475 -v↩    7
     ↪ rho_soil=0.125 > 3layers.nir
```

assuming that $\rho_{\text{leaf}}(\lambda) = 0.475$, $\rho_{\text{soil}}(\lambda) = 0.125$ and $\tau_{\text{leaf}}(\lambda) = \rho_{\text{leaf}}(\lambda)$ at NIR wavelengths. It is possible to perform an equivalent simulation for red wavelengths by entering the following command line:

```
gawk -f 3layers.awk -v rho_leaf=0.04 -v tau_leaf=0.04 -v ↩    8
     ↪rho_soil=0.08 > 3layers.red
```

assuming that $\rho_{\text{leaf}}(\lambda) = 0.04$, $\rho_{\text{soil}}(\lambda) = 0.08$ and $\tau_{\text{leaf}}(\lambda) = \rho_{\text{leaf}}(\lambda)$ in this part of the electromagnetic spectrum.

### 6.5.5   Evaluating the Multiple-Layer Model

Figure 6.16 presents the output generated by Program 6.4, which has been plotted in gnuplot using the following commands, assuming that the data files 3layers.red and 3layers.nir are located in the working directory; otherwise, their full or relative pathnames must be provided:

```
set xlabel "Leaf Area Index (LAI)"                              13
plot '3layers.red' u 1:2 t "Red (0.65 micrometers)", \         14
     '3layers.nir' u 1:2 t "NIR (0.85 micrometers)"            15
```

The main thing to note is that the simulated responses produced by the three-layer model are very similar to the curvi-linear asymptotic relationships observed over real plant canopies, shown in Figure 6.10, in terms of both form and magnitude. Thus, the extra leaf-layer appears to have improved significantly the realism of the model.

It is possible that adding even more leaf layers would enhance the model still further. The code given in Program 6.4 could certainly be amended to explore this possibility. The way in which the problem has been approached, however, means that the complexity of the code grows every time the number of leaf layers in the model is increased. This is because each of the pathways that radiation traverses through the canopy must be stated explicitly. There are a very large number of such pathways through a four-layer model (i.e., three leaf-layers plus the soil substrate), taking into account up to seventh-order multiple scattering. The potential for making mistakes in the formulation of the mathematical model, and in the implementation of this as computer code, therefore grows concomitantly. Consequently, the approach exemplified in this chapter will not be pursued any further. Instead, two other ways of tackling the same problem are explored in the next chapter, both of which enable simulations to be performed for model canopies containing any number of leaf layers and taking into account very high orders of multiple scattering.

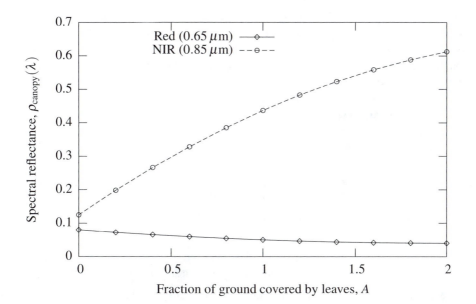

Figure 6.16: Spectral reflectance of a mixed soil and vegetation canopy as a function of LAI, based on the three-layer model given in Program 6.4.

**Exercise 6.2**: Run the `3layers.awk` implementation of the three-layer canopy reflectance model (Program 6.4) using various values of $\tau(\lambda)_{\text{leaf}} \neq \rho_{\text{leaf}}(\lambda)$ over both a bright soil ($\rho_{\text{soil}}(\lambda) > \rho_{\text{leaf}}(\lambda)$) and dark soil background ($\rho_{\text{soil}}(\lambda) < \rho_{\text{leaf}}(\lambda)$). Plot the results using gnuplot.

## 6.6 SUMMARY

A relatively sophisticated model is developed from scratch in this chapter. This model represents the interaction of solar radiation with a plant canopy. In its final form, the model reproduces reasonably well the variation in spectral reflectance as a function of LAI observed for real vegetation canopies. This result does not guarantee, of course, that the model adequately represents the processes that generate the responses observed in nature. It does, however, demonstrate the benefits of starting with a simple model and gradually extending this as required. It also shows that a real-world entity as complex in structure as a plant canopy can be represented quite effectively by a series of plane-parallel leaf-layers suspended above a soil substrate, and that the resultant model can be formulated and implemented using remarkably simple mathematical methods and computational techniques.

It is possible to envisage a number of modifications and extensions that might be made to the multiple-layer model, which would allow it to be used over a much wider range of situations. For instance, a further layer might be incorporated to represent reflection, absorption and transmission by clouds above the vegetation canopy, or snow blanketing the soil substrate. Equally, one might allow each of the leaf layers to cover different fractions of the soil substrate, or the reflectance of the adaxial and abaxial surfaces of the leaf layers to differ. These possibilities, however, are left as an exercise for the reader to explore.

## SUPPORTING RESOURCES

The following resources are provided in the `chapter6` sub-directory of the CD-ROM:

| | |
|---|---|
| `mixture.awk` | gawk program to simulate the spectral reflectance of a linearly mixed soil and vegetation surface |
| `2layer.awk` | gawk program to simulate the spectral reflectance of a two-layer plant canopy |
| `2layr2.awk` | gawk program to simulate the spectral reflectance of a two-layer plant canopy, incorporating fifth-order multiple scattering |
| `3layers.awk` | gawk program to simulate the spectral reflectance of a three-layer plant canopy |
| `interact.bat` | Command line instructions used in this chapter |
| `interact.plt` | gnuplot commands used to generate the figures presented in this chapter |
| `interact.gp` | gnuplot commands used to generate figures presented in this chapter as EPS files |

# Chapter 7

# Analytical and Numerical Solutions

---

**Topics**

- Analytical and numerical solutions to environmental models

- Bouguer's Law and radiation extinction in plant canopies

**Methods and techniques**

- Arrays in gawk

- Iteration and iterative methods

- More control-flow structures in gawk: the `do-while` loop

---

## 7.1 INTRODUCTION

A simple model was developed in the preceding chapter to represent the interaction of solar radiation with a plant canopy. The model canopy initially consisted of a single layer, comprising a linear mixture of leaves and soil. The sophistication of the model was subsequently increased, in stages, by (i) placing the leaves in a separate layer suspended above the soil substrate, (ii) taking into account multiple scattering within the canopy, and (iii) adding an extra leaf layer. These modifications, particularly the last one, improved the ability of the model to replicate the asymptotic relationship observed between spectral reflectance and vegetation amount in real plant canopies at both red and NIR wavelengths. At the same time, however, the modifications increased the complexity of the model and, hence, its computational implementation. This is because the model represents explicitly all of the possible pathways that solar radiation can trace through the plant canopy, which makes it difficult to account for very high order multiple scattering or to add further leaf layers into the model.

Two solutions to these problems are examined in this chapter. These solutions illustrate the principal alternative approaches that are commonly employed to tackle a wide range of complex modeling problems. The first returns to the two-layer case (i.e., a single leaf-layer suspended above a soil substrate) to demonstrate how the contribution of multiple scattering to the total canopy reflectance can be solved by analytical means; that is, by reformulating the model so that it is expressed concisely in terms of a set of mathematical functions and constants. This approach is referred to as an analytical or closed-form solution. The second approach involves modifying the computational model so that it can accommodate any number of leaf layers and take into account very high order multiple scattering. This approach does not attempt to find an analytical solution to the model. It relies, instead, on the computer's ability to solve the problem through "brute force", by performing a pre-defined sequence of numerical operations repeatedly. This approach is known as a numerical solution.

## 7.2  AN EXACT ANALYTICAL SOLUTION TO THE TWO-LAYER MODEL

### 7.2.1  Reformulating the Two-Layer Model

The two-layer model developed in the preceding chapter can be used to simulate the spectral reflectance of a simple plant canopy, taking into account single scattering from the leaf and soil layers, as well as third-order and fifth-order multiple scattering between the layers. In reality, some of the incident solar radiation is scattered more than five times between the leaf layer and the soil substrate before it eventually escapes the canopy. A decision was taken, however, to exclude these higher-order multiple scattering terms from the model formulation based on the results presented in Section 6.4.4, which suggest that fifth-order scattering makes a relatively small contribution to the total canopy reflectance at NIR wavelengths and a negligible one at red wavelengths. Nevertheless, it cannot be assumed that this is always the case; the model was tested using only a limited range of input values. It is possible that high-order multiple scattering makes a more significant contribution to the spectral reflectance of the canopy in other situations. One might imagine that this is the case, for example, when the substrate is highly reflective (e.g., when the substrate is snow).

Although the formulation of the two-layer model used in the preceding chapter does not prevent one from including higher-order scattering terms, the task is made more difficult because of the need to specify explicitly each of the radiation pathways concerned. An alternative formulation of the same model is explored in this section, one that provides an exact analytical solution to the multiple scattering problem. The solution holds only for the special case in which the leaf layer completely covers the ground below ($A = 1$). To simplify the presentation slightly, the terms $R_C$, $R_L$ and $R_S$ are used to denote the spectral reflectance of the canopy, the leaf layer and the soil substrate, respectively, and $T_L$ to denote the spectral transmittance of the leaf layer. These replace the terms $\rho_{\text{canopy}}(\lambda)$, $\rho_{\text{leaf}}(\lambda)$, $\rho_{\text{soil}}(\lambda)$ and $\tau_{\text{leaf}}(\lambda)$, respectively, employed in the preceding chapter.

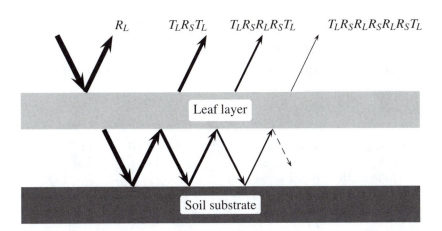

Figure 7.1: Multiple scattering of incident solar radiation in a two-layer plant canopy. $R_L$ and $T_L$ denote the spectral reflectance and transmittance of the leaf layer, respectively, and $R_S$ is the spectral reflectance of the soil substrate.

Figure 7.1 shows a number of different pathways that incident solar radiation can take through the two-layer plant canopy. Note that, while the model assumes radiation travels vertically upward and downward through the canopy, the arrows that represent the different radiation pathways are drawn at oblique angles with respect to the leaf layer and the soil substrate for the sake of diagrammatic clarity. Thus, incident solar radiation can be reflected from the leaf layer ($R_L$) or be scattered numerous times between the leaf layer and the soil substrate before escaping from the canopy. The terms corresponding to third-order ($T_L R_S T_L$), fifth-order ($T_L R_S R_L R_S T_L$) and seventh-order ($T_L R_S R_L R_S R_L R_S T_L$) multiple scattering are indicated in Figure 7.1, but it should be noted that radiation can also be scattered nine times, eleven times and so on, *ad infinitum*. The spectral reflectance of the canopy ($R_C$) is therefore given by the sum of the infinite series presented in Equation 7.1.

$$R_C = R_L + T_L R_S T_L + T_L R_S R_L R_S T_L + T_L R_S R_L R_S R_L R_S T_L + \ldots \quad (7.1)$$

Herein lies the problem: to determine $R_C$ exactly, the sum of this infinite series must be calculated. If the calculation is truncated at a particular order of multiple scattering, as in the preceding chapter, the result is only an approximation to $R_C$, i.e.,

$$R_C \approx R_L + T_L R_S T_L + T_L R_S R_L R_S T_L \quad (7.2)$$

Harte (1988) presents an exact analytical solution to this problem, based on the approach of Rasool and Schneider (1971), which is adopted here. It starts by noting that the second and successive terms of the infinite series presented in Equation 7.1 are products (multiples) of $T_L R_S$ and $T_L$. Thus, Equation 7.1 can be rewritten as

$$R_C = R_L + T_L R_S \left( 1 + R_L R_S + R_L R_S R_L R_S + \ldots \right) T_L \quad (7.3)$$

which can be further simplified to

$$R_C = R_L + T_L R_S \left( 1 + R_L R_S + (R_L R_S)^2 + \ldots + (R_L R_S)^\infty \right) T_L \quad (7.4)$$

Note that the term in parentheses in Equation 7.4 describes a convergent geometric series. In other words, the quotient of successive elements of this term is a constant value (i.e., $R_L R_S$), since $(R_L R_S)^0 = 1$ and $(R_L R_S)^1 = R_L R_S$. This is important because the sum of an infinite geometric series of this form can be expressed as follows:

$$\sum_{i=0}^{\infty} (x)^i = \frac{1}{1-x} \tag{7.5}$$

when the absolute value of $x$ is less than 1; that is, $|x| < 1$ (Harris and Stocker 1998). Now since $R_L$ and $R_S$ are fractional values, their product ($R_L R_S$) must also be less than 1. It follows, therefore, that

$$\sum_{i=0}^{\infty} (R_L R_S)^i = \frac{1}{1-R_L R_S} \tag{7.6}$$

Consequently, we can restate Equation 7.4 as

$$R_C = R_L + T_L R_S \left( \frac{1}{1-R_L R_S} \right) T_L \tag{7.7}$$

which further simplifies to

$$R_C = R_L + \frac{T_L R_S T_L}{1-R_L R_S} \tag{7.8}$$

This represents an exact analytical solution to the two-layer model for the special case in which $A = 1$. This solution accounts for infinite-order scattering of solar radiation between the leaf layer and the soil substrate.

### 7.2.2   Implementing and Running the Exact Analytical Solution

A computational implementation of the exact analytical solution to the two-layer model (Equation 7.8) is given in Program 7.1. The terms $R_C$, $R_L$, $T_L$ and $R_S$ are represented by the variables R_Canopy, R_Leaf, T_Leaf and R_Soil, respectively. As usual, the values of these variables need to be supplied via the command line, using a separate -v flag for each variable. Program 7.1 represents a direct translation of the mathematical formula given in Equation 7.8 into gawk code. Apart from the comments (lines 1 to 14), all of the code is contained within the BEGIN block (lines 16 to 20). This is because the intention is to generate data (i.e., to perform a simulation), rather than to process data contained in a file. Note that a line continuation symbol (\) is used to break a long statement over lines 17 and 18.

Program 7.1 can be run from the command line by typing the following instruction to determine the spectral reflectance of the two-layer canopy at NIR wavelengths, assuming that $R_L = 0.475$, $R_S = 0.125$ and $T_L = R_L$:

```
gawk  -f  analytic.awk  -v  R_Leaf=0.475  -v  T_Leaf=0.475  -v ↵      1
    ↪R_Soil=0.125
```

Program 7.1: analytic.awk

```
# Exact analytical solution to the simple, two-layer (leaf      1
# + soil substrate) model of light interaction with a plant     2
# canopy, for the special case in which the leaf layer          3
# completely covers the soil substrate. Accounts for            4
# infinite-order multiple scattering.                           5
#                                                               6
# Usage: gawk -f analytic.awk -v R_Leaf=value \                 7
#             -v T_Leaf=value -v R_Soil=value [ > output_file ]  8
#                                                               9
# Variables:                                                    10
# R_Leaf     Leaf spectral reflectance                         11
# T_Leaf     Leaf spectral transmittance                       12
# R_Soil     Soil spectral reflectance                         13
# R_Canopy   Total canopy spectral reflectance                 14
#                                                               15
BEGIN{                                                          16
   R_Canopy=R_Leaf+(T_Leaf*T_Leaf*R_Soil) \                    17
      /(1-(R_Leaf*R_Soil));                                    18
   print R_Canopy;                                             19
}                                                               20
```

Similarly, the following command line can be used to calculate the equivalent value at red wavelengths, assuming that $R_L = 0.04$, $R_S = 0.08$ and $T_L = R_L$:

```
gawk -f analytic.awk -v R_Leaf=0.04 -v T_Leaf=0.04 -v ⇨      2
     ⇨R_Soil=0.08
```

### 7.2.3 Evaluating the Exact Analytical Solution

Table 7.1 presents a comparison of the results obtained using alternative formulations and implementations of the two-layer model: (i) twolayr2.awk (Program 6.3, Chapter 6), the approximate solution, taking into account up to fifth-order multiple scattering and (ii) analytic.awk (Program 7.1), the exact analytical solution. For the purpose of comparability, the results derived from both implementations relate to the $A = 1$ case; that is, where the leaf layer completely covers the ground below. Data are presented for a snow substrate and a soil substrate (i.e., high and low reflectance backgrounds, respectively), and for red and NIR wavelengths (i.e., for low and high leaf reflectance, respectively). The values used for the spectral reflectance of the snow substrate are $R_S = 0.9$ at red wavelengths and $R_S = 0.75$ in the NIR.

Table 7.1 suggests that, in most of the cases considered, the inclusion of very high order multiple scattering has relatively little impact on the canopy spectral reflectance predicted by the two-layer model. The exception is the case of the snow substrate in the NIR, where the approximate solution underestimates the exact solution by about 0.034 (3.4%) reflectance.

Table 7.1: Spectral reflectance of a simple plant canopy with either a soil or a snow substrate derived using alternative formulations and implementations of the two-layer model.

| Substrate | $\lambda$ | Parameter Values | | | Formulation/Implementation | |
|---|---|---|---|---|---|---|
| | | $R_L$ | $T_L$ | $R_S$ | twolayr2.awk | analytic.awk |
| Soil | Red | 0.040 | 0.040 | 0.080 | 0.0401 | 0.0401 |
| | NIR | 0.475 | 0.475 | 0.040 | 0.5049 | 0.5050 |
| Snow | Red | 0.040 | 0.040 | 0.900 | 0.0415 | 0.0415 |
| | NIR | 0.475 | 0.475 | 0.750 | 0.7045 | 0.7379 |

Note that $R_L$ and $T_L$ are, respectively, the spectral reflectance and transmittance of the leaf layer, and $R_S$ is the spectral reflectance of the substrate.

More importantly, these results demonstrate that there is frequently more than one solution to an environmental modeling problem. The approximate solution is the more flexible in this instance because it allows canopy reflectance to be modeled as a function of the fraction of ground covered by vegetation. By contrast, the exact analytical solution is limited to the special case in which the leaf layer completely covers the ground below ($A = 1$). Nevertheless, it allows one to gauge the error associated with the approximate solution. It is rare, though, to enjoy the luxury of having both exact and approximate solutions.

> **Exercise 7.1**: Run the exact analytical solution to the two-layer model (Program 7.1) to calculate the canopy spectral reflectance at blue (0.5 $\mu$m) and green (0.55 $\mu$m) wavelengths. Estimate the values of $R_L$ and $R_S$ at these wavelengths from Figure 6.3 (page 145). Assume that $T_L = R_L$.

## 7.3   AN ITERATIVE NUMERICAL SOLUTION TO THE MULTIPLE LEAF-LAYER MODEL

The example discussed in Section 7.2 demonstrates that it is sometimes possible to find an exact analytical solution to a model. Not all models are amenable to this type of solution though; some must be solved numerically. The latter involves a sequence of operations that are performed repeatedly and, each time, applied to the result of the previous iteration (British Computer Society 1998). Sometimes an estimate, or guess, is made regarding the correct solution, to which the iterative procedure is initially applied. Provided that the model has been formulated and implemented correctly, the expectation is that the procedure will move ever closer to the correct solution with each successive iteration. Depending on the nature of the model, and in some cases the accuracy of the initial estimate, the iterative procedure will eventually reach the correct solution or a very close approximation to it. Perhaps the best known example of this approach is Newton's method for calculating $\sqrt{x}$ (Harris and Stocker 1998).

Figure 7.2: Interaction of a photon stream with a multiple leaf-layer canopy (time step 0).

A numerical solution to the light-interaction model, which can accommodate any number of leaf layers and take into account very high order multiple scattering within the canopy, is explored in this section. The computational implementation of this model relies heavily on looping (control-flow) constructs of the type introduced in the preceding chapters. A new control-flow construct, **gawk**'s do while statement, is also introduced here. The circumstances in which this construct or the for loop is most appropriate are examined. Before proceeding to that stage, however, the underlying conceptual model is revisited once more.

### 7.3.1 Revisiting the Conceptual Model

Imagine that solar radiation arrives at the top of a multiple layer plant canopy as a stream of photons; that is, as individual particles of light (Figure 7.2). Assume that each of the leaf layers in the canopy completely covers the soil substrate (equivalent to the $A = 1$ case in the preceding section and chapter). Imagine, also, that one is able to observe the passage of these photons over infinitesimally small intervals in time, as they travel through the canopy. Now consider what one might see.

At first, the photons strike the top leaf-layer (Figure 7.2). As a result, some of the photons are reflected back out to space, some are transmitted downward into the canopy, while the remainder are absorbed by the top leaf-layer (Figure 7.3, top). The photons that are transmitted down through the first leaf-layer are then intercepted by the second leaf-layer. Some of these photons are reflected back up toward the first leaf-layer, some are transmitted down toward the soil substrate, and some are absorbed (Figure 7.3, middle). Eventually, some of the photons reach the soil substrate and are either absorbed by it or are reflected back up through the canopy (Figure 7.3, bottom). At the same time, other photons are scattered upward and downward between the leaf layers. This process continues until, ultimately, the photons are either absorbed by the leaf layers and the soil substrate or they escape from the canopy via

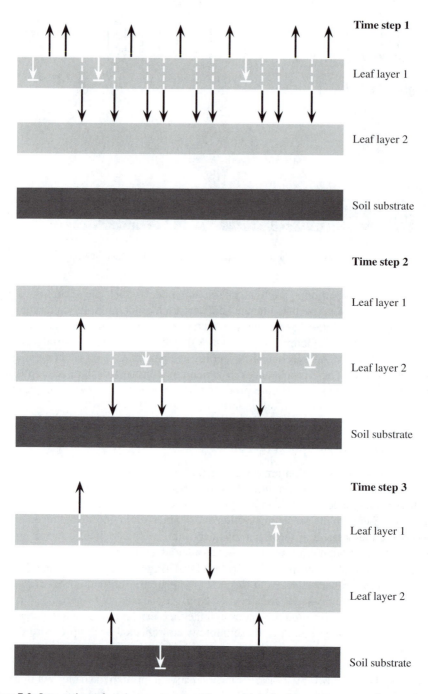

Figure 7.3: Interaction of a photon stream with a multiple leaf-layer canopy at time steps 1 (top), 2 (middle) and 3 (bottom).

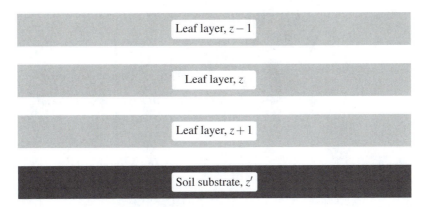

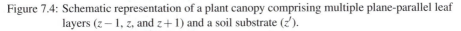

Figure 7.4: Schematic representation of a plant canopy comprising multiple plane-parallel leaf
layers ($z-1$, $z$, and $z+1$) and a soil substrate ($z'$).

the top leaf-layer. The reflectance of the canopy as a whole is therefore given by the
fraction of incident photons that eventually exit the canopy via the top leaf-layer.

Re-stating the conceptual model in this way highlights the following points:

1. There are effectively two "streams" of photons, which travel through the canopy
   in opposite directions — one passes vertically downward into the canopy, the
   other passes vertically upward and eventually out of the canopy.

2. Photons continue to travel in the same direction after interacting with a leaf
   layer if they are transmitted through it.

3. The direction in which a photon travels changes (reverses) if it interacts with
   and is reflected from either a leaf layer or the soil substrate.

4. Photons are lost altogether (the photon stream is attenuated) if they interact
   with and are absorbed by either a leaf layer or the soil substrate.

Importantly, these points are independent of the number of leaf layers in the model
canopy. In other words, our understanding of the nature of the problem has been
further generalized. With this in mind, the mathematical model is now reformulated
for a final time to account for the interaction of solar radiation with a plant canopy
comprising any number of plane-parallel leaf layers.

### 7.3.2 Formulating the Mathematical Model

Consider an individual leaf-layer, $z$, located somewhere within a plant canopy that
comprises several such plane-parallel layers plus a soil substrate, $z'$ (Figure 7.4). To
determine the reflectance of the canopy as a whole, the numbers of photons that are
scattered both upward and downward after interacting with this layer (and each of the
other canopy layers) must be established. The former comprises photons traveling

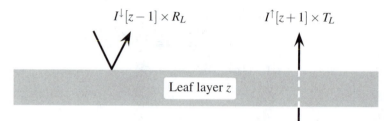

Figure 7.5: Contributions to flux traveling upward from leaf-layer $z$ in a simple, plane-parallel, plant canopy.

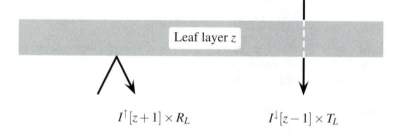

Figure 7.6: Contributions to flux traveling downward from leaf-layer $z$ in a simple, plane-parallel, plant canopy.

upward from the layer below (layer $z + 1$) that are transmitted through layer $z$, as well as photons traveling downward from the layer above (layer $z - 1$) that are reflected upward from layer $z$ (Figure 7.5). The latter consists of photons traveling downward from the layer above ($z - 1$) that are transmitted through layer $z$, plus photons traveling upward from the layer below ($z + 1$) that are reflected downward from layer $z$ (Figure 7.6). The soil substrate (layer $z'$) represents a special case in this context because it contributes only to the upward flux stream (Figure 7.7). Note that the arrows representing the pathways of the reflected flux in Figure 7.5 through Figure 7.7 are drawn at oblique angles with respect to the leaf and soil layers solely for the purpose of diagrammatic clarity; the model assumes that radiation travels either vertically upward or vertically downward only.

The symbols $I^\uparrow[z]$ and $I^\downarrow[z]$ are used here to refer to the photons (flux) traveling upward and downward, respectively, from layer $z$ in the model canopy. Based on the descriptions above, $I^\uparrow[z]$ can therefore be expressed as

$$I^\uparrow[z] = \left( I^\downarrow[z-1] \times R_L \right) + \left( I^\uparrow[z+1] \times T_L \right) \tag{7.9}$$

where $R_L$ and $T_L$ are the spectral reflectance and transmittance, respectively, of leaf layer $z$ (Figure 7.5). For the purpose of this example, it is assumed that each of the

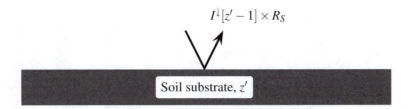

Figure 7.7: Contribution to flux traveling upward from the soil substrate ($z'$) in a simple, plane-parallel, plant canopy.

leaf layers has the same spectral reflectance and transmittance properties, and that the values of these properties are identical for both the upper (adaxial) and lower (abaxial) surfaces of each layer. By the same token, $I^\downarrow[z]$ (Figure 7.6) is given by

$$I^\downarrow[z] = \left(I^\uparrow[z+1] \times R_L\right) + \left(I^\downarrow[z-1] \times T_L\right) \tag{7.10}$$

Moreover, the flux traveling upward from the soil substrate is given by

$$I^\uparrow[z'] = I^\downarrow[z'-1] \times R_S \tag{7.11}$$

where $z' - 1$ is the leaf layer immediately above the soil substrate (Figure 7.7).

It should be evident from Equation 7.9 through Equation 7.11 that the number of photons traveling upward or downward from any given layer in the canopy is partly dependent on the number of photons that are incident on that layer from the layers immediately above and below it. Thus, for example, the values of both $I^\downarrow[z-1]$ and $I^\uparrow[z+1]$ are required to calculate $I^\uparrow[z]$. The problem is that $I^\downarrow[z-1]$ is dependent, in turn, on $I^\uparrow[z]$. Consequently, to determine $I^\downarrow[z-1]$, $I^\uparrow[z]$ must already be known, and the converse is also true. At first glance, the circularity involved in this argument appears to pose an insuperable mathematical conundrum, but it transpires that the problem can be solved simply and effectively using iterative computational techniques, which are demonstrated in the following section.

### 7.3.3 Implementing the Multiple Leaf-Layer Model

There are several important issues that need to be addressed when implementing the multiple layer model given in Equation 7.9 through Equation 7.11. In particular, the program must be able to store and manipulate information on the spectral reflectance and transmittance properties of the leaf layers, as well as the number of photons leaving each layer in both the upward and downward directions. As noted previously, it is assumed that each of the leaf layers has the same spectral reflectance and transmittance properties, and that the adaxial and abaxial surfaces are identical in terms of spectral reflectance and transmittance. Thus, only two gawk variables, R_Leaf and T_Leaf, are required to represent these properties and to store their values. A further variable, R_Soil, is needed to represent the spectral reflectance of the soil substrate.

Representing the number of photons leaving each layer is more problematical. One solution is to use separate variables for each layer; for example, I_Up_Layer1,

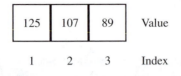

Figure 7.8: Representation of the array I_Up.

I_Down_Layer1, I_Up_Layer2, I_Down_Layer2 and so on, where I_Up_Layer1 denotes the number of photons traveling upward from layer 1, and I_Down_Layer1 refers to the number traveling downward from the same layer. The problem with this approach is that it becomes less efficient and more error-prone as the number of leaf layers increases. Thus, one must keep track of 20 similarly named variables for a model incorporating 10 leaf-layers. More importantly, the objective is to implement a model that can represent a canopy containing any number of leaf layers. This means that one does not know in advance how many leaf layers will be specified and, hence, how many variables are required. A much better solution, therefore, is to employ gawk's array capabilities.

*Using Arrays to Store Collections of Related Data Values*

An array is a computational structure that allows a collection of related data items to be grouped together and referred to by a single identifier or name (British Computer Society 1998). An array can be thought of as a table of cells, or elements, each of which is capable of storing a single numerical value or character string (i.e., text). Each element of an array is uniquely referenced by its index. For example, rather than storing the number of photons passing upward from each leaf layer using separate variables, I_Up_Layer1 = 125, I_Up_Layer2 = 107, I_Up_Layer3 = 89, and so on, it is possible to group these values together and store them in a single array called I_Up (Figure 7.8). In this context, I_Up[1] refers to element 1 of the array, which contains the value 125, I_Up[2] refers to element 2 of the array, which contains the value 107, and I_Up[3] refers to element 3 of the array, which contains the value 89. In the example studied here, the index is used to distinguish the different leaf layers. Note that the array index is enclosed by square brackets ([]).

   Unlike some other computer programming languages, gawk is very flexible about how it sets up and handles arrays. For example, arrays need not be declared before use, nor is there any need to specify in advance the number of elements that are contained within an array (Robbins 2001). Moreover, further elements may be added to an array at any time (i.e., dynamically). The technical details of how this is achieved need not detain us here, but it is useful to note that gawk's arrays are associative (Robbins 2001). In practice, this means that the index of an array is not restricted to positive integer values, they can also be character strings. For example, one could store monthly rainfall data in an array called rainfall using the name of the month as the index (e.g., rainfall[january], rainfall[february], ... ). Of course, one could also have used rainfall[1] to refer to the rainfall in January, rainfall[2] to refer to

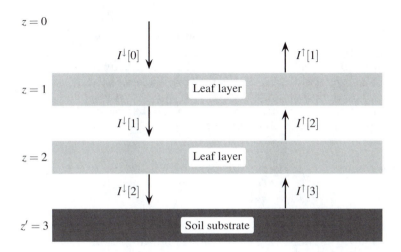

Figure 7.9: Two-stream model of radiation transport through a multiple leaf-layer model of a vegetation canopy (where `layers=2`; see text for explanation).

that in February, and so on, but associative arrays in which the indexes are character strings are arguably more intuitive.

Two other important features of gawk's arrays are (i) that they cannot share the same name as a variable used in the same gawk program and (ii) unless one specifies otherwise, each element of the array is automatically assigned a value of `""`, known as the null (empty) string. When used in mathematical calculations, gawk's null string equates to the value zero (0).

*Putting Arrays into Practice*

An implementation of the multiple leaf-layer model given in Equation 7.9 through Equation 7.11, which makes use of arrays, is presented in Program 7.2. Ignoring comments, this program consists of just six lines of code. These are examined in detail below with the help of Figure 7.9, which illustrates the situation for a three-layer model (two leaf layers and a soil substrate).

Program 7.2 requires the user to enter via the command line the number of leaf layers in the model canopy. This information is stored in the variable `layers`. The user must also supply values for the spectral reflectance and transmittance of the leaf layers (stored in the variables `R_Leaf` and `T_Leaf`, respectively), as well as the spectral reflectance of the soil substrate (`R_Soil`).

The program uses two separate arrays (`I_Up` and `I_Down`; lines 16 and 17) to store data on the upward and downward fluxes of radiation through the model canopy. In this context, `I_Down[0]` denotes downwelling solar radiation incident at the top of the canopy, `I_Down[1]` refers to flux traveling downward from the top leaf-layer, and so on through to `I_Down[layers]`, which indicates flux traveling downward from the bottom leaf-layer. Likewise, `I_Up[1]` refers to flux traveling upward from the top leaf-layer (i.e., flux exiting the canopy), and so on through to `I_Up[layers]`, which

Program 7.2: iterate.awk

```
# Program to calculate the spectral reflectance of a        1
# plant canopy based on a simple two-stream model           2
# (upward and downward fluxes) for a user-specified         3
# number of leaf layers above a soil substrate. The         4
# program requires selected information to be provided      5
# via the command line:                                     6
#                                                           7
# Usage: gawk -f iterate.awk -v R_Leaf=value \              8
#              -v T_Leaf=value -v R_Soil=value \            9
#              -v layers=value [ > outFile ]                10
#                                                           11
# Variables:                                                12
# R_Leaf       Spectral reflectance of each leaf layer      13
# T_Leaf       Spectral transmittance of each leaf layer    14
# R_Soil       Spectral transmittance of the soil substrate 15
# I_Down[z]    Flux traveling downards from layer z         16
# I_Up[z]      Flux traveling upwards from layer z          17
# layers       Number of leaf layers in canopy             18
# iteration    Number of iterations performed               19
#                                                           20
BEGIN {                                                     21
   # Set the total incident solar radiation                22
   I_Down[0]=1;                                             23
                                                            24
   # Outer loop performs iterations                         25
   for (iteration=1;iteration<=20;iteration++){             26
                                                            27
      # Inner loop deals with each leaf layer in turn       28
      for (z=1; z<=layers; z++){                            29
         I_Up[z]=(I_Down[z-1]*R_Leaf)+(I_Up[z+1]*T_Leaf);   30
         I_Down[z]=(I_Up[z+1]*R_Leaf)+(I_Down[z-1]*T_Leaf); 31
      }                                                     32
                                                            33
      # ...and then the soil substrate                      34
      I_Up[layers+1]=I_Down[layers]*R_Soil;                 35
                                                            36
   }                                                        37
                                                            38
   # Finally, print out the number of leaf layers (LAI) and 39
   # the spectral reflectance of the canopy as a whole      40
   printf("%-3i %5.4f\n", layers, I_Up[1]);                 41
                                                            42
}                                                           43
```

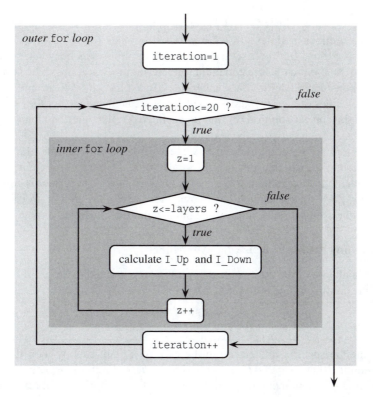

Figure 7.10: Flow-chart representation of the two nested `for` loops used in `iterate.awk` (Program 7.2).

denotes flux traveling upward from the bottom leaf-layer. Finally, `I_Up[layers+1]` indicates flux traveling upward from the soil substrate.

The majority of the code is placed in the `BEGIN` block (lines 21 to 43) because the program is intended to perform simulations, and hence to generate new data, rather than to process data contained in a file. In the `BEGIN` block, the total downwelling solar radiation incident on the top of the canopy is initialized to 1 (`I_Down[0]=1`) on line 23. The objective is to determine what fraction of this eventually exits the canopy (i.e., is reflected by it).

The remainder of the `BEGIN` block consists primarily of two `for` loops, one "nested" within the other (see also Figure 7.10). The inner `for` loop (lines 29 to 32) cycles through each of the leaf layers in turn, from the top (`z=1`) to the bottom (`z=layers`) of the canopy. In so doing it calculates the fraction of incident flux traveling upward and downward from each leaf layer (lines 30 and 31; Equations 7.9 and 7.10). After considering all of the leaf layers, the program moves on to deal with the special case of flux traveling upward from the soil substrate (line 35; Equation 7.11). This allows the incident flux (photons) to traverse its way down through the canopy to the soil substrate and eventually back up and out of the canopy via the top leaf-layer, in a manner analogous to that shown in Figure 7.3.

The outer `for` loop (lines 26 and 37) ensures that the work of the inner `for` loop is repeated a number of times. In other words, the inner `for` loop forms the body of the outer `for` loop. In this case, the number of iterations has been set, somewhat arbitrarily, to 20; a more objective method for determining the appropriate number of iterations is considered later in this chapter (p.190). With each cycle around the outer `for` loop, the values of `I_Up[z]` and `I_Down[z]` are recalculated so that they gradually converge on the correct solution for that layer. In effect, the outer `for` loop accounts for multiple scattering between the leaf layers.

After the outer `for` loop has been traversed for the twentieth time, the program progresses to line 41, which prints out the number of leaf layers in the canopy and the fraction of incident solar radiation exiting the top leaf layer. The former is equivalent to the canopy LAI, since each leaf layer completely covers the area of ground below. The latter is the spectral reflectance of the plant canopy as a whole.

### 7.3.4   Running the Multiple Leaf-Layer Model

The implementation of the multiple leaf-layer model presented in Program 7.2 can be run from the command line by typing

```
gawk -f iterate.awk -v R_Leaf=0.475 -v T_Leaf=0.475 -v ⇨      3
   ⇨R_Soil=0.125 -v layers=3 > layers.nir
```

to determine the spectral reflectance in the NIR of a plant canopy with three leaf-layers, assuming that $R_L = 0.475$, $R_S = 0.125$ and $T_L = R_L$. Similarly, the following command line instruction can be used to calculate the canopy spectral reflectance at red wavelengths, assuming that $R_L = 0.04$, $R_S = 0.08$ and $T_L = R_L$:

```
gawk -f iterate.awk -v R_Leaf=0.04 -v T_Leaf=0.04 -v ⇨        4
   ⇨R_Soil=0.08 -v layers=3 > layers.red
```

The revised model can be run numerous times, varying the value of `layers` on each occasion while keeping the other parameters constant. This allows the spectral reflectance of the plant canopy to be simulated for different numbers of leaf layers and, hence, different LAI. The results of each model run can be added to the end of the files `layers.nir` and `layers.red` by replacing the "greater than" symbol (>) in the example command lines shown above with two "greater than" symbols (>>), which means "append the output to the following file". For example, the following instruction adds the result to the end of the file `layers.nir`:

```
gawk -f iterate.awk -v R_Leaf=0.475 -v T_Leaf=0.475 -v ⇨      5
   ⇨R_Soil=0.125 -v layers=4 >> layers.nir
```

Running the model in this way, and varying the number of leaf layers between 0 and 10, produces the data shown in Table 7.2.

### 7.3.5   Evaluating the Multiple Leaf-Layer Model

Figure 7.11 is a visualization of the data in Table 7.2, generated using the following gnuplot commands:

Table 7.2: Output from `iterate.awk` at red and NIR wavelengths.

| LAI | Canopy Reflectance | |
|---|---|---|
| (Leaf Layers) | Red | NIR |
| 0 | 0.0800 | 0.1250 |
| 1 | 0.0401 | 0.5050 |
| 2 | 0.0401 | 0.6249 |
| 3 | 0.0401 | 0.6755 |
| 4 | 0.0401 | 0.6994 |
| 5 | 0.0401 | 0.7112 |
| 6 | 0.0401 | 0.7169 |
| 7 | 0.0401 | 0.7196 |
| 8 | 0.0401 | 0.7208 |
| 9 | 0.0401 | 0.7212 |
| 10 | 0.0401 | 0.7214 |

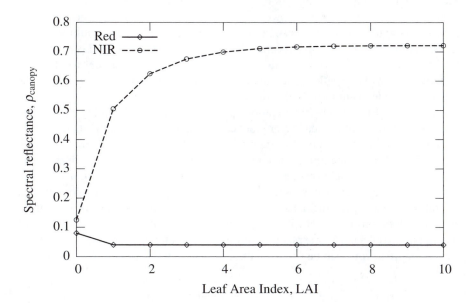

Figure 7.11: Relationship between LAI and canopy spectral reflectance at red and NIR wavelengths predicted by the numerical solution to the multiple leaf-layer model (`iterate.awk`; Program 7.2).

```
set ylabel "Spectral reflectance, rho_canopy"        1
set xlabel "Leaf Area Index, LAI"                     2
set style data linespoints                            3
set key 1.25,0.75                                     4
plot 'layers.red' title "Red" lt 1 lw 2, \            5
     'layers.nir' title "NIR" lt 2 lw 2               6
```

These commands assume that the files `layers.red` and `layers.nir` are located in the
working directory; otherwise, their full or relative pathnames must be provided.

Figure 7.11 suggests that the numerical solution to the multiple leaf-layer model
reproduces the expected relationship between canopy spectral reflectance and LAI
reasonably well (Figure 6.10 on page 156). Moreover, for the single leaf-layer case
(LAI = 1), the results generated by `iterate.awk` (Program 7.2) are identical (at least
to 4 decimal places) to those produced by the exact analytical solution, `analytic.awk`
(Program 7.1), presented in the preceding section (Table 7.1 on page 178).

### 7.3.6   How Many Iterations are Required?

Program 7.2 performs 20 iterations (line 26) to calculate the fraction of incident solar
radiation reflected by the multiple leaf-layer plant canopy. This value was selected
rather arbitrarily; it could equally well have been 10 or 30. In this context, reducing the number of iterations from 20 to 10 would halve the time required to run
the model. This is not a particularly significant consideration here because the time
taken to run Program 7.2 is quite short anyway, even on a low-specification computer.
There are many other instances, however, in which the model run times are very large
(e.g., hours or days). It makes sense in such circumstances to limit the number of iterations, performing no more than is absolutely necessary. The question is, how many
iterations are required? The obvious response is, a sufficient number to obtain an
accurate answer. The challenge, though, is to establish an objective method against
which to evaluate quantitatively this rather subjective criterion.

The effect of the number of iterations performed, when calculating the spectral
reflectance of the multiple leaf-layer plant canopy, can be explored by making some
minor modifications to the final few lines of code in Program 7.2. These are presented
in Program 7.3. First, the `printf` statement (line 40 in Program 7.3) has been moved
inside the outer `for` loop (note the closing brace ending the outer `for` loop on line 41
in Program 7.3; cf. line 37 in Program 7.2), so that it is performed once per iteration.
Second, the `printf` statement has been altered so that the program prints out the
number of iterations performed thus far, as well as the estimated fraction of incident
solar radiation reflected by the plant canopy as a whole after that number of iterations.

The modified code can be run from the command line by typing

```
gawk -f iterate2.awk -v R_Leaf=0.475 -v T_Leaf=0.475 -v ↩     6
    ↪R_Soil=0.125 -v layers=10 > iterate2.nir
```

and

```
gawk -f iterate2.awk -v R_Leaf=0.04 -v T_Leaf=0.04 -v ↩       7
    ↪R_Soil=0.08 -v layers=10 > iterate2.red
```

Program 7.3: iterate2.awk

```
# Program to calculate the spectral reflectance of a          1
# plant canopy, based on a simple two-stream model           2
# (upward and downward fluxes) for a user-specified          3
# number of leaf layers above a soil substrate. The          4
# program requires selected information to be provided       5
# via the command line:                                      6
#                                                            7
# Usage: gawk -f iterate2.awk -v  R_Leaf=value \             8
#                 -v T_Leaf=value -v R_Soil=value \          9
#                 -v layers=value [ > outFile ]             10
#                                                           11
# Variables:                                                12
# R_Leaf      Spectral reflectance of each leaf layer       13
# T_Leaf      Spectral transmittance of each leaf layer     14
# R_Soil      Spectral transmittance of the soil substrate  15
# I_Down[z]   Flux traveling downards from layer z          16
# I_Up[z]     Flux traveling upwards from layer z           17
# layers      Number of leaf layers in canopy              18
# iteration   Number of iterations performed               19
                                                           20
BEGIN {                                                     21
   # Set the total incident solar radiation                22
   I_Down[0]=1;                                             23
                                                           24
   # Outer loop performs iterations                         25
   for (iteration=1;iteration<=20;iteration++){            26
                                                           27
      # Inner loop deals with each of the leaf layers in turn 28
      for (z=1; z<=layers; z++){                           29
         I_Up[z]=(I_Down[z-1]*R_Leaf)+(I_Up[z+1]*T_Leaf);   30
         I_Down[z]=(I_Up[z+1]*R_Leaf)+(I_Down[z-1]*T_Leaf); 31
      }                                                     32
                                                           33
      # ...and then the soil substrate                     34
      I_Up[layers+1]=I_Down[layers]*R_Soil;                35
                                                           36
      # Print out the number of iterations performed thus far 37
      # and the estimated spectral reflectance of the canopy 38
      # as a whole after that number of iterations          39
      printf("%-3i %5.4f\n", iteration, I_Up[1]);          40
   }                                                        41
                                                           42
}                                                           43
```

Table 7.3: Output from `iterate2.awk` (Program 7.3) at red and NIR wavelengths, showing
the predicted canopy spectral reflectance as a function of the number of iterations.

| Iterations | 1 | 2 | 3 | 4 | 5 | 10 | 20 |
|:---:|:---:|:---:|:---:|:---:|:---:|:---:|:---:|
| Red | 0.0400 | 0.0401 | 0.0401 | 0.0401 | 0.0401 | 0.0401 | 0.0401 |
| NIR | 0.4750 | 0.5822 | 0.6305 | 0.6578 | 0.6750 | 0.7091 | 0.7214 |

to determine, respectively, the red and NIR reflectance of a plant canopy consisting
of 10 leaf-layers as a function of the number of iterations performed, based on the
usual values for leaf reflectance, leaf transmittance and soil reflectance. The results
obtained are summarized in Table 7.3.

It is evident from Table 7.3 that the 10 leaf-layer model reaches a stable value
(0.0401) after only two iterations at red wavelengths. In other words, further iterations of the model do not produce an appreciable change in the predicted value of
canopy reflectance in this part of the spectrum. As has already been noted, this is
because the overwhelming majority of incident red radiation is reflected by the top
leaf-layer, the contribution from which is modeled effectively after two iterations.
Moreover, the impact of multiple scattering is negligible at red wavelengths because
of the relatively strong absorption of radiation at these wavelengths by the leaf layers.
By comparison, the predicted canopy reflectance at NIR wavelengths continues to
increase as a function of the number of iterations performed, and does not appear to
reach a stable value within the range examined. Nevertheless, these data describe an
asymptotic relationship, which can be visualized in gnuplot as follows:

```
unset key                                                                    7
set xlabel "Number of iterations"                                            8
plot 'iterate2.nir' w lp                                                      9
```

Thus, the model predicts relatively small changes in the canopy reflectance after 10
or so iterations (Figure 7.12).

These results highlight two important issues: first, the number of iterations needed
to reach a stable solution differs according to the input values presented to the model
(i.e., as a function of wavelength in this instance); second, an objective method of
determining the required number of iterations has yet to be established. These issues
are, of course, interrelated. A solution to both of these problems is examined below.

### 7.3.7  Objective Determination of the Required Number of Iterations

Previously, the number of iterations employed to simulate the spectral reflectance of
the multiple layer plant canopy was "hard wired" into the code (Program 7.2). Thus,
the number of iterations performed was greater than strictly necessary at red wavelengths, while further iterations were required to reach a stable solution in the NIR.
A much better approach is to allow the model to continue running while the result of
the current iteration differs significantly from that of the preceding iteration because
this implies that the model has not yet reached a stable solution. However, once the

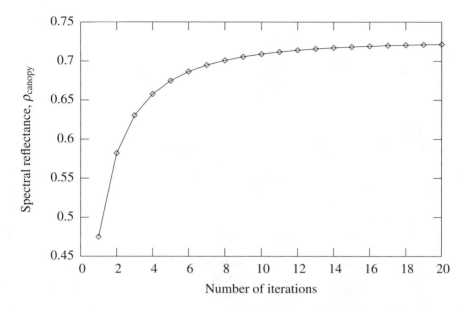

Figure 7.12: Estimated spectral reflectance of a 10 leaf-layer plant canopy at near-infrared wavelengths as a function of the number of model iterations.

difference between the results of successive iterations is less than a threshold value, defined by the user, the iterative process should cease. This is because further iterations are likely to result in only marginal changes to the model output. Put another way, assuming that the threshold value is small, the computational "cost" of performing extra iterations outweighs the potential "benefit" that may be obtained in terms of improvements to the accuracy of the model output.

The required number of iterations is generally not known in advance, so that a for loop, which performs a pre-determined number of iterations, is usually not the most appropriate control-flow structure to use. gawk provides two other control-flow structures that are much better suited to this type of problem: the while and do while loops (Aho *et al.* 1988, Robbins 2001). As their names suggest, these constructs are similar in terms of both syntax and function. The do while loop is used here. The general syntax of the do while statement is as follows (Robbins 2001, Figure 7.13):

```
do
     body
while (condition)
```

Thus, the lines of code in the body are performed repeatedly while the condition is true, otherwise the program moves on to the next line of code. The structure of the do while loop is such that the code in the body is performed at least once.

Program 7.4 presents an alternative implementation of the numerical solution to the multiple leaf-layer light-interaction model. This program uses a do while loop (lines 29 to 39) to control the number of iterations performed, replacing the outer for loop used in Program 7.2. Lines 30 to 38 represent the body of the do while loop.

Program 7.4: iterate3.awk

```
# Program to calculate the spectral reflectance of a          1
# plant canopy, based on a simple two-stream model           2
# (upward and downward fluxes) for a user-specified          3
# number of leaf layers above a soil substrate.              4
#                                                            5
# Usage: gawk -f iterate3.awk -v  R_Leaf=value \             6
#               -v T_Leaf=value -v R_Soil=value \            7
#               -v layers=value -v threshold=value \         8
#               [ > outFile ]                                9
#                                                           10
# Variables:                                                11
# R_Leaf        Spectral reflectance of each leaf layer     12
# T_Leaf        Spectral transmittance of each leaf layer   13
# R_Soil        Spectral transmittance of the soil substrate 14
# I_Down[z]     Flux traveling downwards from layer z       15
# I_Up[z]       Flux traveling upwards from layer z         16
# I_Up_prev     Flux traveling upward from the top leaf-    17
#               layer during the previous iteration          18
# layers        Number of leaf layers in canopy            19
# iteration     Number of iterations performed thus far     20
# threshold     Reflectance threshold to terminate iterations 21
#                                                           22
BEGIN {                                                     23
   # Set the total incident solar radiation                24
   I_Down[0]=1;                                             25
                                                            26
   iteration=0;                                             27
   # Outer do-while loop controls number of iterations     28
   do{                                                      29
      I_Up_prev=I_Up[1];                                    30
      # Inner loop deals with each leaf layer in turn       31
      for (z=1; z<=layers; ++z){                            32
         I_Up[z]=(I_Down[z-1]*R_Leaf)+(I_Up[z+1]*T_Leaf);   33
         I_Down[z]=(I_Up[z+1]*R_Leaf)+(I_Down[z-1]*T_Leaf); 34
      }                                                     35
      # ...and then the soil substrate                     36
      I_Up[layers+1]=I_Down[layers]*R_Soil;                 37
      ++iteration;                                          38
   } while((I_Up[1]-I_Up_prev)>threshold)                  39
                                                            40
   # Print out threshold, iterations and canopy reflectance 41
   printf("%5.4f %3i %5.4f\n", threshold, iteration, I_Up[1]);42
}                                                           43
```

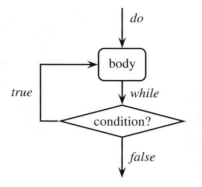

Figure 7.13: Flow-chart representation of the general structure of a do while loop.

Program 7.4 also introduces an additional command-line parameter, threshold (lines 8, 21, 39 and 42). This value, which is measured in terms of the difference in the predicted canopy spectral reflectance between successive iterations of the model, is used to decide when to halt the iterative procedure (line 39). At the beginning of each iteration, the fraction of incident solar radiation exiting the canopy via the top leaf-layer, estimated during the previous iteration, is stored in the variable I_Up_prev (line 30). The upward and downward fluxes from each layer in the canopy are then re-calculated (lines 32 through 37), after which the variable used to count the number of iterations performed thus far is incremented by one (++iteration, line 38). While the difference between the results from the current and previous iterations is larger than the user-defined threshold ((I_Up[1]-I_Up_prev)>threshold; line 39), further iterations of the model are performed (the program loops back from line 39 to line 30). As soon as the difference between successive iterations is equal to or smaller than the threshold value, the do while loop halts. The program then prints out the threshold value, the number of iterations performed and the fraction of incident solar radiation reflected by the canopy (i.e., the flux escaping the top leaf-layer; line 42).

### 7.3.8 Running and Evaluating the Revised Computational Model

Program 7.4 can be run from the command line by typing

```
gawk -f iterate3.awk -v R_Leaf=0.475 -v T_Leaf=0.475 -v ↻      8
    ↻R_Soil=0.125 -v layers=10 -v threshold=0.01 > iterate3↻
    ↻.nir
```

to determine the spectral reflectance of a plant canopy consisting of 10 leaf-layers, based on the usual values of leaf reflectance, leaf transmittance and soil reflectance at NIR wavelengths, and a threshold value of 0.01 (i.e., 1% reflectance). The following command line gives the equivalent result at red wavelengths:

```
gawk -f iterate3.awk -v R_Leaf=0.04 -v T_Leaf=0.04 -v ↻       9
    ↻R_Soil=0.08 -v layers=10 -v threshold=0.01 > iterate3.↻
    ↻red
```

Table 7.4: Output from `iterate3.awk` (Program 7.4) at red and NIR wavelengths, showing the number of iterations performed and the predicted canopy spectral reflectance, $R_C$.

| Reflectance Threshold | | Wavelength | | | |
|---|---|---|---|---|---|
| | | Red | | NIR | |
| Absolute | Percentage | Iterations | $R_C$ | Iterations | $R_C$ |
| 0.1 | 10.00 | 1 | 0.0400 | 3 | 0.6305 |
| 0.05 | 5.00 | 1 | 0.0400 | 3 | 0.6305 |
| 0.01 | 1.00 | 2 | 0.0401 | 7 | 0.6950 |
| 0.001 | 0.10 | 2 | 0.0401 | 16 | 0.7191 |
| 0.0001 | 0.01 | 2 | 0.0401 | 28 | 0.7230 |

The results obtained are summarized in Table 7.4, along with those for further runs of the same model using different threshold values. Table 7.4 demonstrates that the number of iterations required increases as the reflectance threshold becomes more stringent (i.e., smaller). At NIR wavelengths, for instance, 7 iterations are needed for the model output to stabilize to within a threshold of 0.01 (1%) reflectance, while 16 iterations are required for a threshold of 0.001 (0.1%) reflectance.

> **Exercise 7.2**: Run Program 7.4 to simulate the spectral reflectance at NIR wavelengths of a plant canopy comprising 10 leaf-layers and a substrate comprising a layer of snow. Assume that $R_L = 0.475$, $R_L = R_T$ and $R_S = 0.75$. How many iterations are required for the model to stabilize using a reflectance threshold of 0.001?

## 7.4   BOUGUER'S LAW AND THE ATTENUATION COEFFICIENT

In this section, a further modification is made to the implementation of the multiple leaf-layer canopy reflectance model presented previously. This modification allows the model to be used to study the attenuation (extinction) of incident solar radiation as a function of the distance (i.e., the number of leaf layers) that it traverses into the canopy. In general, the amount of solar radiation penetrating a plant canopy varies inversely with depth because of the increasing probability that it is either absorbed or reflected upward by leaf layers higher up the canopy. Thus, for example, the amount of sunlight available to plants found on the forest floor is typically very much smaller than that available to the tallest trees, which form the canopy crown. The resulting vertical distribution of light within the forest has important implications for canopy photosynthesis and productivity, and different species of plants adopt contrasting strategies to compete for the available light resource (Jones 1983, Monteith and Unsworth 1995).

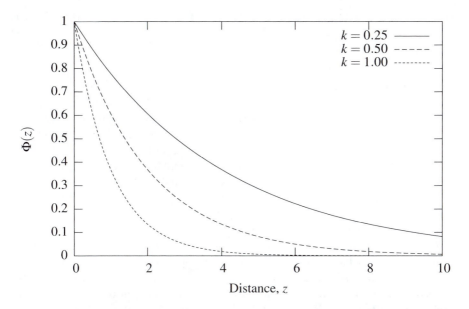

Figure 7.14: Attenuation of incident solar radiation as a function of the distance traversed through a homogeneous turbid medium, according to Bouguer's Law for various values of the attenuation coefficient, $k$.

The extinction of solar radiation within plant canopies is often modeled using Bouguer's Law, also known as Beer's Law or Lambert's Law of Absorption (Campbell and Norman 1998, Glickman 2000). This describes the attenuation of a parallel beam of monochromatic (single wavelength) radiation as it passes through a homogeneous turbid medium, and is given by

$$\Phi(z) = \Phi(0)\exp^{-kz} \tag{7.12}$$

where $z$ is the distance traveled through the medium, $\Phi(0)$ is the flux density at $z = 0$ (before entering the medium), $\Phi(z)$ is the flux density remaining at distance $z$, and $k$ is a constant of proportionality, known as the attenuation or extinction coefficient (Jones 1983, Monteith and Unsworth 1995). The value of the attenuation coefficient, $k$, therefore describes the overall optical density of the medium; that is, the extent to which it is either opaque or translucent. In the context of plant canopies, it provides a single value that can be used to compare the light regimes in different canopies or in the same canopy as it grows and develops over time. Equation 7.12 can be represented graphically in gnuplot using the following commands, which produce Figure 7.14:

```
set dummy z                                      10
phi_0=1.0                                        11
phi(k,z) = phi_0*exp(-k*z)                       12
set key top right                                13
set xlabel "Distance, z"                         14
set ylabel "Phi(z)"                              15
```

```
plot [0:10] phi(0.25,z) w l t "k=0.25", \                      16
             phi(0.50,z) w l t "k=0.50", \                      17
             phi(1.00,z) w l t "k=1.00"                         18
```

Note that line 10 instructs gnuplot to employ a "dummy" variable, z, to denote the *x*-axis. This facility makes it possible to refer to the variable plotted on the *x*-axis (the distance traversed into the canopy) by its conventional symbol, z, which is more convenient in this instance (Williams and Kelly 1998). Also note that the Bouguer Law function is defined in terms of two parameters, k and z, which allows both parameters to be varied when the function is plotted: the value of z varies continuously along the *x*-axis; the value of *k* is specified explicitly on the command line (lines 16 to 18).

Strictly speaking, Bouguer's Law pertains to systems in which the radiation is absorbed without scattering as it passes through a homogeneous turbid medium (Monteith and Unsworth 1995). It is often applied empirically, though, in situations where radiation is typically scattered once only; that is, where the effect of multiple scattering is negligible (Jones 1983, Monteith and Unsworth 1995). In terms of the multiple leaf-layer model used here, this is a reasonable first approximation at red wavelengths because the spectral reflectance and transmittance of leaves and soil are small in this part of the electromagnetic spectrum; however, the assumption of single scattering clearly does not hold at NIR wavelengths. In the remainder of this section, therefore, Bouguer's Law is applied empirically to estimate the attenuation coefficient for the multiple leaf-layer canopy at red wavelengths only. In doing so, it is assumed that the number of leaf layers traversed down from the top of the canopy is a surrogate for the distance parameter, z, in Bouguer's Law (Equation 7.12).

### 7.4.1 Implementing a Computational Model of Bouguer's Law

An implementation of Bouguer's Law is given in Program 7.5 (bouguer.awk). This program is a very slightly modified version of Program 7.4. More specifically, lines 41 through 43 of Program 7.5 report the fraction of incident solar radiation remaining at different levels within the canopy (I_Down[z]). A for loop is used to consider each of the leaf layers in turn, starting at z=0 (at the top of the canopy; see Figure 7.9) and finishing at z=layers (at the bottom of the canopy, beneath the lowest leaf-layer and above the soil substrate).

### 7.4.2 Running and Evaluating the Modified Computational Model

Program 7.5 can be run from the command line by typing the following instruction to determine the attenuation of incident solar radiation as a function of depth into a 10 leaf-layer canopy at red wavelengths, based on a threshold value of 0.01 (1% reflectance) and the usual values of leaf reflectance and transmittance and of soil reflectance:

```
gawk -f bouguer.awk -v R_Leaf=0.04 -v T_Leaf=0.04 -v ⇨       10
    ⇨R_Soil=0.08 -v layers=10 -v threshold=0.01 > bouguer.⇨
    ⇨red
```

Program 7.5: bouguer.awk

```
# Program to calculate the extinction of incident solar       1
# radiation as a function if the number of leaf layers        2
# traversed into a vegetation canopy, based on a simple       3
# two-stream model (upward and downward fluxes).              4
#                                                             5
# Usage: gawk -f bouguer.awk -v  R_Leaf=value \               6
#                  -v T_Leaf=value -v R_Soil=value \          7
#                  -v layers=value -v threshold=value \       8
#                  [ > outFile ]                              9
#                                                            10
# Variables:                                                 11
# R_Leaf        Spectral reflectance of each leaf layer      12
# T_Leaf        Spectral transmittance of each leaf layer    13
# R_Soil        Spectral transmittance of the soil substrate 14
# I_Down[z]     Flux traveling downards from layer z         15
# I_Up[z]       Flux traveling upwards from layer z          16
# I_Up_prev     Flux traveling upward from the top leaf-     17
#               layer during the previous iteration          18
# layers        Number of leaf layers in canopy             19
# iteration     Number of iterations performed thus far      20
# threshold     Reflectance threshold to terminate iterations 21
#                                                            22
BEGIN {                                                      23
   # Set the total incident solar radiation                 24
   I_Down[0]=1;                                              25
   iteration=0;                                              26
   # Outer do-while loop controls number of iterations      27
   do{                                                       28
      I_Up_prev=I_Up[1];                                     29
      # Inner loop deals with each of the leaf layers in turn 30
      for (z=1; z<=layers; ++z){                             31
         I_Up[z]=(I_Down[z-1]*R_Leaf)+(I_Up[z+1]*T_Leaf);    32
         I_Down[z]=(I_Up[z+1]*R_Leaf)+(I_Down[z-1]*T_Leaf);  33
      }                                                      34
      # ...and then the soil substrate                       35
      I_Up[layers+1]=I_Down[layers]*R_Soil;                  36
      ++iteration;                                           37
   } while((I_Up[1]-I_Up_prev)>threshold)                    38
   # Print out the fraction of incident solar radiation      39
   # remaining at different depths (z) within the canopy.    40
   for(z=0; z<=layers; ++z){                                 41
      printf("%3i %5.4f\n", z, I_Down[z]);                   42
   }                                                         43
}                                                            44
```

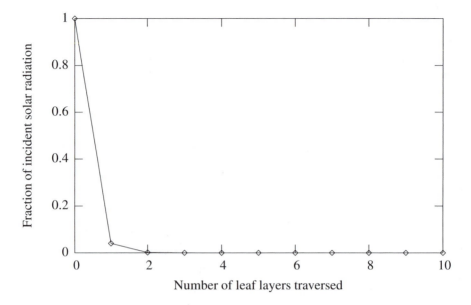

Figure 7.15: Attenuation of incident solar radiation at red wavelengths as a function of the number of leaf-layers traversed down from the top of the canopy.

### 7.4.3   Visualizing the Output

The results obtained using `bouguer.awk` (Program 7.5) can be visualized by typing the following commands in gnuplot assuming that the file `bouguer.red` is located in the working directory:

```
unset key                                                        19
set ylabel "Fraction of incident solar radiation"               20
set xlabel "Number of leaf layers traversed"                    21
plot 'bouguer.red' w lp                                          22
```

The resultant plot is presented in Figure 7.15. This shows the expected negative exponential relationship between the fraction of incident solar radiation remaining and increasing depth into the canopy; thus, the leaf layers at the bottom of the canopy receive considerably less solar radiation than those at the top. Note that over 90% of the incident radiation is attenuated after just one leaf-layer is traversed.

### 7.4.4   Function Fitting in gnuplot

The relationship between radiation extinction and depth into the model canopy can be expressed mathematically by fitting the equation for Bouguer's Law (Equation 7.12) to the data shown in Figure 7.15. The curve-fitting capabilities of gnuplot can be used for this purpose, as demonstrated below.

```
set fit logfile 'bouguer.fit'                                   23
fit phi(k,z) 'bouguer.red' via k                                24
```

```
FIT:      data read from 'bouguer.red'
          #datapoints = 11
          residuals are weighted equally (unit weight)

function used for fitting: phi(k,z)
fitted parameters initialized with current variable values




Iteration 0
WSSR          : 0.128181          delta(WSSR)/WSSR    : 0
delta(WSSR) : 0                   limit for stopping : 1e-05
lambda      : 0.146877

initial set of free parameter values

k               = 1

After 7 iterations the fit converged.
final sum of squares of residuals : 1.33222e-09
rel. change during last iteration : -5.84194e-07

degrees of freedom (ndf) : 10
rms of residuals      (stdfit) = sqrt(WSSR/ndf)      : 1.15422e-05
variance of residuals (reduced chisquare) = WSSR/ndf : 1.33222e-10

Final set of parameters             Asymptotic Standard Error
=======================             ===========================

k               = 3.21639          +/- 0.0002874   (0.008935%)
```

Figure 7.16: Results reported by gnuplot when fitting the Bouguer Law function to the data in the file bouguer.red.

Note that the general form of the function phi(k,z) was defined previously (line 12 on page 197). Recall that this is a negative exponential function, specified in terms of two parameters, k and z, which correspond to the variables $k$ and $z$ in Equation 7.12. Line 23 instructs gnuplot to direct the results of the function-fitting procedure to the file bouguer.fit in the current directory. Line 24 fits the user-defined function phi(k,z) to the data contained in the file bouguer.red by varying the value of the k. Thus, this procedure finds the value of k that results in the best fit between the function phi(k,z) and the data in the file bouguer.red.

The output produced by these gnuplot commands is presented in Figure 7.16, which indicates that the estimated value of the attenuation coefficient, $k$, is 3.21639 at this wavelength. Another value given in the output is an estimate of the uncertainty associated with $k$ ($\pm 0.0002874$), which can be interpreted loosely as describing the confidence limits about $k$ (Williams and Kelly 1998); the smaller this value is, the better the function fits the data. The result obtained here suggests a very good fit. This finding is also demonstrated in Figure 7.17, produced in gnuplot using the following command, which confirms how well the amount of solar radiation penetrating

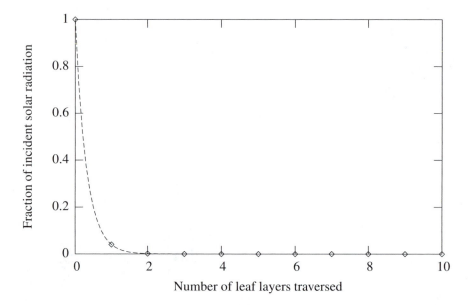

Figure 7.17: Attenuation of incident solar radiation at red wavelengths as a function of the number of leaf-layers traversed downward into a multi-layer plant canopy (points) and the negative exponential function (Bouguer's Law) fitted to these data (dashed lines).

to different depths in the model canopy is characterized by Bouguer's Law:

```
plot 'bouguer.red' w p, phi(k,z) w l                                       25
```

Relationships similar to the one observed in Figure 7.17 are often exploited by manufacturers of sensors designed to estimate, in a non-destructive way, the LAI of plant canopies, particularly arable crops. These instruments are used to measure the incident solar radiation above and below the canopy. By making certain assumptions about the angular distribution of leaves within the canopy, it is possible to derive fairly accurate estimates of LAI from such devices.

## 7.5   SUMMARY

This chapter explores two formulations, and several different implementations, of a model intended to represent the interaction of solar radiation with a plant canopy. The first formulation represents an analytical solution for the special case in which the canopy consists of a single leaf-layer that completely covers the soil substrate below. The second formulation employs a numerical solution to the problems of handling multiple leaf-layers and multiple scattering. Each of these two general approaches has its merits. It is not the intention of this chapter to establish which is "better". In any case, this will vary according to the nature of the problem under investigation. What it demonstrates, however, is that there is rarely only one way of solving an environmental modeling problem.

Much of this chapter is concerned with different computational implementations of Equation 7.9 through Equation 7.11, which underpin the numerical solution to the multiple leaf-layer model. These are employed in Program 7.2 to Program 7.5. It is instructive to note that these "two stream" equations are similar in form to the Kubelka-Munk equations, which describe the transport of solar radiation through a turbid medium that absorbs and scatters radiation (Monteith and Unsworth 1995, Campbell and Norman 1998). The Kubelka-Munk equations are commonly given as follows:

$$-dI^\uparrow = -(S+K)I^\uparrow dz + I^\downarrow S dz \qquad (7.13)$$

$$dI^\downarrow = -(S+K)I^\downarrow dz + I^\uparrow S dz \qquad (7.14)$$

where $z$ is the distance traveled into the turbid medium, $K$ is an absorption coefficient, $S$ is a scattering coefficient, $I^\uparrow$ is the upward traveling flux and $I^\downarrow$ is the downward traveling flux (Monteith and Unsworth 1995). These equations form the basis of many of the more sophisticated models of light interaction with plant canopies that are widely used in studies of ecology, climatology and Earth observation (terrestrial remote sensing) (Goel 1987, Myneni *et al.* 1990, Pinty and Verstraete 1998). Readers wishing to pursue this aspect of environmental modeling further will find that the comprehensive review by Goel (1987) is an excellent starting point.

## SUPPORTING RESOURCES

The following resources are provided in the `chapter7` sub-directory of the CD-ROM:

| | |
|---|---|
| `analytic.awk` | gawk implementation of the exact analytical solution to the two-layer plant canopy model for the special case of $A = 1$ |
| `iterate.awk` | gawk implementation of the iterative numerical solution to the multiple leaf-layer model |
| `iterate2.awk` | modified version of `iterate.awk`, which prints out the number of iterations taken to reach a stable solution |
| `iterate3.awk` | modified version of `iterate2.awk`, which uses a `do while` loop to control the number of iterations |
| `bouguer.awk` | modified version of `iterate3.awk`, which calculates the attenuation of solar radiation as a function of the number of leaf layers traversed into the canopy |
| `iterate.bat` | Command line instructions used in this chapter |
| `interate.plt` | gnuplot commands used to generate the figures presented in this chapter |
| `interate.gp` | gnuplot commands used to generate figures presented in this chapter as EPS files |

# Chapter 8

# Population Dynamics

**Topics**

- Discrete and continuous models of population growth

- Difference equations and differential equations

- Constrained and unconstrained models of population growth

- Chaotic behavior in deterministic systems

- Inter-specific competition

- Predator-prey relationships

**Methods and techniques**

- Control-flow structures in gawk: the `while` loop

- Numerical integration using Euler and Runge-Kutta methods

## 8.1 INTRODUCTION

As a general rule, environmental systems tend to be highly dynamic; that is to say, the system state changes through time, sometimes dramatically so, in response to a set of forcing factors. A good example is the growth or decline of a population, where the system state is defined by the number of individuals of a particular species, typically within a finite geographical area. The species type is not a primary concern here. Instead, consideration is given to generic models that can be applied to populations of human beings, animals, plants, bacteria and a host of other organisms. This generality is desirable because it implies a degree of economy, such that a single model can be

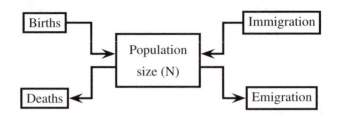

Figure 8.1: Factors leading to a change in the size of a population. The arrows denote the flow of individuals to and from the total population.

used to study several related problems. It is also beneficial because it enables parallels to be explored between environmental systems, so that knowledge of one can be used to help understand the functioning of others.

The models examined in this chapter can be divided into two broad categories: discrete and continuous. The distinction relates to the way in which each treats time as a variable. In discrete models, time proceeds in a series of finite steps that are typically, although not necessarily, of equal length; each step relates to a specific time interval such as an hour, a day, a month or a year. The system state is updated at the end of each time step to account for changes caused by processes operating on the system over the preceding period. In continuous models, by contrast, the system state is updated instantaneously in response to continuous changes in the forcing factors. Discrete and continuous models also differ in terms of the mathematical equations on which they are founded: the former are expressed in terms of difference equations; the latter employ differential equations. These are introduced informally throughout this chapter.

## 8.2    UNCONSTRAINED OR DENSITY-INDEPENDENT GROWTH

### 8.2.1    Development of the Conceptual Model

Various factors cause the population of a particular species in a given area to change in size over time. For instance, new individuals may be added to the population as a result of births or immigration from surrounding areas, while existing individuals may be lost through deaths or emigration to surrounding areas (Figure 8.1). Depending on the balance between these four demographic processes, the population may increase, remain stable or decline over time (Donovan and Welden 2002). Note that in studies of population dynamics it is common to refer to either the population size, which is the number of individuals of a given species, or the population density, which is the population size per unit area.

Following Alstad (2001), four assumptions are made to simplify the development of the initial model of population dynamics developed here, namely that (i) there is no migration of individuals between the study area and the surrounding regions (the study area is isolated in some way) or else the number of immigrants equals the number of emigrants, (ii) reproduction is asexual, (iii) every individual shares the same likelihood of producing offspring and of dying, and (iv) there is an infinite supply of the natural resources (food, water and shelter) required to support the population. The respective implications of these assumptions are that (i) changes to the population size (and density) are controlled solely by births and deaths, (ii) the relative abundance of males and females within the population can be ignored, (iii) the age structure of the population need not be taken into account, and (iv) there are no limits to the growth of the population (Alstad 2001). The last of these is a prerequisite for what is known as unconstrained or density-independent growth, which occurs when the population is unaffected by factors relating to sparsity (e.g., a lack of partners for mating) or over-crowding (e.g., through competition for a finite quantity of natural resources).

### 8.2.2 Formulation, Implementation and Evaluation of the Discrete Model

*Formulation*

In discrete models of population growth it is assumed that individuals are born in a series of cohorts at finite intervals in time (Alstad 2001). Models such as these are appropriate to species that reproduce on a seasonal or an annual basis; this category includes many types of plants, insects and mammals. With each successive time step, therefore, a number of individuals are born and a number die, depending on the size of the population at that point in time. This can be represented mathematically as follows:

$$N_{t+1} = N_t + BN_t - DN_t \tag{8.1}$$

where $N_t$ is the population size (or density) during time step $t$, $N_{t+1}$ is the population during the subsequent time step $(t+1)$, $B$ is the average number of offspring produced by an individual in any one time step (the number of births per capita per time step, where $B \geq 0$) and $D$ is the probability that an individual will die during any given time step (where $0 \leq D \leq 1$). For example, if the population size at time $t$ is 100 ($N_t = 100$), if each individual produces on average two offspring per time step ($B = 2$) and if there is a one in two chance that an individual will die during any given time step ($D = 0.5$), then

$$N_{t+1} = 100 + (2 \times 100) - (0.5 \times 100) = 250 \tag{8.2}$$

Note that Equation 8.1 implicitly assumes that $B$ and $D$ are constant with respect to time, which means that they are independent of the population size (or density), $N_t$. Consequently, Equation 8.1 is known as a density-independent model of population growth (Donovan and Welden 2002).

Equation 8.1 is a form of finite difference equation (FDE), which describes a change (or difference) in the population over a finite interval of time. This is more obvious if Equation 8.1 is rearranged as follows:

$$\begin{aligned} N_{t+1} - N_t &= BN_t - DN_t \\ &= (B-D)N_t \end{aligned} \tag{8.3}$$

where $N_{t+1} - N_t$ is the change in the size of the population between consecutive time steps, $t$ and $t+1$. The difference in the population size is often denoted by $\Delta N$, while the corresponding interval of time is denoted by $\Delta t$. More generally, FDEs can be used to describe the change in any environmental property, not just population size, with respect to finite differences in a given dimension, not just time (e.g., the change in mean air temperature with distance along a geographical axis).

Equation 8.1 can be simplified as follows:

$$N_{t+1} = (1 + B - D)N_t \qquad (8.4)$$

where the term $(1 + B - D)$, often replaced by the symbol $\lambda$, describes the discrete or finite rate of growth in the population. As a result, Equation 8.4 can be restated as

$$N_{t+1} = \lambda N_t \qquad (8.5)$$

and hence

$$\lambda = \frac{N_{t+1}}{N_t} \qquad (8.6)$$

Thus, $\lambda$ describes the proportional change in the population size per unit interval of time (Alstad 2001). In some texts, $\lambda$ is referred to as the Malthusian factor after the Reverend Thomas Malthus (1766–1834) whose *Essay on the Principle of Population*, first published in 1798, explored various issues concerning the growth of the human population and the ability to support it by means of agricultural production.

Assuming that $B$ and $D$ (and hence $\lambda$) remain constant over time and that the initial population size ($N$ at $t = 0$; $N_0$) is known, it is possible to use Equation 8.5 to examine the growth of a population over a number of time steps. For instance, the population size at $t = 1$ is

$$N_1 = \lambda N_0 \qquad (8.7)$$

while at $t = 2$ it is

$$N_2 = \lambda N_1 \qquad (8.8)$$

Substituting Equation 8.7 into Equation 8.8 gives

$$\begin{aligned} N_2 &= \lambda(\lambda N_0) \\ &= \lambda^2 N_0 \end{aligned} \qquad (8.9)$$

Similarly, the population at $t = 3$ is

$$\begin{aligned} N_3 &= \lambda N_2 \\ &= \lambda(\lambda N_1) \\ &= \lambda(\lambda(\lambda N_0)) \\ &= \lambda^3 N_0 \end{aligned} \qquad (8.10)$$

Program 8.1: discrete.awk

```
# Discrete model of unconstrained (density-independent)    1
# population growth (see Eq. 8.11).                         2
#                                                           3
# Usage: gawk -f discrete.awk -v pop_init=value \           4
#              -v lambda=value -v period=value \            5
#              [ > outputFile ]                             6
#                                                           7
# Variables:                                                8
# ----------                                                9
# pop_init  initial population                             10
# lambda    discrete (finite) rate of population growth    11
# period    period (time steps) over which growth modeled  12
#                                                          13
BEGIN{                                                     14
    for(time_step=0;time_step<=period;++time_step){       15
        print time_step, pop_init*(lambda**time_step);    16
    }                                                      17
}                                                          18
```

and so on (Roughgarden 1998). Assuming that the time step is a year, that the initial population is $10,000$ ($N_0 = 10,000$) and that the population grows at a steady rate of 5% per annum ($B - D = 0.05$ and therefore $\lambda = 1.05$), after one year the population size will be $10,500$ ($N_1 = 1.05 \times 10,000 = 10,500$), after two years it will be $11,025$ ($N_2 = 1.05 \times 1.05 \times 10,000 = 11,025$), and so on. Note that this is a geometrical sequence, and populations that behave in this way are said to exhibit geometrical growth.

Equation 8.10 can be generalized to determine the size of a population after an arbitrary number of time steps, $t$,

$$N_t = \lambda^t N_0 \tag{8.11}$$

where $N_0$ is the initial population (at $t = 0$), $N_t$ is the population at time $t$, and $t$ is a positive integer (a whole number). Equation 8.11 describes a population that grows without limit when $\lambda > 1$, that remains constant when $\lambda = 1$ and that declines toward extinction when $\lambda < 1$ (Alstad 2001).

*Implementation*

The next task is to convert the discrete density-independent model of population growth (Equation 8.11) into gawk code. Program 8.1 (discrete.awk) provides an example implementation. Apart from the comments (lines 1 through 12), the code in this program is placed entirely within the BEGIN block (lines 14 to 18) because the aim is to generate data (to perform a simulation) rather than to process data contained in a file. Note that values for each of the main variables — pop_init (the initial size of the population, $N_0$), lambda (the discrete rate of population growth, $\lambda$) and period (the number of time steps over which the population growth is to be simulated, $t$) — are not specified in the body of the code; they have to be supplied via the command line. This makes the code more flexible, permitting the user to experiment with different values of the variables without having to edit the program each time.

The code in the BEGIN block consists of a single for loop (lines 15 to 17). This is used to calculate the size of the population during each time step ($t = 0, 1, 2, \ldots$). The body of the loop (line 16) prints out the current time step, time_step, and the corresponding population size, pop_init*(lambda**time_step). Remember that any number, except zero, raised to the power of zero is equal to 1 ($x^0 = 1$, if $x \neq 0$; n.b. $0^0$ is undefined). Thus, lambda**time_step=1 when time_step=0. As a result, line 16 prints out the initial population size (pop_init) the first time around the loop.

It is instructive to note that Program 8.1 and the mathematical equation on which it is based sometimes generate a fractional number of individuals (if $N_0 = 10$ and $\lambda = 1.05$, then $N_1 = 10 \times 1.05 = 10.5$). In reality, of course, fractional individuals do not exist, and Program 8.1 can be amended to prevent this outcome. The simplest way to do this is to use **gawk**'s int(x) function. This converts its argument to an integer (a whole number) by truncating the fractional component (everything after the decimal point). For example, the result of int(10.6) is 10. Thus, line 16 in Program 8.1 could be amended to read int(pop_init*(lambda^time_step)). Alternatively, the print statement could be replaced by a printf statement, such as printf("%i %.0f\n", time_step, pop_init*(lambda^time_step). The term %.0f in the printf format string instructs gawk to print out the corresponding item in the argument list (pop_init*(lambda^time_step)) as a floating-point number to zero decimal places. This rounds the value of the argument up or down to the nearest integer (values in the range 10.0 to 10.5 are reported as 10, while those in the range 10.51 to 10.99 are output as 11). In the remainder of this section, however, the code given in Program 8.1 is used unamended to perform various simulations.

*Evaluation*

Program 8.1 can be used to examine changes in the size of a population for which births and deaths take place in discrete cohorts at finite intervals in time. By providing the program with appropriate values on the command line, it is possible to explore the impact of different initial population sizes and discrete rates of growth. Assuming that $N_0 = 10$, for example, the growth of the population can be simulated over a period of 50 time steps for each of three different rates of growth ($\lambda = 1.05$, $\lambda = 1.06$ and $\lambda = 1.07$) by typing the following instructions on the command line:

```
gawk -f discrete.awk -v pop_init=10 -v lambda=1.05 -v ↵        1
    ↪period=50 > discr105.dat
gawk -f discrete.awk -v pop_init=10 -v lambda=1.06 -v ↵        2
    ↪period=50 > discr106.dat
gawk -f discrete.awk -v pop_init=10 -v lambda=1.07 -v ↵        3
    ↪period=50 > discr107.dat
```

Note that the result of each simulation is redirected to a separate data file in the working directory, namely discr105.dat, discr106.dat and discr107.dat.

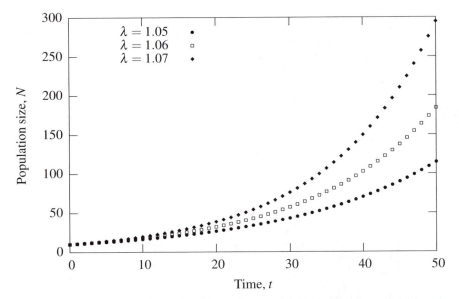

Figure 8.2: Population growth predicted by a discrete density-independent model based on three separate rates of growth ($\lambda = 1.05$, $\lambda = 1.06$ and $\lambda = 1.07$) and a common initial population ($N_0 = 10$).

*Visualization*

The results of the model runs listed above can be visualized by typing the following commands in gnuplot:

```
reset                                                          1
set style data points                                         2
set xlabel "Time, t"                                          3
set ylabel "Population size, N"                               4
set key top left                                              5
plot 'discr105.dat' t "lambda=1.05" lt 3 pt 7, \             6
     'discr106.dat' t "lambda=1.06" lt 3 pt 6, \             7
     'discr107.dat' t "lambda=1.07" lt 3 pt 5               8
```

These commands assume that the relevant data files are located in the working directory. Note that the `points` data style is used here to emphasize the fact that the data pertain to discrete intervals in time (Figure 8.2).

Figure 8.2 demonstrates the geometric pattern of population growth that occurs when $\lambda > 1$. As expected, the population grows more rapidly as the value of $\lambda$ increases. Figure 8.3 shows a similar plot generated using data produced by the same model for $N_0 = 100$ and $\lambda < 1$. In this case, the population declines over time and does so more rapidly as the value of $\lambda$ decreases.

Figure 8.4 and Figure 8.5 show the same two data sets plotted on semi-logarithmic axes (i.e., a logarithmic scale on the $y$-axis and a linear scale on the $x$-axis) having issued the following commands in gnuplot:

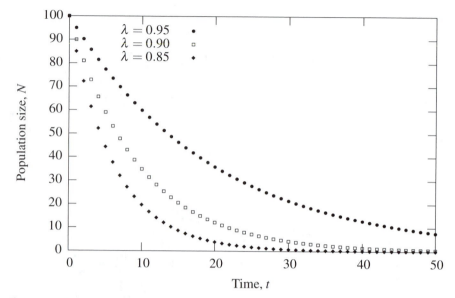

Figure 8.3: Population decline predicted by a discrete, density-independent model based on three separate rates of growth ($\lambda = 0.95$, $\lambda = 0.90$ and $\lambda = 0.85$) and a common initial population ($N_0 = 100$).

```
set logscale y                                                              9
replot                                                                     10
```

These figures suggest that there is a simple linear relationship between the logarithm of $N_t$ and $t$. Taking the natural logarithm (logarithm to the base $e$) of both sides of Equation 8.11 and simplifying, this relationship can be expressed as follows:

$$\begin{aligned} \ln(N_t) &= \ln(\lambda^t N_0) \\ &= \ln(\lambda)t + \ln(N_0) \end{aligned} \tag{8.12}$$

(Roughgarden 1998). This equation has the general form $y = mx + c$. Thus, the slope of the lines in Figure 8.4 and Figure 8.5 is given by $\ln(\lambda)$. The significance of this result is examined in the next section.

Finally, before proceeding any further, the following gnuplot command should be issued to turn off the logarithmic scaling on the $y$-axis in all subsequent plots:

```
unset logscale y                                                           11
```

> **Exercise 8.1**: Reproduce the results shown in Figure 8.3 by running Program 8.1 for $N_0 = 100$, $\lambda = 0.95$, $\lambda = 0.90$ and $\lambda = 0.85$, visualizing the output in gnuplot. Plot the same data on a semi-logarithmic scale by typing set logscale y and then replot in gnuplot to reproduce Figure 8.5. Remember to unset logscale y before producing any further plots.

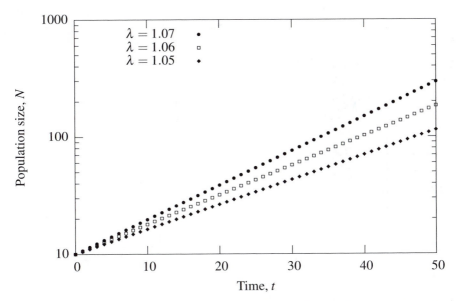

Figure 8.4: Discrete population growth curves from Figure 8.2 plotted on a semi-logarithmic scale.

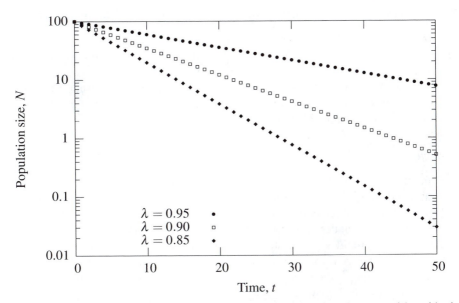

Figure 8.5: Discrete population growth curves from Figure 8.3 plotted on a semi-logarithmic scale.

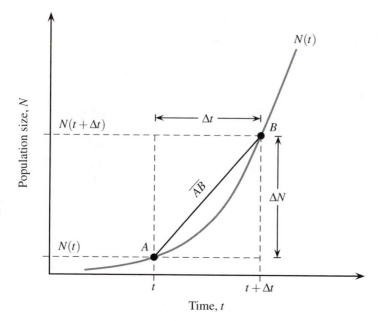

Figure 8.6: Average rate of population growth measured between two census points, $A$ and $B$, expressed as the slope of the secant line $\overline{AB}$. The underlying continuous growth curve, $N(t)$, is shown by the curved gray line.

### 8.2.3 Formulation, Implementation and Evaluation of the Continuous Model

The populations of certain species, including *Homo sapiens* and some bacteria grown *in vitro*, vary in size almost continuously through time, and it is appropriate to use continuous models to describe their growth or decline. Although the mathematical formulation of such models can be derived from first principles (Alstad 2001), this requires knowledge of calculus, which some readers may not possess. Consequently, a less formal, more graphical, approach is adopted here.

*Formulation*

Imagine that, instead of having been produced by a discrete model of population growth, the data in Figure 8.2 come from a series of censuses conducted at regular intervals in time for three continuously varying populations, each of which grows at a different rate. The underlying trends in population growth are therefore continuous, and each census provides a "snapshot" of the population sizes at a particular moment in time. The challenge is to derive a continuous mathematical function that describes the size of each population, and its rate of growth, at any point in time. This is addressed informally, below.

Let $\Delta t$ denote the time interval between successive censuses $A$ and $B$ conducted at times $t$ and $t + \Delta t$, respectively (Figure 8.6). This interval might, for example, be a year, a month or a day. Similarly, let $\Delta N$ denote the amount by which the

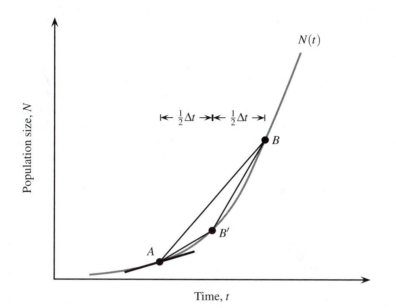

Figure 8.7: Effect of reducing $\Delta t$ on the representation of the continuous population growth curve, $N(t)$. The tangent to $N(t)$ at point $A$ is shown by the bold black line. This represents the rate of change in the population size at that instant in time.

population changes in size over this period of time. The average rate of change in the size of the population between census $A$ and census $B$ is therefore $\Delta N$ divided by $\Delta t$ ($\Delta N/\Delta t$). This is represented graphically by the slope of the (solid black) secant line $\overline{AB}$ shown in Figure 8.6. In the absence of any other information about the way in which the population varies in size between census $A$ and census $B$, the secant line, $\overline{AB}$, represents a first approximation to the shape of the (solid gray) continuous growth curve, $N(t)$, over the corresponding interval of time. It is evident, however, that points along the secant line, $\overline{AB}$, tend to overestimate the population size; that is, they lie above the curve, $N(t)$. This is because the slope of the secant line, $\overline{AB}$, is greater than that of the continuous growth curve, $N(t)$, at point $A$. The latter defines the instantaneous rate of change in the size of the population at time $t$.

Now imagine that another census $B'$ is conducted midway in time between $A$ and $B$; that is, the time interval ($\Delta t$) between successive censuses is halved (Figure 8.7). The resulting secant lines, $\overline{AB'}$ and $\overline{B'B}$, are shorter than $\overline{AB}$, and they provide a better approximation to the shape of the continuous growth curve $N(t)$. Furthermore, the slope of the secant line $\overline{AB'}$ is nearer to that of the continuous growth curve at point $A$. It should be evident that halving the time interval between censuses once again will produce an even more accurate representation of the continuous growth curve. Taking this to the limit, as $\Delta t$ approaches zero ($\Delta t \rightarrow 0$), the resulting secant lines become infinitesimally small so that they effectively define the continuous growth curve. Moreover, as $\Delta t \rightarrow 0$ each secant line is tangent to the continuous growth curve at that point in time, and its slope therefore defines the rate at which the population is growing at that instant. This is expressed more formally as

$$\lim_{\Delta t \to 0} \frac{\Delta N}{\Delta t} = \frac{dN}{dt} = rN \tag{8.13}$$

where $N$ is the population size, $\frac{dN}{dt}$ ("$dN$ by $dt$") is known as the derivative of $N$ with respect to $t$, and $r$ is known as the intrinsic or instantaneous rate of population growth. Thus, $\frac{dN}{dt}$ describes the rate of change in the size of the population at any given moment in time (i.e., the instantaneous rate of change). This is equivalent to the slope of the curve $N(t)$. Note that Equation 8.13 is a form of differential equation.

Equation 8.13 indicates that the rate of change in the size of the population at any given moment in time, $\frac{dN}{dt}$, is equal to the population size, $N$, at that time, $t$, multiplied by the instantaneous rate of growth, $r$. The value of $r$, however, is as yet unknown. To resolve this, recall that the data in Figure 8.2 were transformed into a straight-line relationship by plotting them on semi-logarithmic axes (Figure 8.4). Now imagine, once again, that the data points in Figure 8.4 were obtained from censuses conducted at regular intervals in time ($\Delta t$) for each of the three continuously varying populations. At the limit, when $\Delta t \to 0$, there is an infinite number of census points for each population, such that they effectively describe continuous straight lines. In the preceding section, it was shown that the slope of these lines is $\ln(\lambda)$. It follows, therefore, that

$$r = \ln(\lambda) \tag{8.14}$$

and, hence, that

$$\lambda = e^r \tag{8.15}$$

Substituting Equation 8.15 into Equation 8.11 and modifying the notation slightly to present $N$ as a continuous function of $t$ therefore gives

$$N(t) = e^{rt} N(0) \tag{8.16}$$

where $N(t)$ is the population size at an arbitrary point in time, $t$, $N(0)$ is the initial size of the population (at $t = 0$), and $r$ is the instantaneous rate of population growth. Equation 8.16 indicates that the population size grows (or declines) exponentially with respect to time, depending on the instantaneous rate of growth, $r$.

*Implementation*

Program 8.2 (`continue.awk`) is an implementation of the continuous model of density-independent population growth presented in Equation 8.16. The code is very similar to that used in Program 8.1. The main difference occurs on line 16, where the formula for the continuous model is implemented. Although Program 8.2 uses the same looping structure as Program 8.1 (`discrete.awk`; the discrete model), stepping forward in integer units of time, it is possible to use the continuous model code to predict the population at any point in time. For example, the variable `time` could be incremented in steps of `0.1` by replacing the statement `++time` on line 15 with `time+=0.1`. Finally, note that $\exp(0) = 1$. The first time around the `for` loop, when `time=0`, the program prints out the initial population size given on the command line.

Program 8.2: continue.awk

```
# Continuous model of unconstrained (density-independent)    1
# population growth (see Eq. 8.16).                          2
#                                                            3
# Usage: gawk -f continue.awk -v pop_init=value \            4
#                -v growth_rate=value -v period=value \       5
#                [ > outputFile ]                            6
#                                                            7
# Variables:                                                 8
# ----------                                                 9
# pop_init       initial population                         10
# growth_rate    instantaneous (intrinsic) growth rate      11
# period         period of time over which growth is modeled 12
#                                                           13
BEGIN{                                                      14
    for(time=0;time<=period;++time){                       15
        print time, pop_init*exp(growth_rate*time);        16
    }                                                       17
}                                                           18
```

*Evaluation*

Program 8.2 can be used to simulate continuous population growth over a period of 50 units of time for three different instantaneous rates of growth ($r = 0.05$, $r = 0.06$ and $r = 0.07$), based on an initial population of 10 individuals ($N(0) = 10$), by typing the following instructions on the command line:

```
gawk -f continue.awk -v pop_init=10 -v growth_rate=0.05 -v↷   4
   ↷ period=50 > cont005.dat
gawk -f continue.awk -v pop_init=10 -v growth_rate=0.06 -v↷   5
   ↷ period=50 > cont006.dat
gawk -f continue.awk -v pop_init=10 -v growth_rate=0.07 -v↷   6
   ↷ period=50 > cont007.dat
```

Recall from Equation 8.14 that $r = \ln(\lambda)$, so that the instantaneous growth rates used here are roughly equivalent to the finite growth rates used in the discrete model simulations reported in the preceding section ($\ln(1.05) \approx 0.05$, $\ln(1.06) \approx 0.06$ and $\ln(1.07) \approx 0.07$). Note that the result of each simulation is redirected to a separate data file, namely cont005.dat, cont006.dat and cont007.dat.

*Visualization*

The results of the model runs described above can be visualized by issuing the following set of commands in gnuplot:

```
set style data lines                                        12
plot 'cont005.dat' t "r=0.05", \                            13
```

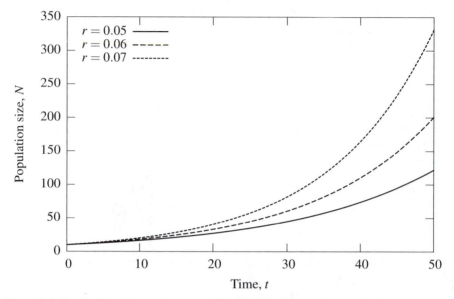

Figure 8.8: Results from the continuous model of density-independent population growth (Program 8.2; continue.awk) for $N_0 = 10$ and $r = 0.05$, $r = 0.06$ and $r = 0.07$.

```
      'cont006.dat'  t  "r=0.06",  \                                    14
      'cont007.dat'  t  "r=0.07"                                        15
```

Note that the lines data style is used here to emphasize the fact that population size is a continuous function of time. The results are presented in Figure 8.8, which can be compared directly to the discrete model case presented in Figure 8.2. Before moving on, it is worth emphasizing once again the difference between the discrete and continuous models of density-independent growth: the latter produces a value of $N$ for any value of $t$ ($t = 1.25$), whereas the former reports values of $N$ for integer values of $t$ only ($t = 1, 2, 3, \dots$).

**Exercise 8.2**: Run Program 8.2 for $N = 100$ and three different values of $r$ ($-0.05$, $-0.1$ and $-0.15$). Visualize the output in gnuplot and compare the resulting plot with Figure 8.3

## 8.3   CONSTRAINED OR DENSITY-DEPENDENT GROWTH

Few populations grow geometrically or exponentially without limit for a long period of time. There are many reasons why this is the case; prime among these are the constraints imposed by the finite pools of natural resources, such as food, water, light and shelter, that are available to support the population. In general, the competition for resources between individuals of the same species, known as intra-specific competition, becomes ever more intense as the population grows in size (Donovan and

Welden 2002). Consequently, the birth and death rates per capita vary according to the size of the population (Alstad 2001). This phenomenon is known as constrained or density-dependent growth because the population growth rate is affected by factors relating to sparsity and over-crowding.

It is possible to derive both discrete and continuous models of density-dependent population growth. In this section, the continuous model is examined first, before moving on to consider a remarkable outcome of the discrete model.

### 8.3.1 Developing the Conceptual Model

One of the earliest attempts to model density-dependent population growth was made by the Belgian mathematician Pierre François Verhulst (1804–1849). In essence, Verhulst hypothesized that there is a limit to the number of individuals that can be supported on a continuing basis by any given environment (Whittaker 1982). He reasoned that this is because of the constraints imposed by factors such as the finite amount of food available for consumption. The theoretical maximum population that can be supported by an environmental system in steady state is known as the carrying capacity (Roughgarden 1998). Verhulst suggested that the rate of growth of a population is dependent on both the population size and the carrying capacity of the environment that it inhabits. For instance, when a population is small in number relative to the carrying capacity, the resources available to each individual are plentiful; as a result, the population is able to grow at a rate largely determined by the intrinsic birth and death rates for that species (i.e., at a rate approximately equal to that of the density-independent model). By contrast, when the population size is equal to the carrying capacity of the environment, the available resources are sufficient to sustain that number of individuals and no more; further growth in the population is therefore inhibited by resource limitations. In these circumstances, the population growth rate tends to zero (Alstad 2001).

### 8.3.2 Continuous Logistic Model

*Formulation*

Verhulst's model of density-dependent growth, also known as the logistic growth model, is usually presented as

$$\frac{dN}{dt} = rN\left(\frac{K-N}{K}\right) \tag{8.17}$$

or, equivalently, as

$$\frac{dN}{dt} = rN\left(1 - \frac{N}{K}\right) \tag{8.18}$$

where $N$ is the population size, $r$ is the instantaneous rate of population growth and $K$ is the carrying capacity (Roughgarden 1998). Equation 8.18 differs from its density-independent counterpart (Equation 8.16) through inclusion of a multiplicative term,

$(K-N)/K$. Alstad (2001) interprets $K-N$ as a measure of the unused resources in the environment and $(K-N)/K$ as the unused fraction of the carrying capacity. Thus, $(K-N)/K \approx 1$, and hence $\frac{dN}{dt} \approx rN$, when $N \approx 0$ (the population is very small); that is to say, the rate of change in the population size is approximately equal to the unconstrained or density-independent case (Equation 8.16). So, when small, the population experiences a period of almost exponential growth. By contrast, $(K-N)/K \approx 0$, and hence $\frac{dN}{dt} \approx 0$, when $N \approx K$ (the population is very large). Consequently, as the population size approaches the carrying capacity, the rate of growth slows to zero. Note that it is theoretically possible for the population size to exceed the carrying capacity $(N > K)$ for short periods of time, whether as a result of immigration or of over-stocking. In these circumstances, $(K-N)/K$, and hence $\frac{dN}{dt}$, becomes negative. As a result, the population decreases in size over time. In effect, therefore, the term $(K-N)/K$ in Equation 8.17 introduces negative feedback between the size of a population and its rate of growth (Alstad 2001).

It is possible to solve Equation 8.18 analytically to derive a formula that expresses $N$ as a continuous function of $t$. The formal derivation of this solution is beyond the scope of this book — interested readers should consult Giordano *et al.* (1997) — but it can be shown that

$$N(t) = \frac{K}{1 + [(K - N(0))/N(0)]e^{-rt}} \tag{8.19}$$

where $N(t)$ is the population at time $t$, $N(0)$ is the initial population (at $t = 0$), $r$ is the instantaneous rate of growth and $K$ is the carrying capacity (Roughgarden 1998).

### Implementation

The continuous logistic model of population growth (Equation 8.19) is implemented in Program 8.3, `cntlogst.awk`. The structure of this code is very similar to that of Program 8.2, `continue.awk`, in that it makes use of a `for` loop contained within the BEGIN block. Two additional variables are employed, namely `carry_cap`, which is used to store the value of the carrying capacity (lines 14 and 19), and `pop_now`, which is used to store the population calculated for time $t$ (lines 12, 19 and 21). This calculation (implementation of Equation 8.19) is performed on lines 19 and 20. These lines of code constitute a single **gawk** statement, which has been split over two lines using the line continuation symbol (\) at the end of line 19. The current time and population size are printed out on line 21. Thus, Program 8.3 reports the size of a continuously varying population at regular intervals in time, controlled by the parameters of the `for` loop.

### Evaluation

Program 8.3 can be used to simulate the density-dependent growth of a continuously varying population, initially consisting of 10 individuals ($N(0) = 10$), over a period of 200 units of time and for three intrinsic rates of growth ($r = 0.05$, $r = 0.06$ and $r = 0.07$) by issuing the following instructions on the command line:

Program 8.3: cntlogst.awk

```
# Continuous-time model of density-dependent (constrained)    1
# population growth, based on the Verhulst (logistic)          2
# equation (see Eq. 8.19).                                     3
#                                                              4
# Usage: gawk -f cntlogst.awk -v pop_init=value \              5
#               -v growth_rate=value -v carry_cap=value \      6
#               -v period=value [ > output_file ]              7
#                                                              8
# Variables:                                                   9
# ----------                                                  10
# pop_init      initial population                            11
# pop_now       current population                            12
# growth_rate   instantaneous (intrinsic) growth rate         13
# carry_cap     carrying capacity                             14
# period        period of time over which growth is modeled   15
                                                              16
BEGIN{                                                        17
  for(time=0;time<=period;++time){                            18
    pop_now=carry_cap/(1+((carry_cap-pop_init)/pop_init)* \   19
      exp(-growth_rate*time));                                20
    print time, pop_now;                                      21
  }                                                           22
}                                                             23
```

```
gawk -f cntlogst.awk -v pop_init=10 -v growth_rate=0.05 -v↳   7
  ↳ carry_cap=1000 -v period=200 > c_log005.dat
gawk -f cntlogst.awk -v pop_init=10 -v growth_rate=0.06 -v↳   8
  ↳ carry_cap=1000 -v period=200 > c_log006.dat
gawk -f cntlogst.awk -v pop_init=10 -v growth_rate=0.07 -v↳   9
  ↳ carry_cap=1000 -v period=200 > c_log007.dat
```

Note that values for the initial population size, carrying capacity and period of time over which the population growth is to be modeled are specified on the command line, and that the result of each simulation is redirected to a separate data file.

*Visualization*

The output from the simulations described above can be visualized in gnuplot by issuing the following set of commands:

```
plot 'c_log005.dat' t "r=0.05" lw 2, \                       16
     'c_log006.dat' t "r=0.06" lw 2, \                       17
     'c_log007.dat' t "r=0.07" lw 2                          18
```

The resultant population growth curves (Figure 8.9) are sigmoidal (S-shaped). To begin with, each population grows at an almost exponential rate (i.e., at a rate that

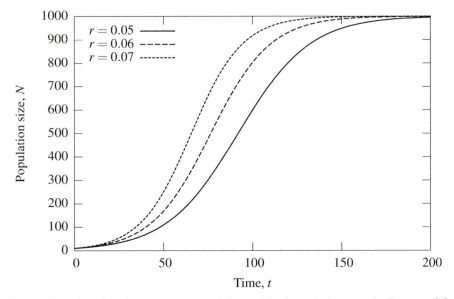

Figure 8.9: Output from the continuous logistic model of population growth (Program 8.3; cntlogst.awk) for $N(0) = 10$, $K = 1000$ and various values of $r$.

is approximately equal to the unconstrained growth model), but as it increases in size the competition for resources becomes more intense and hence its rate of growth decreases. Later, as the size of the population approaches the carrying capacity for the environment, the growth rate decreases to zero, and the population size gradually approaches its steady-state value (Haefner 1996).

The relationship between population size ($N$) and population growth ($\frac{dN}{dt}$) in the logistic model is shown in Figure 8.10. Note that population growth is small when $N$ is small because, despite the fact that the growth rate ($r((K-N)/K)$) is high, the number of individuals available to reproduce is limited. Population growth is also small when $N$ is large because, despite the fact that the population is large, the growth rate is small. Population growth is maximized when $N = K/2$.

---

**Exercise 8.3**: What happens if the initial population exceeds the carrying capacity? Test this by running Program 8.3 for $N(0) = 1000$, $K = 100$ ($N(0) > K$) and $r = 0.07$. Visualize the output in gnuplot.

---

### 8.3.3  Discrete Logistic Model

It is possible to construct a version of the logistic model for species that are born in discrete cohorts separated by finite intervals of time (i.e., a discrete logistic growth model), and one such model is examined here. The primary intention in doing so is to demonstrate the remarkable behavior that this model exhibits under certain con-

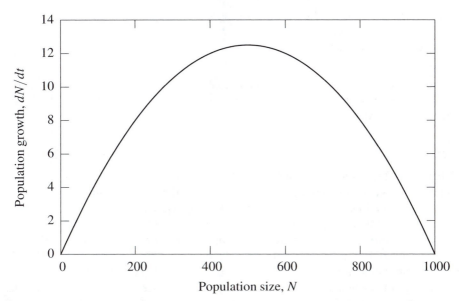

Figure 8.10: Relationship between population size ($N$) and population growth ($dN/dt$) in the continuous density-dependent model of population growth for $N(0) = 10$, $K = 1000$ and $r = 0.05$.

ditions. More specifically, it illustrates how a relatively simple deterministic equation can generate apparently random, and hence unpredictable, output (Gleick 1987, Hall 1991). This type of "chaotic" behavior in population growth models was first demonstrated by Sir Robert May (1974).

*Formulation*

Numerous different formulations of the discrete logistic growth model are reported in the literature. Perhaps the most obvious example is

$$\Delta N_t = RN_t \left( \frac{K - N_t}{K} \right) \tag{8.20}$$

where $\Delta N_t$ is the amount by which the population grows (or declines) during time step $t$, $N_t$ is the population size during time step $t$, $R = B - D$ (the per capita birth rate minus the per capita death rate; see Section 8.2.2) and $K$ is the carrying capacity (Donovan and Welden 2002). This is simply the product of Equation 8.3 (i.e., the discrete version of the unconstrained growth model) and the logistic growth term, $(K - N_t)/K$. The corresponding finite difference equation is, therefore,

$$N_{t+1} = N_t + RN \left( \frac{K - N_t}{K} \right) \tag{8.21}$$

(Roughgarden 1998). A different solution, suggested by May (1976), is

$$N_{t+1} = N_t \exp\left[ r \left( \frac{K - N_t}{K} \right) \right]$$

(8.22)

where $r$ is the instantaneous or intrinsic rate of population growth. This formulation is directly equivalent to the continuous logistic model (Equation 8.17) at the limit, where $\Delta t \to 0$ (Alstad 2001), and is the solution employed here.

### Implementation

Equation 8.22 is implemented in gawk code in Program 8.4, dsclogst.awk. After the usual set of explanatory comments (lines 1 to 14), the program employs the BEGIN block and a for loop to simulate the growth of the population with each successive time step. Each time the program traverses the for loop, the value of the current time step and the corresponding population size are printed out (line 18), after which the population size in the next time step is calculated (line 19). This is repeated until the requisite number of time steps (steps) have elapsed. Note that values must be supplied on the command line for the initial population size (pop), instantaneous rate of population growth (growth_rate), carrying capacity (carry_cap) and number of time steps over which the simulation is to be performed (steps).

### Evaluation

Program 8.4 can be run from the command line as follows:

```
gawk  -f dsclogst.awk  -v pop=10  -v growth_rate=0.05  -v ⇦          10
    ⇦carry_cap=1000  -v steps=200 > d_log005.dat
gawk  -f dsclogst.awk  -v pop=10  -v growth_rate=0.06  -v ⇦          11
    ⇦carry_cap=1000  -v steps=200 > d_log006.dat
gawk  -f dsclogst.awk  -v pop=10  -v growth_rate=0.07  -v ⇦          12
    ⇦carry_cap=1000  -v steps=200 > d_log007.dat
```

These commands assume that the population consists of 10 individuals initially and that the growth of the population is to be modeled over a total period of 200 time steps. Three separate simulations are performed, each for a different (instantaneous) rate of population growth ($r = 0.05$, $r = 0.06$ and $r = 0.07$). The results are redirected into separate data files (d_log005.dat, d_log006.dat and d_log007.dat) in the working directory.

### Visualization

The output from the simulations described above can be visualized in gnuplot by issuing the following set of commands, which produce Figure 8.11:

```
set style data points                                               19
plot 'd_log005.dat' t "r=0.05" w p, \                               20
     'd_log006.dat' t "r=0.06" w p, \                               21
     'd_log007.dat' t "r=0.07" w p                                  22
```

Program 8.4: dsclogst.awk

```
# Discrete-time model of density-dependent (constrained)        1
# population growth, based on the Verhulst (logistic)           2
# equation (see Eq. 8.22).                                      3
#                                                               4
# Usage: gawk -f dsclogst.awk -v pop=value \                    5
#                -v growth_rate=value -v carry_cap=value \      6
#                -v steps=value [ > output_file ]               7
#                                                               8
# Variables:                                                    9
# ----------                                                    10
# pop            population at time step                        11
# growth_rate    instantaneous (intrinsic) growth rate          12
# carry_cap      carrying capacity                               13
# steps          total number of time steps                     14
                                                                15
BEGIN{                                                          16
  for(time_step=0;time_step<=steps;++time_step){                17
    print time_step, pop;                                       18
    pop=pop*exp(growth_rate*(1-pop/carry_cap));                 19
  }                                                             20
}                                                               21
```

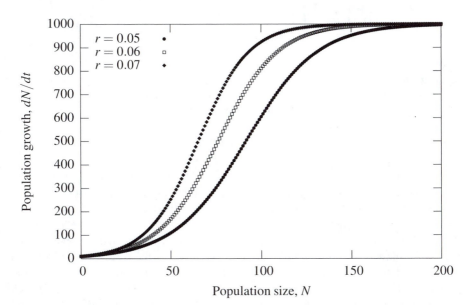

Figure 8.11: Output from the discrete logistic model of population growth (Program 8.4; dsclogst.awk) for $N_0 = 10$, $K = 1000$ and three different values of $r$ (0.05, 0.06 and 0.07).

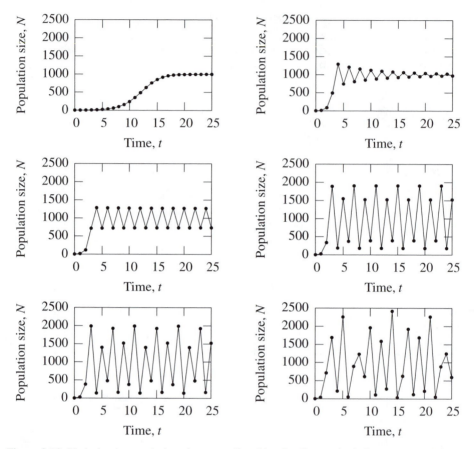

Figure 8.12: Variation in population size as predicted by the discrete logistic growth model as a function of $r$ for $N_0 = 2$ and $K = 1000$. Key: $r = 0.5$ (top left), $r = 1.9$ (top right), $r = 2.05$ (middle left), $r = 2.6$ (middle right), $r = 2.67$ (bottom left) and $r = 3.0$ (bottom right).

Note that the `points` data style has been used to emphasize the fact that the data relate to discrete time steps. It should be evident from Figure 8.11 that the resultant curves are identical in form and magnitude to those in Figure 8.9; that is, the discrete version of the logistic model behaves in the same way as its continuous counterpart for the values of $r$ examined here.

The behavior of the discrete logistic model alters dramatically, however, as the value of $r$ increases. For example, Figure 8.12 presents the results of six different runs of the discrete logistic model (Program 8.4; `dsclogst.awk`) using the same initial population size ($N_0 = 2$; pop) and carrying capacity ($K = 1000$; `carry_cap`) on each occasion, but varying the rate of population growth ($r = 0.5, 1.90, 2.05, 2.6, 2.7$ and $3.0$; `growth_rate`). To make it easier to interpret the resulting patterns, lines have been drawn between consecutive data points. For $r = 0.5$, the population size follows the normal sigmoidal growth curve, reaching a steady-state value (i.e., the carrying

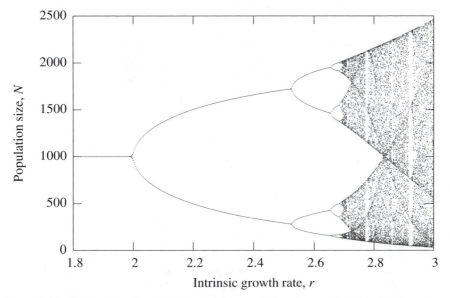

Figure 8.13: "Bifurcation diagram" showing the population sizes predicted by the discrete version of the density-dependent population growth model as a function of the intrinsic rate of increase, *r*.

capacity) after about 20 time steps (Figure 8.12, top left). When the growth rate is increased to $r = 1.9$ the population grows dramatically at first (Figure 8.12, top right), so that by the fourth time step the population consists of almost 1300 individuals. This exceeds the carrying capacity of the environment, so that there are insufficient resources to support all of the individuals. Consequently, the population decreases to around 750 by the fifth time step. This is less than the carrying capacity, however, so the population grows once more to around 1200 by the sixth time step. During subsequent time steps, the population goes through a series of these "boom and bust" cycles, which gradually diminish in magnitude so that the population size stabilizes around a steady-state value (the carrying capacity).

The behavior of the system changes once again when $r > 2$. Here, the population increases in size very dramatically over the first few time steps and then settles into an oscillatory pattern centered around the value of the carrying capacity. For instance, the system state oscillates between two population sizes when $r = 2.05$ (Figure 8.12, middle left), four when $r = 2.6$ (Figure 8.12, middle right) and eight when $r = 2.67$ (Figure 8.12, bottom left). The behavior of the system becomes chaotic when $r \geq 3$ (Figure 8.12, bottom right), with seemingly random fluctuations in the population size from one time step to the next, and the population does not reach an equilibrium size. This behavior is remarkable because it is produced by a simple deterministic equation. May (1991) refers to this as "*deterministic chaos*".

The behavior of the discrete logistic model can be visualized over a range of values of *r*, not just the few selected above. Figure 8.13, for example, was produced by running the model repeatedly for $r = 1.8$ to $r = 3.0$ in steps of 0.0025. On each

Table 8.1: Approximate threshold values for different types of behavior in the discrete logistic population growth model.

| Growth Rate | Model Behavior |
|---|---|
| $0 < r \leq 2.0$ | Reaches stable equilibrium |
| $2.0 < r \leq 2.52$ | Two-point cycle |
| $2.52 < r \leq 2.65$ | Four-point cycle |
| $2.65 < r \leq 2.6825$ | Eight-point cycle |
| $2.6825 < r < 3.0$ | 16-, 32-, 64-point cycles, and so on |
| $r \geq 3.0$ | Chaotic |

occasion the model was run for a total of 500 time steps, and the results for the last 100 of these were saved for each model run in a single data file (Roughgarden 1998). Figure 8.13 indicates that the population size tends toward a single steady-state value for $r < 2.0$. Beyond this it oscillates, first in a two-point cycle, then four-point, eight-point, sixteen-point and so on, as $r$ increases, before breaking into chaotic behavior when $r \geq 3$ (Roughgarden 1998). The values of $r$ that mark the break-points between these different types of behavior are shown in Table 8.1. Note that, despite the apparently increasing disorder above $r = 2.7$, there is a considerable amount of fine structure hidden within this part of Figure 8.13. This is evident in the enlarged section shown in Figure 8.14, which clearly demonstrates that the bifurcation patterns evident in the range $2.0 \leq r \leq 2.8$ are repeated in miniature in the range $2.92 \leq r \leq 2.93$. The fine structure is also present for $r > 3.0$. Thus, although the population size varies chaotically when $r > 3.0$, the fluctuations are not random (Gleick 1987).

One consequence of the chaotic behavior exhibited by the discrete logistic model when $r > 3.0$ is that the model output is highly sensitive to the initial population size, $N_0$ (Roughgarden 1998). This is illustrated in Figure 8.15, which presents the results of two model runs, one using $N_0 = 99$ and the other using $N_0 = 101$, for $r = 3.0$ and $K = 500$. Despite the fact that the two populations differ in size by only two individuals to begin with, which is small in both absolute and relative terms, their trajectories diverge rather rapidly (Figure 8.15). This is significant because there is often great uncertainty over the values that should be used to initialize a model; here, such uncertainty can affect dramatically the nature of the model output.

## 8.4   NUMERICAL INTEGRATION (OR STEPPING) METHODS

In the preceding section, analytical solutions to both the continuous and the discrete versions of the logistic population growth model were obtained from the literature (Equation 8.19 and Equation 8.21, respectively). This was done for the sake of convenience and brevity. Normally, however, analytical solutions have to be derived from the corresponding differential or difference equations. This can be challenging, and it is often the case that an analytical solution cannot be found or else does not exist (Wainwright and Mulligan 2004). In such circumstances, the model has to be solved

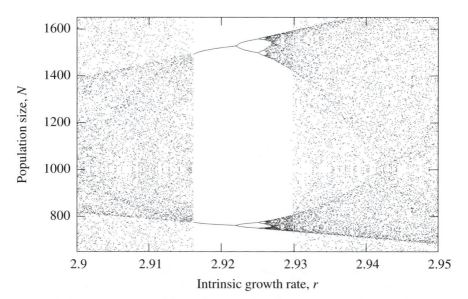

Figure 8.14: Enlarged section of the "bifurcation diagram" presented in Figure 8.13 showing the fine structure present.

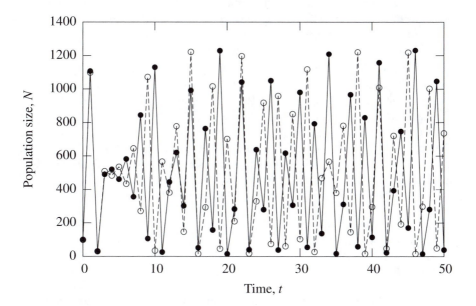

Figure 8.15: Effect of a small difference in $N_0$, $N_0 = 99$ (●) and $N_0 = 102$ (○), on the population dynamics of the discrete logistic model for $r = 3.0$ and $K = 500$.

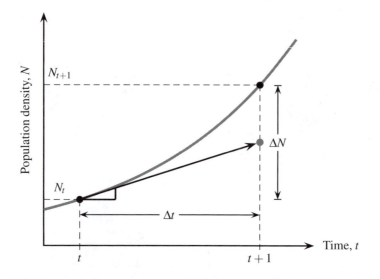

Figure 8.16: Diagrammatic representation of Euler's method of numerical integration used to estimate $N_{t+1}$ based on $N_t$ and $\frac{dN}{dt}$, where $\Delta t = 1$. The solid gray line represents the underlying continuous function $N(t)$.

numerically, using methods similar to those introduced in Chapter 7. In this section, a set of numerical techniques, known formally as numerical integration (Harris and Stocker 1998) and informally as stepping techniques (Borse 1997), is introduced. These methods are used to provide a numerical solution to the differential equations of the continuous logistical population growth model.

Consider, for the moment, Figure 8.16. Imagine that the continuous gray curve represents the underlying, but as yet unknown and unspecified, relationship between population size and time. Suppose that, for some reason, it is not possible to derive by analytical means a continuous mathematical function that describes the form of this relationship, but that the size of the population at time $t$ ($N_t$) and its rate of growth at that instant ($\frac{dN}{dt}$) are known. The challenge is to use this information to predict the population size at other points in time, such as at $t + 1$ ($N_{t+1}$). This is sometimes known as an initial value problem. Two approaches to this type of problem, based on the methods of Euler and Runge-Kutta, are demonstrated below.

### 8.4.1   Euler's Method

*Conceptual Basis and Mathematical Formulation*

The simplest solution to the initial value problem, known as Euler's method (Harris and Stocker 1998), is to take the rate of population growth ($\frac{dN}{dt}$) at time $t$ and to project this forward for a finite period of time ($\Delta t$) to estimate the change in the size of the population ($\Delta N$) over that interval. The continuous function is, thus, approximated

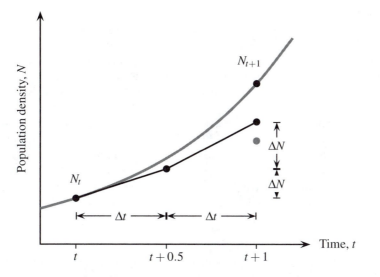

Figure 8.17: Euler's method of numerical integration used to estimate $N_{t+1}$ based on $N_t$ and $\frac{dN}{dt}$, where $\Delta t = 0.5$. The gray dot is the value of $N_{t+1}$ estimated using $\Delta t = 1$.

by a straight line segment (Figure 8.16). This can be expressed formally as

$$\Delta N \approx \frac{dN}{dt} \cdot \Delta t \tag{8.23}$$

and hence

$$N_{t+1} \approx N_t + \frac{dN}{dt} \cdot \Delta t \tag{8.24}$$

where $N_t$ is the population size at time $t$, $\frac{dN}{dt}$ is the rate of population growth at that moment and $N_{t+1}$ is the population size at time $t + 1$.

The weakness of this very basic approach is immediately apparent from Figure 8.16. Specifically, if Euler's method is used to project the population size forward over the entire period between time $t$ and time $t + 1$ ($\Delta t = 1$) it often yields a poor estimate of the actual population at time $t + 1$ ($N_{t+1}$). In this particular example it underestimates substantially the true value. This is because the derivative, $\frac{dN}{dt}$, of the underlying function, $N(t)$, varies continuously with time (the slope of the curve $N(t)$ varies as a function of $t$).

The situation can often be improved by reducing $\Delta t$, the time period over which the population growth is projected. Figure 8.17, for example, illustrates what happens when $\Delta t$ is halved ($\Delta t = 0.5$), so that the population size at time $t + 1$ ($N_{t+1}$) is estimated in two steps. First, the rate of population growth at time $t$ is used to estimate the change in the size of the population up to time $t + 0.5$. The predicted population size at time $t + 0.5$ is then employed to calculate the rate of population growth at that instant, and this is subsequently used to estimate the population at time $t + 1$.

In the example shown in Figure 8.17, halving $\Delta t$ increases the accuracy with which the population size at $t + 1$ is estimated. Despite this, the result is still a poor approximation to the actual population size at that moment in time ($N_{t+1}$). The accuracy of the estimate can be improved by reducing $\Delta t$ further, increasing the number of steps used to calculate $N_{t+1}$. Theoretically, at the limit, $\Delta t \rightarrow 0$, Euler's method yields the exact analytical solution (Borse 1997). There are, however, two important practical constraints. First, decreasing $\Delta t$ and increasing the number of steps affects the computational load; the more steps, the longer it takes to compute the result. Second, computers do not store numbers with infinite precision, and the resulting rounding errors ultimately limit the level of accuracy that can be attained (Borse 1997). The most appropriate value of $\Delta t$ to use for any given model is often, therefore, a matter of trial-and-error. This value can be established by progressively reducing $\Delta t$ until further reductions no longer yield significant changes in the model output.

*Implementation*

Program 8.5 illustrates how Euler's method can be applied to the continuous logistic model of population growth. Apart from the comments (lines 1 to 14), the code is entirely contained in the BEGIN block (lines 16 to 31), because the intention is to generate data as part of a simulation. In this particular example, values for most of the main parameters — that is, the initial population size ($N_0$, pop), the intrinsic growth rate ($r$, growth_rate), the carrying capacity ($K$, carry_cap) and the time period over which the population growth is to be modeled ($t$, stop_time) — are hard-wired into the code (lines 17 to 21), rather than being provided on the command line, for the sake of convenience. A while loop (lines 23 to 30) is used to perform a sequence of actions (lines 24 to 29) repeatedly while the current time (time; initialized to zero on line 20) is less than or equal to the time at which the simulations are meant to stop (stop_time). The first of these actions is to print out the values of current time and the population size (line 24). Line 27 then increments the current time by $\Delta t$, the value of which is entered via the command line. The estimated change in the size of the population over this finite interval of time ($\Delta N$; delta_pop) is calculated on line 28. This represents an implementation of Equation 8.23, where $\frac{dN}{dt} = rN\left(1 - \frac{N}{K}\right)$ (Equation 8.18). Finally, the updated population size is calculated on line 29 (Equation 8.24).

Program 8.5 can be run from the command line as follows, for three different values of $\Delta t$ (1.0, 0.5 and 0.1):

```
gawk -f euler.awk -v dt=1.0 > euler1.dat            13
gawk -f euler.awk -v dt=0.5 > euler05.dat           14
gawk -f euler.awk -v dt=0.1 > euler01.dat           15
```

Note that the results are redirected to three separate data files, namely euler1.dat, euler05.dat and euler01.dat, which are located in the working directory.

*Visualization and Evaluation*

The results obtained by running Program 8.5 for $\Delta t = 1.0$, 0.5 and 0.1 are visualized in Figure 8.18, which was produced by typing the following commands in gnuplot:

Program 8.5: euler.awk

```
# Continuous-time model of density-dependent (constrained)    1
# population growth, based on the Verhulst (logistic)         2
# equation (see Eq. 22), solved using Euler's method of       3
# numerical integration.                                      4
#                                                             5
# Usage: gawk -f euler.awk -v dt=value [ > output_file ]      6
#                                                             7
# Variables:                                                  8
# ----------                                                  9
# pop              population at time step                    10
# growth_rate      instantaneous (intrinsic) growth rate      11
# carry_cap        carrying capacity                          12
# time             current time                               13
# stop_time        time at which simulation should stop       14
# dt               integration time step (1/steps)            15
                                                              16
BEGIN{                                                        17
  pop=10.0;                                                   18
  growth_rate=0.5;                                            19
  carry_cap=1000;                                             20
  time=0;                                                     21
  stop_time=10;                                               22
                                                              23
  while(time<=stop_time){                                     24
    printf("%.1f\t%.3f\n", time, pop);                        25
                                                              26
    # Numerical integration using Euler's method              27
    time+=dt;                                                 28
    delta_pop=growth_rate*pop*(1-(pop/carry_cap))*dt;         29
    pop+=delta_pop;                                           30
  }                                                           31
}                                                             32
```

```
N=10.0                                                        23
r=0.5                                                         24
K=1000.0                                                      25
set dummy t                                                   26
pop(t)=K/(1+((K-N)/N)*exp(-r*t))                              27
plot pop(t) t "Analytical solution" w l lw 2, \              28
     'euler1.dat' t "Euler: dt=1.0" w lp pt 7, \             29
     'euler05.dat' every 2 t "Euler: dt=0.5" w lp pt 8, \    30
     'euler01.dat' every 10 t "Euler: dt=0.1" w lp pt 10     31
```

These commands assume that the data files euler1.dat, euler05.dat and euler01.dat are located in the working directory; otherwise, their full or relative pathnames must be provided. Lines 23 to 27 define the parameters and the equation of the analytical

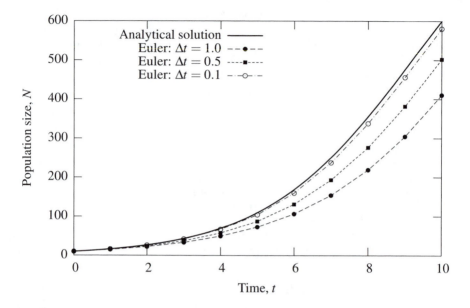

Figure 8.18: Analytical and numerical (Euler's method) solutions to the continuous logistic
model of population growth for $r = 0.5$, $N_0 = 10$ and $\Delta t = 1, 0.5$ and $0.1$.

solution to the continuous logistic model (Equation 8.19). These commands should
be familiar from the preceding chapter (Section 7.4). The plot is created on lines 28
to 31 using a single gnuplot command, which is split over four lines of code by the
line continuation symbol (\) on lines 28 through 30. Note that the keyword every on
lines 30 and 31 instructs gnuplot to plot every $n^{th}$ value in the named data set. For
example, the file euler05.dat contains estimates of the population size at $t = 0, 0.5$,
1.0, 1.5, and so on. By selecting every second record of data from this file (every 2),
only the results for $t = 0, 1.0$, and so on, are plotted. This allows the results of the
three model runs to be directly compared.

It is clear from Figure 8.18 that the numerical solution gets closer to the analytical
solution as $\Delta t$ becomes smaller. Table 8.2 presents similar results, examining the
difference at $t = 10$ between the exact analytical solution ($N_{10} = 599.86$) and the
numerical solution based on Euler's method, for different values of $\Delta t$. Note that the
percentage difference between the analytical and numerical solutions decreases by an
order of magnitude as $\Delta t$ is reduced by a factor of 10. At the same time, however, the
total number of computational steps required to calculate $N_{10}$ increases by an order
of magnitude. Thus, there is a trade-off between computation time and accuracy.

### 8.4.2   Runge-Kutta Methods

*Conceptual Basis and Mathematical Formulation*

The main advantages of Euler's method are that it is relatively simple to program
and fast to compute (Harris and Stocker 1998). It is unreliable in many instances,

Table 8.2: Difference at $t = 10$ between the analytical solution to the continuous logistic model for $r = 0.5$, $K = 1000$ and $N_0 = 10$ and the numerical solution based on Euler's method for different values of $\Delta t$.

| $\Delta t$ | Steps | $N_{10}$ | % Difference |
|---|---|---|---|
| 1.0 | 1 | 410.548 | 31.559 |
| 0.5 | 2 | 502.277 | 16.267 |
| 0.1 | 10 | 580.571 | 3.216 |
| 0.01 | 100 | 597.945 | 0.319 |
| 0.001 | 1000 | 599.668 | 0.032 |
| 0.0001 | 10000 | 599.840 | 0.003 |

however, and so is rarely used in practice. Nevertheless, it is important because it forms the basis for understanding virtually all other methods of numerical integration (Borse 1997). Among the most widely used of these are the mid-point method (also known as the modified Euler method) and the method of Runge-Kutta (Borse 1997, Harris and Stocker 1998, Wainwright and Mulligan 2004). The mid-point method is typically more accurate than the classical Euler method; this is because it uses the slope of the continuous function $N(t)$ in the middle of the interval ($\frac{dN}{dt}$ at $t + 0.5$; Figure 8.19), which is normally a better approximation to the average rate of change over the whole interval (Borse 1997, Harris and Stocker 1998). The method of Runge-Kutta takes this a stage further; the slope that is used to predict the population size at the end of the time interval is a weighted average of the slope at the start of the interval (at $t$) and at several other points between $t$ and $t + 1$.

The fourth-order Runge-Kutta method is used in this section. This method offers a good compromise between programming effort, computation time and numerical accuracy, and can be expressed as

$$N_{t+1} \approx N_t + \frac{1}{6} \left( k_1 + 2k_2 + 2k_3 + k_4 \right) \Delta t \tag{8.25}$$

where

$$
\begin{aligned}
k_1 &= f'(t, N_t) \\
k_2 &= f'(t + 0.5\Delta t, N_t + 0.5k_1\Delta t) \\
k_3 &= f'(t + 0.5\Delta t, N_t + 0.5k_2\Delta t) \\
k_4 &= f'(t + \Delta t, N_t + k_3\Delta t)
\end{aligned}
\tag{8.26}
$$

and where $f'(t_i, N_i)$ is the derivative of the function $N(t)$ at $t$, which is simply a different way of writing $\frac{dN}{dt}$ (Harris and Stocker 1998). Despite the rather daunting appearance of these equations, the calculations involved are quite simple and their subsequent implementation in gawk code is relatively straightforward.

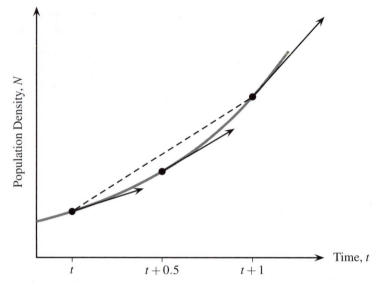

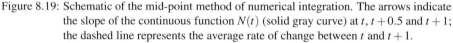

Figure 8.19: Schematic of the mid-point method of numerical integration. The arrows indicate the slope of the continuous function $N(t)$ (solid gray curve) at $t$, $t + 0.5$ and $t + 1$; the dashed line represents the average rate of change between $t$ and $t + 1$.

*Implementation*

Program 8.6, rk4.awk, illustrates the use of the fourth-order Runge-Kutta method to solve numerically the continuous logistic model of population growth. The first 21 lines of this program are identical to those of Program 8.5, euler.awk, which used Euler's method of numerical integration, and they perform a similar function. The values of the Runge-Kutta parameters, $k_1$, $k_2$, $k_3$ and $k_4$ (Equation 8.26), are calculated on lines 29 to 35. Note that lines 29, 31, 33 and 35 use the differential equation form of the continuous logistic model (Equation 8.18). The $k$ parameters are used, in turn, to calculate values for the variables pop_k1, pop_k2 and pop_k3, which are estimates of the population size at intermediate points between time $t$ and time $t + 1$. These variables are employed on line 36 to estimate the change in the population size over the period $t$ to $t + 1$ (Equation 8.25). Finally, the estimated population size at time $t + 1$ is calculated on line 37.

Program 8.6 can be run from the command line as follows:

```
gawk -f rk4.awk -v dt=0.1 > rk4.dat                                    16
```

assuming that $\Delta t = 0.1$. The results are redirected to the file rk4.dat.

*Visualization*

The output from Program 8.6 can be visualized in gnuplot as follows (Figure 8.20):

```
plot pop(t) t "Analytical solution" w l lw 2, \                        32
     'rk4.dat' every 10  t "rk4: dt=0.1" w p pt 7 ps 2                 33
```

Program 8.6: rk4.awk

```
# Continuous-time model of density-dependent (constrained)    1
# population growth, based on the Verhulst (logistic)          2
# equation (see Eq. 22), solved using fourth-order            3
# Runge-Kutta numerical integration.                          4
#                                                             5
# Usage: gawk -f rk4.awk -v dt=value [ > output_file ]        6
#                                                             7
# Variables:                                                  8
# ----------                                                  9
# pop            population at time step                      10
# growth_rate    instantaneous (intrinsic) growth rate        11
# carry_cap      carrying capacity                            12
# time           current time                                 13
# stop_time      time at which simulation should stop         14
# dt             integration step size                        15
#                                                             16
BEGIN{                                                        17
  pop=10.0;                                                   18
  carry_cap=1000;                                             19
  growth_rate=0.5;                                            20
  time=0;                                                     21
  stop_time=10;                                               22
                                                             23
  while(time<=stop_time){                                    24
    printf("%f\t%f\n", time, pop);                           25
                                                             26
    # Fourth-order Runge-Kutta method                        27
    #(after Harris and Stocker, 1998)                        28
    k1=growth_rate*pop*(1-(pop/carry_cap));                  29
    pop_k1=pop+k1*dt/2.0;                                    30
    k2=growth_rate*pop_k1*(1-(pop_k1/carry_cap));           31
    pop_k2=pop+k2*dt/2.0;                                    32
    k3=growth_rate*pop_k2*(1-(pop_k2/carry_cap));           33
    pop_k3=pop+k3*dt;                                        34
    k4=growth_rate*pop_k3*(1-(pop_k3/carry_cap));           35
    pop+=(1.0/6.0)*(k1+2.0*k2+2.0*k3+k4)*dt;                36
    time+=dt;                                                37
  }                                                          38
}                                                            39
```

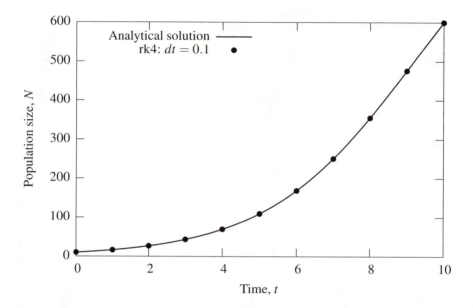

Figure 8.20: Results of the analytical (solid line) and fourth-order Runge-Kutta numerical
(dots) solutions to the continuous logistic model of population growth for $r = 0.05$, $N(0) = 10$ and $dt = 0.1$.

Recall that the function pop(t) on line 32 was defined previously and is the analytical
solution to the continuous logistic model. It is evident that the fourth-order Runge-
Kutta method provides a close approximation to the analytical solution in this case.

## 8.5  INTER-SPECIFIC COMPETITION

### 8.5.1  Conceptual Basis and Mathematical Formulation

The models that have been considered thus far have been concerned with the growth
or decline of a single species taken in isolation. It is extremely rare, however, for just
one species to have exclusive access to the resources of a particular environment; it is
much more common for these resources to be shared, often unequally, between two
or more species. Individuals must therefore compete for the available resources, not
only with members of their own species (intra-specific competition), but also with
members of other species (inter-specific competition). Ultimately, the competitive
interaction between two species produces one of two outcomes: sustained coexistence
of the two species or the demise of the less competitive species (Alstad 2001).

The competition between two species can be represented in simple terms by extending the continuous logistic model of population growth, which accounted for
intra-specific competition (Section 8.3; Equation 8.18), as follows:

$$\frac{dN_1}{dt} = r_1 N_1 \left( \frac{K_1 - (N_1 + N_2)}{K_1} \right) \tag{8.27}$$

where $N_1$ is the population size of species 1, $N_2$ is the population size of species 2, $r_1$ is the intrinsic growth rate of species 1 and $K_1$ is the carrying capacity of the environment for that species. Thus, an increase in the population of species 2 ($N_2$) will decrease the amount of resources that remains available to species 1 because the combined populations of species 1 and 2 ($N_1 + N_2$) are subtracted from $K_1$ in the negative feedback term of Equation 8.27. An equivalent formulation can be derived for species 2:

$$\frac{dN_2}{dt} = r_2 N_2 \left( \frac{K_2 - (N_1 + N_2)}{K_2} \right) \tag{8.28}$$

where $r_2$ is the intrinsic growth rate of species 2 and $K_2$ is the carrying capacity of the environment for that species.

Equation 8.27 and Equation 8.28 suggest that the competition between the two species is simple, direct and brutal. An increase in the population of species 1 means that the resources remaining available to support the growth of species 2 are reduced accordingly. It is highly unusual, however, for two species to make identical demands on an environment; for instance, their diets may differ (Alstad 2001). If the type of food that the two species consume differs only slightly, the competition between them for this resource will be intense. If their diets differ substantially, however, the level of inter-specific competition will be much less pronounced. This can be represented by introducing a scaling factor, or competition coefficient, into Equation 8.27 as follows:

$$\frac{dN_1}{dt} = r_1 N_1 \left( \frac{K_1 - (N_1 + \alpha N_2)}{K_1} \right) \tag{8.29}$$

where $\alpha$ is a factor that describes the effect of competition on a member of species 1 caused by members of species 2 (Roughgarden 1998). For example, if $\alpha = 1$ an extra individual of species 2 has exactly the same effect on species 1 as another member of species 1 (i.e., inter-specific competition is as intense as intra-specific competition); if $\alpha > 1$, an extra individual of species 2 has a greater effect on species 1 than the introduction of another member of species 1 (i.e., inter-specific competition is more intense than intra-specific competition); if $\alpha < 1$, an extra individual of species 2 has a smaller effect on species 1 than the introduction of another member of species 1 (i.e., inter-specific competition is less intense than intra-specific competition); if $\alpha = 0$, there is no inter-specific competition and the two populations grow independently, according to separate continuous logistic equations (Roughgarden 1998).

A similar competition coefficient, $\beta$, can be introduced to Equation 8.28:

$$\frac{dN_2}{dt} = r_2 N_2 \left( \frac{K_2 - (N_2 + \beta N_1)}{K_2} \right) \tag{8.30}$$

where $\beta$ describes the effect of competition on a member of species 2 caused by members of species 1. In most cases, the values of $\alpha$ and $\beta$ differ so that the effect of competition between species 1 and 2 is unequal and not reciprocal (Alstad 2001).

Equation 8.29 and Equation 8.30 are known as the Lotka-Volterra competition equations, named after the two scientists, Alfred Lotka and Vito Volterra, who developed them independently in the 1920s (Lotka 1925, Volterra 1926). Note that there

```
pop_1              10                                          1
pop_2              20                                          2
carry_1            1000                                        3
carry_2            750                                         4
alpha              0.7                                         5
beta               0.6                                         6
growth_1           0.5                                         7
growth_2           0.75                                        8
stop_time          100                                         9
delta_t            0.1                                         10
```

Figure 8.21: Example data file, `params1.dat`, containing the set of parameter values that are required as input to Program 8.7, `compete.awk`.

is no closed-form analytical solution to these differential equations and so numerical integration techniques must be employed.

## 8.5.2   Implementation

Program 8.7 presents an implementation of the Lotka-Volterra competition equations (Equation 8.29 and Equation 8.30) using Euler's method of numerical integration. The first 18 lines of this code are comments, which outline the purpose of the program, explain how it should be run via the command line, and list the primary variables that it employs. Lines 20 to 29 employ a group of pattern-action rules to read the parameter values that are required by the model ($N_1$, $N_2$, $K_1$, $K_2$, $\alpha$, $\beta$, $r_1$, $r_2$, $t$ and $\Delta t$) from an input data file. Given the number of parameter values that have to be specified, this approach is more convenient than entering the values via the command line and more flexible than stating them explicitly within the program itself. An example of the corresponding data file is given in Figure 8.21. Note that the first field of each record in the data file contains the name of the variable in Program 8.7 associated with that parameter; the second field specifies its value. The pattern-action rules read these values from the data file and store them using the appropriate variables.

Lines 31 through 47 of the program contain an END block, which is used here to perform the model simulations (i.e., to generate data) after the parameter values have been read from the input data file. A while loop (lines 33 through 46) is employed inside the END block to model the growth of the two populations over the specified period of time (i.e., while time is less than or equal to stop_time). The current time (time) is initially set to zero (line 32). The first action of the while loop is to print out the current time and the population size for species 1 and 2 (line 34). The current time is then increased by a small amount $\Delta t$ (delta_t; line 36), and the change in the size of each population over this interval of time is calculated using Equation 8.29 and Equation 8.30 (lines 38 to 39 and 41 to 42). These values are then used to calculate the two population sizes at the new point in time (lines 44 and 45). The updated time and population sizes are printed out at the start of the next iteration of the while loop.

Program 8.7: compete.awk

```
# Simple model of population growth with inter-specific    1
# competition described by the Lokta-Volterra equations    2
# (Eqs. 4 and 5), solved using Euler's method of           3
# numerical integration.                                   4
#                                                          5
# Usage: gawk -f compete.awk params.dat [ > outputFile ]   6
#                                                          7
# Variables                                                8
# ------------                                             9
# pop_1, pop_2 Initial population density, species 1 and 2 10
# carry_1      Carrying capacity, species 1               11
# carry_2      Carrying capacity, species 2               12
# alpha, beta  Competition coefficients, species 1 and 2  13
# growth_1     Growth rate, species 1                     14
# growth_2     Growth rate, species 2                     15
# time         Current time                               16
# stop_time    Time period after which simulation should stop 17
# dt           Time step for numerical integration        18
                                                          19
(NR==1){pop_1=$2;}                                        20
(NR==2){pop_2=$2;}                                        21
(NR==3){carry_1=$2;}                                      22
(NR==4){carry_2=$2;}                                      23
(NR==5){alpha=$2;}                                        24
(NR==6){beta=$2;}                                         25
(NR==7){growth_1=$2;}                                     26
(NR==8){growth_2=$2;}                                     27
(NR==9){stop_time=$2;}                                    28
(NR==10){dt=$2;}                                          29
                                                          30
END{                                                      31
    time=0;                                               32
    while(time<=stop_time){                               33
        print time, pop_1, pop_2;                         34
        # Increment time by dt                            35
        time+=dt;                                         36
        # Calculate change in population species 1 (Eq. 4) 37
        delta_pop_1=dt*growth_1*pop_1* \                  38
            ((carry_1-pop_1-(alpha*pop_2))/carry_1);      39
        # Calculate change in population species 2 (Eq. 5) 40
        delta_pop_2=dt*growth_2*pop_2* \                  41
            ((carry_2-pop_2-(beta*pop_1))/carry_2);       42
        # Calculate new population sizes species 1 and 2  43
        pop_1+=delta_pop_1;                               44
        pop_2+=delta_pop_2;                               45
    }                                                     46
}                                                         47
```

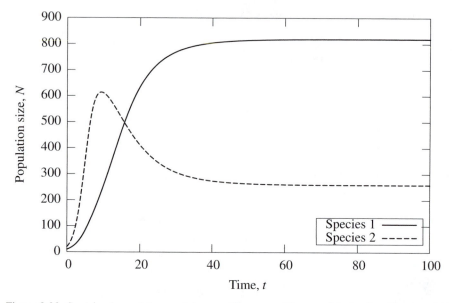

Figure 8.22: Sustained coexistence: inter-specific competition modeled using Program 8.7, compete.awk, for $N_1 = 10$ and $N_2 = 20$ initially, $r_1 = 0.5$, $r_2 = 0.75$, $\alpha = 0.7$, $\beta = 0.6$, $K_1 = 1000$ and $K_2 = 750$.

### 8.5.3  Running the Model

Program 8.7 can be run from the command line as follows:

```
gawk -f compete.awk params1.dat > compete1.dat
```
17

This command assumes that $N_1 = 10$ and $N_2 = 20$ initially, and that $r_1 = 0.5$, $r_2 = 0.75$, $\alpha = 0.7$, $\beta = 0.6$, $K_1 = 1000$ and $K_2 = 750$ (Figure 8.21). Thus, species 2 has the higher intrinsic rate of growth (0.75 as opposed to 0.5 for species 1), but a lower carrying capacity in this environment (750 as opposed to 1000 for species 1). The competition coefficients $\alpha$ and $\beta$ are similar, although the relative impact of inter-specific competition is greater on species 1 than it is on species 2 ($\alpha > \beta$). Finally, the growth of both populations is simulated over a period of 100 units of time and for $\Delta t = 0.1$. The output from the model is redirected to the file compete1.dat.

### 8.5.4  Visualization

The output from Program 8.7 can be visualized by typing the following commands in gnuplot (Figure 8.22):

```
set style data lines                                                    34
set key bottom right box                                                35
plot 'compete1.dat' u 1:2 t "Species 1" lw 2, \                         36
     'compete1.dat' u 1:3 t "Species 2" lw 2                            37
```

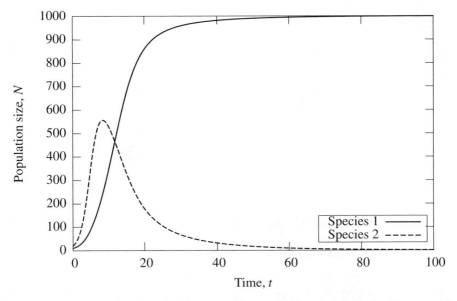

Figure 8.23: Demise of one species: inter-specific competition modeled using Program 8.7, `compete.awk`, for $N_1 = 10$ and $N_2 = 20$ initially, and $r_1 = 0.5$, $r_2 = 0.75$, $\alpha = 0.5$, $\beta = 0.8$, $K_1 = 1000$ and $K_2 = 750$.

These commands assume that the data file `compete1.dat` is located in the working directory; otherwise, its full or relative pathname must be supplied.

Figure 8.22 indicates that the population of species 2 initially grows in number faster than that of species 1. This is because of the higher intrinsic rate of growth of species 2 ($r_2 > r_1$) and the competitive impact that it has on species 1 ($\alpha = 0.7$). Despite this, the population of species 1 continues to grow in size and exceeds that of species 2 after $t = 16$, primarily because of the environment's greater carrying capacity for species 1. The increased competition resulting from the large number of individuals of species 1 causes the population of species 2 to decrease in number. By $t = 60$, however, the two populations settle into a state of sustained coexistence in which $N_1 \approx 819$ and $N_2 \approx 259$. Note that the size of each population under these steady-state conditions is smaller than the carrying capacity of the environment for either species in isolation ($K_1 = 1000$ and $K_2 = 750$).

The Lotka-Volterra equations do not always produce conditions of sustained coexistence. Figure 8.23, for example, shows the results of a further simulation using Program 8.7 in which species 2 eventually becomes extinct as a result of competition with species 1. This simulation was performed using slightly different values of the competition coefficients ($\alpha = 0.5$ and $\beta = 0.8$) compared to the preceding example, but the values of the other variables are unchanged (Figure 8.24).

```
gawk -f compete.awk params2.dat > compete2.dat
```

18

```
pop1            10                                          1
pop2            20                                          2
carry1          1000                                        3
carry2          750                                         4
alpha           0.5                                         5
beta            0.8                                         6
growth1         0.5                                         7
growth2         0.75                                        8
stop_time       100                                         9
delta_t         0.1                                        10
```

Figure 8.24: Second example data file, `params2.dat`, containing the set of parameter values that are required as input to Program 8.7, `compete.awk`.

## 8.6   PREDATOR-PREY RELATIONSHIPS

### 8.6.1   Conceptual Basis and Mathematical Formulation

The final model of population growth examined in this chapter considers the case in which one species preys upon another. All other things being equal, one might expect that the predator species will generally become more numerous as the population of the prey species increases in number because of the greater level of resources available to support the predator population. As the number of predators increases, however, more prey are likely to be consumed, which may lead to a decline in the population of the prey species. This may, in turn, lead to a reduction in the number of predators and eventually, perhaps, to the recovery of the prey species population. There is, thus, a fine balance and interrelationship between the populations of the two species, which is another form of inter-specific competition.

One of the simplest mathematical models of predator-prey relationships is based, once again, on the pioneering work of Lotka (1925) and Volterra (1926), and is given by the following pair of differential equations:

$$\frac{dN_1}{dt} = rN_1 - aN_1N_2 \tag{8.31}$$

and

$$\frac{dN_2}{dt} = abN_1N_2 - dN_2 \tag{8.32}$$

where $N_1$ is the population size of the prey species, $N_2$ is the population size of the predator species, $r$ is the intrinsic rate of growth of the prey species, $d$ is the death rate of the predators, $a$ is a coefficient that describes the rate of success with which predators capture prey and $b$ is a coefficient relating the number of prey consumed by each predator to the number of predator births (Roughgarden 1998). There are several important assumptions that underpin these equations. The first is that, in the absence of predation, the growth of the prey species is unconstrained (density-independent), so that $dN_1/dt = rN_1$ when $N_2 = 0$ (the prey species increases in number exponentially over time when there are no predators; Equation 8.31). The second assumption, which is implicit, is that the predators encounter the prey in a random fashion and that the number of encounters increases in proportion to the number of predators and prey ($N_1 N_2$). Only a fraction of encounters results in a predator capturing its prey, and hence in the death of the prey ($a N_1 N_2$; Equation 8.31). The third assumption is that the predators are never satiated; that is, they consume as many prey as they can capture. Finally, it is assumed that the number of predator births is proportional to the number of prey consumed ($ab N_1 N_2$; Equation 8.32), and that a fixed proportion of the predator population dies at any given point in time ($d N_2$; Equation 8.32).

### 8.6.2 Implementation

Program 8.8, `predprey.awk`, presents an implementation in gawk of the Lotka-Volterra predator-prey model (Equation 8.31 and Equation 8.32). The program employs fourth-order Runge-Kutta numerical integration to solve this pair of differential equations because a closed-form analytical solution does not exist.

The first six lines of the program consist of comments that outline the purpose of the code and that show how it should be run via the command line. The bulk of the code is contained within the BEGIN block (lines 8 to 43) because the intention is to generate data via a simulation model. For the sake of simplicity, the values of the main variables have been hard-wired into the program (lines 9 to 16), although they could equally well have been entered on the command line or via an input data file. A `while` loop is used to simulate the growth of the predator and prey populations over time, continuing while the current time (`time`) is less than or equal to the specified point in time when the simulation should stop (`stop_time`). The first action within the `while` loop is to print out the current time and the sizes of the prey and predator populations (line 19). The current time is then incremented by a small amount, $\Delta t$ (`delta_t`; line 21). The size of each population at the new point in time is recalculated (lines 24, 26, 28, 30, 32, 34, 36 and 40 for the prey species and lines 25, 27, 29, 31, 33, 35, 37 and 41 for the predator species) using fourth-order Runge-Kutta numerical integration to solve Equation 8.31 and Equation 8.32. The updated population sizes are printed out in the next iteration of the `while` loop.

### 8.6.3 Running the Model

Program 8.8 can be run from the command line as follows:

```
gawk -f predprey.awk -v dt=0.01 > predprey.dat
```
*19*

Remember that the values of the main variables are hard-wired in the code and hence do not have to be supplied on the command line. Thus, the population of each species initially consists of 20 individuals ($N_1 = 20$ and $N_2 = 20$), the prey species exhibits a relatively high intrinsic rate of growth ($r = 0.9$) and 60% of the predator population dies in each time step ($d = 0.6$). Moreover, on average the predators are successful in capturing the prey once in every 10 encounters ($a = 0.1$) and produce an offspring for every two prey that they consume ($b = 0.5$). Note that output from the computational model is redirected to the file `predprey.dat` in the working directory.

Program 8.8: predprey.awk

```
# A simple model of predator-prey interaction based          1
# on the Lotka-Volterra equations (see Equations 8.31 and    2
# 8.32), solved using fourth-order Runge-Kutta numerical     3
# integration techniques.                                    4
#                                                            5
# Usage: gawk -f predprey.awk -v dt=value [ > outputFile ]   6
                                                            7
BEGIN{                                                       8
   prey=20.0;           # Population density, prey (N_1)     9
   pred=20.0;           # Population density, predator (N_2) 10
   death=0.6;           # Death rate, predator species (d)   11
   p_coeff=0.1;         # Coefficient of predation (a)       12
   p_effic=0.5;         # Predator efficiency (b)            13
   growth_rate=0.9;     # Growth rate, prey species (r)      14
   time_stop=60;        # Time at which simulation should stop (t)15
   time=0;              # Current time                       16
                                                            17
   while(time<=time_stop){                                   18
      print time, prey, pred;                                19
                                                            20
      time+=dt;                                              21
                                                            22
      # Prey species (Eq. 8.31), predator species (Eq. 8.32) 23
      k1_prey=(growth_rate*prey)-(p_coeff*prey*pred);        24
      k1_pred=(p_effic*p_coeff*prey*pred)-(death*pred);      25
      prey_1=prey+k1_prey*dt/2;                              26
      pred_1=pred+k1_pred*dt/2;                              27
      k2_prey=(growth_rate*prey_1)-(p_coeff*prey_1*pred_1);  28
      k2_pred=(p_effic*p_coeff*prey1*pred_1)-(death*pred_1); 29
      prey_2=prey+k2_prey*dt/2;                              30
      pred_2=pred+k2_pred*dt/2;                              31
      k3_prey=(growth_rate*prey_2)-(p_coeff*prey_2*pred_2);  32
      k3_pred=(p_effic*p_coeff*prey_2*pred_2)-(death*pred_2);33
      prey_3=prey+k3_prey*dt;                                34
      pred_3=pred+k3_pred*dt;                                35
      k4_prey=(growth_rate*prey_3)-(p_coeff*prey_3*pred_3);  36
      k4_pred=(p_effic*p_coeff*prey_3*pred_3)-(death*pred_3);37
                                                            38
      # Calculate revised populations                        39
      prey+=(1/6)*(k1_prey+2*k2_prey+2*k3_prey+k4_prey)*dt;  40
      pred+=(1/6)*(k1_pred+2*k2_pred+2*k3_pred+k4_pred)*dt;  41
   }                                                          42
}                                                            43
```

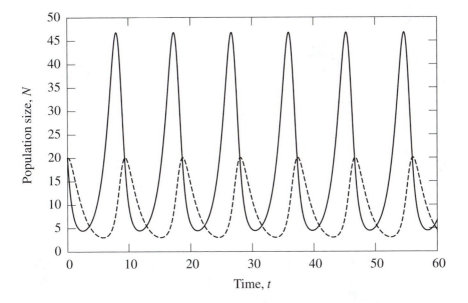

Figure 8.25: Output from the Lotka-Volterra predator-prey equations (solid line = prey species; dashed line = predator species).

### 8.6.4 Visualization

The output produced by Program 8.8, `predprey.awk`, can be visualized (Figure 8.25) by typing the following commands in gnuplot:

```
unset key                                          38
plot 'predprey.dat' u 1:2 lw 2, \                  39
     'predprey.dat' u 1:3 lw 2                      40
```

These commands assume that the file `predprey.dat` is located in the working directory; otherwise, its full or relative pathname must be supplied. Note that this file contains three fields of data: the current time, the population size of the prey species and the population size of the predator species.

Figure 8.25 demonstrates that the Lotka-Volterra predator-prey model produces regular oscillations in the sizes of the predator and prey populations, based on the parameter values used in Program 8.8. The oscillations are slightly out of phase, so that an increase in the size of the prey population is followed, after a short lag, by a corresponding increase in the predator population. As the predator population grows, the number of prey consumed rises and hence the prey population starts to fall. The prey population eventually becomes too small to support the number of predators and, after another short lag, the predator population begins to decrease. The reduced number of predators means that the prey population flourishes once more, and the whole cycle begins again.

The same information can be represented in the form of a phase-plane diagram (Alstad 2001), where the sizes of the prey and predator populations are plotted on the

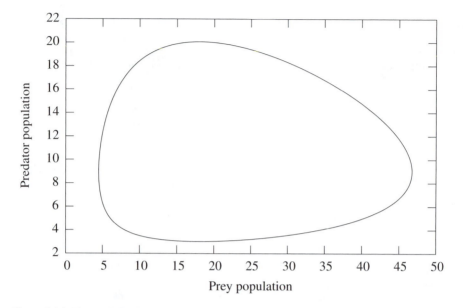

Figure 8.26: Phase-plane diagram showing the hysteresis loop between the sizes of the predator and prey populations predicted by the Lotka-Volterra model (Program 8.8).

$x$- and $y$-axes, respectively. This can be achieved in gnuplot as follows:

```
set xlabel "Prey population"                                    41
set ylabel "Predator population"                               42
plot 'predprey.dat' u 2:3 lw 2                                 43
```

These commands produce Figure 8.26. Notice that the repeated oscillations seen in Figure 8.25 trace a closed loop in Figure 8.26. This is sometimes known as a hysteresis loop because the change in the predator population lags somewhat behind the change in the prey population that causes it. Figure 8.26 and Figure 8.25 also demonstrate that the predator and prey species continue to coexist in this realization of the model, with neither forcing the other into extinction (Roughgarden 1998); other values of the model parameters may, however, produce very different outcomes.

**Exercise 8.4**: Experiment with Program 8.8, changing the values of the main variables ($N_1$, $N_2$, $r$, $d$, $a$ and $b$) in the code. Which conditions lead to coexistence of the predator and the prey populations and which, if any, lead to the demise of one or both of them? Plot the results in gnuplot.

## 8.7  SUMMARY

This chapter examines a range of models of population growth, starting with a simple model of unconstrained (density-independent) growth, progressing through to models of constrained (density-dependent) growth, to ones that represent competition for resources between species (inter-specific competition) and the relationships between predator and prey species. Each is comparatively simple, in the sense that few real populations behave exactly in the way that the models describe. Nevertheless, each provides important insights that help to understand the functioning of environmental systems. Readers who wish to explore population growth models further are encouraged to consult the following sources: Roughgarden (1998), Wilson (2000), Alstad (2001), May (2001), and Donovan and Welden (2002).

In the following chapter, a population growth model, similar to the ones explored here, is used to simulate the growth of two species of daisy on an imaginary planet, known as Daisyworld. The two types of daisy compete with one another for the available resources. Their growth rates are also controlled by the temperature of the planet, which is, in turn, a function of the amount of solar radiation that the planet receives.

## SUPPORTING RESOURCES

The following resources are provided in the `chapter8` sub-directory of the CD-ROM:

| | |
|---|---|
| `discrete.awk` | gawk implementation of the discrete model of unconstrained (density-independent) population growth. |
| `continue.awk` | gawk implementation of the continuous model of unconstrained (density-independent) population growth. |
| `cntlogst.awk` | gawk implementation of the continuous model of constrained (density-dependent) population growth, based on the Verhulst (logistic) equation. |
| `dsclogst.awk` | gawk implementation of the discrete model of constrained (density-dependent) population growth, based on the Verhulst (logistic) equation. |
| `euler.awk` | gawk implementation of the continuous model of constrained (density-dependent) population growth, based on the Verhulst (logistic) equation, solved using Euler's method. |
| `rk4.awk` | gawk implementation of the continuous model of constrained (density-dependent) population growth, based on the Verhulst (logistic) equation, solved using the fourth-order Runge-Kutta method. |
| `compete.awk` | gawk implementation of the continuous model of inter-specific competition, described by the Lotka-Volterra equations, solved using Euler's method. |
| `predprey.awk` | gawk implementation of the continuous model of predator-prey interaction, described by the Lotka-Volterra equations, solved using the fourth-order Runge-Kutta method. |

| | |
|---|---|
| `params1.dat` | Example parameter value file required as input to `compete.awk`. |
| `params2.dat` | Example parameter value file required as input to `compete.awk`. |
| `populate.bat` | Command line instructions used in this chapter. |
| `populate.plt` | gnuplot commands used to generate the figures presented in this chapter. |
| `populate.gp` | gnuplot commands used to generate figures presented in this chapter as EPS files. |

# Chapter 9

# Biospheric Feedback on Daisyworld

**Topics**

- Daisyworld model and the Gaia hypotheses

- Biospheric feedback and steady-state conditions

- Perturbing Daisyworld by increasing solar luminosity

- Exploring the impact of biodiversity

**Methods and techniques**

- Modularizing gawk code with user-defined functions

- Sensitivity analysis

## 9.1   INTRODUCTION

This chapter examines a model of the feedback mechanisms that exist between the biota (the living organisms) and the abiotic environment of an imaginary planet, known as Daisyworld. The model combines many of the elements examined in the preceding four chapters. For instance, it explores the growth in the populations of two species of daisy over time (Chapter 8) as a function of planetary temperature. The latter is, in turn, partly controlled by changes in the amount of solar radiation incident on the planet's surface (Chapter 5). The two species of daisy differ solely with respect to color: one is dark, the other is light, compared to the soil substrate in which they

Table 9.1: Multiple Gaia hypotheses (after Kirchner 2002).

| Hypothesis | Description |
|---|---|
| Influential Gaia | Biota collectively significantly affect planet's abiotic environment |
| Co-evolutionary Gaia | Evolution of biota and abiotic environment are closely coupled |
| Homeostatic Gaia | Biota act to stabilize abiotic environment (negative feedback loops dominate) |
| Geophysiological Gaia | Biosphere operates as a single, giant, organism |
| Optimizing Gaia | Biota optimize the abiotic environment for their own benefit |
| Gaia as a metaphor | — |

grow. Thus, the model is also concerned with the interaction between solar radiation and plant canopies, or, more specifically, the fractions of incident radiation that are reflected or absorbed by the daisies (Chapter 6 and Chapter 7).

The Daisyworld model was developed by Watson and Lovelock (1983) largely in response to criticism of Lovelock's Gaia hypothesis (named after the goddess of the Earth in classical Greek mythology), which suggests that Earth behaves like a self-regulating super-organism in which the biota and abiotic environment interact to maintain conditions that are suitable for, and adapted to, the continued existence of life (Lovelock 1995b). One of the main criticisms of the Gaia hypothesis is that it is teleological; that is to say, it implies that the biota are imbued with foresight or a sense of purpose, which they employ to achieve a specific goal (Doolittle 1981, Dawkins 1982). Lovelock (1995a) strongly contests this assertion and, together with Andrew Watson, he developed the Daisyworld model to show how the biota might regulate their abiotic environment without recourse to foresight or planning, through a combination of positive and negative feedback mechanisms (Saunders 1994, Lovelock 1995a). Despite this, a number of criticisms of the Gaia hypothesis remain, the most serious of which are that it is untestable (Kirchner 1989) and unfalsifiably vague (Kirchner 1990). The second of these two criticisms arises partly because, over time, Lovelock and co-workers have expressed the Gaia hypothesis in a number of different ways (Kirchner 2002). These variants are summarized in Table 9.1.

Setting aside the merits and deficiencies of the Gaia hypothesis (or hypotheses), which it was originally developed to defend, the Daisyworld model has considerable value in its own right as a heuristic tool. While the model does not purport to show how the biotic and abiotic components of Earth's biosphere are actually linked, it suggests a way in which they might interact so that the biota exert a degree of control over their abiotic environment (Hardisty et al. 1993). The model also displays certain properties that may be relevant to the way in which Earth's climate and biosphere have evolved. For these reasons, it is examined in detailed in this chapter.

## 9.2 DESCRIPTION AND ASSUMPTIONS OF THE CONCEPTUAL MODEL

Daisyworld is an imaginary planet illuminated by a distant sun (Watson and Love-lock 1983, Hardisty *et al.* 1993). Its atmosphere is cloudless and contains a negligible quantity of greenhouse gases. The only forms of life on the planet are two species of daisy, which differ solely in terms of color: one species is dark, the other is light, compared to the soil substrate in which they grow. For the sake of convenience, the dark and light species are referred to as "black" and "white" daisies, respectively, hereafter; by implication the soil is "gray". As a consequence of their physical char-acteristics (i.e., their color), the black daisies reflect less of the incident solar radiation than does the soil substrate, and hence they absorb more; in contrast, the white daisies reflect more and absorb less solar radiation than does the soil substrate.

To simplify matters still further, Daisyworld is assumed to be completely flat. This characteristic has an important ramification: the amount of solar radiation that is incident on the planet is uniform across the whole of its surface; by contrast, for a spherical planet, such as Earth, it varies from equator to pole (Watson and Love-lock 1983). It is also implicitly assumed that Daisyworld traces a circular orbit around its sun and that the planet's surface is always normal to the incident solar radiation (i.e., the angle of obliquity is zero). Thus, Daisyworld experiences neither diurnal variation nor seasonal changes in the amount of incident solar radiation (Hardisty *et al.* 1993). Over time, however, Daisyworld's sun gradually becomes brighter (its luminosity increases), which is typical of a main sequence star, such as Earth's sun (Saunders 1994). As a result, there is a concomitant increase in the amount of solar radiation incident on the planet's surface, which has important consequences for the climate of Daisyworld.

A fraction of the planet's surface is suitable for daisy growth. The availability of this fertile land is one of three factors that control the rates at which the daisy populations grow; the other two factors are the fraction of daisies that die during a given period of time and the rate at which new daisies appear (are born) over the same interval of time. The death rate is assumed to be fixed, independent of the population size, and the daisies that die decompose very rapidly to become soil. By contrast, the birth rate varies according to temperature: the optimum temperature for daisy growth is 22.5 °C, but growth can occur at any temperature in the range of 5 °C to 40 °C (22.5 °C ± 17.5 °C). The amount of fertile land available for daisy growth is, of course, limited; hence, there is both intra-specific and inter-specific competition among the black and white daisies for this resource.

The temperature of Daisyworld as a whole (i.e., the average global temperature) is dependent on the amount of incident solar radiation and the albedo of the planet surface. The latter is, in turn, the sum of the albedo values of the various materials that cover the planet's surface (black daisies, white daisies and bare soil) weighted by the fraction of ground covered by each material type (its relative areal extent). The surface temperature differs somewhat from one location to another, however, due to local differences in the planet's albedo. Thus, the temperature is lower than the global average over patches of white daisies because they reflect more and, hence, absorb less of the incident solar radiation than either the soil substrate or the black daisies.

Similarly, the temperature is higher than the global average over patches of black daisies because they reflect less and, hence, absorb more of the incoming radiation than either the soil substrate or the white daisies.

## 9.3   FORMULATING THE MATHEMATICAL MODEL

The processes and interactions that are described verbally in the preceding section are represented graphically in Figure 9.1. This diagram provides a useful overview of how Daisyworld functions as a system; it highlights the various feedback loops, which cause the system to behave in a complex non-linear fashion. The next step is to develop a mathematical model of the system. This model is formulated below.

It is convenient to start by representing the amount of solar energy incident on Daisyworld. As noted previously, this quantity varies with time and is independent of the other state variables in the system (it is an exogenous variable). Consequently, it is drawn outside the main system box in Figure 9.1, which is indicated by the dashed line. Watson and Lovelock (1983) express the amount of solar energy incident on the surface of Daisyworld as the product of the solar constant, $S$ (see also Section 5.4.1), and a factor, $L$, which describes the relative luminosity (or brightness) of the sun. They give the value of $S$ as $9.17 \times 10^5$ erg·cm$^{-2}$·s$^{-1}$, which is equivalent to $917$ W·m$^{-2}$. The luminosity factor of Daisyworld's sun would therefore need to be about 1.5 to produce the same level of exo-atmospheric irradiance as Earth's sun ($S \times L = 917$ W·m$^{-2} \times 1.5 \approx 1380$ W·m$^{-2}$; see also Equation 5.7 on page 123).

The amount of solar radiation incident on Daisyworld is critically important to the system as a whole because of the direct effect that it has on the average temperature of the planet, $T_{global}$. $T_{global}$ is also dependent on the fraction of incident solar radiation that is reflected from the planet surface and, hence, the fraction that it absorbs: the greater the fraction that is reflected, the less is absorbed; the smaller the fraction that is absorbed, the less the planet surface heats up. The fraction of incident solar radiation that is reflected by the planet as a whole is given by the global albedo, $A_{global}$. By implication, the fraction of incident solar radiation that is absorbed by Daisyworld is $1 - A_{global}$ (i.e., everything except that which is reflected). Thus, the total amount of radiation absorbed by Daisyworld is $S \times L \times (1 - A_{global})$.

Assuming that Daisyworld behaves like a blackbody radiator, the amount of solar energy that it absorbs and the amount that it emits must be equal. Consequently, the Stefan-Boltzmann equation, which was introduced in Chapter 5 (Equation 5.3 on page 122), can be used to estimate the average temperature of the planet. Remember that the Stefan-Boltzmann equation relates the energy emitted by a blackbody radiator to its temperature, as follows:

$$M = \sigma T^4 \tag{9.1}$$

where $M$ is the radiation emitted by the object (W·m$^{-2}$), $T$ is its temperature in kelvin ($K$) and $\sigma$ is the Stefan-Boltzmann constant ($5.67 \times 10^{-8}$ W·m$^{-2}$·K). Equation 9.1 can be rearranged to express the temperature of a blackbody in terms of the amount of

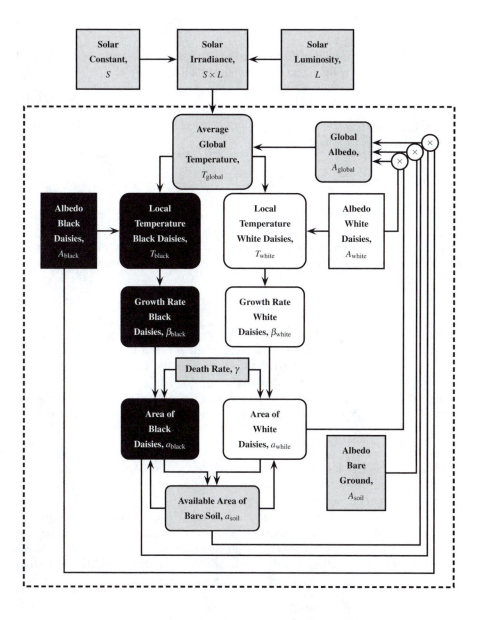

Figure 9.1: Diagrammatic representation of the Daisyworld model.

energy that it emits, as follows:

$$T = \left(\frac{M}{\sigma}\right)^{0.25} \tag{9.2}$$

It has already been shown that the amount of solar radiation absorbed by Daisyworld and hence, as a blackbody radiator, the amount that it emits is

$$M = S \times L \times (1 - A_{global}) \tag{9.3}$$

Substituting Equation 9.3 into Equation 9.2, and subtracting 273.2 to convert from kelvin into degrees Celsius, gives

$$T_{global} = \left(\frac{SL(1 - A_{global})}{\sigma}\right)^{0.25} - 273.2 \tag{9.4}$$

which describes the average global temperature of Daisyworld.

As noted earlier, the average global albedo of Daisyworld, $A_{global}$, varies as a function of the fraction of the planet surface that is covered by black daisies, by white daises and by bare soil, as well as their respective albedo values. This can be expressed as follows:

$$A_{global} = (a_{soil}A_{soil}) + (a_{black}A_{black}) + (a_{white}A_{white}) \tag{9.5}$$

where $a_{soil}$, $a_{black}$ and $a_{white}$ are the fractional areas of the planet surface covered by bare soil, black daisies and white daisies, respectively, and where $A_{soil}$, $A_{black}$ and $A_{white}$ are the corresponding albedo values. Watson and Lovelock (1983) use the following initial values for these variables: $A_{soil} = 0.5$, $A_{black} = 0.25$, $A_{white} = 0.75$, $a_{black} = 0.01$ and $a_{white} = 0.01$. Finally, the fraction of the planet surface that is not covered by daisies, but which is suitable for daisy growth (i.e., fertile bare soil), is

$$a_{soil} = a_{suit} - (a_{black} + a_{white}) \tag{9.6}$$

where $a_{suit}$ is the fraction of the planet surface that is suitable for daisy growth. In this chapter, it is assumed that $a_{suit} = 1.0$. If a value less than one is used for $a_{suit}$, Equation 9.5 must be modified accordingly.

Watson and Lovelock (1983) suggest that the local temperature of an area covered by either black daises or white daisies can be expressed as a function of the difference between the local albedo and the average global albedo, which causes local variations in the rate of surface heating, and the rate at which heat energy is redistributed from warmer to cooler areas. The relevant formulation is as follows:

$$T_{black} = \left(q'(A_{global} - A_{black}) + T_{global}\right) \tag{9.7}$$

$$T_{white} = \left(q'(A_{global} - A_{white}) + T_{global}\right) \tag{9.8}$$

where $T_{black}$ and $T_{white}$ are the local temperatures over areas of black daisies and white daisies, respectively, and $q'$ is a factor that describes the redistribution (i.e., the

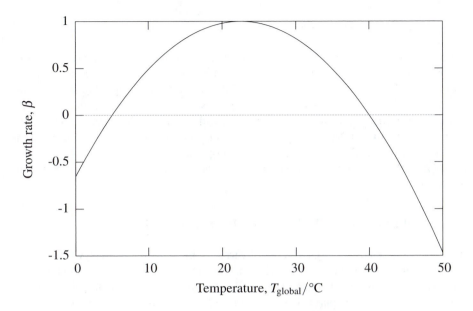

Figure 9.2: Parabolic relationship between daisy growth rate and local temperature.

conduction) of thermal energy from warmer to cooler areas. Thus, if $q' = 0$, the local temperatures are always equal to the global average temperature because there is perfect conduction of thermal energy from warmer locations to cooler ones. Watson and Lovelock (1983) suggest that the value of $q'$ should be less than $0.2SL/\sigma$, and they employ $q' = 20$ in their own simulations.

The intrinsic (instantaneous) growth rates of the black daisies and the white daisies are assumed to be simple parabolic functions of local temperature, as follows:

$$\beta_{\text{black}} = 1 - 0.003265 \left(22.5 - T_{\text{black}}\right)^2 \tag{9.9}$$

$$\beta_{\text{white}} = 1 - 0.003265 \left(22.5 - T_{\text{white}}\right)^2 \tag{9.10}$$

where $\beta_{\text{black}}$ and $\beta_{\text{white}}$ are the growth rates for the black daisies and the white daisies, respectively (Figure 9.2). The numerical constants (1, 0.003265 and 22.5) used in these equations are such that daisy growth occurs when the local temperature lies in the range of 5 °C to 40 °C, and is maximized when the local temperature is 22.5 °C. Outside this range, Equation 9.9 and Equation 9.10 result in a negative growth rate; that is, a decline in the population for that species of daisy.

As the populations of the two species of daisy grow (or decline) over time, the fractional area of the planet surface that each occupies changes. This is expressed by the following pair of differential equations:

$$\frac{da_{\text{black}}}{dt} = \left(a_{\text{black}} \left(a_{\text{soil}} \beta_{\text{black}} - \gamma\right)\right) \tag{9.11}$$

$$\frac{da_{\text{white}}}{dt} = (a_{\text{white}} (a_{\text{soil}}\beta_{\text{white}} - \gamma)) \tag{9.12}$$

where $\gamma$ is the death rate for both types of daisy. These equations encapsulate both inter-specific and intra-specific competition. Thus, the growth rate for each species is multiplied by the amount of fertile land that remains unoccupied ($a_{\text{soil}}\beta_{\text{black}}$ and $a_{\text{soil}}\beta_{\text{white}}$), for which daisies from each species must compete with daisies of the same species (inter-specific competition) and with daisies of the other species (intra-specific competition). Note, however, that the death rate, $\gamma$, is assumed to be identical for both species and is independent of the population size. Watson and Lovelock (1983) employ $\gamma = 0.3$ in their simulations; that is, 30% of the population dies in each time step.

## 9.4   IMPLEMENTING THE COMPUTATIONAL MODEL

In view of the relatively large number of variables involved in the Daisyworld model, the first step taken here is to summarize each of the variables in a table, together with their corresponding mathematical symbols, the names used to identify them in the gawk code and, where appropriate, their units of measurement (Table 9.2).

Four separate implementations of the computational model are examined below. In the first, the luminosity, $L$, of Daisyworld's sun is held constant through time. The intention is to examine how Daisyworld responds to a fixed level of external forcing and, in particular, to explore the steady-state behavior of the system under these conditions. In the second implementation, the solar luminosity is increased in steps of 0.025 from $L = 0.5$ to $L = 1.7$. At each step, the system is allowed to run until it reaches steady state before proceeding to the next value of $L$. The objective of this implementation is to examine how the system responds to a gradual change in external forcing. In the third implementation, an extra species of daisy is added to the model to investigate the effect of increasing biodiversity on Daisyworld. Finally, the fourth implementation replaces elements of the code from the third implementation with user-defined functions. The intention here is to show how the code can be made modular so that, in this particular example, the model can be extended to simulate the effect of having a very large number of daisy species, each with a different albedo.

### 9.4.1   Implementation 1: Constant Solar Luminosity

*Implementation*

Program 9.1, `daisy1.awk`, is an implementation of the Daisyworld model for a fixed (user-defined) value of solar luminosity. The code is contained entirely within the BEGIN block (lines 1 to 46) because the intention is to generate data from the model. Inside the BEGIN block, lines 4 to 13 initialize the values of the main variables used in the program, rather than requiring them to be entered via the command line. Thus, the fraction of the planet's surface that is initially covered by black daisies is set to 0.01 (i.e., 1% of the planet surface; line 4); the same value is used to initialize the fractional area covered by white daisies (line 5). These values provide a starting point from

Table 9.2: List of parameters and variables in the Daisyworld model and their implementation in the corresponding gawk code (Program 9.1).

| Entity | Symbol | Variable name | Units |
|---|---|---|---|
| Emitted flux density | $M$ | – | $\text{W}\cdot\text{m}^{-2}$ |
| Stefan-Boltzmann constant | $\sigma$ | Stefan | $\text{W}\cdot\text{m}^{-2}\cdot\text{K}^{-4}$ |
| Solar constant | $S$ | solar_const | $\text{W}\cdot\text{m}^{-2}$ |
| Solar luminosity | $L$ | luminosity | – |
| Global albedo | $A_{\text{global}}$ | global_albedo | – |
| Soil albedo | $A_{\text{soil}}$ | albedo_soil | – |
| Albedo of black daisies | $A_{\text{black}}$ | albedo_black | – |
| Albedo of white daisies | $A_{\text{white}}$ | albedo_white | – |
| Fraction of planet suitable for daisy growth | $a_{\text{suit}}$ | area_suit | – |
| Fraction of planet covered by bare soil | $a_{\text{soil}}$ | area_soil | – |
| Fraction of planet covered by black daisies | $a_{\text{black}}$ | area_black | – |
| Fraction of planet covered by white daisies | $a_{\text{white}}$ | area_white | – |
| Global temperature (average) | $T_{\text{global}}$ | global_temp | °C |
| Local temperature (black daisies) | $T_{\text{black}}$ | temp_black | °C |
| Local temperature (white daisies) | $T_{\text{white}}$ | temp_white | °C |
| Energy diffusion factor | $q'$ | qfactor | – |
| Growth rate of black daisies | $\beta_{\text{black}}$ | growth_black | – |
| Growth rate of white daisies | $\beta_{\text{white}}$ | growth_white | – |

Program 9.1: daisy1.awk

```
BEGIN {                                                              1
                                                                    2
  # Initialize main variables                                      3
  area_black    = 0.01;                                            4
  area_white    = 0.01;                                            5
  area_suit     = 1.0;                                             6
  solar_const   = 917.0;                                           7
  Stefan_Boltz  = 5.67E-08;                                        8
  albedo_soil   = 0.5;                                             9
  albedo_black  = 0.25;                                            10
  albedo_white  = 0.75;                                            11
  death_rate    = 0.3;                                             12
  q_factor      = 20.0;                                            13
                                                                    14
  # Run model for 100 time steps                                  15
  for(time=0;time<=100;++time){                                    16
    # Equation 9.6                                                 17
    area_soil=area_suit-(area_black+area_white);                   18
    # Equation 9.5                                                 19
    albedo_global=(area_soil*albedo_soil)+ \                       20
      (area_black*albedo_black)+(area_white*albedo_white);         21
    # Equation 9.4                                                 22
    temp_global=(((solar_const*luminosity* \                       23
      (1-albedo_global))/Stefan_Boltz)**0.25)-273;                 24
    # Equations 9.7 and 9.8                                        25
    temp_black=((q_factor*(albedo_global-albedo_black))+ \         26
      temp_global);                                                27
    temp_white=((q_factor*(albedo_global-albedo_white))+ \        28
      temp_global);                                                29
    # Print current time, global temperature and daisy areas      30
    printf("%3i %7.4f %6.4f %6.4f\n", \                            31
      time, temp_global, area_black, area_white);                  32
    # Equations 9.9 and 9.10                                       33
    growth_black=(1-(0.003265*((22.5-temp_black)**2)));            34
    growth_white=(1-(0.003265*((22.5-temp_white)**2)));            35
    # Update daisy area for next time step                         36
    # Equations 9.11 and 9.12                                      37
    area_black+=area_black*((area_soil*growth_black)- \           38
      death_rate);                                                 39
    area_white+=area_white*((area_soil*growth_white)- \           40
      death_rate);                                                 41
    # Do not allow daisies to become extinct                      42
    if(area_black<0.01){area_black=0.01};                          43
    if(area_white<0.01){area_white=0.01};                          44
  }                                                                45
                                                                    46
}                                                                  47
```

which the daisies can grow. In this implementation, the entire planet is assumed to be suitable for daisy growth (line 6). The solar constant is given as $917\,\text{W·m}^{-2}$ (line 7) and the value of the Stefan-Boltzmann constant ($\sigma$; Equation 9.1) is set to $5.67 \times 10^{-8}$ (line 8); note that the latter could have been coded as `Stefan_Boltz = 0.0000000567`, but it is more convenient when dealing with very large or very small numbers, such as this, to use the equivalent scientific notation (`Stefan_Boltz = 5.67E-8`). The albedo values of the soil, black daisies and white daisies are initialized on lines 9 through 11; the values employed here are the same as those used by Watson and Lovelock (1983). Thus, the soil reflects half of the incident solar radiation (and, hence, absorbs the other half), while the black daisies and white daisies reflect one quarter and three quarters, respectively. This means that the black daisies are not perfectly black. Instead, they are a dark shade of gray. Similarly, the white daisies are not perfectly white, but are a pale shade of gray. Finally, the death rate of the daisies, $\gamma$, is set to 0.3 (line 12) and the heat diffusion factor, $q'$, is fixed at 20.0 (line 13), in common with Watson and Lovelock (1983).

The bulk of the code is contained within a `for` loop (lines 16 to 45). This is used to run the model for 101 iterations; each iteration representing a discrete time-step ($t = 0$ to $t = 100$ inclusive). This is intended to allow the model to reach steady-state conditions, if they exist. The lines of code inside the `for` loop implement the mathematical equations presented in the preceding section. Thus, for example, line 18 corresponds to Equation 9.6. In many cases, the gawk statements are quite long; they have therefore been split over two lines of code using the line continuation symbol (\), and the continuation line is indented by one tab stop. For instance, lines 20 and 21 represent a single gawk statement, implementing Equation 9.5.

In general, the translation from mathematical equations to gawk code is fairly straightforward and self-evident. There are, however, two particular points that are worth noting. First, in Equation 9.4 a number is raised to the power of 0.25; this is implemented in Program 9.1 using one of gawk's exponentiation symbols, `**` (line 24). Second, the multiplication operator, which is implicit in the mathematical equations (the term $a_{soil}A_{soil}$ in Equation 9.6), must be made explicit in the gawk code (`area_soil*albedo_soil`; line 20).

The order in which the equations are implemented in Program 9.1 is different from the sequence in which they were introduced in the previous section. The reason for this is that the result of one equation typically forms the input to another. Thus, the average global albedo (`albedo_global`; lines 20 and 21) cannot be established until the area available for daisy growth (`area_soil`) has been determined (line 18). Similarly, the average global temperature (lines 23 and 24) can only be calculated once the average global albedo is known. The average global temperature is, in turn, required to calculate the local temperatures over the patches of black daisies and white daisies (lines 26 to 29).

A `printf` statement is used on lines 31 and 32 of the program to print out the average global temperature, and the fractional areas covered by the black daisies and the white daisies, at the current point in time. The growth rates of the black daisies and the white daisies are calculated on lines 34 and 35, respectively, based on the corresponding local temperature values. The rates of growth are used, in turn, to

```
 0  26.8742  0.0100  0.0100
 1  26.7692  0.0140  0.0168
 2  26.5566  0.0195  0.0280
 3  26.1617  0.0273  0.0462
 4  25.4810  0.0380  0.0749
 5  24.4002  0.0530  0.1182
 6  22.8705  0.0735  0.1782
 7  21.0655  0.1002  0.2507
 8  19.4890  0.1312  0.3211
 9  18.6609  0.1617  0.3720
10  18.5866  0.1874  0.3995
11  18.9154  0.2074  0.4115
12  19.3701  0.2227  0.4155
13  19.8215  0.2344  0.4160
14  20.2216  0.2432  0.4149
```

Figure 9.3: First 15 lines of the file `daisy1.dat`.

determine the change in the fractional area of the planet's surface covered by each species of daisy (lines 38 to 41), and hence to calculate revised values of the relevant variables (`area_black` and `area_white`) for the start of the next time step (the next iteration around the `for` loop). Finally, lines 42 and 43 of the code ensure that neither species of daisy becomes extinct. Thus, the fractional area covered by either species is reset to 0.01 (1% of the planet's surface) if its actual area falls below this value. These lines of code do not have direct counterparts in the mathematical model, but they provide a simple constraint, suggested by Watson and Lovelock (1983), on the dynamics of the Daisyworld system. They imply that there are always small refugia within which the daisy species can survive, even under the harshest conditions.

*Evaluation*

Program 9.1 can be used to examine the behavior of Daisyworld under conditions of constant solar luminosity, as follows:

```
gawk -f daisy1.awk -v luminosity=1.0 > daisy1.dat
```
                                                                                    *1*

In this example, the variable that records the solar luminosity is assigned a value of 1.0 on the command line and the output from the model is redirected to the file `daisy1.dat` in the working directory (Figure 9.3). The model can be run using other values of solar luminosity by changing the above command line accordingly.

*Visualization*

The data produced by `daisy1.awk` (Program 9.1) can be visualized by typing the following commands in gnuplot, assuming that the file `daisy1.dat` is located in the working directory:

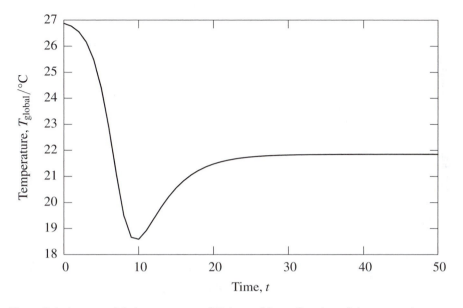

Figure 9.4: Average global temperature of Daisyworld as a function of time, assuming a constant solar luminosity ($L = 1.0$).

```
reset                                                        1
set style data lines                                         2
set ylabel "Temperature, T_global/deg C"                     3
set xlabel "Time, t"                                         4
unset key                                                    5
set xrange [0:50]                                            6
plot 'daisy1.dat' u 1:2 lw 2                                 7
```

These commands, which should be familiar from the preceding chapters, produce a plot of the average global temperature of Daisyworld as a function of time, under conditions of constant solar luminosity ($L = 1.0$). The resulting plot (Figure 9.4) indicates that the temperature of the planet starts at a little less than 27 °C, declines to about 18.5 °C by $t = 10$, and ultimately reaches a steady state of just under 22 °C (i.e., very close to the optimum temperature for daisy growth) after $t = 30$.

To understand why the planet behaves in this way, it is instructive to examine how the fractional area covered by each type of daisy varies as a function of time. This can be explored in gnuplot by entering the following commands:

```
set ylabel "Fractional area, a"                              8
set format y "%4.2f"                                         9
set key bottom right box                                     10
plot 'daisy1.dat' u 1:3 t "Black daisies" lw 2, \            11
     'daisy1.dat' u 1:4 t "White daisies" lw 2               12
```

Note that line 9, above, specifies the format used for the tic-mark labels on the $y$-axis; that is, four digit floating-point numbers reported to two decimal places (e.g., 0.25).

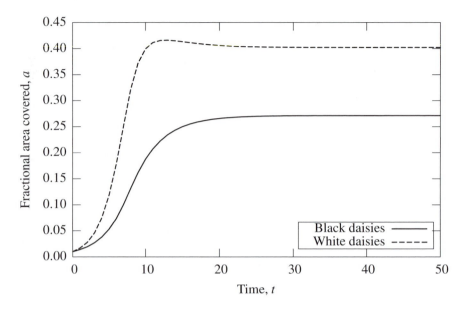

Figure 9.5: Fractional area of Daisyworld covered by black daisies and white daisies as a function of time, assuming a constant solar luminosity ($L = 1.0$).

The resulting plot indicates that both species of daisy increase in area very rapidly during the first 10 or so time steps (Figure 9.5). Initially, the white daisies grow faster because they reflect more of the incident solar radiation; consequently, the local temperature over these patches is cooler than over the rest of the planet, and hence it is closer to the optimum for daisy growth. As the fractional area covered by white daisies increases, the global average temperature of Daisyworld decreases (i.e., the white daisies exert a negative feedback effect on the temperature of the planet). As the planet becomes cooler, the growth rate of the black daisies increases and the fractional area covered by this species increases correspondingly. This has the effect of warming the planet because the black daisies absorb more of the incident solar radiation (i.e., the black daisies exert a positive feedback effect on the temperature of the planet). By about $t = 30$, the relative warming effect of the black daisies and the relative cooling effect of the white daisies counterbalance one another, so that the average global temperature and the fractional areas covered by the black daisies and white daisies stabilize with respect to time (i.e., the planet enters a steady state).

Program 9.1 can be run using different values of solar luminosity to demonstrate that steady-state behavior is the norm, rather than the exception, on Daisyworld (see Figure 9.6). Note that the average temperature of the planet in steady state decreases as the solar luminosity increases over the range of the values considered here.

**Exercise 9.1**: Show how the fractional areas of black daisies and white daisies vary as a function of time for $L = 0.8, 0.9, 1.0, 1.1$ and $1.2$.

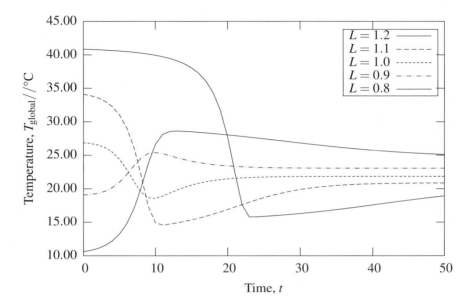

Figure 9.6: Global average temperature versus time for $L = 0.8, 0.9, 1.0, 1.1$ and $1.2$.

### 9.4.2 Implementation 2: Increasing Solar Luminosity

The main focus of the paper by Watson and Lovelock (1983) is the way in which Daisyworld responds to a gradual increase in solar luminosity as a function of time. This behavior mimics that of main sequence stars, such as Earth's sun, which tend to increase in brightness with age (Saunders 1994), and it has important consequences for the climate of Daisyworld. Note that the planet is allowed to reach steady state each time, before the luminosity is increased again (Watson and Lovelock 1983).

*Implementation*

Program 9.2, `daisy2.awk`, is an implementation of the Daisyworld model in which the solar luminosity, $L$, is gradually increased in steps of 0.025 from 0.5 to 1.7. As in the previous implementation, the code is contained entirely within the BEGIN block (lines 1 to 46), because the intention is to generate data (i.e., to perform a simulation) using the model. For the sake of convenience, lines 3 to 13 initialize the values of the main variables rather than requiring them to be entered via the command line. An additional variable, `threshold`, is initialized on line 13, the purpose of which is outlined below. A `for` loop (lines 15 to 45) is used to run the model for several different values of solar luminosity. Inside this outer loop, a `do-while` loop (lines 18 to 42) is used to run the model through time until the system approaches steady state for the current value of solar luminosity. In this example, steady state is deemed to have been reached when the global average temperature of Daisyworld changes by less than a user-specified amount (line 42) between successive iterations of the `do while` loop; here, the threshold value is set to 0.01 °C (line 13). The absolute temperature

Program 9.2: daisy2.awk

```
BEGIN {                                                              1
  # Initialize main variables                                       2
  area_black    = 0.01;                                             3
  area_white    = 0.01;                                             4
  area_suit     = 1.0;                                              5
  solar_const   = 917;                                              6
  Stefan_Boltz  = 5.67E-08;                                         7
  albedo_soil   = 0.5;                                              8
  albedo_black  = 0.25;                                             9
  albedo_white  = 0.75;                                            10
  death_rate    = 0.3;                                             11
  q_factor      = 20;                                              12
  threshold     = 0.01;                                            13
                                                                   14
  for(luminosity=0.5;luminosity<=1.7;luminosity+=0.025){           15
    # Run model until it reaches steady state for                  16
    # current solar luminosity                                     17
    do{                                                            18
      prev_temp_global=temp_global;                                19
      area_avail=area_suit-(area_black+area_white);                20
      albedo_global=(area_avail*albedo_soil)+ \                    21
        (area_black*albedo_black)+(area_white*albedo_white);       22
      temp_global=(((solar_const*luminosity* \                     23
        (1-albedo_global))/Stefan_Boltz)**0.25)-273;               24
      temp_black=((q_factor*(albedo_global-albedo_black))+ \       25
        temp_global);                                              26
      temp_white=((q_factor*(albedo_global-albedo_white))+ \       27
        temp_global);                                              28
      growth_black=(1-(0.003265*((22.5-temp_black)**2)));          29
      growth_white=(1-(0.003265*((22.5-temp_white)**2)));          30
      area_black+=area_black*((area_avail*growth_black)- \         31
        death_rate);                                               32
      area_white+=area_white*((area_avail*growth_white)- \         33
        death_rate);                                               34
      # Do not allow daisies to become extinct                     35
      if(area_black<0.01){area_black=0.01};                        36
      if(area_white<0.01){area_white=0.01};                        37
      # Check whether global average temperature for current       38
      # and previous model run differ by more than threshold       39
      temp_difference=temp_global-prev_temp_global                 40
      if(temp_difference<=0){temp_difference*=-1}                  41
    } while(temp_difference>threshold)                             42
    printf("%5.3f %7.4f %6.4f %6.4f\n", \                          43
      luminosity, temp_global, area_black, area_white);            44
  }                                                                45
}                                                                  46
```

difference between successive iterations of the do while loop is calculated on lines 40 and 41; the temperature of the previous iteration is recorded on line 19. Further iterations are performed for the current solar luminosity value while the absolute temperature difference remains greater than the threshold value (line 42). Once the temperature difference between two successive iterations is less than or equal to the threshold value, the program exits the do while loop and prints out the current values of the solar luminosity, global average temperature and the fractional area of the planet surface covered by black daisies and by white daisies, respectively (lines 43 and 44). The program then returns to the for loop, incrementing the solar luminosity by 0.025, before entering the do while loop again.

## Evaluation

Program 9.2 can be run from the command line as follows:

```
gawk -f daisy2.awk > daisy2.dat
```
2

Note that the output from the model is redirected to the file daisy2.dat in the working directory. It is also instructive to run the same model a second time without daisies, to examine the effect that the biota have on the evolution of Daisyworld's temperature as a function of increasing solar luminosity. This can be achieved most simply by commenting out (i.e., placing a # symbol at the start of) lines 3, 4, 36 and 37 in Program 9.2, which has the effect of setting the area covered by each species of daisy to zero. The revised program can then be saved in a new file, nolife.awk, and run as follows:

```
gawk -f nolife.awk > nolife.dat
```
3

## Visualization

The two data sets produced above, daisy2.dat and nolife.dat, can be visualized together by typing the following commands in gnuplot:

```
set format y "%g"                                        13
set xrange [0.5:1.7]                                     14
set xzeroaxis                                            15
set key top left Left box                                16
set ylabel "Temperature, T_global/deg C"                17
set xlabel "Solar luminosity, L"                         18
plot 'nolife.dat' u 1:2 t "No biota" lw 2, \             19
     'daisy2.dat' u 1:2 t "With biota" lw 2              20
```

These commands assume that both data files are located in the working directory; otherwise, their full or relative pathnames must be provided. Note that line 13 sets the format of the labels placed beside the tic-marks on the $y$-axis to standard scientific notation, line 14 restricts the values displayed on the $x$-axis (i.e., solar luminosity) to the range 0.5 to 1.7, and line 15 draws a line at $y = 0$ marking the $x$-axis. This set of commands produces the plot shown in Figure 9.7.

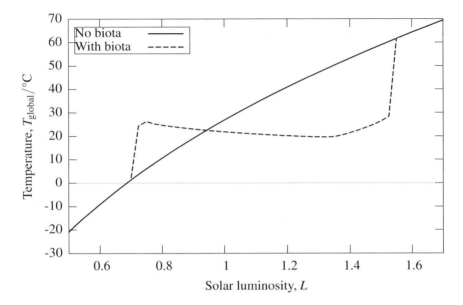

Figure 9.7: Global average temperature of Daisyworld, with and without biota, as a function of solar luminosity.

Figure 9.7 demonstrates the behavior of the Daisyworld model with and without biota (daisies). In the absence of biota (solid line), the temperature of the planet rises monotonically with increasing solar luminosity. With the biota present in the model, the planet exhibits a different response (dashed line). The daisies begin to exert an influence on the climate of Daisyworld when the solar luminosity rises above 0.7. At this point, the daisies cause the planet to warm much more rapidly than is the case when the biota are absent, and the global average temperature rises to a peak of just over 26 °C at $L = 0.75$. Subsequent increases in solar luminosity up to $L = 1.325$ are accompanied by a gradual reduction in the global average temperature; that is, the presence of daisies causes the planet to cool slightly. Thus, the effect of the daisies is to stabilize the temperature of the planet over a wide range of solar luminosity values. When the solar luminosity exceeds 1.325, however, the temperature of Daisyworld starts to increase gradually once more until $L = 1.55$ when it rises very suddenly, eventually merging with the temperature curve for the planet without life.

To understand why Daisyworld behaves in this way, it is instructive to examine how the fractional area of the planet's surface covered by each type of daisy varies as a function of solar luminosity. This can be achieved by entering the following commands in gnuplot, assuming that `daisy2.dat` is located in the working directory:

```
unset key                                                        21
set ylabel "Fractional area covered, a"                          22
plot 'daisy2.dat' u 1:3 lw 2, \                                  23
     'daisy2.dat' u 1:4 lw 2                                      24
```

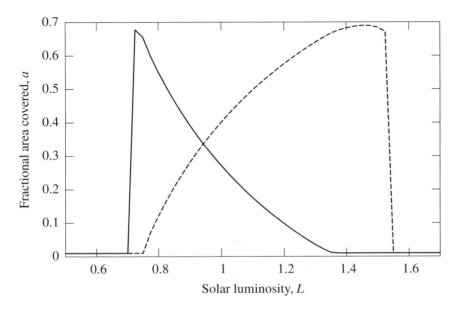

Figure 9.8: Solar luminosity versus the fractional area of Daisyworld covered by black (solid line) and white (dashed line) daisies.

The resulting plot (Figure 9.8) indicates that the black daisies grow very rapidly at first, covering approximately two-thirds of the planet's surface by $L = 0.725$; they then gradually die back, declining almost to zero by $L = 1.375$. The white daisies, by contrast, grow more slowly at first, eventually reaching a peak at $L = 1.45$, before dying back very rapidly by $L = 1.55$.

The results obtained using this implementation of the Daisyworld model can be interpreted as follows. Initially, when Daisyworld's sun is very weak (i.e., the solar luminosity is low; $L \leq 0.7$), the planet is cold and neither species of daisy is able to grow. Gradually, though, as the solar luminosity increases, the temperature of the planet rises to the point where daisy growth commences. At this stage, Daisyworld's climate favors the growth of the black daisies. This is because they absorb more of the incident solar radiation and, as a result, the local temperature over the patches of black daisies is somewhat higher than the global average. The temperature of the planet as a whole, however, remains too low to permit the growth of white daisies and, in the absence of inter-specific competition, the black daisies expand to cover the majority of the planet's surface. In doing so, they raise the temperature of the planet to the point at which the white daisies start to grow and, beyond this, the black daisies begin to die back. The white daisies grow slowly at first, partly because they have to compete with the black daisies for the limited amount of fertile land that remains unoccupied. As the area covered by white daisies increases so does the planetary albedo. This has the effect of cooling the planet slightly because a greater fraction of the incident solar radiation is reflected back out to space. For a while the two species of daisy co-exist. Moreover, the delicate balance between the relative warming effect of the black daisies and the corresponding cooling effect of the white ones causes the temperature of the planet to stabilize, despite the continuing increases in solar luminosity. Watson and Lovelock (1983) describe this situation as one of biological homeostasis. Eventually, though, the increasing solar luminosity causes the temperature of Daisyworld to rise above the maximum that the daisies can tolerate, and the populations of both species crash.

> **Exercise 9.2**: Modify Program 9.2, daisy2.awk, so that there is only one species of daisy present (black or white). Save the file as daisy2x.awk. Run the revised program from the command line. Create plots of the global average temperature of Daisyworld and the fractional area covered by daisies as a function of solar luminosity. Compare the results with those of the two-species model. Repeat the exercise for the other species.

### 9.4.3   Implementation 3: Exploring the Impact of Biodiversity

In the third implementation of the Daisyworld model examined here, an additional species of daisy is introduced. This minor modification forms the basis of a simple experiment that is designed to evaluate the impact of biodiversity on the behavior of the model system. Specifically, the experiment tests the extent to which the system becomes more or less stable when a third species of daisy is added.

*Implementation*

Adding an extra species of daisy to the computational model is a relatively straightforward task; it involves a small number of modifications to the previous program, daisy2.awk. These modifications are presented in Program 9.3, daisy3.awk.

One change is the inclusion of two further variables, area_gray and albedo_gray, which are used to store the fractional area of ground covered by gray daisies and their albedo, respectively (lines 4 and 11). Note that the albedo of the gray daisies is given as 0.5 in this case, which is the same as that of the soil substrate. The equations used to calculate the area of land available for further daisy growth (Equation 9.6; lines 22 and 23) and to determine the global albedo (Equation 9.5; lines 24 to 27) have been amended to take into account the gray daisies. Further lines of code have also been added to calculate the local temperature of the gray daisies (lines 32 and 33), their rate of growth (line 37), the resulting change in the fractional area of the planet's surface that they cover (lines 41 and 42), and to prevent the gray daisies from becoming extinct (line 47). Finally, the print statement (lines 54 to 56) has been amended so that it also outputs the fractional area covered by the gray daisies.

*Evaluation*

Program 9.3 can be run from the command line as follows:

```
gawk -f daisy3.awk > daisy3.dat
```

*4*

Program 9.3: daisy3.awk (lines 1 to 44)

```
BEGIN {                                                             1
  # Initialize main variables                                      2
  area_black   = 0.01;                                             3
  area_gray    = 0.01;                                             4
  area_white   = 0.01;                                             5
  area_suit    = 1.0;                                              6
  solar_const  = 917;                                              7
  Stefan_Boltz = 5.67E-08;                                         8
  albedo_soil  = 0.5;                                              9
  albedo_black = 0.25;                                            10
  albedo_gray  = 0.5;                                             11
  albedo_white = 0.75;                                            12
  death_rate   = 0.3;                                             13
  q_factor     = 20;                                              14
  threshold    = 0.01;                                            15
                                                                  16
  for(luminosity=0.5;luminosity<=1.7;luminosity+=0.025){          17
    # Run model until it reaches steady state for                 18
    # current solar luminosity                                    19
    do{                                                           20
      prev_temp_global=temp_global;                               21
      area_avail=area_suit- \                                     22
        (area_black+area_white+area_gray);                        23
      albedo_global= (area_avail*albedo_soil)+ \                  24
        (area_black*albedo_black)+ \                              25
        (area_white*albedo_white) + \                             26
        (area_gray*albedo_gray);                                  27
      temp_global=(((solar_const*luminosity* \                    28
        (1-albedo_global))/Stefan_Boltz)**0.25)-273;              29
      temp_black=((q_factor*(albedo_global-albedo_black))+ \      30
        temp_global);                                             31
      temp_gray=((q_factor*(albedo_global-albedo_gray))+ \        32
        temp_global);                                             33
      temp_white=((q_factor*(albedo_global-albedo_white))+ \      34
        temp_global);                                             35
      growth_black=(1-(0.003265*((22.5-temp_black)**2)));         36
      growth_gray=(1-(0.003265*((22.5-temp_gray)**2)));           37
      growth_white=(1-(0.003265*((22.5-temp_white)**2)));         38
      area_black+=area_black*((area_avail*growth_black)- \        39
        death_rate);                                              40
      area_gray+=area_gray*((area_avail*growth_gray)- \           41
        death_rate);                                              42
      area_white+=area_white*((area_avail*growth_white)- \        43
        death_rate);                                              44
```

Program 9.3 (continued): daisy3.awk (lines 45 to 58)

```
# Do not allow daisies to become extinct              45
if(area_black<0.01){area_black=0.01};                 46
if(area_gray<0.01){area_gray=0.01};                   47
if(area_white<0.01){area_white=0.01};                 48
# Check whether global average temperature for current 49
# and previous model run differ by more than threshold 50
temp_difference=temp_global-prev_temp_global          51
if(temp_difference<0){temp_difference*=-1}            52
} while(temp_difference>threshold)                     53
printf("%5.3f %7.4f %6.4f %6.4f %6.4f\n", \           54
   luminosity, temp_global, area_black, \             55
   area_gray, area_white);                            56
}                                                      57
}                                                      58
```

*Visualization*

The output produced by the three-species version of the Daisyworld model (Program 9.3; `daisy3.awk`) can be visualized in gnuplot, where it can also be compared to the results obtained using the two-species model (Program 9.2; `daisy2.awk`) and the no biota model (`nolife.awk`), by typing the following commands (Figure 9.9):

```
set key top left box                                  25
set ylabel "Temperature, T_global/deg C"              26
plot 'nolife.dat' u 1:2 t "No biota" lt 0 lw 2, \     27
   'daisy2.dat' u 1:2 t "2 species" lt 1 lw 2, \      28
   'daisy3.dat' u 1:2 t "3 species" lt 2 lw 2         29
```

These commands assume that all three of the data files, `nolife.dat`, `daisy2.dat` and `daisy3.dat`, are located in the working directory. Similarly, the fractional area of the planet's surface that is covered by each of the three daisy species can be visualized as follows:

```
unset key                                             30
set ylabel "Fractional area, a"                       31
plot 'daisy3.dat' u 1:3 t "Black daisies" lw 2, \     32
   'daisy3.dat' u 1:4 t "Gray daisies" lw 2, \        33
   'daisy3.dat' u 1:5 t "White daisies" lw 2          34
```

These commands produce the plot shown in Figure 9.10.

Figure 9.9 indicates that the relationship between global average temperature and solar luminosity predicted by the three-species model is similar to that predicted by the two-species model. The temperature of Daisyworld is arguably very slightly more stable for solar luminosity values between 0.725 and 1.5 in the three-species model, with the exception of a small perturbation around $L = 1.0$, but the difference is small and the evidence is, at best, inconclusive. The impact of an increase in biodiversity

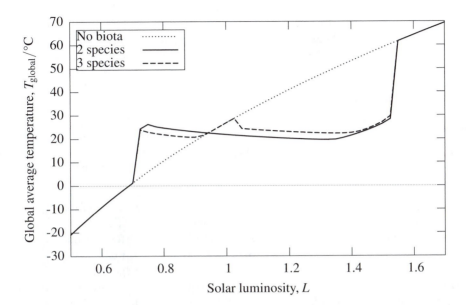

Figure 9.9: Variation in the global average temperature of Daisyworld as a function of solar luminosity. Comparison of results obtained using the two-species and three-species versions of the model.

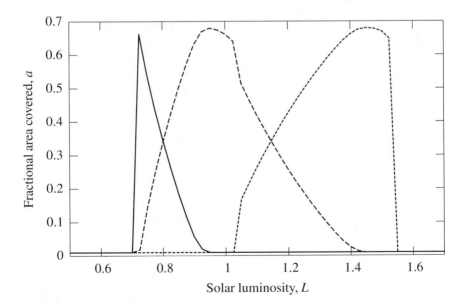

Figure 9.10: Fractional area of the planet surface covered by black (solid line), gray (long-dash line) and white (short-dash line) daisies as a function of relative solar luminosity.

on the global average temperature of Daisyworld (i.e., from two to three species of daisy) therefore appears to be minimal. However, close inspection of Figure 9.10 reveals that there are only ever two species of daisy that cover more than the specified minimum fraction (0.01) of the planet's surface for any given value of solar luminosity (i.e., the white daisies and gray daisies when $0.725 \leq L \leq 0.925$, and the gray daisies and black daisies when $1.025 \leq L \leq 1.525$). It is possible that this is an artifact of the specific value used for the albedo of the gray daisies in this simulation, which is the same as that of the soil substrate; this is left as an exercise for the reader to investigate (see Exercise 9.3, below). Alternatively, it is possible that the full impact of biodiversity becomes apparent only when a larger number of daisy species is modeled. This possibility is examined in the following section.

---

**Exercise 9.3**: Modify Program 9.3, `daisy3.awk`, so that the albedo of the gray daisies is specified on the command line. Save the file as `daisy3x.awk`. Run the revised program from the command line using a different value for the albedo of the gray daisies ($0.3 \leq A_g \leq 0.7$) each time. Redirect the output to separate data files. Create plots of the global average temperature of Daisyworld, and of the fractional areas covered by each species of daisy, as a function of solar luminosity. Compare the results with those from `daisy3.awk`.

---

### 9.4.4    Implementation 4: Modularizing the Code with User-Defined Functions

Modifying Program 9.2 (`daisy2.awk`) so that it incorporates an extra species of daisy is a relatively straightforward task. It requires only a few minor changes to the code, including additional lines to calculate the local temperature and the growth rate of the gray daisies, and the fractional area of Daisyworld that they cover (Program 9.3; `daisy3.awk`). This process is readily achieved using a text editor, by copying and pasting the equivalent lines of code for the black daisies and then editing these lines so that occurrences of the word `black` are replaced by the word `gray`. This approach is reasonably acceptable for small programs, but it is problematic for longer and more complex ones. Imagine, for example, having to modify Program 9.3 so that it simulates 10 or more species of daisy. There is considerable potential for making mistakes in these circumstances and "bugs" (coding errors) inevitably creep in. User-defined functions represent a much more elegant solution to this problem. They also make the code modular; that is to say, they help to divide the code into a number of smaller modules, which are more manageable individually and can be re-used.

*User-Defined Functions*

A user-defined function consists of a number of gawk statements, which perform a specific task, encapsulated in a named block of code. User-defined functions offer several important benefits: they help to modularize a program by breaking up the code into a collection of discrete building blocks; they can reduce the length of a program and improve its readability because they can be used many times in the same program without having to repeat the corresponding lines of code; and they foster the re-use of common functions in different programs, either by cutting and pasting the function definitions between program files or, more effectively, through the use of a function "library" (a separate file containing a number of frequently employed user-defined functions).

A user-defined function typically receives a number of inputs, which are known as the "arguments" to the function, and it manipulates these to generate an output. The general syntax of a function definition, which commences with the keyword function, is as follows:

```
function name_of_function(list_of_parameters)
{
    body_of_function
}
```

Function definitions can appear before, after or between the BEGIN, main and END pattern-action blocks, but may not be placed inside them. The name of a user-defined function can comprise any sequence of letters, digits and underscores, except that it may not start with a digit (Robbins 2001). The convention used here is to begin function names with the prefix fn_ to distinguish them from variables and arrays; for instance, a function to calculate the growth rate of daisies might be called fn_growth. Note that, in general, the same name cannot be used for a variable and a function, a function and an array, or an array and a variable in a single gawk program. Moreover, there must not be any blank spaces between the name of the function and the left parenthesis of its parameter list. The parameter list specifies the items of data on which the function operates; there may be zero or more items in this list. Where there is more than one item in the list, the individual items are separated by commas. The number of items in the list and the order in which they are listed are significant, as is demonstrated later in this section. The body of the function contains one or more gawk statements, which specify the actions to be performed; these define what the function does.

A user-defined function is said to be "called" when it is used within the BEGIN, main or END pattern-action blocks of a program. Note that it is not necessary to define a function prior to the point at which it is called in a program, because gawk parses all of the code before executing it (Robbins 2001). When a function is called, the values given in the parameter list are passed to, and stored in, the corresponding variables used in the body of the function; this is sometimes referred to as "calling by value". The actions specified in the body of the function are then performed on these values. In many instances, the result of a user-defined function is returned to the main part of the gawk program so that, for example, it can be employed in subsequent calculations. On other occasions, a user-defined function may simply perform an action, such as printing out a number, rather than returning a value; this is the equivalent of a void function in the C programming language or a procedure in Pascal (Robbins 2001).

Figure 9.11 provides a schematic representation of how a user-defined function is specified and employed in a gawk program. Note that in this example the function, fn_name, is defined after the END block, and that the parameter list consists of two

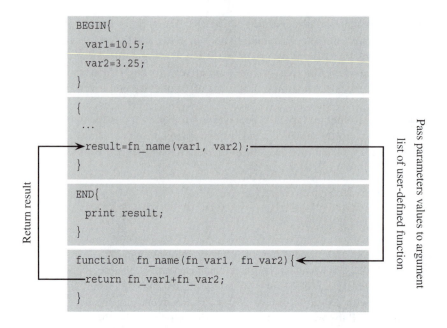

Figure 9.11: Schematic representation of a user-defined function in gawk. The function
fn_name is defined at the end of the program, and is "called" from the main
block.

arguments, fn_var1 and fn_var2, separated by a comma. The function is called in
the main pattern-action block of the program, where it is passed a copy of the values held by the variables var1 and var2; these variables are initialized in the BEGIN
block (var1=10.5 and var2=3.25). The values passed to the function are stored in
the variables fn_var1 and fn_var2, respectively; that is, fn_var1 receives the value
of var1 (fn_var1=10.5) and fn_var2 receives the value of var2 (fn_var2=3.25). Note
that fn_var1 and fn_var2 are referred to as local variables, since their values are only
known within the user-defined function and they cease to exist once the function is
exited, in contrast to global variables, such as var1 and var2, whose values are known
globally and persist throughout the program (Robbins 2001). The values of the variables fn_var1 and fn_var2 are manipulated by the body of the function, which sums
their values in this example. The result is returned to the main pattern-action block
of the program where it is stored in the variable result (result=13.75). The value of
result is printed out in the END block.

*Re-Implementation of the Daisyworld Model with User-Defined Functions*

Program 9.4, daisy4.awk, re-implements the three-species Daisyworld model with a
number of user-defined functions, fn_temp, fn_growth and fn_area; this program is
otherwise identical to Program 9.3, daisy3.awk. Each function is called three times,
once for each species of daisy (white, gray and black). The first function, fn_temp,
which is defined on lines 63 to 67, calculates the local temperatures over each of the

Program 9.4: daisy4.awk (lines 1 to 43)

```
BEGIN {                                                              1
                                                                    2
  # Initialize main variables                                       3
  area_black   = 0.01;                                              4
  area_gray    = 0.01;                                              5
  area_white   = 0.01;                                              6
  area_suit    = 1.0;                                               7
  solar_const  = 917;                                               8
  Stefan       = 5.67E-08;                                          9
  albedo_soil  = 0.5;                                              10
  albedo_black = 0.25;                                             11
  albedo_gray  = 0.5;                                              12
  albedo_white = 0.75;                                             13
  death_rate   = 0.3;                                              14
  q_factor     = 20;                                               15
  threshold    = 0.01;                                             16
                                                                   17
  for(luminosity=0.5;luminosity<=1.7;luminosity+=0.025){           18
    # Run model until it reaches steady state for                  19
    # current solar luminosity                                     20
    do{                                                            21
      prev_temp_global=temp_global;                                22
                                                                   23
      area_avail=area_suit- \                                      24
        (area_black+area_white+area_gray);                         25
      albedo_global=(area_avail*albedo_soil)+ \                    26
        (area_black*albedo_black)+ \                               27
        (area_white*albedo_white) + \                              28
        (area_gray*albedo_gray);                                   29
      temp_global=(((solar_const*luminosity* \                     30
        (1-albedo_global))/Stefan)**0.25)-273;                     31
                                                                   32
      temp_black=fn_temp(albedo_black);                            33
      temp_gray=fn_temp(albedo_gray);                              34
      temp_white=fn_temp(albedo_white);                            35
                                                                   36
      growth_black=fn_growth(temp_black);                          37
      growth_gray=fn_growth(temp_gray);                            38
      growth_white=fn_growth(temp_white);                          39
                                                                   40
      area_black+=fn_area(area_black,growth_black);                41
      area_gray+=fn_area(area_gray,growth_gray);                   42
      area_white+=fn_area(area_white,growth_white);                43
```

Program 9.4 (continued): daisy4.awk (lines 44 to 80)

```
                                                                    44
    # Do not allow daisies to become extinct                       45
    if(area_black<0.01){area_black=0.01};                          46
    if(area_gray<0.01){area_gray=0.01};                            47
    if(area_white<0.01){area_white=0.01};                          48
                                                                    49
    # Check whether global average temperature for current         50
    # and previous model run differ by more than threshold         51
    temp_difference=temp_global-prev_temp_global                   52
    if(temp_difference<0){temp_difference*=-1}                     53
  } while(temp_difference>threshold)                                54
                                                                    55
  printf("%5.3f %7.4f %6.4f %6.4f %6.4f\n", \                      56
    luminosity, temp_global, area_black, \                         57
    area_gray, area_white);                                        58
  }                                                                 59
}                                                                   60
                                                                    61
# Function to calculate local temperatures                         62
function fn_temp(albedo_daisy)                                      63
{                                                                   64
  return ((q_factor*(albedo_global-albedo_daisy))+ \              65
    temp_global);                                                  66
}                                                                   67
                                                                    68
# Function to calculate daisy growth rates                         69
function fn_growth(temp_daisy)                                      70
{                                                                   71
  return (1-(0.003265*((22.5-temp_daisy)**2)));                   72
}                                                                   73
                                                                    74
# Function to calculate change in daisy area                       75
function fn_area(area_daisy,growth_daisy)                           76
{                                                                   77
  return (area_daisy*((area_avail*growth_daisy)- \                 78
    death_rate));                                                  79
}                                                                   80
```

three species of daisy; it accepts a single argument, the albedo of the selected species, albedo_daisy, and returns the local temperature over that species; it also employs three global variables, q_factor, albedo_global and temp_global, which are either initialized or calculated in the BEGIN block. The function is called on lines 33 to 35, in the BEGIN block. On each occasion, the function is passed the albedo value of a particular species of daisy and the result is stored in the corresponding variable for the local temperature of that species. The second function, fn_growth, which is defined on lines 70 to 73, operates in a similar manner; it accepts a single argument, the temperature of the selected species of daisy, temp_daisy, and returns the rate of growth for that species. This function is called on lines 37 to 39, in the BEGIN block. The third function, fn_area, which is defined on lines 76 to 80, accepts two arguments, the fractional area of Daisyworld covered by the chosen species of daisy, area_daisy, and the growth rate of that species, growth_daisy; it also uses two global variables, area_avail and death_rate, which are either initialized or calculated elsewhere in the program. The values of these variables are used to calculate the amount by which the area covered by that species of daisy changes during the current time step. This function is called on lines 41 to 43, in the BEGIN block.

## *Evaluation*

Program 9.4 can be run from the command line as follows:

```
gawk  -f  daisy4.awk  >  daisy4.dat
```
6

Note that the output from the model is redirected to a file, daisy4.dat, in the working directory. The contents of this file should be identical to those of daisy3.dat because the programs used to generate them are functionally equivalent.

---

**Exercise 9.4**: Verify that Program 9.4, daisy4.awk, produces the same results as Program 9.3, daisy3.awk, by plotting the output from both in gnuplot. Modify daisy4.awk so that it simulates conditions on Daisyworld in the presence of five species of daisy, each with a different albedo in the range $0.25 \leq A \leq 0.75$. Save the revised program as daisy4x.awk and run this from the command line, redirecting the output to a data file, daisy4x.dat. Create a plot of the global average temperature of Daisyworld versus solar luminosity. Compare the results with the three-species, two-species and no biota models.

---

## *Revised Implementation using Separate Program Files*

Earlier, it was noted that user-defined functions can be placed in a file separate from the program that calls them. This approach allows commonly used functions to be employed in many different programs without having to incorporate the corresponding code explicitly in each one. This feature helps to keep the main program files short and manageable; it also encourages the re-use of tried and tested code. While a

Program 9.5: daisyvar.awk

```
BEGIN {                                              1
    # Initialize main variables                      2
    area_black    = 0.01;                            3
    area_gray     = 0.01;                            4
    area_white    = 0.01;                            5
    area_suit     = 1.0;                             6
    solar_const   = 917;                             7
    Stefan        = 5.67E-08;                        8
    albedo_soil   = 0.5;                             9
    albedo_black  = 0.25;                            10
    albedo_gray   = 0.5;                             11
    albedo_white  = 0.75;                            12
    death_rate    = 0.3;                             13
    q_factor      = 20;                              14
    threshold     = 0.01;                            15
}                                                    16
```

separate function library is not really necessary in the case of the Daisyworld model, which is a relatively short program and where the user-defined functions are highly specific to this particular piece of code, the approach is nevertheless adopted here to illustrate how it is implemented.

Thus, Program 9.4, daisy4.awk, can be split into three separate files, daisyvar.awk (Program 9.5), daisy5.awk (Program 9.6) and daisyfns.awk (Program 9.7). The first of these contains the code initializing the main variables used in the model, the second contains the bulk of the code, including the calls to the user-defined functions, and the third contains the definitions of the user-defined functions. Note that two of these files consist of BEGIN blocks, namely daisyvar.awk (Program 9.5) and daisy5.awk (Program 9.6). gawk will happily parse two or more BEGIN blocks (and, indeed, two or more main and END blocks), although the order in which they are listed on the command line is significant, as is shown below.

*Re-Evaluation*

Program 9.6 and Program 9.7 can be run together via the command line as follows:

```
gawk -f daisyvar.awk -f daisy5.awk -f daisyfns.awk > ⇨      7
    ⇨daisy5.dat
```

This should produce the same result as Program 9.4, daisy4.awk. Note that the name of each of the three program files is preceded by a separate command-line switch, -f, which indicates that the corresponding file contains code to be interpreted, rather than data to be processed. As noted above, the order in which the program files are given on the command line is significant: the file containing the variable initialization statements (daisyvar.awk) must, in this instance, be placed before the code in which the variables are first used (daisy5.awk); otherwise, the variables will be uninitialized

Program 9.6: daisy5.awk

```
BEGIN {                                                                       1
  for(luminosity=0.5;luminosity<=1.7;luminosity+=0.025){                      2
    # Run model until it reaches steady state for                             3
    # current solar luminosity                                                4
    do{                                                                       5
      prev_temp_global=temp_global;                                           6
                                                                              7
      area_avail=area_suit- \                                                 8
        (area_black+area_white+area_gray);                                    9
      albedo_global=(area_avail*albedo_soil)+ \                              10
        (area_black*albedo_black)+ \                                         11
        (area_white*albedo_white) + \                                        12
        (area_gray*albedo_gray);                                             13
      temp_global=(((solar_const*luminosity* \                              14
        (1-albedo_global))/Stefan)**0.25)-273;                              15
                                                                             16
      temp_black=fn_temp(albedo_black);                                     17
      temp_gray=fn_temp(albedo_gray);                                       18
      temp_white=fn_temp(albedo_white);                                     19
                                                                             20
      growth_black=fn_growth(temp_black);                                   21
      growth_gray=fn_growth(temp_gray);                                     22
      growth_white=fn_growth(temp_white);                                   23
                                                                             24
      area_black+=fn_area(area_black,growth_black);                         25
      area_gray+=fn_area(area_gray,growth_gray);                            26
      area_white+=fn_area(area_white,growth_white);                         27
                                                                             28
      # Do not allow daisies to become extinct                              29
      if(area_black<0.01){area_black=0.01};                                 30
      if(area_gray<0.01){area_gray=0.01};                                   31
      if(area_white<0.01){area_white=0.01};                                 32
                                                                             33
      # Check whether global average temperature for current                34
      # and previous model run differ by more than threshold                35
      temp_difference=temp_global-prev_temp_global                          36
      if(temp_difference<0){temp_difference*=-1}                            37
    } while(temp_difference>threshold)                                       38
                                                                             39
    printf("%5.3f %7.4f %6.4f %6.4f %6.4f\n", \                             40
      luminosity, temp_global, area_black, \                                41
      area_gray, area_white);                                               42
  }                                                                          43
}                                                                            44
```

Program 9.7: daisyfns.awk

```
# Function to calculate local temperatures              1
function fn_temp(albedo_daisy)                          2
{                                                       3
  return ((q_factor*(albedo_global-albedo_daisy))+ \    4
    temp_global);                                       5
}                                                       6
                                                        7
# Function to calculate daisy growth rates             8
function fn_growth(temp_daisy)                          9
{                                                      10
  return (1-(0.003265*((22.5-temp_daisy)**2)));       11
}                                                      12
                                                       13
# Function to calculate change in daisy area           14
function fn_area(area_daisy,growth_daisy)              15
{                                                      16
  return (area_daisy*((area_avail*growth_daisy)- \     17
    death_rate));                                      18
}                                                      19
```

(i.e., they will be assumed to have the value zero or contain the null string) when the code is first run, which is not what is intended here.

---

**Exercise 9.5**: Which of the following command lines produces an error, and why?

```
gawk -f daisy5.awk -f daisyfns.awk -f daisyvar.awk
gawk -f daisyvar.awk -f daisyfns.awk -f daisy5.awk
gawk -f daisyfns.awk -f daisyvar.awk -f daisy5.awk
```

What is wrong with the following command line?

```
gawk -f daisyvar.awk daisyfns.awk daisy5.awk
```

---

## 9.5  SENSITIVITY ANALYSIS AND UNCERTAINTY ANALYSIS

It is important to investigate the extent to which the results generated by a model are sensitive to the parameter values with which it is initialized. If a small change in the value of an input parameter produces a large change in the values output from the model, the model is said to be sensitive to the parameter and the parameter is said to have a high influence on the model (Ford 1999). Conversely, if a large change in the value of an input parameter produces a small change in the values output from the model, the model is said to be insensitive to the parameter and the parameter is said to have a low influence on the model. The process of establishing the sensitivity of a model to its input parameters is known as sensitivity analysis (Saltelli *et al.* 2000, Saltelli *et al.* 2004). Knowledge gained by performing a sensitivity analysis of a model can help to elucidate the way in which the modeled system functions and to identify those parameters of the model whose values need to be specified most accurately. Ultimately, the aim of sensitivity analysis is to focus attention on the critical parts of the model and the environmental system that it purports to represent.

In practice, sensitivity analysis is usually performed by perturbing the values of the model parameters in known amounts and measuring the effects that these variations have on the model outputs. The simplest method is to vary the value of one parameter at a time, while the values of the other parameters are held constant. This approach, known as one-at-a-time (OAT) or univariate sensitivity analysis, is used to quantify the influence that each parameter exerts on the model and is the method adopted here. It has a significant limitation, however, because it cannot describe the sensitivity of the model to simultaneous changes in two or more parameters, each of which may have a limited influence when considered in isolation, but which combine to produce major changes in the model output. This problem requires the use of more sophisticated multivariate sensitivity analysis techniques, such as Monte Carlo simulation with simple random or Latin Hypercube sampling, which are beyond the scope of this book (Saltelli *et al.* 2000, Saltelli *et al.* 2004).

Figure 9.12 and Figure 9.13 show the results of two separate OAT sensitivity analyses performed on the three-species implementation of the Daisyworld model, `daisy4.awk`. In the first of these, the model is run a number of times, each time using a different value for the albedo of the black daisies ($0.1 \leq A_{black} \leq 0.45$), while the values of the other parameters are held constant (Figure 9.12); in the second, it is the albedo of the white daisies that is varied ($0.55 \leq A_{white} \leq 0.90$), while the other parameters are fixed (Figure 9.13). Figure 9.12 suggests that the value used to represent the albedo of the black daisies has a small but significant impact on the form of the relationship between solar luminosity and the globally averaged temperature of Daisyworld when the solar luminosity is between 0.7 and 0.9; in general terms, the planet warms up faster as the albedo of the black daisies decreases. This parameter has little impact on the temperature of Daisyworld, however, when the solar luminosity exceeds 0.9. By contrast, the albedo of the white daisies has a significant influence on the temperature of Daisyworld when the solar luminosity exceeds 1.0, but very little effect below this value.

The sensitivity analyses outlined above can also be used to explore the relationship between the albedo of the daisy species and the fractional area of the planet surface that they occupy. Thus, Figure 9.14 and Figure 9.15 show that the albedo of the black daisies has a significant influence on the fractional areas covered by both the black daisies and the white daisies at different solar luminosities.

Uncertainty analysis is the corollary of sensitivity analysis. It examines the propagation of uncertainty in the values of the model parameters (and sometimes in the structure and formulation of the model) through to uncertainty in the model outputs. For instance, the exact value of a parameter may not be known. Instead, it may be

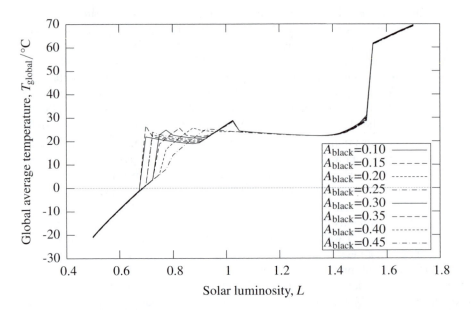

Figure 9.12: Sensitivity analysis of the three-species Daisyworld model, `daisy4.awk`, showing variation in the globally averaged temperature of the planet as a function of the albedo of the black daisies ($0.10 \le A_{\text{black}} \le 0.45$).

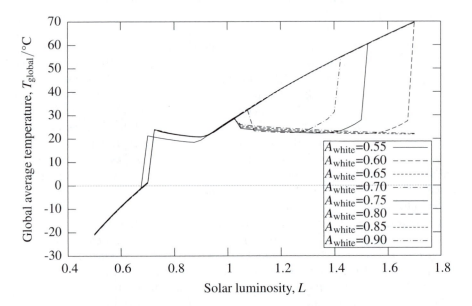

Figure 9.13: Sensitivity analysis of the three-species Daisyworld model, `daisy4.awk`, showing variation in the globally averaged temperature of the planet as a function of the albedo of the white daisies ($0.55 \le A_{\text{white}} \le 0.90$).

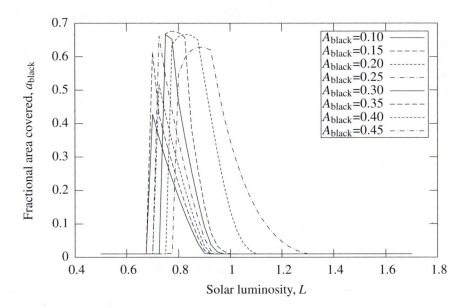

Figure 9.14: Sensitivity analysis of the three-species Daisyworld model, `daisy4.awk`, showing variation in the fractional area of the planet covered by black daisies as a function of the albedo of the black daisies ($0.10 \leq A_{\text{black}} \leq 0.45$).

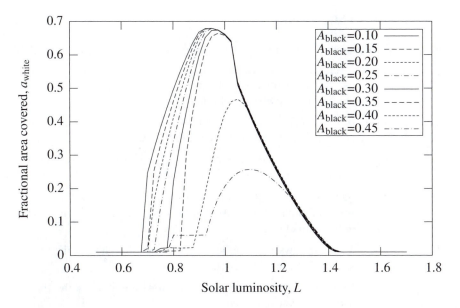

Figure 9.15: Sensitivity analysis of the three-species Daisyworld model, `daisy4.awk`, showing variation in the fractional area of the planet covered by white daisies as a function of the albedo of the black daisies ($0.10 \leq A_{\text{black}} \leq 0.45$).

thought to lie somewhere between fixed upper and lower limits, or perhaps to vary about a mean value according to a given standard deviation. The input value for this parameter is therefore constrained, but not prescribed. The uncertainty in the value that should be used for this parameter results in variation (uncertainty) in the model outputs, which can be quantified. In crude terms, therefore, the modeler wishes to know whether a 10% difference in the value of the parameter causes a 1%, 10% or 100% change in the model outputs. This type of uncertainty analysis is not performed here, but is left as an exercise for the interested reader.

## 9.6   SUMMARY

This chapter examines a model of biospheric feedback between the biota (different species of daisy) on an imaginary planet, known as Daisyworld, and their abiotic environment (Watson and Lovelock 1983). The model combines many of the topics covered in the preceding chapters, including constrained population growth and inter-specific competition (Chapter 8), and the interaction of solar radiation with various surface materials (Chapter 5 through Chapter 7). Several implementations of the Daisyworld model are presented, including one in which the solar luminosity is held constant and others in which it gradually increases over time. The sensitivity of the model to the number of daisy species (biodiversity) and to variations in the input parameter values is also explored. The computational model is also modularized with user-defined functions.

## SUPPORTING RESOURCES

The following resources are provided in the `chapter9` sub-directory of the CD-ROM:

| | |
|---|---|
| daisy1.awk | **gawk** implementation of the two-species Daisyworld model for conditions of constant solar luminosity. |
| daisy2.awk | **gawk** implementation of the two-species Daisyworld model for conditions of increasing solar luminosity. |
| daisy3.awk | **gawk** implementation of the three-species Daisyworld model for conditions of increasing solar luminosity. |
| daisy4.awk | **gawk** implementation of the two-species Daisyworld model for conditions of increasing solar luminosity, employing user-defined functions. |
| daisyvar.awk | Modular **gawk** program to initialize the variables for the three-species Daisyworld model. |
| daisyfns.awk | Library of user-defined **gawk** functions required by the three-species Daisyworld model. |

| | |
|---|---|
| `daisy5.awk` | Modular gawk program implementing the three-species Daisyworld model under conditions of increasing solar luminosity, employing user-defined functions. Also requires `daisyvar.awk` and `daisyfns.awk`. |
| `daisy.bat` | Command line instructions used in this chapter. |
| `daisy.plt` | gnuplot commands used to generate the figures presented in this chapter. |
| `daisy.gp` | gnuplot commands used to generate figures presented in this chapter as EPS files. |

# Chapter 10

# Modeling Incident Solar Radiation and Hydrological Networks over Natural Terrain

**Topics**

- Calculating the gradient and aspect of terrain slopes from a DEM

- Estimating the direct, diffuse and total solar irradiance on sloping terrain

- Modeling hydrological networks (local drainage direction) using a DEM

**Methods and techniques**

- Processing 2D arrays in gawk

- Plotting vectors and arrows in 2D and 3D using gnuplot

## 10.1 INTRODUCTION

The Daisyworld model, which is explored in the preceding chapter, considers a highly idealized planet, the surface of which is assumed to be flat and level. This assumption simplifies the model in various ways, the most important of which is that radiation from Daisyworld's sun is incident on every part of the planet surface at the same angle (i.e., normal [perpendicular] to the surface). The same assumption cannot be applied generically to models of solar radiation interaction with Earth's land surface, which is rarely flat or level over significant distances, but slopes at different angles (gradients) and in different directions (aspects) according to the local terrain. As a result, those parts of the land surface that face toward the sun tend to receive more solar radiation than those that face away. More generally, the amount of solar radiation received at Earth's surface varies according to the slope of the terrain and the angle of the sun. The latter is, of course, a function of the latitude of the site, the time of day and the day of year, as is demonstrated in Chapter 5. The analysis begun in Chapter 5 is extended here to produce a model that simulates the amount of solar radiation received by the undulating terrain around Llyn Efyrnwy.

Terrain slope also exerts an effect on other environmental processes, perhaps the most obvious example of which is rainfall run-off (Beven 2001). Thus, the route that precipitation takes from the point at which it falls on the ground, via a sequence of streams and rivers, to the sea or to a lake is primarily controlled by the slope of the local terrain. It is relatively easy to derive comprehensive spatial data on terrain slope (i.e., gradient and aspect) from a DEM. This information can be used, in turn, to model the hydrological network (i.e., the local drainage direction) of an area (Moore 1996). From this, it is possible to simulate the hydrological response of the study area, as well as other properties related to soil erosion and sediment transport (Wainwright and Mulligan 2004). The second part of this chapter therefore introduces a model to predict the hydrological network for the Llyn Efyrnwy area, based on the DEM introduced in Chapter 2. The structure of the network is compared to the "blue line" features (i.e., streams and rivers) identified in the corresponding 1:50,000 scale OS topographic map.

## 10.2   VISUALIZING DIGITAL ELEVATION DATA AS AN ARRAY

Recall from Chapter 2 that the DEM of Llyn Efyrnwy, eyfynwy.dem, consists of a $51 \times 51$ element grid of elevation values sampled at regular spatial intervals (100 m) on the ground. In addition to the 3D visualizations presented in Chapter 2 (Figure 2.18, page 50), these data can be viewed as a form of planimetric map by issuing the following commands in gnuplot:

```
reset                                                                          1
unset key                                                                      2
set xlabel "Easting, m"                                                        3
set ylabel "Northing, m"                                                       4
set xtics 295000, 1000, 300000                                                 5
set ytics 315000, 1000, 320000                                                 6
set dgrid3d 51,51,16                                                           7
set size square                                                                8
set tics out                                                                   9
set style line 1 lt -1 lw 0.01                                                10
set palette gray gamma 1                                                      11
set pm3d at s hidden3d 1 map                                                  12
set colorbox vertical user origin 0.8,0.185 size 0.04,0.655                   13
set cblabel "Elevation, m"                                                    14
splot 'efyrnwy.dem' u 1:2:3 w pm3d                                            15
```

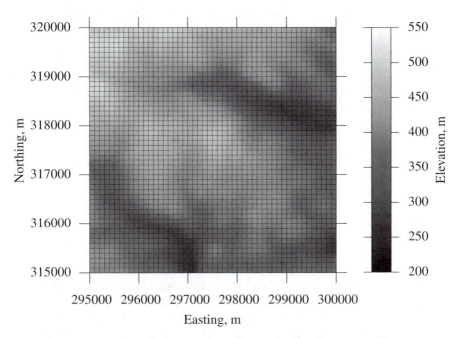

Figure 10.1: Planimetric visualization of the terrain elevation at Llyn Efyrnwy.

Most of these instructions have been introduced in the preceding chapters and so should be familiar by now. They produce the plot shown in Figure 10.1. Lines 1 to 15 instruct gnuplot to create a solid surface model rendered as a grayscale (line 11) planimetric map (line 12). The gray tones in the plot represent the elevation values, with black representing the lowest points in the DEM and white representing the highest ones. Line 9 instructs gnuplot to place the tic-marks facing outward and line 10 specifies the particular line type used to mark the data grid. Lines 13 and 14 specify the position and label, respectively, of the continuous grayscale key, which appears on the right-hand side of the plot. Some experimentation with the parameter values used to control the position of the key is normally required to obtain an aesthetically pleasing result. The significance of the resulting plot (Figure 10.1) is that it highlights the fact that the DEM is a regular rectangular grid of elevation values. It is important to note, however, that the elevation values relate to the intersections on the grid, rather than the grayscale rendered cells that gnuplot produces in this visualization.

## 10.3   HANDLING MULTI-DIMENSIONAL ARRAYS IN gawk

Although gnuplot can be used to display data organized in the form of a grid or array, gawk is needed to transform the data into the required format (e.g., to calculate the gradient and aspect of the terrain from the original elevation values). The methods used to handle arrays in gawk are introduced in Chapter 7 (page 184). The arrays used in that chapter are, however, one-dimensional (i.e., they store a list of values). By con-

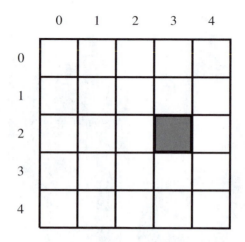

Figure 10.2: Graphical representation of a 2D array, data. The shaded cell is data[2,3].

trast, the data examined in this chapter are inherently multi-dimensional. The dimensions are defined by the Easting, Northing and elevation of the data points. All three dimensions are measured in meters relative to ODN, using the OS coordinate system. gawk handles multi-dimensional arrays such as this by using a separate index for each dimension. Thus, while each element in a 1D array is referenced by a single unique index (data[1], data[2], data[3], ... , data[x]), the elements of a 2D array are identified by two indexes separated by a comma (e.g., data[1,1], data[1,2], data[1,3], ... , data[2,1], ... , data[y,x]). The way in which gawk stores and processes the indexes is different from that of other programming languages (Robbins 2001), but the details need not concern us here.

Figure 10.2 is a graphical representation of a 2D array, data. The array contains 25 elements, which are arranged in a $5 \times 5$ element grid or matrix. It is usual for the first index of an array, such as this, to denote the row number, referenced from the top downward, and for the second index to indicate the column number, referenced from left to right. In some computer programming languages, such as C, the numbers used for the indexes start at zero (0); in others, such as FORTRAN, they begin at one (1). These two systems are commonly referred to as zero-based and one-based arrays, respectively. gawk is essentially agnostic in this respect, so that either system can be used. Hence, data[2,3] refers to the shaded cell in Figure 10.2.

Observant readers will notice that the array indexing system described above is different from the grid referencing system used by the OS, which is normally given in the following order: Easting (equivalent to column number), measured east to west, and Northing (equivalent to row number), measured south to north (i.e., referenced from the bottom up; see, for example, Figure 10.1). gawk is capable of coping with either of these array indexing systems. The onus is placed on the person who implements the computational model to make clear which system is being used and to be consistent in this respect, otherwise all manner of unintended consequences, errors and misunderstandings may arise.

Program 10.1: readarr.awk

```
# Reads digital elevation data stored in the form        1
# of triplets (Easting, Northing, elevation) into        2
# a 2D GAWK array.                                        3
#                                                         4
# Usage: gawk -f readarr.awk input_file                  5
#                                                         6
# Variables:                                              7
# ----------                                              8
# E          Easting                                      9
# N          Northing                                    10
# elevation Array in which DEM data are stored           11
                                                         12
{                                                        13
  E=$1;                   # Read in Easting              14
  N=$2;                   # Read in Northing             15
                                                         16
  elevation[E,N]=$3;  # Read elevation data into array   17
}                                                        18
```

Program 10.1, readarr.awk, illustrates how digital elevation data held in the file efyrnwy.dem (Figure 2.12, page 44) can be read into a 2D gawk array. Apart from the comments, the program is contained entirely within the main pattern-action block (lines 13 to 18). The program takes the Easting and Northing values from the first and second fields of the current data record and stores these in the variables E and N, respectively (lines 14 and 15). These variables are then employed as the indexes to a 2D array, elevation, which is used to hold the corresponding elevation value, read from the third field of each record (line 17). Although the program does nothing with the data in the array, it illustrates how an array can be populated from a data file. This simple program forms the basis of a more extensive program presented in the next section, which calculates the gradient and aspect of every cell in a DEM array.

## 10.4    DETERMINING TERRAIN GRADIENT AND ASPECT

### 10.4.1    Formulation

Various methods have been developed to calculate gradient and aspect values from digital elevation data stored in a 2D array (Burrough and McDonnell 1998). The method used here is that of Zevenbergen and Thorne (1987), which represents a good compromise between accuracy and computational simplicity (Skidmore 1989). This method calculates the gradient and aspect of each cell in an array (elevation[E,N]) by measuring the difference in elevation between the cells immediately to the north and south of it (elevation[E,N+$\Delta$] and elevation[E,N-$\Delta$], respectively; where $\Delta$ is the cell size in meters) and between the cells immediately to the east and west of it (elevation[E-$\Delta$,N] and elevation[E+$\Delta$,N], respectively; Figure 10.3). These values

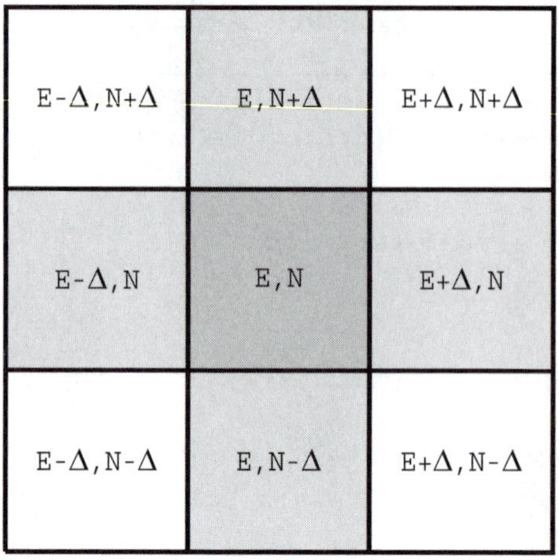

| E-Δ,N+Δ | E,N+Δ | E+Δ,N+Δ |
|---------|-------|---------|
| E-Δ,N   | E,N   | E+Δ,N   |
| E-Δ,N-Δ | E,N-Δ | E+Δ,N-Δ |

Figure 10.3: Relative indexing used to calculate the local gradient and aspect of a given cell, elevation[E,N], in an array containing digital elevation data, where E and N are the Easting and Northing, respectively, of the cell and where Δ is the cell size.

are used to calculate the gradient of the central cell in the north-south (NS) and east-west (EW) directions. The EW gradient can be expressed more formally as follows:

$$\text{gradEW} = \frac{\text{elevation}[E - \Delta, N] - \text{elevation}[E + \Delta, N]}{2\Delta} \tag{10.1}$$

Similarly, the NS gradient is given by Equation 10.2.

$$\text{gradNS} = \frac{\text{elevation}[E, N + \Delta] - \text{elevation}[E, N - \Delta]}{2\Delta} \tag{10.2}$$

The overall gradient of the cell, measured in arc degrees (°), can be computed from these two values using Equation 10.3.

$$\theta = \tan^{-1}\left(\sqrt{\text{gradEW}^2 + \text{gradNS}^2}\right) \tag{10.3}$$

Finally, the aspect of the cell relative to true north is given by Equation 10.4.

$$\phi = 180 + \tan^{-1}\left(-\frac{\text{gradEW}}{\text{gradNS}}\right) \tag{10.4}$$

### 10.4.2   Implementation

Equation 10.1 through Equation 10.4 are implemented in Program 10.2, gradasp.awk. The code used in this program can be interpreted as follows.

Lines 1 to 24 contain a series of comments that explain the purpose of the program and how it should be run from the command line, and that list the main variables used in the code. The BEGIN block (lines 26 to 29) initializes two variables, PI and rad2deg:

Program 10.2: gradasp.awk (lines 1 to 45; comments, BEGIN block and main pattern-action block).

```
# Takes a 2D DEM and for each cell determines the local      1
# gradient and aspect using the method of Zevenbergen        2
# and Thorne (1987). Prints out two files, each             3
# containing the coordinates of the cell concerned          4
# and either (i) its gradient or (ii) its aspect.           5
#                                                            6
# Usage: gawk -f gradasp.awk -v size=value input_file       7
#               [> output_file]                             8
#                                                            9
# Variables:                                                10
# ----------                                                11
# min_E       eastern-most Easting in DEM                   12
# max_E       western-most Easting in DEM                   13
# min_N       southern-most Northing in DEM                 14
# min_N       northern-most Northing in DEM                 15
# this_E      Easting of current cell                       16
# this_N      Northing of current cell                      17
# size        DEM cell size (metres)                        18
# gradNS      Gradient of current cell measured North-South 19
# gradEW      Gradient of current cell measured East-West   20
# gradient    Gradient of current cell                      21
# aspect      Aspect of current cell                        22
# elevation   2D array storing the DEM                      23
# rad2deg     Conversion factor - radians to degrees        24
                                                            25
BEGIN{                                                      26
  PI=3.14159263359;                                         27
  rad2deg=360.0/(2*PI);                                     28
}                                                           29
                                                            30
{                                                           31
  E=$1;                    # Reading in Easting             32
  N=$2;                    # Reading in Northing            33
                                                            34
  # Initialize minimum and maximum Easting and Northing     35
  # of the DEM                                              36
  if(NR==1){min_E=E; max_E=E; min_N=N; max_N=N}            37
                                                            38
  if(E<min_E){min_E=E}    # Determine minimum and           39
  if(E>max_E){max_E=E}    # maximum Eastings and            40
  if(N<min_N){min_N=N}    # Northings (i.e., limits)        41
  if(N>max_N){max_N=N}    # of DEM                          42
                                                            43
  elevation[E,N]=$3;      # Read elevation data into array  44
}                                                           45
```

Program 10.1 (continued): gradasp.awk (lines 46 to 77; END block).

```
                                                                    46
END{                                                                47
  # Pass across DEM by Easting                                      48
  for(this_E=min_E+size; this_E<max_E; this_E+=size){               49
                                                                    50
    # Pass up through DEM by Northing                               51
    for(this_N=min_N+size; this_N<max_N; this_N+=size){             52
                                                                    53
      # Calculate Zevenbergen and Thorne's (1987) parameters        54
      gradEW=(elevation[this_E-size,this_N]- \                       55
             elevation[this_E+size,this_N])/(2*size);                56
      gradNS=(elevation[this_E,this_N+size]- \                       57
             elevation[this_E,this_N-size])/(2*size);                58
                                                                    59
      # Calculate the gradient (degrees)                            60
      gradient[this_E,this_N]=rad2deg*\                              61
             (gradNS*gradNS + gradEW*gradEW)**0.5;                   62
      # Print this value out to a file                              63
      printf("%6i %6i %5.2f\n", this_E, this_N,                     64
        gradient[this_E,this_N]) > "gradient.dem";                  65
                                                                    66
      # Calculate the aspect (degrees)                              67
      aspect=180+rad2deg*atan2(-gradEW,gradNS);                     68
      # Print this value out to a file                              69
      printf("%6i %6i %3i\n", this_E, this_N,                       70
          aspect) > "aspect.dem";                                  71
                                                                    72
    }                                                               73
                                                                    74
  }                                                                 75
                                                                    76
}                                                                   77
```

the former should need no explanation; the latter is a factor used to convert angular values from radians to degrees.

The main pattern-action block (lines 31 to 45) reads the digital elevation data from the input file into a 2D array, elevation. The Easting and Northing values (fields 1 and 2 of the data file) are used as the indexes to the array (lines 32, 33 and 44), and the corresponding elevation values are read into the array from the third field of the file (line 44). The code on lines 37 and 39 to 42 determines the minimum and maximum Eastings and Northings of the data set. This information is used later on in the program, in the END block.

The gradient and aspect of each cell in the array, elevation, is calculated in the END block (lines 47 to 77). This is achieved using two for loops to navigate a cell at a time through the array, east to west and south to north: the outer loop (lines 49 to 75) controls the Easting; the inner loop (lines 52 to 73) controls the Northing. The variables this_E and this_N are used to store the location of the current cell for which gradient and aspect values are calculated. Note that the for loops start one cell in from the eastern-most edge (min_E+size; line 49) and one cell up from the southern-most edge (min_N+size; line 52) of the array. This is because the gradient and aspect values of a given cell are calculated with reference to the cells immediately above and below it, as well as to either side of it. Clearly, this is not possible for the cells around the edge of the array. For the same reason, the for loops stop processing just before they reach the northern-most and western-most edges of the array (this_E<max_E, line 49, and this_N<max_N, line 52). Consequently, the data files for gradient and aspect that are output from this program are two columns and two rows smaller than the input DEM (i.e., they are both $49 \times 49$ element grids). This feature is illustrated in Figure 10.4.

Inside the for loops, lines 55 and 56 calculate the EW gradient of the current cell (gradEW; Equation 10.1), lines 57 and 58 calculate its NS gradient (gradNS; Equation 10.2) and lines 61 and 62 calculate its overall gradient (gradient; Equation 10.3). The latter value is printed out on lines 64 and 65, along with the Easting and the Northing of the cell to which it relates. Note that the output is written to a file called gradient.dem in the working directory, rather than being sent to the computer screen. This outcome is achieved using the redirection symbol and the appropriate file name enclosed in double quotation marks (> "gradient.dem"; line 65). The output file is opened once only while the program is running, such that the Easting, Northing and gradient values for subsequent cells are appended to the end of the named file (Robbins 2001).

Finally, the aspect (or azimuth) of the cell is calculated on line 68. Note that the built-in function used by **gawk** to calculate the inverse tangent of an angle, $\tan^{-1}$, is atan2; there is no atan function (Robbins 2001). The output from this function is given in radians, and the result is converted into arc degrees through multiplication by the conversion factor rad2deg. To convert this angle into arc degrees relative to due north, 180° is added to the result. Lines 70 and 71 write the Easting, Northing and aspect of the current cell to a file called aspect.dem, which is created in the working directory, using the redirection method described above.

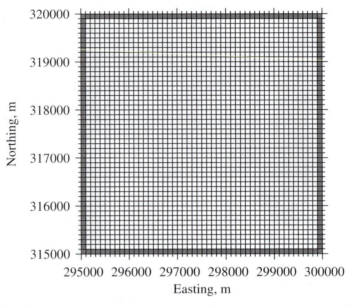

Figure 10.4: Relationship between the size of the data array given as input to Program 10.2, gradasp.awk, (gray and white cells combined) and the output data arrays that it produces (white cells only).

### 10.4.3   Evaluation

Program 10.2 can be run from the command line as follows:

```
gawk -f gradasp.awk -v size=100 efyrnwy.dem                            1
```

Note that the grid cell size, size, measured in meters, is specified on the command line. As noted previously, the program creates two output files, gradient.dem and aspect.dem, in the working directory. The former contains the gradient (measured in arc degrees; °) of each cell in the input DEM; the latter contains the corresponding aspect values (measured in arc degrees relative to true north). Note that the names of the output data files do not have to provided on the command line because these are specified in the program itself (lines 65 and 73 of Program 10.2, gradasp.awk).

### 10.4.4   Visualization

The gradient data produced by Program 10.2 can be visualized in gnuplot as follows:

```
set xrange [295000:300000]                                            16
set yrange [320000:315000]                                            17
set cblabel "Gradient, arc degrees"                                   18
splot 'gradient.dem' u 1:2:3 w pm3d                                    19
```

These instructions generate the plot shown in Figure 10.5. The aspect data can be similarly visualized by entering the following gnuplot commands:

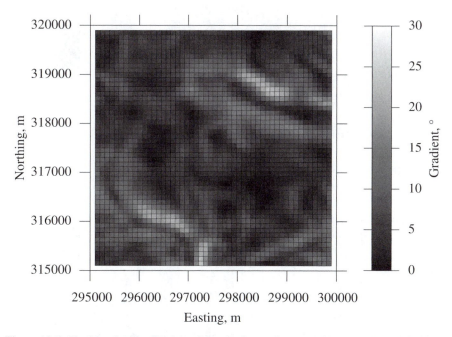

Figure 10.5: Planimetric visualization of the terrain gradient (arc degrees) at Llyn Efyrnwy.

```
set cbrange [0:360]                              20
set cbtics 0,30,360                              21
set cblabel "Aspect, arc degrees"               22
set palette gray rgbformulae -13,-13,-13        23
splot 'aspect.dem' u 1:2:3 w pm3d               24
```

These commands produce the plot shown in Figure 10.6. Note that the instruction on line 20 sets the range of values to be plotted on the continuous grayscale key, limiting this to 0 to 360. Line 21 controls the tic-marks placed beside the key. Line 22 specifies the text used to label the key. Line 23 instructs gnuplot to use a grayscale palette that runs from black (0°) to white (180°) and back to black again (360°), rather than the normal linear grayscale palette (white to black); this command ensures that north-facing slopes are shaded dark gray through to black, while the south-facing ones are shaded light gray through to white.

Figure 10.5 and Figure 10.6 assist in the visual interpretation of the terrain around Llyn Efyrnwy, especially Figure 10.6, which is similar in appearance to a shaded relief map. The value of the data used to create these figures is exploited more fully in the following section, in which they are used to calculate the direct, diffuse and total (global) solar irradiance incident on every cell in the DEM. Information of this type is important to various types of study concerned with the surface radiation regime, including those attempting to model the surface energy budget (Monteith and Unsworth 1995, Campbell and Norman 1998) or the growth of crops (Russell *et al.* 1989, Bouman 1992, Goudriaan and van Laar 1994).

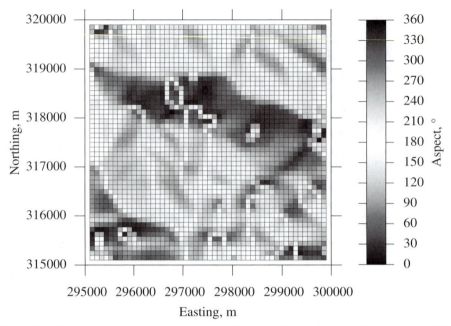

Figure 10.6: Planimetric visualization of the terrain aspect (arc degrees relative to true north) at Llyn Efyrnwy.

## 10.5 SOLAR IRRADIANCE ON SLOPING TERRAIN

Chapter 5 presents an analysis of the temporal variation in solar radiation incident at Llyn Efyrnwy based on a computational model and measurements made *in situ*. The computational model employed in that context calculates both the direct and the diffuse solar radiation incident on a horizontal element of Earth's surface using the mathematical formulations given by Campbell and Norman (1998), which are adapted from those of Liu and Jordan (1960). In this section, the analysis is extended to simulate the amount of direct and diffuse solar radiation incident on the undulating terrain around Llyn Efyrnwy, taking into account the zenith and azimuth angles of the sun and the slope (gradient and aspect) of the terrain.

### 10.5.1 Formulation

To calculate the solar radiation incident on an inclined surface, the angle between the sun and the surface normal must be known. This angle is given by Equation 10.5.

$$\zeta = \cos^{-1} \left( \cos\theta \cos\Psi + \sin\theta \sin\Psi \cos(\phi - \phi_s) \right) \tag{10.5}$$

where $\zeta$ is the angle between the normal to the inclined surface and the direction of the sun, $\theta$ is the gradient of the surface slope, $\Psi$ is the solar zenith angle, $\phi$ is the aspect (azimuth angle) of the surface slope, and $\phi_s$ is the azimuth angle of the sun (Iqbal 1983, Campbell and Norman 1998, Twidell and Weir 2006). The equations used to calculate the gradient and aspect (azimuth) of a sloping terrain facet are presented in the previous section, while those used to calculate the solar zenith angle are presented in Chapter 5. The solar azimuth angle, $\phi_s$, measured clockwise with respect to due north, is given by Equation 10.6 for the morning (for hour angles $\leq 0$) and by Equation 10.7 for the afternoon (for hour angles $> 0$).

$$\phi_s = 180 - \cos^{-1}\left(\frac{-(\sin\delta - \cos\Psi\sin\phi_L)}{\cos\phi_L\sin\Psi}\right) \tag{10.6}$$

$$\phi_s = 180 + \cos^{-1}\left(\frac{-(\sin\delta - \cos\Psi\sin\phi_L)}{\cos\phi_L\sin\Psi}\right) \tag{10.7}$$

where $\delta$ is the solar declination angle and $\theta_L$ is the latitude of the site (Campbell and Norman 1998). Note that the equation used to calculate the solar declination angle is presented in Chapter 5 (Equation 5.13).

The direct solar irradiance on an inclined surface can be calculated by modifying Equation 5.8 as follows:

$$I_{\text{direct}} = E_0 \tau^m \cos\zeta \tag{10.8}$$

where $E_0$ is the total exo-atmospheric solar irradiance (Equation 5.7; assumed to be 1380 W·m$^{-2}$ here), $\tau$ is the atmospheric transmittance (assumed to be 0.7 here), and $m$ is the atmospheric air mass (Equation 5.11). Likewise, the diffuse solar irradiance on an inclined surface can be approximated by modifying Equation 5.9 as follows:

$$I_{\text{diffuse}} = \left(\frac{1 + \cos\theta}{2}\right) 0.3 \left(1 - \tau^m\right) E_0 \tag{10.9}$$

The term $(1 + \cos\theta)/2$, often referred to as the sky view factor, describes the fraction of the sky hemisphere that is visible from a point on the surface. As the gradient of the terrain, $\theta$, increases, the fraction of the sky hemisphere visible from that point decreases and the amount of diffuse solar radiation it receives reduces concomitantly. Note that this is a relatively simple model of diffuse irradiance on a sloping surface. More sophisticated models also take into account the solar radiation reflected from adjacent terrain, which contributes to the diffuse radiation received at a given point. This phenomenon is sometimes known as the adjacency effect (Vermote *et al.* 1997), but its inclusion is beyond the scope of the discussion presented here.

### 10.5.2   Implementation

Program 10.3, `demirrad.awk`, implements Equation 10.5 through Equation 10.9 in a computational model. This program produces three output files, each of which relates to a different component of the incident solar radiation (i.e., direct, diffuse and total). The program is quite long, but it can be broken down into a number of discrete sections to help understand how it functions.

Lines 1 to 45 are comments, which describe what the program does, explain how it is operated from the command line, and list the primary variables employed in the

Program 10.3: demirrad.awk (lines 1 to 45; comments).

```
# Takes a 2D DEM in the form Easting, Northing, elevation      1
# and determines the local gradient and aspect of             2
# each cell using the method of Zevenbergen and               3
# Thorne (1987). It then uses this information to             4
# calculate the direct, diffuse and total solar irradiance   5
# on each cell given prescribed values of total exo-         6
# atmospheric solar irradiance, atmospheric transmittance    7
# and atmospheric pressue (at the altitude of the site and   8
# at sea level), and user-defined values of the latitude     9
# of the site, the day of year and the solar hour angle.     10
#                                                             11
# Usage: gawk -f dsirrad.awk -v size=value -v latitude=value  12
#                -v DOY=value -v hour_angle=value input_file  13
#                [> output_file]                              14
#                                                             15
# Variables:                                                  16
# ----------                                                  17
# E_0        Exo-atmospheric solar irradiance (W.m^{-2})      18
# tau        Atmospheric transmittance                       19
# p_alt      Atmospheric pressure at average altitude (mbar) 20
# p_sea      Atmospheric pressure at sea level (mbar)        21
# rad2deg    Conversion factor - radians to degrees          22
# deg2rad    Conversion factor - degrees to radians          23
# latitude   Latitude of site (initially in degrees)         24
# hour_angle Solar hour angle at site (initially degrees)    25
# zenith     Solar zenith angle (radians)                    26
# azimuth    Solar azimuth angle (radians)                   27
# declinatn  Solar declination angle (degrees)               28
# air_mass   Atmospheric air mass                            29
# zeta       Angle between sun and surface normal (radians)  30
# I_direct   Direct solar irradiance on surface (W.m^{-2})   31
# I_diffuse  Diffuse solar irradiance on surface (W.m^{-2})  32
# I_total    Total solar irradiance on surface (W.m^{-2})    33
# sky_view   Sky view factor                                 34
# min_E      eastern-most Easting in DEM (metres)            35
# max_E      western-most Easting in DEM (metres)            36
# min_N      southern-most Northing in DEM (metres)          37
# min_N      northern-most Northing in DEM (metres)          38
# this_E     Easting of current cell (metres)                39
# this_N     Northing of current cell (metres)               40
# size       DEM cell size (metres)                          41
# grad_NS    N-Sg radient of current cell (degrees)          42
# grad_EW    E-W gradient of current cell (degrees)          43
# gradient   Gradient of current cell (degrees)              44
# aspect     Aspect of current cell (degrees)                45
```

Program 10.2 (continued): demirrad.awk (lines 46 to 88; BEGIN block).

```
                                                                          46
BEGIN{                                                                    47
    # Initialize key variables                                           48
    E_0=1380.0;                                                          49
    tau=0.7;                                                              50
    p_alt=1000.0;                                                        51
    p_sea=1013.0;                                                        52
    PI=3.14159263359;                                                    53
    rad2deg=360/(2*3.1415927);                                          54
    deg2rad=1/rad2deg;                                                   55
                                                                          56
    # Convert latitude and hour angle values provided                    57
    # via command line from degrees to radians                           58
    latitude*=deg2rad;                                                  59
    hour_angle*=deg2rad;                                                60
                                                                          61
    # Calculate solar declination angle (radians)                       62
    declinatn=(-23.4*deg2rad)* \                                        63
        cos(deg2rad*(360*(DOY+10)/365));                               64
                                                                          65
    # Calculate solar zenith angle (radians)                            66
    if(hour_angle<(deg2rad*90)  hour_angle>(-90*deg2rad)){             67
        zenith=fn_acos(sin(latitude)*sin(declinatn)+\                  68
            cos(latitude)*cos(declinatn)*cos(hour_angle));             69
    } else {                                                            70
        zenith=fn_acos(sin(latitude)*sin(declinatn)-\                  71
            cos(latitude)*cos(declinatn)*cos(hour_angle));             72
    }                                                                   73
                                                                          74
    # Calculate solar azimuth angle (radians)                           75
    if(hour_angle>0){                                                   76
        azimuth=(deg2rad*180)+fn_acos(-1*(sin(declinatn)-\             77
            cos(zenith)*sin(latitude))/\                               78
(cos(latitude)*sin(zenith)));                                          79
    } else {                                                            80
        azimuth=(deg2rad*180)-fn_acos(-1*(sin(declinatn)-\             81
            cos(zenith)*sin(latitude))/\                               82
(cos(latitude)*sin(zenith)));                                          83
    }                                                                   84
                                                                          85
    # Calculate atmospheric air mass                                    86
    air_mass=(p_alt/p_sea)/cos(zenith);                                87
}                                                                       88
```

Program 10.2 (continued): demirrad.awk (lines 89 to 131; main pattern-action and END blocks).

```
                                                                                89
{                                                                               90
   # Read in Easting and Northing of current cell from DEM                      91
   E=$1;                                                                         92
   N=$2;                                                                         93
                                                                                94
   # Initialize min. and max. Easting and Northing of DEM                       95
   if(NR==1){min_E=E; max_E=E; min_N=N; max_N=N}                                 96
                                                                                97
   # Determine min. and max. Eastings and Northings of DEM                       98
   if(E<min_E){min_E=E}                                                          99
   if(E>max_E){max_E=E}                                                         100
   if(N<min_N){min_N=N}                                                         101
   if(N>max_N){max_N=N}                                                         102
                                                                               103
   # Read elevation data into array (dem)                                       104
   dem[E,N]=$3;                                                                  105
}                                                                              106
                                                                               107
END{                                                                           108
   # Pass across DEM by Easting                                                 109
   for(this_E=min_E+size;this_E<max_E;this_E+=size){                           110
      # Pass up through DEM by Northing                                         111
      for(this_N=min_N+size;this_N<max_N;this_N+=size){                        112
                                                                               113
         # Calculate Zevenbergen and Thorne's parameters                       114
         grad_EW=(dem[this_E-size,this_N]- \                                    115
                 dem[this_E+size,this_N])/(2*size);                             116
         grad_NS=(-dem[this_E,this_N-size]+ \                                   117
                 dem[this_E,this_N+size])/(2*size);                             118
                                                                               119
         # Calculate the gradient (radians)                                     120
         gradient=deg2rad*(360/(2*PI))*(grad_NS*grad_NS +\                      121
            grad_EW*grad_EW)**0.5;                                              122
                                                                               123
         # Calculate the aspect (radians)                                       124
         aspect=PI+(atan2(-grad_EW,grad_NS));                                   125
                                                                               126
         # Calculate angle of incidence of solar radiation                      127
         # for the inclined surface                                             128
         cos_zeta=(cos(gradient)*\                                              129
              cos(zenith))+(sin(gradient)*\                                     130
              sin(zenith)*cos(azimuth-aspect));                                 131
```

Program 10.2 (continued): demirrad.awk (lines 132 to 176; END block (continued)).

```
                                                                        132
        # Calculate the direct solar irradiance                        133
        I_direct=(E_0)*(tau**air_mass)*cos_zeta;                        134
        if(I_direct<0){I_direct=0}                                      135
                                                                        136
        # Print out the Easting, Northing and direct solar             137
        # irradiance for this cell                                     138
        printf("%6d %6d %6.1f\n", this_E, this_N,                       139
            I_direct) > "efyrnwy.dir";                                  140
                                                                        141
        # Calculate the diffuse solar irradiance                       142
        sky_view=0.5*(1+cos(gradient));                                 143
        I_diffuse=sky_view*0.3*(1-(tau**air_mass))*\                    144
            (E_0)*cos(zenith);                                          145
        if(I_diffuse<0){I_diffuse=0}                                    146
                                                                        147
        # Print out the Easting, Northing and diffuse solar            148
        # irradiance for this cell                                     149
        printf("%6d %6d %6.1f\n", this_E, this_N,                       150
            I_diffuse) > "efyrnwy.dif";                                 151
                                                                        152
        # Calculate global solar irradiance on surface                 153
        I_global=I_direct+I_diffuse;                                    154
        if(I_global<0){I_global=0}                                      155
                                                                        156
        # Print out the Easting, Northing and diffuse solar            157
        # irradiance for this cell                                     158
        printf("%6d %6d %6.1f\n", this_E, this_N,                       159
            I_global) > "efyrnwy.tot";                                  160
                                                                        161
    }                                                                   162
  }                                                                     163
}                                                                       164
                                                                        165
function fn_abs(x){                                                     166
    if(x<0){return -1*x} else {return x}                               167
}                                                                       168
                                                                        169
function fn_acos(x){                                                    170
    if(x*x<=1){                                                         171
        return 1.570796 - atan2(x/sqrt(1-x*x),1);                       172
    } else {                                                            173
        return 0;                                                       174
    }                                                                   175
}                                                                       176
```

code. Lines 47 to 88 contain code placed in the BEGIN block. These lines initialize several of the major variables (lines 49 to 55), convert some of the variables from arc degrees into radians (lines 59 and 60), and calculate the solar declination angle (lines 63 and 64), solar zenith angle (lines 67 to 73), solar azimuth angle (lines 76 to 84) and the atmospheric air mass (line 87). Much of this code is taken directly from Program 5.2, solarrad.awk, which is presented in Chapter 5. Lines 76 to 84 are new, however, and they implement Equation 10.6 and Equation 10.7, respectively.

Lines 90 to 106 contain code placed in the main pattern-action block. This code is identical to the main block of Program 10.2, gradasp.awk, given earlier in this chapter, and it performs the same purpose, namely to read the digital elevation data from the input data file into the 2D array, dem.

Lines 108 to 164 contain code placed in the END block. These lines of code represent an extension to the END block presented in Program 10.2, gradasp.awk. Thus, they not only calculate the gradient and aspect (azimuth) of each cell in the array (lines 110 to 125), they also determine the angle between the sun and the surface normal for that cell (Equation 10.5; lines 129 to 131) and the amount of direct solar radiation (Equation 10.9; lines 134 and 135), diffuse solar radiation (Equation 10.8; lines 143 to 146) and total solar radiation (Equation 10.9; lines 154 to 155) incident upon it. The latter three sets of values are written to separate files on lines 139 and 140, lines 150 and 151, and lines 159 and 160, respectively.

A number of the calculations performed by the code outlined above make use of two user-defined functions, which are also included in Program 10.3. The first of these determines the absolute value of a given input (fn_abs; lines 166 to 168); this function is used to convert negative values into positive ones. The second calculates the inverse cosine of an angle (fn_acos; lines 170 to 176), because gawk does not provide a built-in function that serves this purpose.

### 10.5.3   Evaluation

Program 10.3 is a fairly sophisticated computational model, which brings together most of the programming techniques that have been introduced throughout this book. To run this program from the command line, the user must provide a number of input values, including the cell size of the DEM, the latitude of the study site, the day of year and the time of day, in addition to the name of the file that contains the digital elevation data on which the program operates. For example, to determine the direct, diffuse and total solar radiation incident on the terrain around Llyn Efyrnwy (size=100 and latitude=52.756) at solar noon (hour_angle=0) on June 21 (the northern summer solstice; DoY=172), the appropriate command line is as follows:

```
gawk  -f demirrad.awk  -v size=100  -v latitude=52.756  -v DOY↵   2
      ↳=172  -v hour_angle=0 efyrnwy.dem
```

Remember that the names of the three output data files do not have to be specified on the command line because these details are given explicitly in the program (lines 139 and 140, 150 and 151, and 159 and 160).

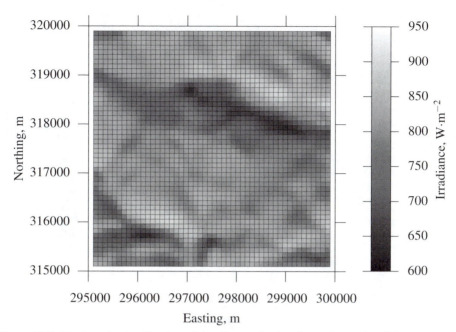

Figure 10.7: Planimetric visualization of the direct solar irradiance for each cell in the Llyn Efyrnwy DEM at solar noon on DoY = 172 (June 21).

### 10.5.4 Visualization

The direct solar irradiance data simulated using Program 10.3 can be visualized in gnuplot with the following commands:

```
set palette gray                              25
set zrange [*:*]                              26
set cbrange [*:*]                             27
set cbtics autofreq                           28
set cblabel "Irradiance W.m^{-2}"             29
splot 'efyrnwy.dir' u 1:2:3 w pm3d            30
```

These instructions produce the plot shown in Figure 10.7. Note that the instruction on line 25, above, resets the standard grayscale palette (black through to white); lines 26 and 27 reset the range of values that are plotted on the $z$-axis and on the grayscale key, respectively, so that they are determined automatically by gnuplot; line 28 instructs gnuplot to determine automatically the number of tic-marks plotted on the grayscale key; line 29 specifies the text used to label the key; and line 30 plots the direct solar irradiance data. It is evident from Figure 10.7 that the predicted values of direct solar irradiance range from about $600\,\mathrm{W\cdot m^{-2}}$ on some north-facing slopes to just under $950\,\mathrm{W\cdot m^{-2}}$ on some south-facing ones. This difference can be seen most clearly in the relatively steep valley running roughly WNW–ESE between Northings 318000 and 319000 on the eastern side of the study area. This valley contains the Afon Conwy, a large stream that drains into the Afon Efyrnwy below Llyn Efyrnwy.

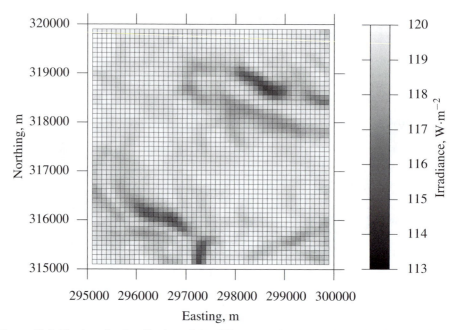

Figure 10.8: Planimetric visualization of the diffuse solar irradiance for each cell in the Llyn Efyrnwy DEM at solar noon on DoY = 172 (June 21).

Similar plots can also be constructed for the diffuse solar irradiance (Figure 10.8) and the total solar irradiance (Figure 10.9), respectively, using the following gnuplot commands:

```
splot 'efyrnwy.dif' u 1:2:3 w pm3d                                    31
```

and

```
splot 'efyrnwy.tot' u 1:2:3 w pm3d                                    32
```

Note that the estimated values of diffuse solar irradiance are smaller in absolute terms and more limited in their range ($113\,\mathrm{W \cdot m^{-2}}$ to $120\,\mathrm{W \cdot m^{-2}}$) than those of the direct solar irradiance. Also, some of the lowest values are recorded on the south-facing slopes of the Afon Conwy valley because of their steep gradient and, hence, restricted sky view factor.

**Exercise 10.1**: Using Program 10.3, `demirrad.awk`, simulate the direct, diffuse and total solar irradiance on the terrain at Llyn Efyrnwy at 8 am on June 21 and at solar noon on January 1. Use gnuplot to visualize the results obtained from each of these experiments.

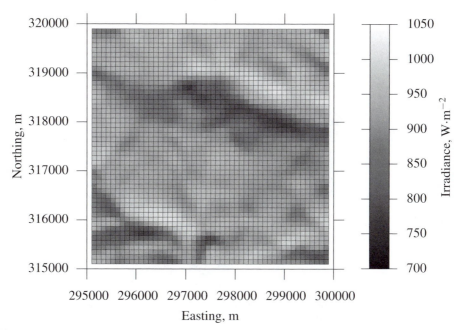

Figure 10.9: Planimetric visualization of the total (global) solar irradiance for each cell in the Llyn Efyrnwy DEM at solar noon on DoY $= 172$ (June 21).

## 10.6  MODELING HYDROLOGICAL NETWORKS

Information on drainage basins, such as the catchment boundaries (or watersheds) and the spatial location and topological connectivity of stream and river networks, is important to many studies in hydrology and geomorphology (Burrough and Mc-Donnell 1998, Beven 2001, Mulligan 2004). Some of the information required by these studies can be obtained directly from automated analyses of DEMs including, most notably, the flow paths that precipitation falling on the terrain subsequently trace across the land surface (Burrough and McDonnell 1998, Mulligan 2004). Knowledge of these pathways can be used, in turn, to model the movement and accumulation of water, soil and environmental contaminants over the terrain.

Various methods have been developed that can be used to model flow paths across a DEM, the simplest of which is known as the D8 or "eight-point pour" algorithm (Moore 1996, Burrough and McDonnell 1998, Mulligan 2004). This algorithm makes the assumption that water flows from each cell in the DEM to one, and only one, of its eight immediate neighbors, whichever defines the direction of the steepest downhill gradient (Figure 10.10). The D8 algorithm has a number of limitations for hydrological modeling, the most notable of which is that flow is restricted to the eight cardinal (north, south, east and west) and ordinal (north-east, south-east, south-west and north-west) directions. A corollary of this is that flow dispersion is not taken into account. Nevertheless, the D8 algorithm provides a useful starting point for modeling hydrological networks and is the method employed here.

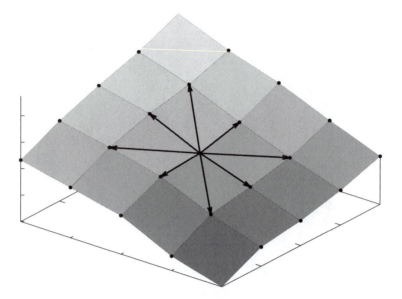

Figure 10.10: Potential directions in which water may flow across a DEM in the D8 or "eight-point pour" algorithm.

Figure 10.11 illustrates the application of the D8 algorithm to an example DEM. The number in each cell of the array represents its elevation in meters. The arrows show the direction of steepest downhill gradient from each cell to one of its eight immediate neighbors. Collectively, the arrows define the flow paths across the DEM, which are also known as local drainage direction (LDD) vectors.

### 10.6.1  Implementation

Program 10.4, d8ldd.awk, presents an implementation in gawk of the D8 method for determining the local drainage directions across a DEM. This program shares many common elements with Program 10.2 and Program 10.4, including a set of explanatory comments (lines 1 to 27), the main pattern-action block (lines 29 to 45), and the two for loops in the END block, which are used to traverse east to west and south to north across the DEM array (lines 49 and 51). d8ldd.awk differs from the preceding two programs, however, in terms of how it processes the data stored in the array, dem. Lines 56, 58 and 59 define two further nested for loops, which are used to examine the elevation values in a 3 × 3 cell window centered on the current cell in the array. The elevation of the central cell in the window is compared with those of its eight neighbors to establish which of these results in the steepest downhill gradient (lines 62 to 74). The Easting and Northing of that cell, and its gradient with respect to the central cell, are recorded (lines 70 to 72). The program prints out the Easting and Northing of the current cell (i.e., the central cell in the 3 × 3 window, from which flow occurs), the relative coordinates of its steepest downhill neighbor (i.e., the cell into which flow occurs) and their respective elevation values (lines 82 to 85).

Program 10.4: d8ldd.awk (lines 1 to 45; comments and main pattern-action block).

```
# Implementation of the D8 ("eight point pour") algorithm.      1
# Derives local drainage direction (LDD) from a DEM (see        2
# [Burrough and McDonnell 1998]). Prints out coordinates of     3
# each cell, the relative coordinates to the neighboring        4
# cell in the direction of steepest downhill descent, plus      5
# the elevations of these two cells.                            6
#                                                               7
# Usage: gawk -f d8ldd.awk -v size=value input [> output]       8
#                                                               9
# Variables:                                                   10
# ----------                                                   11
# size       DEM cell size (meters)                            12
# min_E      western-most Easting in DEM (meters)              13
# max_E      eastern-most Easting in DEM (meters)              14
# min_N      northern-most Northing in DEM (meters)            15
# min_N      southern-most Northing in DEM (meters)            16
# this_E     Easting of current cell (meters)                  17
# this_N     Northing of current cell (meters)                 18
# offset_E   Offset (Easting) from current cell (meters)       19
# offset_N   Offset (Northing) from current cell (meters)      20
# steep_E    Offset (Easting) to neighboring cell with         21
#            steepest gradient (meters)                        22
# steep_N    Offset (Northing) to neighboring cell with        23
#            steepest gradient (meters)                        24
# gradient   Gradient between the current cell and one of      25
#            its eight immediate neighbors (degrees)           26
# steepest   Steepest gradient in neighborhood (degrees)       27
                                                               28
{                                                              29
   # Read in Easting and Northing of current cell from DEM     30
   E=$1;                                                       31
   N=$2;                                                       32
                                                               33
   # Initialize min. and max. Easting and Northing of DEM      34
   if(NR==1){min_E=E; max_E=E; min_N=N; max_N=N}              35
                                                               36
   # Determine min. and max. Eastings and Northings of DEM     37
   if(E<min_E){min_E=E}                                        38
   if(E>max_E){max_E=E}                                        39
   if(N<min_N){min_N=N}                                        40
   if(N>max_N){max_N=N}                                        41
                                                               42
   # Read elevation data into array (DEM)                      43
   dem[E,N]=$3;                                                44
}                                                              45
```

Program 10.2 (continued): d8ldd.awk (lines 46 to 89; END block).

```
                                                                              46
END {                                                                         47
    # Pass across DEM by Easting                                              48
    for(this_E=min_E+size;this_E<max_E;this_E+=size){                         49
        # Pass up through DEM by Northing                                     50
        for(this_N=min_N+size;this_N<max_N;this_N+=size){                     51
                                                                              52
            # Initialize steep downhill descent                              53
            steepest=-999;                                                    54
            # Loop through eight neighboring cells, by Easting               55
            for(offset_E= -size;offset_E<=size;offset_E+=size){               56
                # ...and now loop by Northing                               57
                for(offset_N= -size;offset_N<=size;                          58
                    offset_N+=size){                                         59
                                                                              60
                    # Ignore center cell, from which flow occurs            61
                    if(!((offset_E==0) && (offset_N==0))){                    62
                        gradient=dem[this_E,this_N]-\                         63
                    dem[this_E+offset_E,this_N+offset_N];                     64
                        gradient/=sqrt((offset_E*offset_E)+\                  65
                    (offset_N*offset_N));                                     66
                        if(gradient>steepest){                               67
                            # Steepest downhill gradient detected.           68
                            # Record gradient and direction.                 69
                            steepest=gradient;                               70
                            steep_E=offset_E;                                71
                            steep_N=offset_N;                                72
                        }           # End of inner 'if' block                73
                    }               # End of outer 'if' block                74
                                                                              75
                }                   # End of inner 'for' loop                76
            }                       # End of outer 'for' loop                77
                                                                              78
            # Print coordinates of cell from which flow                     79
            # occurs and relative coordinates of cell into                  80
            # which flow occurs, plus their elevations.                     81
            printf("%6i %6i %6i %6i %6.1f %6.1f\n",\                          82
                this_E, this_N, steep_E, steep_N, \                          83
                dem[this_E,this_N], \                                        84
                dem[this_E+steep_E,this_N+steep_N]);                         85
                                                                              86
        }                           # End Northing 'for' loop                87
    }                               # End Easting 'for loop                   88
}                                   # End 'END' block                         89
```

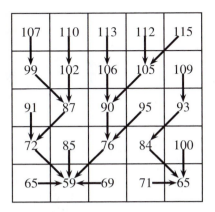

Figure 10.11: Example of a small DEM array showing the LDD vectors derived using the D8 ("eight-point pour") algorithm.

## 10.6.2   Evaluation

Program 10.4 can be run from the command line as follows:

```
gawk -f d8ldd.awk -v size=100 efyrnwy.dem > efyrnwy.ld8
```
*3*

Note that the DEM cell size (100 m) is given on the command line (`size=100`) and that the results are redirected to the file `efyrnwy.ld8` located in the working directory.

## 10.6.3   Visualization

The data generated using Program 10.4 can be visualized using the following gnuplot command:

```
plot 'efyrnwy.dd8' u 1:2:3:4 w vector nohead
```
*33*

This instruction produces the plot shown in Figure 10.12. The `vector` data style used here draws vectors between, in this case, $E$, $N$ and $E + \Delta E$, $N + \Delta N$; it requires four columns of data, namely $E$, $N$, $\Delta E$ and $\Delta N$, which are provided in the first four fields of the input data file `efyrnwy.ld8`. The `nohead` directive instructs gnuplot not to draw an arrowhead at the end of the vector, which simplifies the resulting figure in this instance. Note that it is also possible to superimpose the "blue line" features (i.e., the stream and river networks derived from the 1:50,000 scale OS topographic map), which are contained in the file `rivers.dat`, on the LDD vectors as follows:

```
plot 'efyrnwy.dd8' u 1:2:3:4 w vector nohead,\
     'rivers.dat' w l lt 3 lw 2
```
*34*
*35*

These instructions produce the plot shown in Figure 10.13, which shows a high degree of correspondence between the LDD vectors derived from the Llyn Efyrnwy DEM and the "blue line" features extracted from the OS topographic map of this area. The blue line features are clearly fewer in number and more limited in extent than the LDD vectors, owing to the selection, abstraction and generalization processes that

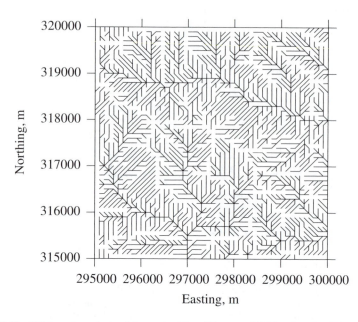

Figure 10.12: LDD vectors derived from the Llyn Efyrnwy DEM using the D8 ("eight-point pour") algorithm.

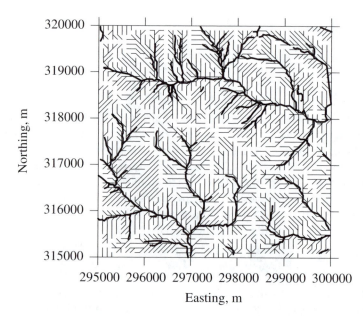

Figure 10.13: LDD vectors derived from the Llyn Efyrnwy DEM using the D8 ("eight-point pour") algorithm with OS stream network superimposed.

are inherent in map production (Buttenfield and McMaster 1991), but all of the major hydrological routes are successfully identified.

### 10.6.4   Modified Implementation

It is possible to make a few small modifications to the END block of Program 10.4 so that the LDD vectors it creates can be visualized in 3D using gnuplot's facility for drawing arrows with arbitrary start and end points in terms of the *x-* , *y-* and *z-* axes. The revised code is given in Program 10.5, d8arrows.awk, of which only the modified END block is presented here. Program 10.5 makes use of a further variable, arrow_number, which is initialized on line 49, to number the arrows that represent the LDD vectors. The number of each arrow and its start and end points in terms of Easting, Northing and elevation are printed out on lines 81 to 86. The arrow number is incremented on line 87, before the main for loops move on to consider the next cell in the DEM array.

### 10.6.5   Evaluation of the Modified Implementation

Program 10.5 can be run from the command line as follows:

```
gawk -f d8arrows.awk -v size=100 efyrnwy.dem > efyrnwy.arr
```
4

Once again, the DEM cell size (100 m) is given on the command line (size=100) and the results are redirected to a file, efyrnwy.arr, in the working directory. An extract from the output file, efyrnwy.arr, is presented in Figure 10.14.

### 10.6.6   Visualization of the LDD Vectors on a Solid Surface Model

The output from Program 10.5 can be visualized in gnuplot by issuing the following commands, most of which have been used elsewhere in this chapter:

```
load 'efyrnwy.arr'                                                      36
unset key                                                              37
set style line 1 lt -1 lw 0.01                                         38
set dgrid3d 51,51,16                                                   39
set pm3d at s hidden3d 1                                               40
set palette gray                                                       41
set colorbox vertical user origin 0.8,0.185 size 0.04,0.655           42
set cblabel "Elevation, m"                                            43
set view 45,45,1,1                                                     44
set surface                                                            45
splot 'efyrnwy.dem' w pm3d                                            46
```

These commands produce the plot shown in Figure 10.15, which gives a powerful impression of the flow paths that water takes across the terrain around Llyn Efyrnwy. Line 36 loads the set arrow instructions, which are contained in the file efyrnwy.arr, into gnuplot. The syntax of these instructions is set arrow tag from x1,y1,z1 to x2,y2,z2 nohead front, where tag is an integer that identifies a particular arrow, x1,y1,z1 denotes the start point of that arrow, x2,y2,z2 indicates its end point, nohead

Program 10.5: d8arrows.awk (lines 47 to 90 only; END block).

```
END{                                                                           47
    # Number of arrow used in 3D visualization                                 48
    arrow_number=1;                                                            49
    # Pass across DEM by Easting                                               50
    for(this_E=min_E+size;this_E<max_E;this_E+=size){                          51
        # Pass up through DEM by Northing                                      52
        for(this_N=min_N+size;this_N<max_N;this_N+=size){                      53
            # Initialize steep downhill descent                                54
            steepest=-999;                                                     55
            # Loop through eight neighboring cells, by Easting                 56
            for(offset_E= -size;offset_E<=size;offset_E+=size){                57
                # ...and now loop by Northing                                  58
                for(offset_N= -size;offset_N<=size;                            59
                    offset_N+=size){                                           60
                    # Ignore center cell, from which flow occurs               61
                    if(!((offset_E==0) && (offset_N==0))){                     62
                        gradient=dem[this_E,this_N]-\                          63
                    dem[this_E+offset_E,this_N+offset_N];                      64
                        gradient/=sqrt((offset_E*offset_E)+\                   65
                    (offset_N*offset_N));                                      66
                        if(gradient>steepest){                                 67
                            # Steepest downhill gradient detected.             68
                            # Record gradient and direction.                   69
                            steepest=gradient;                                 70
                            steep_E=offset_E;                                  71
                            steep_N=offset_N;                                  72
                        }               # End of inner "if" block              73
                    }                   # End of outer "if" block              74
                }                       # End of inner "for" loop              75
            }                           # End of outer "for" loop              76
                                                                               77
            # Print coordinates of cell from which flow                       78
            # occurs and relative coordinates of cell into                    79
            # which flow occurs, plus their elevations.                       80
            printf("set arrow %i from %6i,%6i,%6.1f to \                       81
                %6i,%6i,%6.1f", arrow_number, \                               82
                this_E, this_N, dem[this_E,this_N], \                         83
                this_E+steep_E, this_N+steep_N, \                             84
                dem[this_E+steep_E,this_N+steep_N]);                          85
            printf(" nohead front lt 1 lw 2.0\n");                            86
            ++arrow_number;     # Increment arrow number                      87
        }                       # End Northing "for" loop                     88
    }                           # End Easting "for" loop                      89
}                               # End "END" block                             90
```

```
set arrow 1  from 295100,315100, 408.0 to 295100,315000, 401.0 nohead front lt 1 lw 2.0
set arrow 2  from 295100,315200, 416.0 to 295100,315100, 408.0 nohead front lt 1 lw 2.0
set arrow 3  from 295100,315300, 421.0 to 295200,315400, 412.0 nohead front lt 1 lw 2.0
set arrow 4  from 295100,315400, 419.0 to 295200,315500, 403.0 nohead front lt 1 lw 2.0
set arrow 5  from 295100,315500, 408.0 to 295100,315600, 395.0 nohead front lt 1 lw 2.0
set arrow 6  from 295100,315600, 395.0 to 295100,315700, 382.0 nohead front lt 1 lw 2.0
set arrow 7  from 295100,315700, 382.0 to 295100,315800, 361.0 nohead front lt 1 lw 2.0
set arrow 8  from 295100,315800, 361.0 to 295200,315800, 357.0 nohead front lt 1 lw 2.0
set arrow 9  from 295100,315900, 385.0 to 295100,315800, 361.0 nohead front lt 1 lw 2.0
set arrow 10 from 295100,316000, 396.0 to 295100,315900, 376.0 nohead front lt 1 lw 2.0
set arrow 11 from 295100,316100, 400.0 to 295200,316200, 383.0 nohead front lt 1 lw 2.0
set arrow 12 from 295100,316200, 399.0 to 295200,316300, 374.0 nohead front lt 1 lw 2.0
set arrow 13 from 295100,316300, 398.0 to 295200,316300, 374.0 nohead front lt 1 lw 2.0
set arrow 14 from 295100,316400, 394.0 to 295200,316400, 365.0 nohead front lt 1 lw 2.0
set arrow 15 from 295100,316500, 386.0 to 295200,316600, 349.0 nohead front lt 1 lw 2.0
set arrow 16 from 295100,316600, 382.0 to 295200,316600, 349.0 nohead front lt 1 lw 2.0
set arrow 17 from 295100,316700, 369.0 to 295200,316600, 349.0 nohead front lt 1 lw 2.0
set arrow 18 from 295100,316800, 370.0 to 295200,316800, 353.0 nohead front lt 1 lw 2.0
set arrow 19 from 295100,316900, 368.0 to 295200,317000, 336.0 nohead front lt 1 lw 2.0
set arrow 20 from 295100,317000, 359.0 to 295200,317000, 336.0 nohead front lt 1 lw 2.0
set arrow 21 from 295100,317100, 347.0 to 295200,317000, 336.0 nohead front lt 1 lw 2.0
set arrow 22 from 295100,317200, 374.0 to 295100,317100, 347.0 nohead front lt 1 lw 2.0
set arrow 23 from 295100,317300, 388.0 to 295100,317200, 374.0 nohead front lt 1 lw 2.0
set arrow 24 from 295100,317400, 401.0 to 295100,317300, 388.0 nohead front lt 1 lw 2.0
set arrow 25 from 295100,317500, 417.0 to 295100,317400, 401.0 nohead front lt 1 lw 2.0
```

Figure 10.14: Extract from the data file efyrnwy.arr produced by Program 10.5, d8arrows.awk (first 25 records).

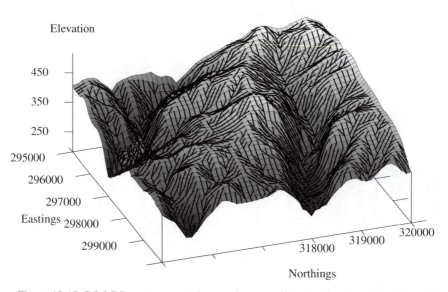

Figure 10.15: D8 LDD vectors superimposed onto a 3D visualization of the Llyn Efyrnwy DEM.

specifies that the arrow should be drawn without an arrowhead (i.e., as a straight line segment) and front instructs gnuplot to plot the arrow on top of the 3D surface. The resulting arrows are superimposed on the solid surface plot of the Llyn Efyrnwy DEM, which is produced by the commands on lines 37 to 46.

## 10.7   SUMMARY AND FURTHER DIRECTIONS

This chapter provides an introduction to some of the methods that can be employed to model the amount of direct, diffuse and total solar radiation received on sloping terrain, and to simulate hydrological networks (i.e., local drainage directions), using digital elevation data stored in a 2D array. Many of the numerical and computational techniques introduced in the preceding chapters are combined here to address these challenges.  In one sense, therefore, this chapter is the culmination of the various topics covered throughout the book. Viewed from a wider perspective, however, it is only just the beginning, an entry point to the field of environmental modeling. There are many routes that the reader might wish to explore from here, according to his or her areas of interest.  These routes could involve computational implementations of environmental models expressed in mathematical form, which are published in the scientific literature, or perhaps the development of new mathematical models *ab initio*. The skills developed throughout this book should enable the reader to engage in either of these activities.

Beyond this point, the reader should consult more widely the scientific literature concerned with different aspects of environmental modeling. Excellent overviews are presented by Jakeman *et al.* (1993), Ford (1999), Deaton and Winebrake (2000) and

Wainwright and Mulligan (2004). Two texts by Harte (1988, 2001) are also highly recommended; these books cover a wide range of environmental problems and they adopt a didactic approach which enhances the reader's skills in developing conceptual models and in formulating these models in mathematical terms.

Issues concerning the computational implementation of a range of ecological and evolutionary models are presented in Wilson (2000) and Donovan and Welden (2001, 2002). The first of these three books uses the C programming language; the second and third employ standard spreadsheet packages. All three books present material which is complementary to that introduced here.

Further coverage of hydrological modeling can be found in Abbott and Refsgaard (1996) and Beven (1998, 2001). These books cover several hydrological models and their applications, and provide sufficient detail to help the reader construct computational implementations of the models concerned.

Over the last 40 years, the computational models designed to represent various aspects of Earth's climate system have become ever more sophisticated and, hence, increasingly complex. An excellent introduction to this evolving field is provided by McGuffie and Henderson-Sellers (2005), who cover topics ranging from basic energy balance models and Daisyworld, through Earth system models of intermediate complexity (EMIC), to coupled climate system models.

Increasingly, given the explicit spatial nature of many environmental problems, many types of environmental models are being implemented in the framework of a GIS. In this context, instructive overviews of the various applications of GIS to problems in environmental modeling are given in a number of texts, including those by Goodchild *et al.* (1996), Clarke *et al.* (2002) and Brimicombe (2003).

The field of environmental modeling is also considered in a number of academic journals, which present research papers on various aspects of environmental modeling and software development. The main journals include *Geographical and Environmental Modelling* (Taylor & Francis), *Environmental Modelling and Software* (Elsevier) and *Environmental Modeling and Assessment* (Springer). The reader may wish to consult these and other related journals for research and review articles on specific topics of interest, and to be informed of the latest developments in the field.

Finally, whichever of these or other directions the reader subsequently decides to take, the skills acquired from this book should make the way ahead more accessible and hopefully more enjoyable. Good luck! *Pob lwc*!

## SUPPORTING RESOURCES

The following resources are provided in the `chapter10` sub-directory of the CD-ROM:

| | |
|---|---|
| `readarr.awk` | Program to read digital elevation data triplets (Easting, Northing, elevation) into a 2D gawk array. |
| `gradasp.awk` | gawk program to calculate terrain slope (gradient and aspect) from digital elevation data. |

| | |
|---|---|
| demirrad.awk | gawk program to calculate the direct, diffuse and total solar irradiance on each cell in a DEM. |
| d8ldd.awk | gawk program to determine the local drainage direction for each cell in a DEM. |
| rivers.dat | Data file containing $E, N$ coordinates of the "blue line" features (i.e., streams and rivers) extracted from the 1:50,000 scale OS topographic map covering Llyn Efyrnwy. |
| dem.bat | Command line instructions used in this chapter. |
| dem.plt | gnuplot commands used to generate the figures presented in this chapter. |
| dem.gp | gnuplot commands used to generate figures presented in this chapter as EPS files. |

# Appendix A

# Installing and Running the Software

## A.1 Introduction

Both of the software packages used in this book — gawk and gnuplot — are "free"; that is, they are covered by the GNU's Not Unix (GNU) General Public License (Appendix B) and by the gnuplot license (Appendix C), respectively. These licenses allow the programs to be copied and distributed freely to others, which means that they can be placed on any number of computers (provided, of course, that one has right of access to do so). You are strongly encouraged to read the terms of the licenses, which are given in full in Appendix B and Appendix C.

This appendix provides instructions on how to install the gawk and gnuplot software on a personal computer running either GNU/Linux or Microsoft Windows. This is preceded by a brief introduction to some basic computing concepts.

## A.2 Some Basic Computing Concepts

### A.2.1 Operating System

An operating system (OS) is a software program or collection of software programs that controls the overall operation of a computer (British Computer Society 1998). The OS performs a number of essential tasks, including the management of the computer's hardware (its memory, disk drives, keyboard, mouse and VDU) and the provision of basic services to other software programs and to the computer user (access to files, network connections, scheduling and switching rapidly between different tasks [multi-tasking], and graphical and audio operations). The part of the OS that handles access to files and programs stored on the computer's disk drives is sometimes known as the disk operating system (DOS).

A computer user normally interacts with the OS in one of three ways: (i) by typing instructions on the keyboard, which are entered and interpreted via the command-

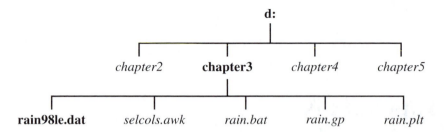

Figure A.1: Graphical representation of some files and directories on the CD-ROM, illustrating the full path to the file `rain98le.dat` under Microsoft Windows (`d:/chapter3/rain98le.dat`).

line interface (CLI); (ii) by point-and-click methods associated with a graphical user interface (GUI); or (iii) by voice control. Each of these approaches has its strengths and weaknesses. For the most part, the exercises in this book make use of the CLI approach.

### A.2.2    Files, Directories, Paths and File Systems

In computing terms, a file is a collection of related data, which is treated as a single entity and assigned a name with which it is referenced. A file can contain textual, numerical, graphical or audio data, a software program, or various other types of information. A file can be stored either temporarily in the computer's memory, while the computer is switched on, or more permanently on one of its disk drives.

Depending on which OS is installed on the computer, the name of a file may be limited to just a few letters (a ... z, A ... z) and digits (0 ... 9) or it may be possible to use a long sequence of letters, digits, blank spaces and other characters (e.g., `;$%()-@^_{}`) to describe the contents and purpose of the file. On many OSs, file names are prohibited from containing certain characters (`/\?*:|"<>`). One of the most basic, and therefore ubiquitous, file naming conventions is the so-called 8.3 system. In the 8.3 system, the filename consists of a base name of up to eight characters (a ... z, 0 ... 9, - and _) and an optional extension composed of a period (.) and up to three further characters; the filename extension is frequently used to indicate the nature and contents of the file (`.dat`, `.txt`, `.doc`). The 8.3 system is case insensitive, so that `MyFile.dat` and `myfile.dat` refer to the same file. This system is employed by Microsoft DOS, versions of Microsoft Windows prior to Windows 95, and the ISO 9660:1999 file system used on many CD-ROMs. It is also the file naming convention employed throughout this book.

A file system is used to organize the files stored on a computer disk drive so that they are uniquely referenced and can be accessed rapidly. Most file systems are hierarchical, with files organized into named directories (also known as folders). Each directory can store a collection of files, as well as other directories, which are known as sub-directories (Figure A.1 and Figure A.2). In the 8.3 file system, directory and sub-directory names can consist of up to eight alphanumeric characters (a ... z, 0 ... 9, - and _). Sub-directories can contain further sub-directories. Thus, the whole file

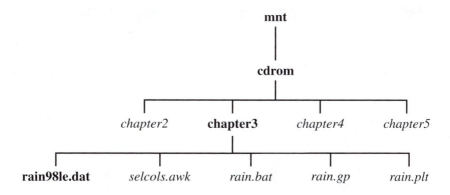

Figure A.2: Graphical representation of some files and directories on the CD-ROM, illustrating the full path to the file `rain98le.dat` under GNU/Linux (`/mnt/cdrom/chapter3 /rain98le.dat`).

system extends in a branching tree structure from the master or root directory. The collection of all directories and sub-directories on a given disk drive is often referred to as a directory tree, and a part of the tree is sometimes known as a sub-tree (British Computer Society 1998).

The location of a given file or directory in the file system is described by its path. A path can be either absolute or relative. The absolute (or full) pathname of a file or directory typically describes its location relative to the root directory of the disk drive on which it is stored. For instance, the full pathname of the file `rain98le.dat` on the CD-ROM may be `d:\chapter3\rain98le.dat` under Microsoft Windows (Figure A.1) and `/mnt/cdrom/chapter3rain98le.dat` under GNU/Linux (Figure A.2). The relative pathname usually describes the location of a file or directory with reference to the current working directory. Thus, assuming that the user is currently working in the `chapter4` directory on the CD-ROM, the relative pathname of the file `rain98le.dat` is `..\chapter3\rain98le.dat` under Microsoft Windows (Figure A.1) and `../chapter3rain98le.dat` under GNU/Linux (Figure A.2), where the symbol `..` means "move up one level in the directory tree".

## A.3  Installing and Running gnuplot

gnuplot is an interactive data and function plotting utility designed for use on personal computers running GNU/Linux, UNIX, MacOS X or Microsoft Windows, in addition to a number of other operating systems. It supports many types of 2D and 3D plots (using lines, points, boxes, contours, vector fields and surfaces), as well as various other specialized plot types. gnuplot can direct its output to the computer screen, pen plotters, printers, and various forms of graphics file (EMF, LATEX, JPEG, PNG, PS, PDF and SVG).

The current official version of gnuplot, used throughout this book, is version 4.0, which was released on April 15, 2004. Copies of the software intended for use on computers running either GNU/Linux or Microsoft Windows are included on the

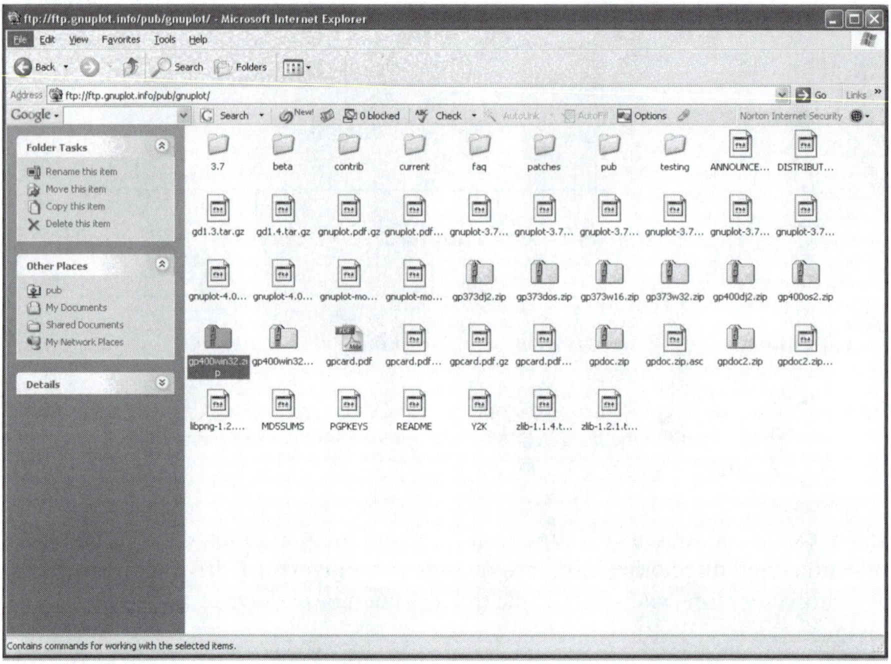

Figure A.3: The gnuplot download page viewed in a standard web browser (ftp://ftp.gnuplot.
info/pub/gnuplot/).

CD-ROM; copies for these and other operating systems can also be obtained from
the gnuplot homepage (http://www.gnuplot.info). Note that a new version of gnuplot
(v4.1) is currently under development. This version can also be downloaded from the
gnuplot homepage.

The remainder of this section describes how to download, install and run gnuplot
on a personal computer running either GNU/Linux or a recent version of Microsoft
Windows.

### A.3.1   Instructions for Microsoft Windows

1. a) Installing from the gnuplot web site

   It is recommended that you download the latest version of the software from the
   gnuplot web site, assuming that you have access to a reasonably fast network
   connection. Using a web browser, go to ftp://ftp.gnuplot.info/pub/gnuplot/ and
   locate the file marked `gp400win32.zip` (Figure A.3). This file is a compressed
   (zipped) archive, approximately 2.9 MB in size, which contains the gnuplot
   software, documentation, examples and associated files. Save this file on your
   hard disk, either by dragging and dropping it onto the Microsoft Windows desktop or by right-clicking the relevant icon and then selecting the `Save as...` option.

(a) Extracting the archive

(b) File extraction wizard.

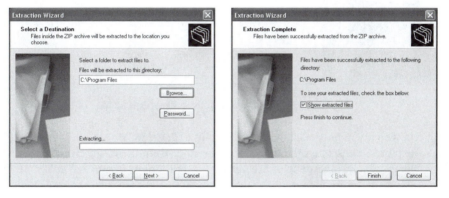

(c) Selecting where to extract the archive.

(d) Completing the extraction procedure.

Figure A.4: Installing gnuplot on Microsoft Windows XP. See text for details.

b) Installing from the CD-ROM

The zip archive gp400win32.zip is also available on the CD-ROM, and can be found in the software directory (d:\software\gp400win32.zip).

2. To unpack the archive, right-click on the gp400win32.zip icon and select the Extract All... option (Figure A.4a), which will start the file extraction wizard (Figure A.4b). Continue by clicking the Next > button.

3. You should now instruct the computer where to install the software. Using the Browse... button, select the C:\Program Files folder (Figure A.4c) if you have permission to do so; otherwise, accept the default location suggested by the installation wizard or choose another folder to which you have access. Now click Next > (Figure A.4c) and then Finish (Figure A.4d).

4. At this point, you should be presented with a file explorer, open at the folder

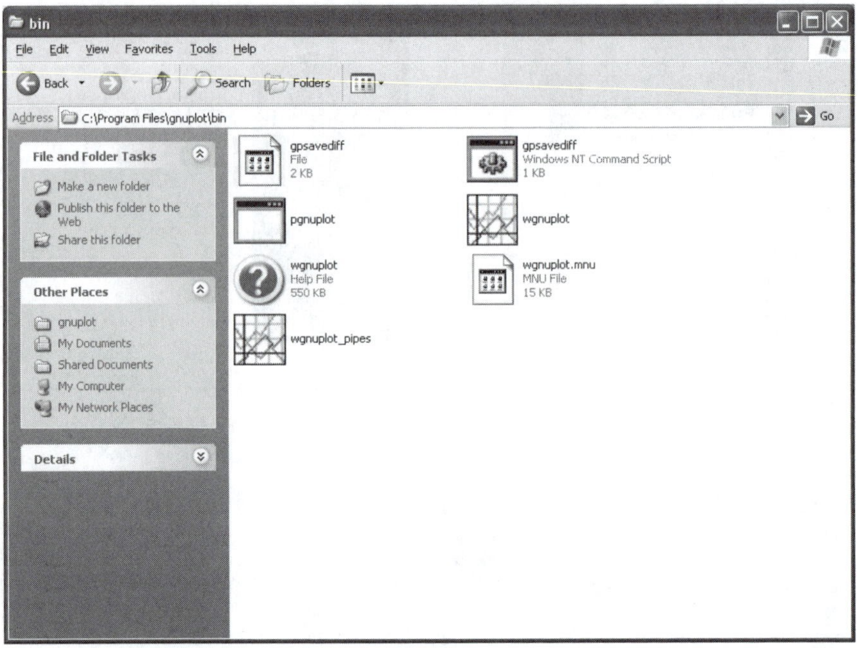

Figure A.5: Folder containing the gnuplot executable.

in which you unpacked the gnuplot archive (`C:\Program Files`). Using the file explorer, click open the `gnuplot` folder and, from there, the `bin` folder. The latter contains the gnuplot executable (i.e., program; Figure A.5). Locate the `wgnuplot` program icon, then right-click on this to create a shortcut. Drag the shortcut icon onto the Microsoft Windows desktop.

5. To start gnuplot, double-click the shortcut icon on the Microsoft Windows desktop to open the interface shown in Figure A.6. gnuplot is now ready to run.

### A.3.2   Instructions for GNU/Linux

1. A version of gnuplot is included with most major distributions of GNU/Linux, in which case it is possible to start gnuplot by opening a GNU/Linux console, or an xterm, and then typing `gnuplot`. This should present you with a command-line interface similar to that shown in Figure A.7.

2. If gnuplot is already installed on your system, check which version is running. Figure A.7, for instance, indicates that the system concerned is using version 4.0 of the software. If your system reports an older version, especially if it precedes version 3.7, you should consider installing a newer one.

3. a) Installing from an RPM

   The procedure used to install gnuplot on GNU/Linux systems differs slightly from one distribution to another. Many distributions make use of the RPM

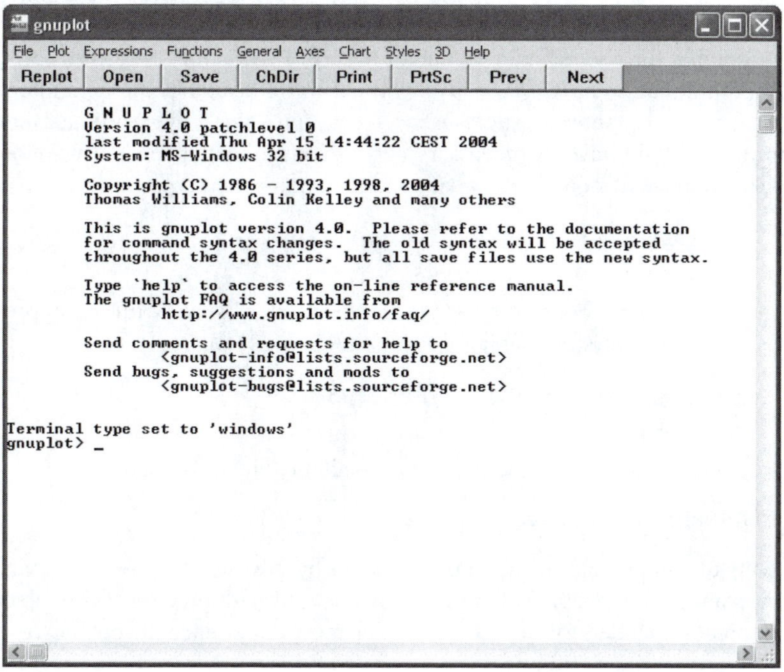

Figure A.6: GUI version of gnuplot for Microsoft Windows.

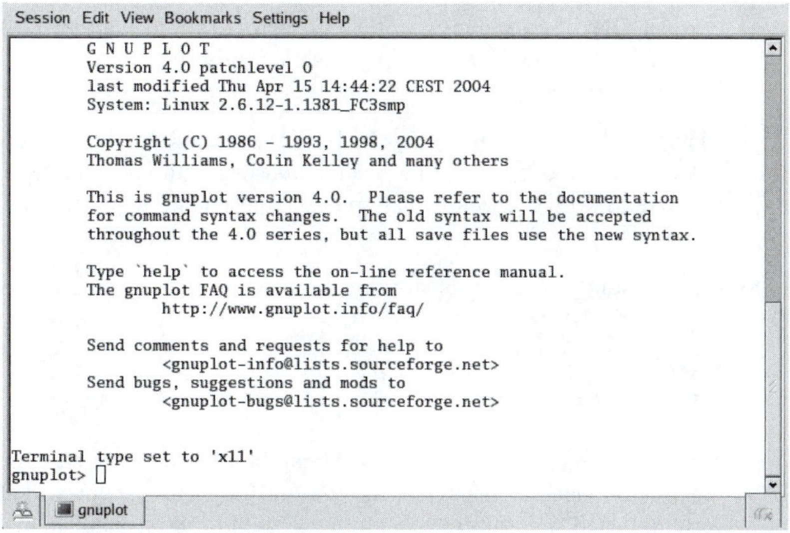

Figure A.7: Command-line interface for gnuplot running under GNU/Linux.

Package Manager (RPM) installation system, however, which is a convenient solution in most cases. To use this facility, search for the most recent version of gnuplot for your system on one of the RPM repository web sites (e.g., http://rpmfind.net/linux/RPM/) and download this file onto your computer. Log onto your computer as super-user (if you don't know what this is, then you probably don't have the necessary permissions to install new software on your system; instead, consult the system administrator) and type

```
rpm -i gnuplot-4.x.x-x.xxxx.rpm
```

in a GNU/Linux console or xterm, replacing x.x-x.xxx with the appropriate values for the version you have downloaded, for example,

```
rpm -i gnuplot-4.0.0-5.i586.rpm
```

Log out as super-user. gnuplot is now available on your system.

b) Installing from source

gnuplot can also be installed on a system by downloading and compiling the relevant source code. Although this is not a difficult procedure, it is generally not recommended for novice users. For more experienced users, however, the instructions are as follows:

- Download the latest version of the source code from the gnuplot web site (ftp.gnuplot.info/pub/gnuplot/gnuplot-4.0.0.tar.gz).

- Open a GNU/Linux console or xterm and extract the gnuplot archive by typing `tar zxvf gnuplot-4.0.0.tar.gz` (or something similar, if you have downloaded a slightly different version of the software).

- Change to the directory into which you extracted the gnuplot archive (`cd gnuplot-4.0.0`). Read the installation instructions contained in the file called INSTALL, checking any special requirements appropriate to your system.

- Compile and install the software by typing the following commands:

```
./configure                                            1
make                                                   2
make install                                           3
```

Note that numerous different options can be passed to the compiler via the `./configure` command (see the INSTALL file), and that you will almost certainly need super-user access to the system to install gnuplot in either `/usr/bin` or `/usr/local/bin`.

- gnuplot should now be available on your system.

## A.4   Installing and Running gawk

gawk is a high-level programming language that, among many other things, can be used to read, manipulate, reformat and output data contained in ASCII text files in a quick, convenient and relatively easy manner. There are several versions of awk: the one used here is the GNU Project (http://www.gnu.org/) implementation, known as gawk (Dougherty 1996, Robbins 2001).

Version 3.1.3 of gawk, which was released in June 2004, is used throughout this book. Copies of this software, designed for use with either GNU/Linux or Microsoft Windows, are included on the CD-ROM; copies for these and other operating systems can also be obtained from the gawk homepage (http://www.gnu.org/software/gawk/gawk.html).

The remainder of this section describes how to download, install and run gawk on a personal computer running either the GNU/Linux operating system or a recent version of Microsoft Windows.

### A.4.1   Instructions for Microsoft Windows

1.  a) Installing from the GnuWin32 web site

    It is recommended that you download gawk from the GnuWin32 web site (http://gnuwin32.sourceforge.net/), if you have access to a reasonably fast network connection. To do so, point your web browser at http://gnuwin32.sourceforge.net/packages.html. This web page lists a number of software tools released under GPL and similar "open source" licenses, which are commonly available on GNU/Linux and UNIX systems, that have been ported for use on 32-bit computers running Microsoft Windows.

    Locate the entry for gawk (marked "Gawk 3.1.3 pattern scanning and processing") and click on the link marked Setup on this line. This will direct you to a list of servers from which the software can be obtained. Choose the server closest to you geographically. You should then be presented with a pop-up window indicating the name of the file that you are about to download (e.g., gawk-3.1.3-2.exe), the size of the file (e.g., 6.35 MB), and three buttons that determine how or whether to proceed. Click on the button marked Run to start installing the software on your computer. You may be asked to confirm, via another pop-up window, that the installation procedure should indeed now run, in which case you should once again click the button marked Run.

    You should now follow the instructions presented by the installation wizard (Figure A.8 and Figure A.9). In most cases, all that is required is to accept the default suggestions, and to click the Next button to proceed to the subsequent step. There are, however, two important exceptions to this general rule. The first is that you will be asked to confirm whether or not you agree to the terms and conditions of the gnuplot software license (given in full in Appendix C). To proceed you must click on the I accept the agreement option (Figure A.8b). The second is that you will be asked to identify where on your system the

(a) Starting the installing wizard.

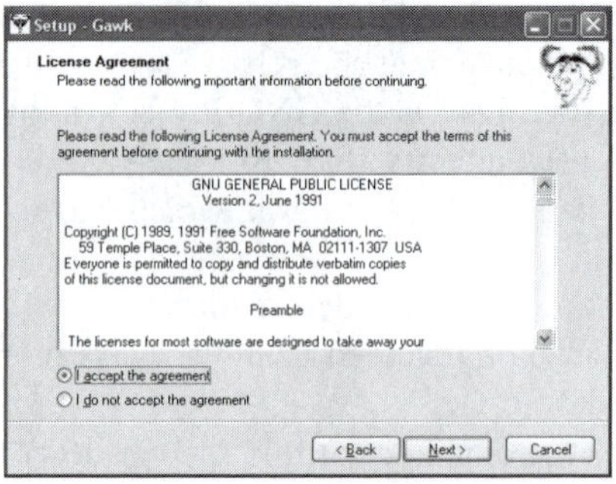

(b) Accepting the license agreement.

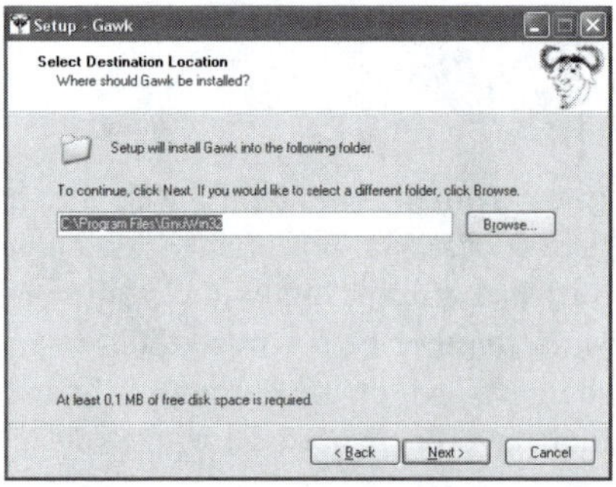

(c) Deciding where to install.

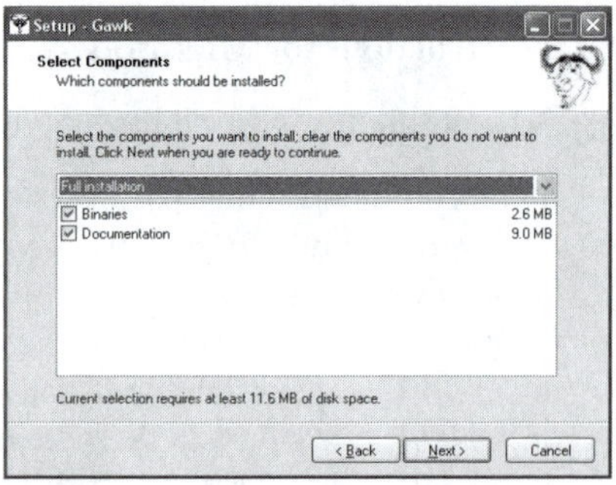

(d) Selecting the components to install.

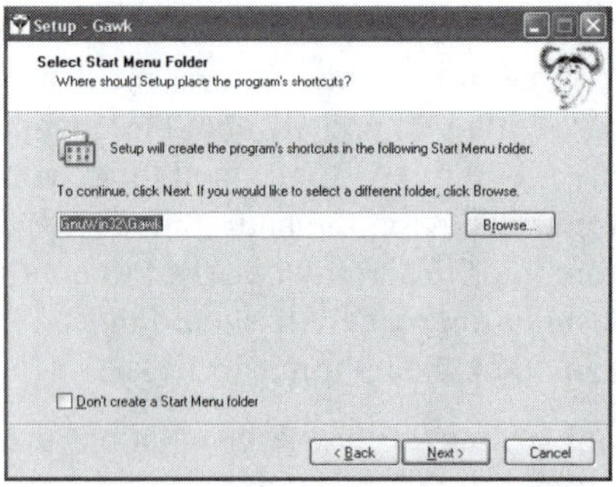

(e) Additional installation tasks.

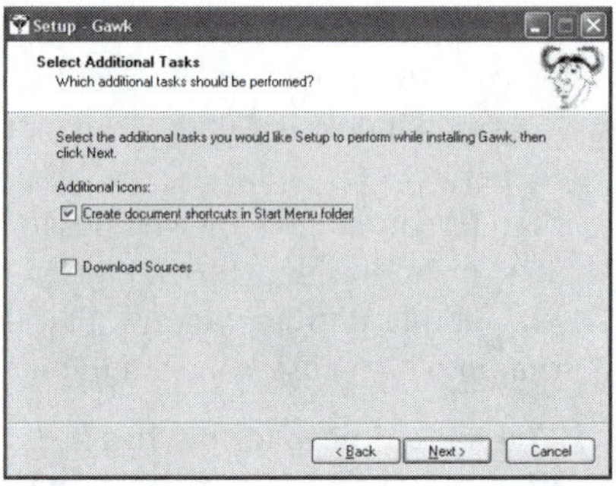

(f) Starting the installation procedure.

Figure A.8: Some of the steps involved in installing gawk on Microsoft Windows.

Figure A.9: Further steps involved in installing gawk on Microsoft Windows.

software should be installed: it is recommended that you click on the Browse button and select the C:\Program Files\GnuWin32 folder (Figure A.8c) if you have permission to do so; otherwise, accept the default location suggested by the installation wizard or choose another folder to which you have access.

b) Installing from the CD-ROM

The program used to set up gawk on your computer (gawk-3.1.3-2.exe) is also provided in the software directory on the CD-ROM (d:\software\gawk -3.1.3-2.exe). Use a file explorer to locate this file, double-click on its icon, and follow the instructions presented by the installation wizard, described above.

2. gawk is a command-line tool (i.e., it does not have a GUI that makes use of point-and-click techniques). To use gawk, therefore, one must run the CLI. In Microsoft Windows XP, the CLI (known in the past as the DOS prompt) can be found via Start→All Programs→Accessories→Command Prompt.

3. Before using gawk, however, Microsoft Windows must be told where to find the gawk utility (gawk.exe). The most convenient way of doing this is to add the folder in which gawk.exe is installed to the environment variable known as the Path. This can be achieved by right-clicking on the My Computer icon in the Start menu (Figure A.10a), and from there selecting the Properties option. Select the Advanced tab in the Properties window, and then click on the Environment Variables button. A new window should appear (Figure A.10b). Search in the lower pane of this window, marked System variables, for the Path entry. Highlight this line in the textbox and then click on the Edit button. This process should produce a further pop-up window entitled Edit System Variable (Figure A.10c). In the text entry box that is marked Variable *value*:, add the following text to the end of the current path ;C:\ Program Files\GnuWin32\bin (or another pathname, if you have installed gawk elsewhere on your system) and then click the OK button.

4. The gawk utility should now be available on your system. To test this, open a command-line window (see above) and type the command gawk. You should be presented with a response similar to that shown in Figure A.11.

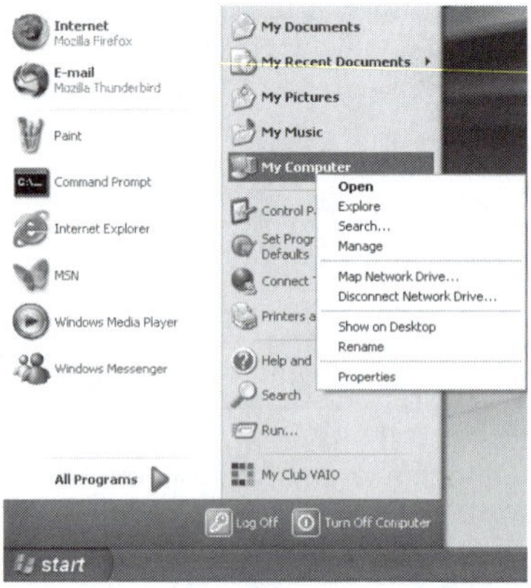

(a) Accessing the Windows environment variables.

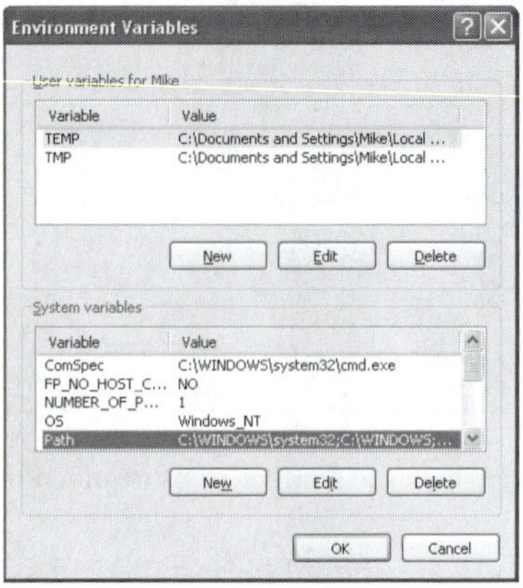

(b) The `Path` environment variable.

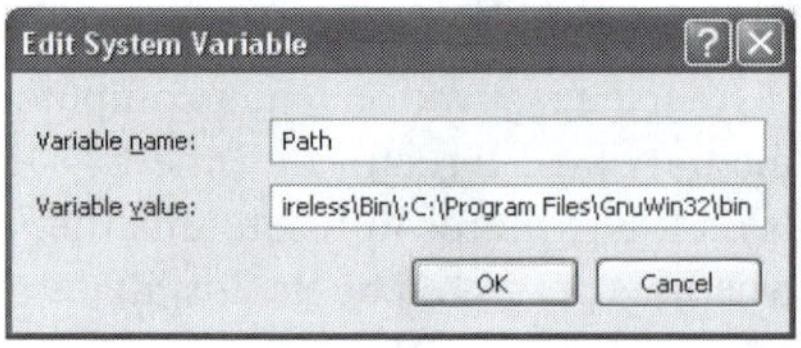

(c) Modifying the `Path`.

Figure A.10: Instructing Microsoft Windows where to find gawk on the system.

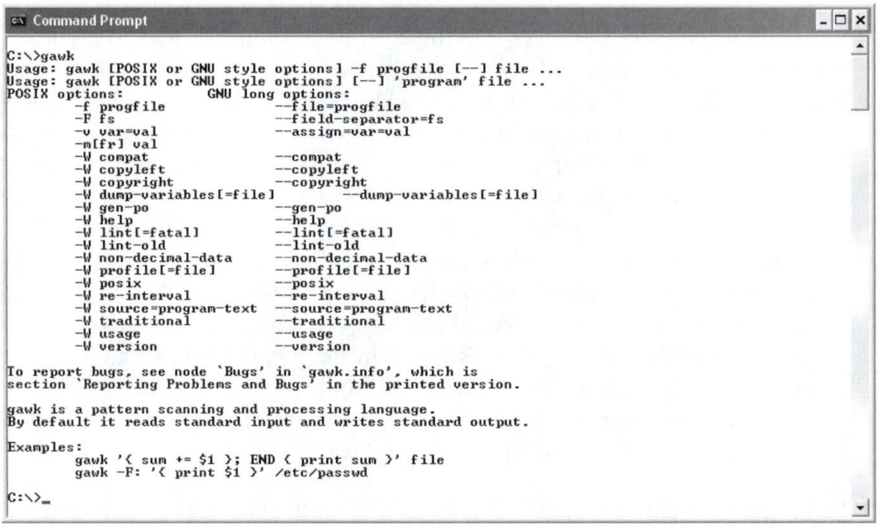

Figure A.11: Running gawk under Microsoft Windows.

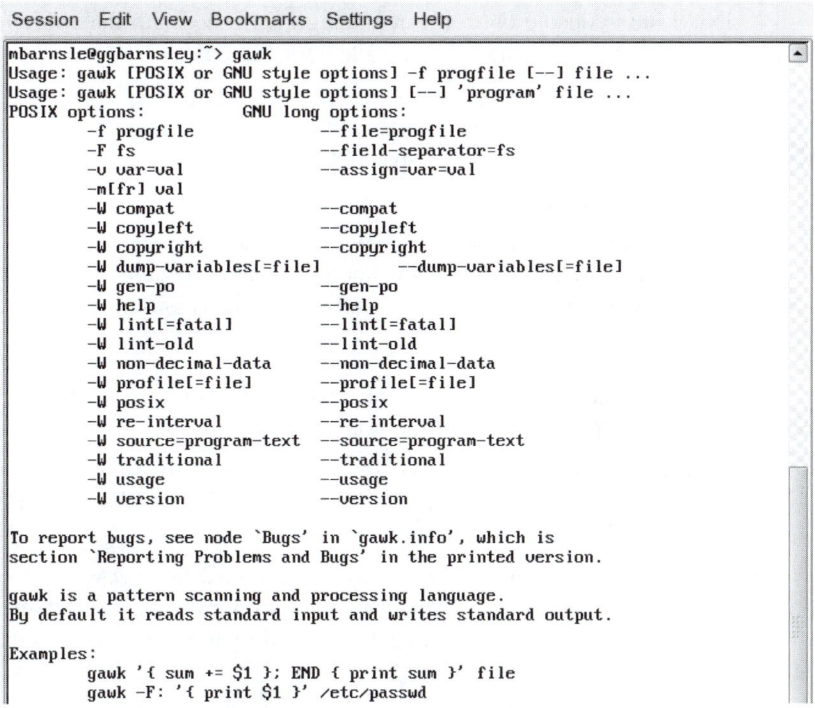

```
Session  Edit  View  Bookmarks  Settings  Help

mbarnsle@ggbarnsley:~> gawk
Usage: gawk [POSIX or GNU style options] -f progfile [--] file ...
Usage: gawk [POSIX or GNU style options] [--] 'program' file ...
POSIX options:          GNU long options:
        -f progfile             --file=progfile
        -F fs                   --field-separator=fs
        -v var=val              --assign=var=val
        -m[fr] val
        -W compat               --compat
        -W copyleft             --copyleft
        -W copyright            --copyright
        -W dump-variables[=file]        --dump-variables[=file]
        -W gen-po               --gen-po
        -W help                 --help
        -W lint[=fatal]         --lint[=fatal]
        -W lint-old             --lint-old
        -W non-decimal-data     --non-decimal-data
        -W profile[=file]       --profile[=file]
        -W posix                --posix
        -W re-interval          --re-interval
        -W source=program-text  --source=program-text
        -W traditional          --traditional
        -W usage                --usage
        -W version              --version

To report bugs, see node `Bugs' in `gawk.info', which is
section `Reporting Problems and Bugs' in the printed version.

gawk is a pattern scanning and processing language.
By default it reads standard input and writes standard output.

Examples:
        gawk '{ sum += $1 }; END { print sum }' file
        gawk -F: '{ print $1 }' /etc/passwd
```

Figure A.12: gawk running in a GNU/Linux console.

## A.4.2  Instructions for GNU/Linux

1.  A version of gawk is included with most major distributions of GNU/Linux, in which case it is possible to use the gawk utility by typing `gawk` in a GNU/Linux console or xterm. Running gawk in this way without further instructions causes it to print out basic information on how it should be used (Figure A.12). The version of gawk available on the system can be checked by typing `gawk --version` on the command line.

2.  If gawk is not already available on your system, you can install it in one of two ways described below.

    a) Installing from an RPM

    The procedure used to install gawk from an RPM on GNU/Linux systems is much the same as that described for gnuplot in Section A.3.2. Search for the most recent version of gawk for your system on one of the RPM repository web sites (e.g., http://rpmfind.net/linux/RPM/) and download this file onto your computer. Log onto your computer as super-user (if you don't know what this is, then you probably don't have the necessary permissions to install new software on your system; instead, consult the system administrator) and type

    `rpm -i gawk-3.1.x.xxxx.rpm`

in a GNU/Linux console or xterm, replacing x.x.x.xxx with the appropriate values for the version you have downloaded, for example,

```
rpm -i gawk-3.1.5.i586.rpm
```

Log out as super-user. gawk is now available on your system.

b) Installing from source

gawk can also be installed on a system by downloading and compiling the relevant source code. Although this is not a difficult procedure, it is generally not recommended for novice users. For more experienced users, however, the instructions are as follows:

- Download the latest version of the source code from the gawk web site (ftp.gnu.org/pub/gawk-3.1.5.tar.gz).

- Open a GNU/Linux console or xterm and extract the gawk archive by typing tar zxvf gawk-3.1.5.tar.gz (or something similar, if you have downloaded a slightly different version of the software).

- Change to the directory into which you extracted the gawk archive; for instance, cd gawk-3.1.5. Read the installation instructions contained in the file called INSTALL, checking any special requirements appropriate to your system.

- Compile and install the software by typing the following commands:

```
./configure                                                  1
make                                                         2
make install                                                 3
```

- Note that numerous different options can be passed to the compiler via the ./configure command (see the INSTALL file), and that you will almost certainly need super-user access to the system to install gawk in either /usr/bin or /usr/local/bin.

- gawk should now be available on your system.

# Appendix B

# GNU General Public License

Version 2, June 1991

Copyright (©) 1989, 1991 Free Software Foundation, Inc.

59 Temple Place – Suite 330, Boston, MA 02111-1307, USA

## B.1 PREAMBLE

The licenses for most software are designed to take away your freedom to share and change it. By contrast, the GNU General Public License is intended to guarantee your freedom to share and change free software—to make sure the software is free for all its users. This General Public License applies to most of the Free Software Foundation's software and to any other program whose authors commit to using it. (Some other Free Software Foundation software is covered by the GNU Library General Public License instead.) You can apply it to your programs, too.

When we speak of free software, we are referring to freedom, not price. Our General Public Licenses are designed to make sure that you have the freedom to distribute copies of free software (and charge for this service if you wish), that you receive source code or can get it if you want it, that you can change the software or use pieces of it in new free programs; and that you know you can do these things.

To protect your rights, we need to make restrictions that forbid anyone to deny you these rights or to ask you to surrender the rights. These restrictions translate to certain responsibilities for you if you distribute copies of the software, or if you modify it.

For example, if you distribute copies of such a program, whether gratis or for a fee, you must give the recipients all the rights that you have. You must make sure that they, too, receive or can get the source code. And you must show them these terms so they know their rights.

We protect your rights with two steps: (1) copyright the software, and (2) offer you this license which gives you legal permission to copy, distribute and/or modify the software.

Also, for each author's protection and ours, we want to make certain that everyone understands that there is no warranty for this free software. If the software is modified by someone else and passed on, we want its recipients to know that what they have is not the original, so that any problems introduced by others will not reflect on the original authors' reputations.

Finally, any free program is threatened constantly by software patents. We wish to avoid the danger that redistributors of a free program will individually obtain patent licenses, in effect making the program proprietary. To prevent this, we have made it clear that any patent must be licensed for everyone's free use or not licensed at all.

The precise terms and conditions for copying, distribution and modification follow.

## B.2  TERMS AND CONDITIONS FOR COPYING, DISTRIBUTION AND MODIFICATION

1. This License applies to any program or other work which contains a notice placed by the copyright holder saying it may be distributed under the terms of this General Public License. The "Program", below, refers to any such program or work, and a "work based on the Program" means either the Program or any derivative work under copyright law; that is to say, a work containing the Program or a portion of it, either verbatim or with modifications and/or translated into another language. (Hereinafter, translation is included without limitation in the term "modification".) Each licensee is addressed as "you".

    Activities other than copying, distribution and modification are not covered by this License; they are outside its scope. The act of running the Program is not restricted, and the output from the Program is covered only if its contents constitute a work based on the Program (independent of having been made by running the Program). Whether that is true depends on what the Program does.

2. You may copy and distribute verbatim copies of the Program's source code as you receive it, in any medium, provided that you conspicuously and appropriately publish on each copy an appropriate copyright notice and disclaimer of warranty; keep intact all the notices that refer to this License and to the absence of any warranty; and give any other recipients of the Program a copy of this License along with the Program.

    You may charge a fee for the physical act of transferring a copy, and you may at your option offer warranty protection in exchange for a fee.

3. You may modify your copy or copies of the Program or any portion of it, thus forming a work based on the Program, and copy and distribute such modifica-

tions or work under the terms of Section 1 above, provided that you also meet all of these conditions:

(a) You must cause the modified files to carry prominent notices stating that you changed the files and the date of any change.

(b) You must cause any work that you distribute or publish, that in whole or in part contains or is derived from the Program or any part thereof, to be licensed as a whole at no charge to all third parties under the terms of this License.

If the modified program normally reads commands interactively when run, you must cause it, when started running for such interactive use in the most ordinary way, to print or display an announcement including an appropriate copyright notice and a notice that there is no warranty (or else, saying that you provide a warranty) and that users may redistribute the program under these conditions, and telling the user how to view a copy of this License. (Exception: if the Program itself is interactive but does not normally print such an announcement, your work based on the Program is not required to print an announcement.)

These requirements apply to the modified work as a whole. If identifiable sections of that work are not derived from the Program, and can be reasonably considered independent and separate works in themselves, then this License, and its terms, do not apply to those sections when you distribute them as separate works. But when you distribute the same sections as part of a whole which is a work based on the Program, the distribution of the whole must be on the terms of this License, whose permissions for other licensees extend to the entire whole, and thus to each and every part regardless of who wrote it.

Thus, it is not the intent of this section to claim rights or contest your rights to work written entirely by you; rather, the intent is to exercise the right to control the distribution of derivative or collective works based on the Program.

In addition, mere aggregation of another work not based on the Program with the Program (or with a work based on the Program) on a volume of a storage or distribution medium does not bring the other work under the scope of this License.

4. You may copy and distribute the Program (or a work based on it, under Section 2) in object code or executable form under the terms of Sections 1 and 2 above provided that you also do one of the following:

(a) Accompany it with the complete corresponding machine-readable source code, which must be distributed under the terms of Sections 1 and 2 above on a medium customarily used for software interchange; or,

(b) Accompany it with a written offer, valid for at least three years, to give any third party, for a charge no more than your cost of physically performing source distribution, a complete machine-readable copy of the corresponding source code, to be distributed under the terms of Sections 1 and 2 above on a medium customarily used for software interchange; or,

(c) Accompany it with the information you received as to the offer to distribute corresponding source code. (This alternative is allowed only for non-commercial distribution and only if you received the program in object code or executable form with such an offer, in accord with Subsection b above.)

The source code for a work means the preferred form of the work for making modifications to it. For an executable work, complete source code means all the source code for all modules it contains, plus any associated interface definition files, plus the scripts used to control compilation and installation of the executable. However, as a special exception, the source code distributed need not include anything that is normally distributed (in either source or binary form) with the major components (compiler, kernel, and so on) of the operating system on which the executable runs, unless that component itself accompanies the executable.

If distribution of executable or object code is made by offering access to copy from a designated place, then offering equivalent access to copy the source code from the same place counts as distribution of the source code, even though third parties are not compelled to copy the source along with the object code.

5. You may not copy, modify, sublicense, or distribute the Program except as expressly provided under this License. Any attempt otherwise to copy, modify, sublicense or distribute the Program is void, and will automatically terminate your rights under this License. However, parties who have received copies, or rights, from you under this License will not have their licenses terminated so long as such parties remain in full compliance.

6. You are not required to accept this License, since you have not signed it. However, nothing else grants you permission to modify or distribute the Program or its derivative works. These actions are prohibited by law if you do not accept this License. Therefore, by modifying or distributing the Program (or any work based on the Program), you indicate your acceptance of this License to do so, and all its terms and conditions for copying, distributing or modifying the Program or works based on it.

7. Each time you redistribute the Program (or any work based on the Program), the recipient automatically receives a license from the original licensor to copy, distribute or modify the Program subject to these terms and conditions. You may not impose any further restrictions on the recipients' exercise of the rights granted herein. You are not responsible for enforcing compliance by third parties to this License.

8. If, as a consequence of a court judgment or allegation of patent infringement or for any other reason (not limited to patent issues), conditions are imposed on you (whether by court order, agreement or otherwise) that contradict the conditions of this License, they do not excuse you from the conditions of this License. If you cannot distribute so as to satisfy simultaneously your obligations

under this License and any other pertinent obligations, then as a consequence you may not distribute the Program at all. For example, if a patent license would not permit royalty-free redistribution of the Program by all those who receive copies directly or indirectly through you, then the only way you could satisfy both it and this License would be to refrain entirely from distribution of the Program.

If any portion of this section is held invalid or unenforceable under any particular circumstance, the balance of the section is intended to apply and the section as a whole is intended to apply in other circumstances.

It is not the purpose of this section to induce you to infringe any patents or other property right claims or to contest validity of any such claims; this section has the sole purpose of protecting the integrity of the free software distribution system, which is implemented by public license practices. Many people have made generous contributions to the wide range of software distributed through that system in reliance on consistent application of that system; it is up to the author/donor to decide if he or she is willing to distribute software through any other system and a licensee cannot impose that choice.

This section is intended to make thoroughly clear what is believed to be a consequence of the rest of this License.

9. If the distribution and/or use of the Program is restricted in certain countries either by patents or by copyrighted interfaces, the original copyright holder who places the Program under this License may add an explicit geographical distribution limitation excluding those countries, so that distribution is permitted only in or among countries not thus excluded. In such case, this License incorporates the limitation as if written in the body of this License.

10. The Free Software Foundation may publish revised and/or new versions of the General Public License from time to time. Such new versions will be similar in spirit to the present version, but may differ in detail to address new problems or concerns.

    Each version is given a distinguishing version number. If the Program specifies a version number of this License which applies to it and "any later version", you have the option of following the terms and conditions either of that version or of any later version published by the Free Software Foundation. If the Program does not specify a version number of this License, you may choose any version ever published by the Free Software Foundation.

11. If you wish to incorporate parts of the Program into other free programs whose distribution conditions are different, write to the author to ask for permission. For software which is copyrighted by the Free Software Foundation, write to the Free Software Foundation; we sometimes make exceptions for this. Our decision will be guided by the two goals of preserving the free status of all derivatives of our free software and of promoting the sharing and reuse of software generally.

## B.3 NO WARRANTY

1. Because the program is licensed free of charge, there is no warranty for the program, to the extent permitted by applicable law. Except when otherwise stated in writing the copyright holders and/or other parties provide the program "as is" without warranty of any kind, either expressed or implied, including, but not limited to, the implied warranties of merchantability and fitness for a particular purpose. The entire risk as to the quality and performance of the program is with you. Should the program prove defective, you assume the cost of all necessary servicing, repair or correction.

2. In no event unless required by applicable law or agreed to in writing will any copyright holder, or any other party who may modify and/or redistribute the program as permitted above, be liable to you for damages, including any general, special, incidental or consequential damages arising out of the use or inability to use the program (including but not limited to loss of data or data being rendered inaccurate or losses sustained by you or third parties or a failure of the program to operate with any other programs), even if such holder or other party has been advised of the possibility of such damages.

END OF TERMS AND CONDITIONS

# Appendix C

# Gnuplot License

# Appendix D

# Standards

## D.1 International Standard Date and Time Notation

The International Organization for Standardization has published ISO 8610:2004, an international standard for the numerical representation of information on dates and times (http://www.iso.org/iso/en/prods-services/popstds/datesandtime.html). The aim of ISO 8610:2004 is to remove the potential for ambiguity and confusion that arises when different national standards are employed. Extracts from ISO 8610:2004 are given in Table D.1, where YYYY denotes the four-digit calendar year, MM is the two-digit ordinal number of the calendar month (01 refers to January and 12 refers to December), DD is the two-digit ordinal number of the day of the month (01 to 31), hh is the two-digit hour (00 to 23), mm is the two-digit minute (00 to 59), ss is the two-digit second (00 to 59) and .s is one or more digits denoting a decimal fraction of a second. Note that an uppercase T, known as the time designator, is used to separate the date and time components. Unless a specific time zone designator is employed, times are expressed in Coordinated Universal Time (UTC).

Table D.1: ISO 8601:2004 date and time notations.

| Format | Notation | Example |
|--------|----------|---------|
| Basic | YYYYMM | 200512 |
| | YYYYMMDD | 20051224 |
| | YYYYMMDDThhmm | 20051224T2359 |
| | YYYYMMDDThhmmss.s | 20051224T2359.9 |
| Extended | YYYY-MM | 2005-12 |
| | YYYY-MM-DD | 2005-12-24 |
| | YYYY-MM-DDThh:mm | 2005-12-24T23:59 |
| | YYYY-MM-DDThh:mm:ss.s | 2005-12-24T23:59:59.9 |

Table D.2: SI Base Units

| Base Quantity | SI Base Unit | |
|---|---|---|
| | Name | Symbol |
| length | meter | m |
| mass | kilogram | km |
| time | second | s |
| electric current | ampere | A |
| thermodynamic temperature | kelvin | K |
| amount of substance | mole | mol |
| luminous intensity | candela | cd |

Two versions of the standard notation are recognized by ISO 8610:2004 (Table D.1). The first is known as the "basic" format, where the components of the date and time information are simply concatenated together (e.g., `20061224T235959`, meaning one second before midnight on December 24, 2006). The second is referred to as the "extended" format, where hyphens (`-`) are used to separate the year, month and day components of the date, and colons (`:`) are used to separate the hour, minute and second of the time (e.g., `2006-12-24T23:59:59`). The former offers the advantage of compactness; the latter is arguably easier for humans to read.

## D.2 SI Units (Systéme International d'Unités)

The International System of Units is an international system for measurement that is used in the fields of science and commerce (Taylor 1995). Universally referred to as SI units (from the French *Systéme International d'Unités*), the system was first established in 1960. SI units are divided into three broad categories, namely (i) base units, (ii) derived units and (iii) supplementary units. The SI base units are listed in Table D.2, while examples of the derived and supplementary units are given in Table D.3. The International System of Units also defines SI prefixes that describe common multiples of the SI units (Table D.4).

There are rules and conventions for the proper usage of SI units in text documents and scientific figures (Taylor 1995). In particular, statements of the form "$10\,m \times 5\,m$" are preferred to "$10 \times 5\,m$" because it makes it clear that the value 10 is measured in units of meters. Similarly, when referring to a range of values (in this example, temperature measured in degree Celsius), the form "$20\,°C$ to $30\,°C$" is preferred to "$20\,°C–30\,°C$" or "20 to $30\,°C$"; the form "$(20$ to $30)°C$" is also acceptable. The SI documentation also specifies the formats that should be used to represent unit symbols and unit names. For example, the following forms are all acceptable: "$kg/m^3$", "$kg·m^{-3}$" or "kilogram per cubic meter". Further information on these and other aspects of the SI units and nomenclature can be found in Taylor (1995).

Table D.3: SI derived and supplementary units

| Derived Quantity | Name | Symbol | Expressed in Base SI Units | Expressed in SI Base units |
|---|---|---|---|---|
| area | square meters | | $m^2$ | $m^2$ |
| volume | cubic meters | | $m^3$ | $m^3$ |
| speed, velocity | meters per second | | $m \cdot s^{-1}$ | $m \cdot s^{-1}$ |
| plane angle | radian | rad | $m \cdot m^{-1}$ | $m \cdot m^{-1}$ |
| solid angle | steradian | sr | $m^2 \cdot m^{-2}$ | $m^2 \cdot m^{-2}$ |
| frequency | hertz | Hz | $s^{-1}$ | $s^{-1}$ |
| force | newton | N | $m \cdot kg$ | $m \cdot kg$ |
| pressure | pascal | Pa | $N \cdot m^{-2}$ | $m^{-1} \cdot kg \cdot s^{-2}$ |
| energy | joule | J | $N \cdot m$ | $m^2 \cdot kg \cdot s^{-2}$ |
| power, radiant flux | watt | W | $J \cdot s^{-1}$ | $m^2 \cdot kg \cdot s^{-3}$ |
| Celsius temperature | degrees Celsius | °C | — | $s \cdot A$ |
| heat flux density, irradiance | watt per square meter | | $W \cdot m^{-2}$ | — |
| radiant intensity | watt per steradian | | $W \cdot sr^{-1}$ | — |
| radiance | watt per square meter per steradian | | $W \cdot m^{-2} \cdot sr^{-1}$ | — |

Table D.4: SI prefixes

| Factor | Name | Symbol |
|--------|-------|--------|
| $10^{24}$ | yotta | Y |
| $10^{21}$ | zetta | Z |
| $10^{18}$ | exa | E |
| $10^{15}$ | peta | P |
| $10^{12}$ | tera | T |
| $10^{9}$ | giga | G |
| $10^{6}$ | mega | M |
| $10^{3}$ | kilo | k |
| $10^{2}$ | hecto | h |
| $10^{1}$ | deka | da |
| $10^{-1}$ | deci | d |
| $10^{-2}$ | centi | c |
| $10^{-3}$ | milli | m |
| $10^{-6}$ | micro | $\mu$ |
| $10^{-9}$ | nano | n |
| $10^{-12}$ | pico | p |
| $10^{-15}$ | femto | f |
| $10^{-18}$ | atto | a |
| $10^{-21}$ | zepto | z |
| $10^{-24}$ | yocto | y |

# Appendix E

# Solutions to Exercises

**Exercise 2.1**

```
plot 'le98temp.dat' using 3:1                                            1
```

**Exercise 2.2**

```
reset                                                                    1
unset key                                                                2
set style data points                                                    3
set xlabel "Easting, m"                                                  4
set ylabel "Northing, m"                                                 5
set zlabel "Elevation, m"                                                6
set xtics 295000, 1000, 300000                                          7
set ytics 315000, 1000, 320000                                          8
set ztics 250, 100, 550                                                  9
splot 'efyrnwy.dem' u 1:2:3                                             10
pause -1 "Press the return key to continue"                            11
                                                                        12
set view 45, 45, 1, 1                                                  13
replot                                                                  14
pause -1 "Press the return key to continue"                            15
                                                                        16
set view 45, 60, 1, 1                                                  17
replot                                                                  18
pause -1 "Press the return key to continue"                            19
                                                                        20
set view 60, 60, 1, 1                                                  21
replot                                                                  22
pause -1  "Press the return key to continue"                           23
```

## Exercise 2.3

```
reset                                                               1
unset key                                                           2
set style data lines                                                3
set xlabel "Easting, m"                                             4
set ylabel "Northing, m"                                            5
set zlabel "Elevation, m"                                           6
set xtics 295000, 1000, 300000                                      7
set ytics 315000, 1000, 320000                                      8
set ztics 250, 100, 550                                             9
set dgrid3d 51,51,16                                               10
set style line 9                                                   11
set pm3d at s hidden3d 9                                           12
splot 'efyrnwy.dem' u 1:2:3 with pm3d                             13
pause -1 "Press the return key to continue"                       14
                                                                   15
set pm3d at b                                                      16
splot 'efyrnwy.dem' u 1:2:3 with pm3d                             17
pause -1 "Press the return key to continue"                       18
                                                                   19
set pm3d at t                                                      20
replot                                                             21
pause -1 "Press the return key to continue"                       22
                                                                   23
set pm3d at bst                                                    24
replot                                                             25
pause -1 "Press the return key to continue"                       26
```

## Exercise 2.4

```
reset                                                               1
unset key                                                           2
set xlabel "Easting, m"                                             3
set ylabel "Northing, m"                                            4
set zlabel "Elevation, m"                                           5
set xtics 295000, 1000, 300000                                      6
set ytics 315000, 1000, 320000                                      7
set style data lines                                                8
set dgrid3d 51,51,16                                                9
unset surface                                                      10
set contour base                                                   11
set cntrparam levels discrete 300, 325, 350, 375, 400, \          12
    425, 450, 475, 500                                             13
set terminal push                                                  14
set terminal table                                                 15
set output 'contours.dat'                                          16
splot 'efyrnwy.dem' using 1:2:3                                    17
set output                                                         18
```

```
set terminal pop                                          19
plot 'contours.dat' index 0 u 1:2, \                      20
     'contours.dat' index 1 u 1:2, \                       21
     'contours.dat' index 2 u 1:2, \                       22
     'contours.dat' index 3 u 1:2, \                       23
     'contours.dat' index 4 u 1:2, \                       24
     'contours.dat' index 5 u 1:2, \                       25
     'contours.dat' index 6 u 1:2, \                       26
     'contours.dat' index 7 u 1:2, \                       27
     'contours.dat' index 8 u 1:2                          28
```

## Exercise 2.5

```
reset                                                     1
set key top right box                                     2
set xdata time                                            3
set timefmt "%Y%m%d%H%M"                                  4
set format x "%b"                                         5
set xlabel "Month, 1998"                                  6
set xrange ["199801010000":"199901010000"]               7
set ylabel "Temperature, degrees Celsius"                 8
set yrange [-5:30]                                        9
set y2label "Precipitation, mm"                           10
set ytics nomirror                                        11
set y2tics nomirror                                       12
set term png                                              13
set output "myplot2.png"                                  14
plot 'le98temp.dat' u 2:3 t "Maximum temperature" w lp, \ 15
     'le98rain.dat' u 2:3 axes x1y2 t "Precipitation" w i 16
set output                                                17
```

## Exercise 3.1

```
gawk '{print $4, $5, $6, $7, $9/10}' rain98le.dat          1
```

```
YEAR MONTH DAY END_HOUR 0                                  1
1998 1 1 900 0                                             2
1998 1 1 2100 14.6                                         3
1998 1 2 900 5.2                                           4
1998 1 2 2100 1.6                                          5

1998 12 29 2100 0.2                                        713
1998 12 30 900 2.4                                         714
1998 12 30 2100 0                                          715
1998 12 31 900 0                                           716
1998 12 31 2100 0.8                                        717
```

```
gawk '{print $4 $5 $6 $7 $9/10}' rain98le.dat              2
```

```
YEARMONTHDAYEND_HOUR0                                                      1
1998119000                                                                2
199811210014.6                                                            3
1998129005.2                                                              4
19981221001.6                                                             5

1998122921000.2                                                          713
199812309002.4                                                          714
1998123021000                                                           715
199812319000                                                           716
1998123121000.8                                                         717
```

## Exercise 3.2

```
gawk  -f selcols3.awk rain98le.dat                                         1
```

```
199811900 0.000000                                                        1
1998112100  14.600000                                                     2
199812900 5.200000                                                        3
1998122100  1.600000                                                      4
199813900 35.800000                                                       5

19981230900 2.400000                                                     713
199812302100 0.000000                                                    714
19981231900 0.000000                                                     715
199812312100 0.800000                                                    716
```

Program E.1: ex3-2.awk

```
(NR >1){printf("%i%i%i%i %f", $4, $5, $6, $7, $9/10.0)}                    1
```

```
199811900 0.0000001998112100  14.600000199812900  5.200000..
```

## Exercise 3.3

Program E.2: seltemp.awk

```
(NR >1){                                                                   1
  printf("%3i %04i%02i%02iT%04i %7.2f %7.2f\n", \                          2
    NR, $4, $5, $6, $7, $9/10.0, $10/10.0);                                3
}                                                                          4
```

```
gawk  -f  seltemp.awk  temp98le.dat > temp98le.out                              1
```

```
    2  19980101T0900      1.80      -0.30                                        1
    3  19980101T2100      7.70       1.60                                        2
    4  19980102T0900      6.80       3.90                                        3
    5  19980102T2100      6.50       3.80                                        4
    6  19980103T0900      9.70       4.80                                        5

  714  19981230T0900      9.00       6.10                                      713
  715  19981230T2100      7.40       4.10                                      714
  716  19981231T0900      5.40       3.70                                      715
  717  19981231T2100      7.00       4.80                                      716
```

```
reset                                                                           1
set xdata time                                                                  2
set timefmt "%Y%m%dT%H%M"                                                        3
set format x "%d/%m"                                                             4
set xlabel "Day/Month, 1998"                                                     5
set ylabel "Temperature, t/degree Celsius"                                       6
set xrange ["19980401T0000":"19980701T0000"]                                     7
set yrange [-5:30]                                                               8
set key top right box                                                           9
set data style linespoints                                                      10
plot 'le98temp.out' u 2:3 t "Maximum temperature", \                           11
     'le98temp.out' u 2:4 t "Minimum temperature"                              12
```

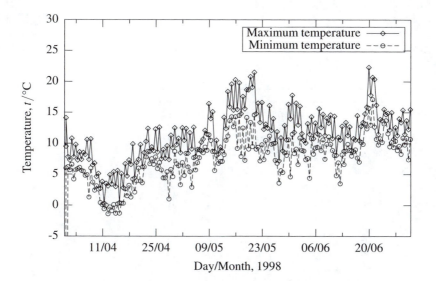

Figure E.1: Output from Exercise 3.3

## Exercise 4.1

Program E.3: winddir.awk

```
(NR>1  && $9!=-999){                                    1
  if($9>345   $9<=15){++N}                              2
  if($9>15  && $9<=45){++NNE}                           3
  if($9>45  && $9<=75){++ENE}                           4
  if($9>75  && $9<=105){++E}                            5
  if($9>105  && $9<=135){++ESE}                         6
  if($9>135  && $9<=165){++SSE}                         7
  if($9>165  && $9<=195){++S}                           8
  if($9>195  && $9<=225){++SSW}                         9
  if($9>225  && $9<=255){++WSW}                         10
  if($9>255  && $9<=285){++W}                           11
  if($9>285  && $9<=315){++WNW}                         12
  if($9>315  && $9<=345){++NNW}                         13
  ++num_obs;                                            14
}                                                       15
END{                                                    16
  printf("%3i %5.2f\n", 0, N/num_obs);                  17
  printf("%3i %5.2f\n", 30, NNE/num_obs);               18
  printf("%3i %5.2f\n", 60, ENE/num_obs);               19
  printf("%3i %5.2f\n", 90, E/num_obs);                 20
  printf("%3i %5.2f\n", 120, ESE/num_obs);              21
  printf("%3i %5.2f\n", 150, SSE/num_obs);              22
  printf("%3i %5.2f\n", 180, S/num_obs);                23
  printf("%3i %5.2f\n", 210, SSW/num_obs);              24
  printf("%3i %5.2f\n", 240, WSW/num_obs);              25
  printf("%3i %5.2f\n", 270, W/num_obs);                26
  printf("%3i %5.2f\n", 300, WNW/num_obs);              27
  printf("%3i %5.2f\n", 330, NNW/num_obs);              28
}                                                       29
```

```
gawk -f winddir.awk wind98le.dat > wind98le.dir          1
```

```
# Create a standard x-y plot                                              1
unset key                                                                 2
set xlabel "Wind direction (degrees relative to north)"                   3
set ylabel "Relative frequency"                                           4
plot 'wind98le.dir' w i lw 2                                              5
pause -1 "Press the return key to continue"                               6
# Create a polar plot                                                     7
set angles degrees        # Angles measured in degrees                    8
set polar                 # Produce a polar plot                          9
set grid polar 15.0       # Mark grid at 15 degree intervals              10
unset border              # Turn off normal borders                       11
set xtics axis            # Plot the xtics along the x-axis               12
set ytics axis            # Plot the ytics along the y=axis               13
```

```
set xrange [-0.30:0.30]                                          14
set yrange [-0.30:0.30]                                          15
set size square            # Make the output square             16
plot 'wind98le.dir' u (90-$1):($2) w i lw 2                      17
```

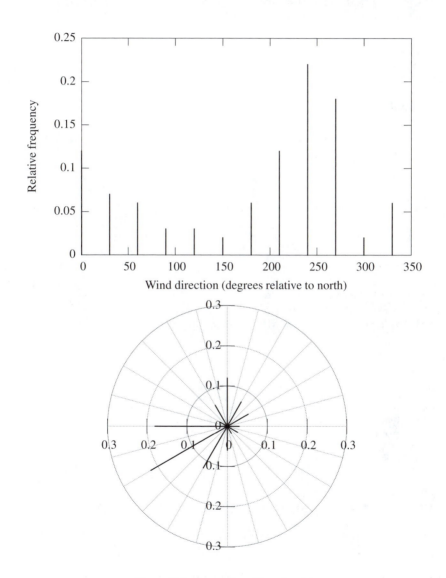

Figure E.2: Output from Exercise 4.1

## Exercise 4.2

```
phi_prime(u_prime,c,k)=exp((-1.0*(u_prime/c)**k))        1
freq_cut_in=phi_prime(3.0,5.32883,1.55037)              2
freq_cut_out=phi_prime(20.0,5.32883,1.55037)            3
print freq_cut_in-freq_cut_out                          4
```

```
0.662989225124349
```

## Exercise 4.3

```
unset key                                                           1
set xtics auto                                                      2
set xrange [0:20]                                                   3
rho=1.225               # Air density kg/m^3 at sea level           4
radius=3.0              # Rotor radius in meters                    5
area=pi*(radius**2)     # Area swept by turbine rotors              6
Cp=0.4                  # Power coefficient                         7
Ng=0.75                 # Generator efficiency                      8
Nb=0.9                  # Mechanical efficiency                     9
set dummy u                                                         10
Power(u)=area*(rho*(u**3)/2)/1000                                   11
set xlabel "Wind speed, u_0/m.s^{-1}"                              12
set ylabel "Wind power density, P/kW.m^{-2}"                       13
plot Power(u)*Cp*Ng*Nb/area lw 2                                    14
```

## Exercise 4.4

Program E.4: meanpwr.awk

```
{                                                                   1
  sum_power+=$2;                                                    2
  ++sum_hours;                                                      3
}                                                                   4
                                                                    5
END{                                                                6
  printf("Mean output while WECS active: %6.4f kW.h\n", \           7
    sum_power/sum_hours);                                           8
  printf("Mean output throughout year:   %6.4f kW.h\n", \          9
    sum_power/(365*24));                                            10
}                                                                   11
```

```
gawk -f meanpwr.awk wind98le.pwr > meanpwr.out
```

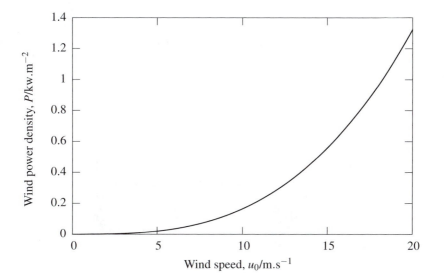

Figure E.3: Output from Exercise 4.3

```
Mean output while WECS active: 1.9322 kW.h
Mean output throughout year:   1.1856 kW.h
```

## Exercise 5.1

Program E.5: ex5-1.awk

```
(NR >1 && $7 ==1200 && $9 >=0){               1
   printf ("%4i%02i%02iT%04i  %4i\n", $4, $5, $6, $7, $9);   2
}                                              3
```

```
gawk  -f ex5 -1.awk radt98le.dat > ex5 -1.dat
```

```
reset                                          1
unset key                                      2
set xdata time                                 3
set timefmt "%Y%m%dT%H%M"                      4
set format x "%b"                              5
set xlabel "Month, 1998"                       6
set ylabel "Solar irradiance, W.h.m^-2"        7
set xrange ["19980101T1200":"19990101T1200"]   8
plot 'ex5 -1.dat' u 1:2 w i                    9
```

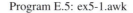

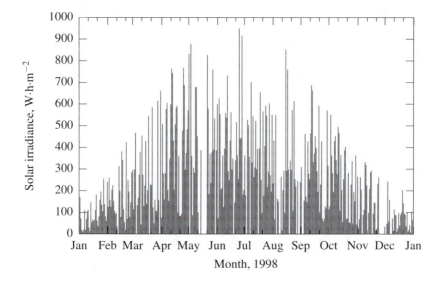

Figure E.4: Output from Exercise 5.1

## Exercise 5.2

Program E.6: ex5-2.awk

```
BEGIN{                                                              1
    latitude=52.756;           # Latitude (degrees)                2
    E_0=1380.0;                # Exo-atmos. solar irradiance        3
    tau=0.7;                   # Atmos. transmittance              4
    p_alt=1000;                # Atmos. pressure (altitude)        5
    p_sea=1013;                # Atmos. pressure (sea level)       6
    DOY=172;                   # Day of year (June 21)             7
                                                                   8
    deg2rad=(2*3.1415927)/360; # Degrees to radians                9
                                                                  10
    latitude*=deg2rad;         # Latitude in radians              11
                                                                  12
    # Solar declination angle                                     13
    declination = (-23.4*deg2rad)* \                              14
        cos(deg2rad*(360*(DOY+10)/365));                          15
                                                                  16
    # for-loop to cycle through 24 hours in a day                 17
    for(hour_angle=-180;hour_angle<=180;hour_angle+=15){          18
        # Cosine of the solar zenith angle                        19
        cos_zenith=sin(latitude)*sin(declination)+ \              20
            cos(latitude)*cos(declination)*\                      21
    cos(hour_angle*deg2rad);                                      22
                                                                  23
```

```
    # Atmospheric air mass                                          24
    air_mass=(p_alt/p_sea)/cos_zenith;                              25
                                                                    26
    # Direct, diffuse and global solar irradiance                  27
    I_direct=(E_0)*(tau**air_mass)*cos_zenith;                      28
    I_diffuse=0.3*(1-(tau**air_mass))*(E_0)*cos_zenith;             29
    I_global=I_direct+I_diffuse;                                    30
                                                                    31
    hour=12+(hour_angle/15.0);                                      32
                                                                    33
    # Output results                                               34
    if(I_global>=0){                                               35
       printf("%4i %7.3f %7.3f %7.3f\n", \                         36
          hour, I_global, I_direct, I_diffuse);                    37
    }                                                              38
  }                                                                39
}                                                                  40
```

```
gawk -f ex5-2.awk > ex5-2.out
```

```
set xdata time                                                     1
set timefmt "%Y%m%dT%H%M"                                          2
set xrange ["19980621T0000":"19980622T0000"]                      3
set x2range [0:24]                                                4
set key top left Left box                                          5
set xlabel "Time"                                                 6
set xtics nomirror                                                7
set format x "%H:%M"                                              8
set ylabel "Global solar irradiance, W.m^2"                      9
plot 'radt98le.out' u 1:2 t "Measured" w lp lw 2 pt 7, \         10
     'ex5-2.out' axes x2y1 t "Modeled" w lp lw 2 pt 7            11
```

## Exercise 6.1

### Program E.7: ex6-1.awk

```
# Simple model of solar radiation interaction with a mixed    1
# soil, vegetation and snow surface. The program calculates   2
# spectral reflectance of a surface covered by specified      3
# areal fractions of soil, vegetation (leaves) and snow       4
#                                                             5
# Usage: gawk -f mixture.awk -v rho_leaf=value \              6
#              -v rho_soil=value -v rho_snow \                7
#              [ > output_file ]                              8
#                                                             9
```

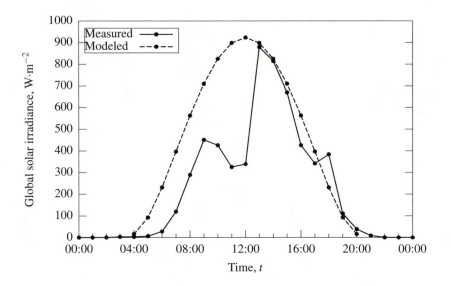

Figure E.5: Output from Exercise 5.2

```
# Variables:                                                      10
# area_leaf   Fractional area covered by leaves                   11
# rho_leaf    Leaf spectral reflectance                           12
# rho_soil    Soil spectral reflectance                           13
# rho_surface Average spectral reflectance of surface             14
                                                                  15
BEGIN{                                                            16
   for(area_leaf=0;area_leaf<=1;area_leaf+=0.1){                  17
      for(area_snow=0;area_snow<=1-area_leaf;area_snow+=0.1){     18
         area_soil=1-(area_leaf+area_snow);                       19
         rho_surface=(area_leaf*rho_leaf)+ \                      20
            area_snow*rho_snow + \                                21
            (area_soil*rho_soil);                                 22
            printf("%3.1f %3.1f %3.1f %4.2f\n", \                 23
         area_leaf, area_snow, area_soil, rho_surface);           24
      }                                                           25
   }                                                              26
}                                                                 27
```

```
gawk -f ex6-1.awk -v rho_leaf=0.475 -v rho_snow=0.75 -v ↳    1
   ↳rho_soil=0.125 > ex6-1.out
```

```
0.0 0.0 1.0 0.12                                                  1
0.0 0.1 0.9 0.19                                                  2
0.0 0.2 0.8 0.25                                                  3
0.0 0.3 0.7 0.31                                                  4
0.0 0.4 0.6 0.38                                                  5
```

```
0.8  0.0  0.2  0.40                                                   61
0.8  0.1  0.1  0.47                                                   62
0.8  0.2  0.0  0.53                                                   63
0.9  0.0  0.1  0.44                                                   64
0.9  0.1  0.0  0.50                                                   65
1.0  0.0  0.0  0.47                                                   66
```

## Exercise 6.2

```
gawk  -f 3layers.awk  -v rho_leaf=0.15  -v tau_leaf=0.07  -v ↩  1
   ↪rho_soil=0.05 > ex6-2a.dat
gawk  -f 3layers.awk  -v rho_leaf=0.07  -v tau_leaf=0.15  -v ↩  2
   ↪rho_soil=0.05 > ex6-2b.dat
gawk  -f 3layers.awk  -v rho_leaf=0.15  -v tau_leaf=0.07  -v ↩  3
   ↪rho_soil=0.20 > ex6-2c.dat
gawk  -f 3layers.awk  -v rho_leaf=0.07  -v tau_leaf=0.15  -v ↩  4
   ↪rho_soil=0.20 > ex6-2d.dat
gawk  -f 3layers.awk  -v rho_leaf=0.0  -v tau_leaf=0.475  -v ↩  5
   ↪rho_soil=0.125 > ex6-2e.dat
gawk  -f 3layers.awk  -v rho_leaf=0.475  -v tau_leaf=0.0  -v ↩  6
   ↪rho_soil=0.125 > ex6-2f.dat
```

```
reset                                                                 1
set xlabel "Leaf Area Index, LAI"                                     2
set ylabel "Spectral reflectance, rho_canopy"                         3
set style data linespoints                                            4
set key bottom right                                                  5
set yrange [0:*]                                                      6
plot 'ex6-2a.dat' \                                                   7
    t "rho_leaf=0.15, tau_leaf=0.07, rho_soil=0.05", \                8
    'ex6-2b.dat' \                                                    9
    t "rho_leaf=0.07, tau_leaf=0.15, rho_soil=0.05", \               10
    'ex6-2c.dat' \                                                   11
    t "rho_leaf=0.15, tau_leaf=0.07, rho_soil=0.20", \               12
    'ex6-2c.dat' \                                                   13
    t "rho_leaf=0.07, tau_leaf=0.15, rho_soil=0.20"                  14
pause -1                                                             15
                                                                    16
set key top right                                                   17
plot 'ex6-2a.dat' \                                                 18
    t "rho_leaf=0.0, tau_leaf=0.475, rho_soil=0.125", \             19
    'ex6-2d.dat' \                                                  20
    t "rho_leaf=0.475, tau_leaf=0.0, rho_soil=0.125", \             21
    'ex6-2e.dat' \                                                  22
    t "rho_leaf=0.0, tau_leaf=0.0, rho_soil=0.125"                  23
pause -1                                                             24
```

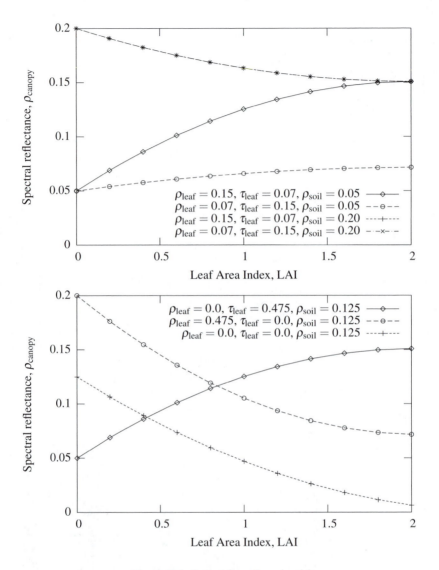

Figure E.6: Output from Exercise 6.2

## Exercise 7.1

```
gawk -f analytic.awk -v R_Leaf=0.07 -v T_Leaf=0.07 -v ↩     1
   ↪R_Soil=0.06
gawk -f analytic.awk -v R_Leaf=0.18 -v T_Leaf=0.18 -v ↩     2
   ↪R_Soil=0.05
```

```
0.0702952
0.181635
```

The spectral reflectance of the vegetation canopy at blue wavelengths predicted by the analytical model is approximately 0.703, while at green wavelengths it is roughly 0.182.

**Exercise 7.2**

```
gawk -f iterate3.awk -v R_Leaf=0.475 -v T_Leaf=0.475 -v ⇨     1
    ⇨R_Soil=0.75 -v layers=10 -v threshold=0.001
```

```
0.0010  16  0.7192
```

The numerical model, iterate3.awk, takes 16 iterations to reach a stable solution given input values of $R_L = 0.475$, $T_L = R_L$, $R_S = 0.75$, 10 leaf-layers and a threshold value of 0.001.

**Exercise 8.1**

```
gawk -f discrete.awk -v pop_init=100 -v lambda=0.95 -v ⇨     1
    ⇨period=50 > ex8-1a.dat
gawk -f discrete.awk -v pop_init=100 -v lambda=0.90 -v ⇨     2
    ⇨period=50 > ex8-1b.dat
gawk -f discrete.awk -v pop_init=100 -v lambda=0.85 -v ⇨     3
    ⇨period=50 > ex8-1c.dat
```

```
reset                                           1
set xlabel "Time, t"                            2
set ylabel "Population size, N"                 3
set style data points                           4
set key top right box                           5
                                                6
plot 'ex8-1a.dat' t "lambda=0.95", \            7
     'ex8-1b.dat' t "lambda=0.90", \            8
     'ex8-1c.dat' t "lambda=0.85"               9
pause -1                                         10
                                                11
set key bottom left box                         12
set logscale y                                  13
replot                                          14
pause -1                                         15
```

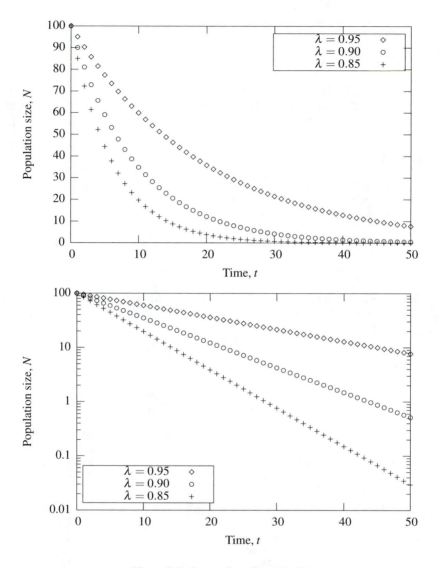

Figure E.7: Ouptut from Exercise 8.1

## Exercise 8.2

```
gawk -f continue.awk -v pop_init=100 -v growth_rate=-0.05↩    1
    ↪ -v period=50 > ex8-2a.dat
gawk -f continue.awk -v pop_init=100 -v growth_rate=-0.10↩    2
    ↪ -v period=50 > ex8-2b.dat
gawk -f continue.awk -v pop_init=100 -v growth_rate=-0.15↩    3
    ↪ -v period=50 > ex8-2c.dat
```

```
reset                                            1
set xlabel "Time, t"                             2
set ylabel "Population size, N"                  3
set style data lines                             4
set key top right box                            5
                                                 6
plot 'ex8-2a.dat' t "r=-0.05", \                 7
     'ex8-2b.dat' t "r=-0.10", \                 8
     'ex8-2c.dat' t "r=-0.15"                    9
pause -1                                         10
                                                 11
set key bottom left box                          12
set logscale y                                   13
replot                                           14
pause -1                                         15
```

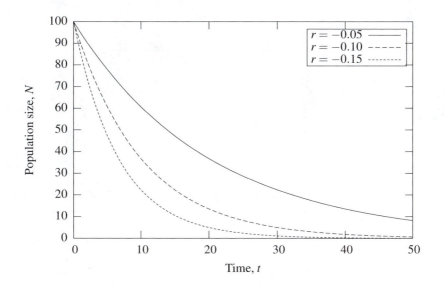

Figure E.8: First plot from Exercise 8.2

## Exercise 8.3

```
gawk -f cntlogst.awk -v pop_init=1000 -v growth_rate=0.07↩  1
    ↪ -v carry_cap=100 -v period=50 > ex8-3.dat
```

```
reset                                            1
set xlabel "Time, t"                             2
set ylabel "Population size, N"                  3
set style data lines                             4
```

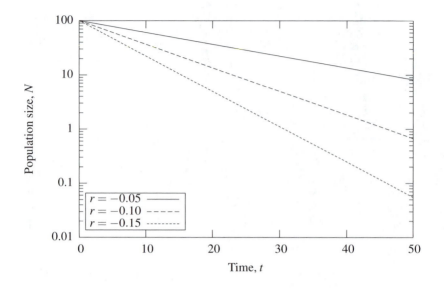

Figure E.9: Second plot from Exercise 8.2

```
set key top right box                                              5
plot 'ex8-3.dat' t "r=0.07, K=100"                                 6
pause -1                                                           7
```

**Exercise 8.4**

This exercise is left to the discretion of the reader to experiment freely with the predator-prey model and its outputs.

**Exercise 9.1**

```
gawk -f daisy1.awk -v luminosity=0.8 > ex9-1_08.dat              1
gawk -f daisy1.awk -v luminosity=0.9 > ex9-1_09.dat              2
gawk -f daisy1.awk -v luminosity=1.0 > ex9-1_10.dat              3
gawk -f daisy1.awk -v luminosity=1.1 > ex9-1_11.dat              4
gawk -f daisy1.awk -v luminosity=1.2 > ex9-1_12.dat              5
```

```
reset                                                             1
set yrange [0:0.7]                                                2
set style data lines                                              3
set key top right box                                             4
set xlabel "Time, t"                                              5
set ylabel "Fractional area covered by black daisies"            6
plot 'ex9-1_08.dat' u 1:3 t "L=0.8", \                            7
     'ex9-1_09.dat' u 1:3 t "L=0.9", \                            8
```

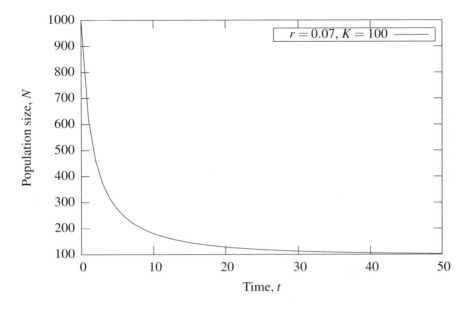

Figure E.10: Output from Exercise 8.3

```
      'ex9-1_10.dat' u 1:3 t "L=1.0", \              9
      'ex9-1_11.dat' u 1:3 t "L=1.1", \              10
      'ex9-1_12.dat' u 1:3 t "L=1.2"                 11
pause -1                                             12
                                                     13
set ylabel "Fractional area covered by white daisies"  14
plot 'ex9-1_08.dat' u 1:4 t "L=0.8", \              15
      'ex9-1_09.dat' u 1:4 t "L=0.9", \              16
      'ex9-1_10.dat' u 1:4 t "L=1.0", \              17
      'ex9-1_11.dat' u 1:4 t "L=1.1", \              18
      'ex9-1_12.dat' u 1:4 t "L=1.2"                 19
pause -1                                             20
```

**Exercise 9.2**

Program E.8: daisy2x.awk

```
BEGIN {                                              1
  # Initialize main variables                        2
  area_daisy    = 0.01;                              3
  area_suit     = 1.0;                               4
  solar_const   = 917;                               5
  Stefan_Boltz  = 5.67E-08;                          6
  albedo_soil   = 0.5;                               7
  death_rate    = 0.3;                               8
```

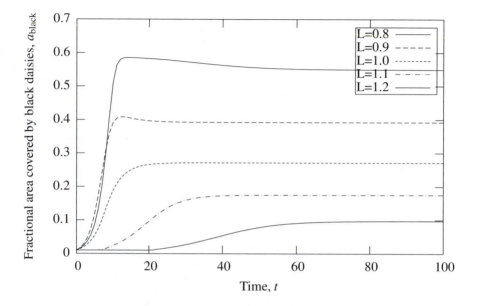

Figure E.11: First plot from Exercise 9.1

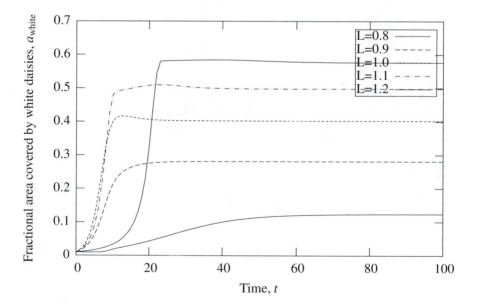

Figure E.12: Second plot from Exercise 9.1

```
q_factor      = 20;                                                    9
threshold     = 0.01;                                                  10
                                                                       11
for(luminosity=0.5;luminosity<=1.7;luminosity+=0.025){                 12
  # Run model until it reaches steady state for                        13
  # current solar luminosity                                           14
  do{                                                                  15
    prev_temp_global=temp_global;                                      16
    area_avail=area_suit-area_daisy;                                   17
    albedo_global=(area_avail*albedo_soil)+ \                          18
      (area_daisy*albedo_daisy);                                       19
    temp_global=(((solar_const*luminosity* \                          20
      (1-albedo_global))/Stefan_Boltz)**0.25)-273;                     21
    temp_daisy=((q_factor*(albedo_global-albedo_daisy))+ \22
      temp_global);                                                    23
    growth_daisy=(1-(0.003265*((22.5-temp_daisy)**2)));                24
    area_daisy+=area_daisy*((area_avail*growth_daisy)- \               25
      death_rate);                                                     26
    # Do not allow daisies to become extinct                           27
    if(area_daisy<0.01){area_daisy=0.01};                              28
    # Check whether global average temperature for                     29
    # current and previous run differ by > threshold                   30
    temp_difference=temp_global-prev_temp_global                       31
    if(temp_difference<=0){temp_difference*=-1}                        32
  } while(temp_difference>threshold)                                   33
  printf("%5.3f %7.4f %6.4f\n", \                                      34
    luminosity, temp_global, area_daisy);                              35
  }                                                                    36
}                                                                      37
```

```
gawk -f daisy2x.awk -v albedo_daisy=0.25 > ex9-2blk.dat               1
gawk -f daisy2x.awk -v albedo_daisy=0.75 > ex9-2wht.dat               2
```

```
reset                                                                  1
set key top left Left box                                              2
set xlabel "Solar luminosity, L"                                       3
set style data lines                                                   4
                                                                       5
set ylabel "Globally averaged temperature, T/deg C"                   6
plot 'ex9-2blk.dat' u 1:2 t "Black daisies only", \                    7
     'ex9-2wht.dat' u 1:2 t "White daisies only", \                    8
     'daisy2.dat' u 1:2 t "Black and white daisies"                    9
pause -1                                                               10
                                                                       11
set yrange [0:0.8]                                                     12
set ylabel "Fractional area covered, a"                               13
plot 'ex9-2blk.dat' u 1:3 t "Black daisies only", \                   14
     'ex9-2wht.dat' u 1:3 t "White daisies only"                      15
pause -1                                                               16
```

# Exercise 9.3

Program E.9: daisy3x.awk

```
BEGIN {                                                                    1
  # Initialize main variables                                             2
  area_black    = 0.01;                                                   3
  area_gray     = 0.01;                                                   4
  area_white    = 0.01;                                                   5
  area_suit     = 1.0;                                                    6
  solar_const   = 917;                                                    7
  Stefan_Boltz  = 5.67E-08;                                              8
  albedo_soil   = 0.5;                                                    9
  albedo_black  = 0.25;                                                  10
  # albedo_gray = 0.5;                                                   11
  albedo_white  = 0.75;                                                  12
  death_rate    = 0.3;                                                   13
  q_factor      = 20;                                                    14
  threshold     = 0.01;                                                  15
                                                                         16
  for(luminosity=0.5;luminosity<=1.7;luminosity+=0.025){                 17
    # Run model until it reaches steady state for                       18
    # current solar luminosity                                          19
    do{                                                                 20
      prev_temp_global=temp_global;                                     21
      area_avail=area_suit- \                                           22
        (area_black+area_white+area_gray);                              23
      albedo_global= (area_avail*albedo_soil)+ \                        24
        (area_black*albedo_black)+ \                                    25
        (area_white*albedo_white) + \                                   26
        (area_gray*albedo_gray);                                        27
      temp_global=(((solar_const*luminosity* \                          28
        (1-albedo_global))/Stefan_Boltz)**0.25)-273;                    29
      temp_black=((q_factor*(albedo_global-albedo_black))+ \            30
        temp_global);                                                   31
      temp_gray=((q_factor*(albedo_global-albedo_gray))+ \             32
        temp_global);                                                   33
      temp_white=((q_factor*(albedo_global-albedo_white))+ \           34
        temp_global);                                                   35
      growth_black=(1-(0.003265*((22.5-temp_black)**2)));               36
      growth_gray=(1-(0.003265*((22.5-temp_gray)**2)));                 37
      growth_white=(1-(0.003265*((22.5-temp_white)**2)));              38
      area_black+=area_black*((area_avail*growth_black)- \              39
        death_rate);                                                    40
      area_gray+=area_gray*((area_avail*growth_gray)- \                41
        death_rate);                                                    42
      area_white+=area_white*((area_avail*growth_white)- \             43
        death_rate);                                                    44
      # Do not allow daisies to become extinct                          45
```

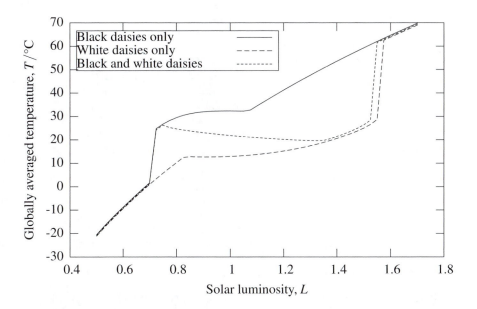

Figure E.13: First plot from Exercise 9.2

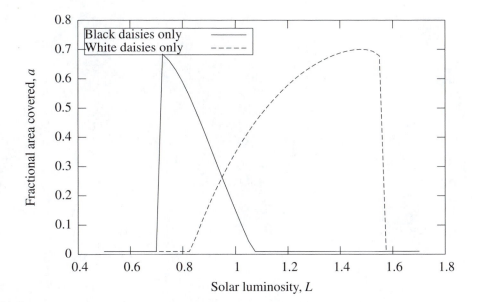

Figure E.14: Second plot from Exercise 9.2

```
        if(area_black<0.01){area_black=0.01};                46
        if(area_gray<0.01){area_gray=0.01};                  47
        if(area_white<0.01){area_white=0.01};                48
        # Check whether global average temperature for       49
        # current and previous run differ by > threshold     50
        temp_difference=temp_global-prev_temp_global         51
        if(temp_difference<0){temp_difference*=-1}           52
    } while(temp_difference>threshold)                       53
    printf("%5.3f %7.4f %6.4f %6.4f %6.4f\n", \              54
        luminosity, temp_global, area_black, \               55
        area_gray, area_white);                              56
  }                                                          57
}                                                            58
```

```
gawk  -f daisy3x.awk  -v albedo_gray=0.3  > ex9-3_03.dat     1
gawk  -f daisy3x.awk  -v albedo_gray=0.4  > ex9-3_04.dat     2
gawk  -f daisy3x.awk  -v albedo_gray=0.5  > ex9-3_05.dat     3
gawk  -f daisy3x.awk  -v albedo_gray=0.6  > ex9-3_06.dat     4
gawk  -f daisy3x.awk  -v albedo_gray=0.7  > ex9-3_07.dat     5
```

```
reset                                                         1
set style data lines                                         2
set key top left box                                         3
set xlabel "Solar luminosity, L"                             4
set ylabel "Globally averaged temperature, T_global/degC"    5
plot 'ex9-3_03.dat' t "a_gray=0.3", \                        6
     'ex9-3_04.dat' t "a_gray=0.4", \                        7
     'ex9-3_05.dat' t "a_gray=0.5", \                        8
     'ex9-3_06.dat' t "a_gray=0.6", \                        9
     'ex9-3_07.dat' t "a_gray=0.7"                          10
pause -1                                                     11
                                                            12
set key top right box                                       13
set ylabel "Fractional area of black daisies, a_black"      14
plot 'ex9-3_03.dat' u 1:3 t "a_gray=0.3", \                 15
     'ex9-3_04.dat' u 1:3 t "a_gray=0.4", \                 16
     'ex9-3_05.dat' u 1:3 t "a_gray=0.5", \                 17
     'ex9-3_06.dat' u 1:3 t "a_gray=0.6", \                 18
     'ex9-3_07.dat' u 1:3 t "a_gray=0.7"                    19
pause -1                                                     20
                                                            21
set key top left box                                        22
set ylabel "Fractional area of white daisies, a_white"      23
plot 'ex9-3_03.dat' u 1:4 t "a_gray=0.3", \                 24
     'ex9-3_04.dat' u 1:4 t "a_gray=0.4", \                 25
     'ex9-3_05.dat' u 1:4 t "a_gray=0.5", \                 26
     'ex9-3_06.dat' u 1:4 t "a_gray=0.6", \                 27
     'ex9-3_07.dat' u 1:4 t "a_gray=0.7"                    28
pause -1                                                     29
```

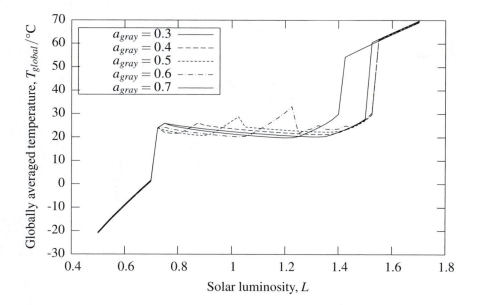

Figure E.15: First plot from Exercise 9.3

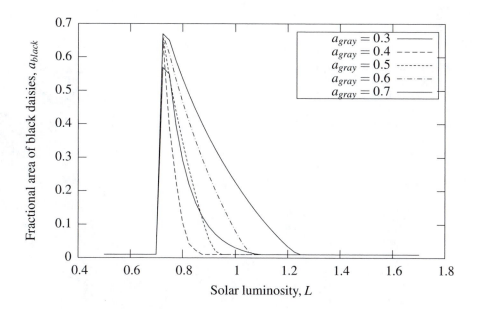

Figure E.16: Second plot from Exercise 9.3

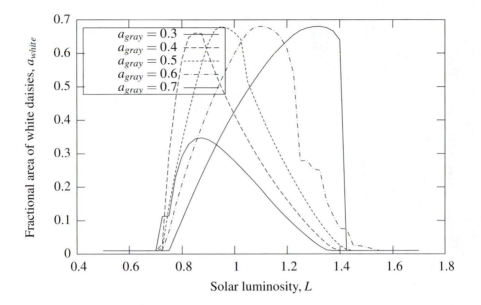

Figure E.17: Third plot from Exercise 9.3

## Exercise 9.4

Program E.10: daisy4x.awk

```
BEGIN {                                                  1
                                                         2
    # Initialize main variables                          3
    area_black      = 0.01;                               4
    area_dkgray     = 0.01;                               5
    area_gray       = 0.01;                               6
    area_palegray   = 0.01;                               7
    area_white      = 0.01;                               8
    area_suit       = 1.0;                                9
    solar_const     = 917;                                10
    Stefan          = 5.67E-08;                           11
    albedo_soil     = 0.5;                                12
    albedo_black    = 0.22;                               13
    albedo_dkgray   = 0.375;                              14
    albedo_gray     = 0.5;                                15
    albedo_palegray = 0.675;                              16
    albedo_white    = 0.75;                               17
    death_rate      = 0.3;                                18
    q_factor        = 20;                                 19
    threshold       = 0.01;                               20
                                                         21
```

```
for (luminosity=0.5; luminosity<=1.7; luminosity+=0.025) {      22
  # Run model until it reaches steady state for                 23
  # current solar luminosity                                    24
  do {                                                          25
    prev_temp_global=temp_global;                              26
                                                               27
    area_avail=area_suit- \                                    28
      (area_black+area_dkgray+area_gray+\                      29
      area_palegray+area_white);                               30
    albedo_global=(area_avail*albedo_soil)+ \                  31
      (area_black*albedo_black)+ \                             32
      (area_dkgray*albedo_dkgray)+ \                           33
      (area_gray*albedo_gray)+ \                               34
      (area_palegray*albedo_palegray)+ \                       35
      (area_white*albedo_white);                               36
    temp_global=(((solar_const*luminosity* \                   37
      (1-albedo_global))/Stefan)**0.25)-273;                   38
                                                               39
    temp_black=fn_temp(albedo_black);                          40
    temp_dkgray=fn_temp(albedo_dkgray);                        41
    temp_gray=fn_temp(albedo_gray);                            42
    temp_palegray=fn_temp(albedo_palegray);                    43
    temp_white=fn_temp(albedo_white);                          44
                                                               45
    growth_black=fn_growth(temp_black);                        46
    growth_dkgray=fn_growth(temp_dkgray);                      47
    growth_gray=fn_growth(temp_gray);                          48
    growth_palegray=fn_growth(temp_palegray);                  49
    growth_white=fn_growth(temp_white);                        50
                                                               51
    area_black+=fn_area(area_black,growth_black);              52
    area_dkgray+=fn_area(area_dkgray,growth_dkgray);           53
    area_gray+=fn_area(area_gray,growth_gray);                 54
    area_palegray+=fn_area(area_palegray,growth_palegray);     55
    area_white+=fn_area(area_white,growth_white);              56
                                                               57
    # Do not allow daisies to become extinct                   58
    if (area_black<0.01) {area_black=0.01};                    59
    if (area_dkgray<0.01) {area_dkgray=0.01};                  60
    if (area_gray<0.01) {area_gray=0.01};                      61
    if (area_palegray<0.01) {area_palegray=0.01};              62
    if (area_white<0.01) {area_white=0.01};                    63
                                                               64
    # Check whether global average temperature for             65
    # current and previous model run differ by more            66
    # than threshold                                           67
    temp_difference=temp_global-prev_temp_global               68
    if (temp_difference<0) {temp_difference*=-1}               69
  } while (temp_difference>threshold)                          70
```

```
                                                                        71
  printf("%5.3f %7.4f %6.4f %6.4f %6.4f %6.4f %6.4f\n", \           72
    luminosity, temp_global, area_black, \                         73
    area_dkgray, area_gray, \                                      74
    area_palegray, area_white);                                    75
  }                                                                  76
}                                                                    77
                                                                     78
# Function to calculate local temperatures                          79
function fn_temp(albedo_daisy)                                       80
{                                                                    81
  return ((q_factor*(albedo_global-albedo_daisy))+ \               82
    temp_global);                                                   83
}                                                                    84
                                                                     85
# Function to calculate daisy growth rates                          86
function fn_growth(temp_daisy)                                       87
{                                                                    88
  return (1-(0.003265*((22.5-temp_daisy)**2)));                    89
}                                                                    90
                                                                     91
# Function to calculate change in daisy area                        92
function fn_area(area_daisy,growth_daisy)                            93
{                                                                    94
  return (area_daisy*((area_avail*growth_daisy)- \                 95
    death_rate));                                                   96
}                                                                    97
```

```
gawk -f daisy4x.awk > daisy4x.dat                                   1
```

```
reset                                                              1
set style data lines                                               2
set key top left box                                               3
set xlabel "Solar luminosity, L"                                   4
                                                                   5
set ylabel "Globally averaged temperature, T_global/degC"          6
plot 'daisy4x.dat'  t "5 daisy species", \                         7
     'daisy3.dat'  t "3 daisy species", \                          8
     'daisy2.dat'  t "2 daisy species", \                          9
     'nolife.dat'  t "0 daisy species"                             10
pause -1                                                            11
```

## Exercise 9.5

The command line `gawk -f daisy5.awk -f daisyfns.awk -f daisyvar.awk` results in a "division by zero" error, because the values of the main variables are initialized after the main pattern-action block.

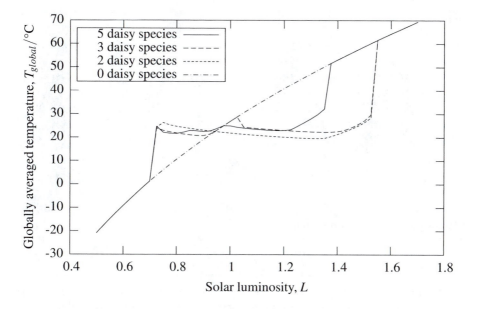

Figure E.18: Output from Exercise 9.4

The command line `gawk -f daisyvar.awk daisyfns.awk daisy5.awk` produces no output because it treats the files `daisyfns.awk` and `daisy5.awk` as input data files (cf. program files), since neither is preceded by a `-f` command-line switch. gawk therefore interprets the rules in `daisyvar.awk`, which initializes several variables. This program file contains no instructions on how to process the "data" contained in the files `daisyfns.awk` and `daisy5.awk`, however, so gawk produces no output.

## Exercise 10.1

To simulate the direct, diffuse and total solar irradiance on the terrain around Llyn Efyrnwy at 8 am on June 21:

```
gawk -f demirrad.awk -v size=100 -v latitude=52.756 -v ↵      1
    ↳DOY=172 -v hour_angle=-60 efyrnwy.dem
```

To visualize the results:

```
reset                                  1
unset key                              2
set xlabel "Easting, m"                3
set ylabel "Northing, m"               4
set xtics 295000, 1000, 300000         5
set ytics 315000, 1000, 320000         6
set xrange [295000:300000]             7
set yrange [320000:315000]             8
set dgrid3d 51,51,16                   9
set size square                        10
```

```
set tics out                                                          11
set style line 1 lt -1 lw 0.01                                        12
set palette gray gamma 1                                              13
set pm3d at s hidden3d 1 map                                          14
set colorbox vertical user origin 0.8,0.185 size 0.04,0.655           15
set cblabel "Irradiance, W.m^{-2}"                                    16
splot 'efyrnwy.tot' u 1:2:3 w pm3d                                    17
pause -1                                                              18
splot 'efyrnwy.dif' u 1:2:3 w pm3d                                    19
pause -1                                                              20
splot 'efyrnwy.tot' u 1:2:3 w pm3d                                    21
pause -1                                                              22
```

To simulate the direct, diffuse and total solar irradiance on the terrain around Llyn Efyrnwy at solar noon on January 1:

```
gawk -f demirrad.awk -v size=100 -v latitude=52.756 -v ↩       2
   ↪DOY=1 -v hour_angle=0 efyrnwy.dem
```

To visualize the results, repeat the gnuplot commands shown above. Figure E.19 shows the total solar irradiance estimated by the second of these two simulations.

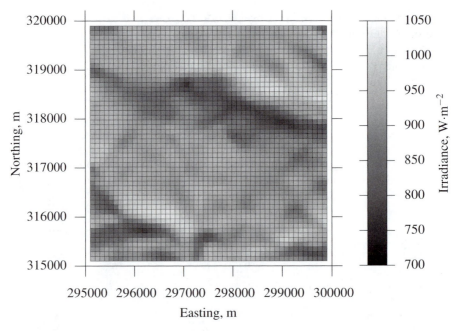

Figure E.19: Output from Exercise 10.1

# Appendix F

# Acronyms and Abbreviations

| | |
|---|---|
| **6S** | Second Simulation of the Satellite Signal in the Solar Spectrum |
| **ASCII** | American Standard Code for Information Interchange |
| **AWS** | automatic weather station |
| **awk** | Aho, Weinberger and Kernighan |
| **BADC** | British Atmospheric Data Centre |
| **CD-ROM** | Compact Disc Read-Only Memory |
| **CH$_4$** | methane |
| **CLI** | command-line interface |
| **CO$_2$** | carbon dioxide |
| **CSV** | comma-separated value |
| **DD** | decimal degrees |
| **DEM** | digital elevation model |
| **DGVM** | dynamic global vegetation model |
| **DMS** | degrees, minutes and seconds |
| **DOS** | disk operating system |
| **EMF** | Enhanced Metafile Format |
| **EMIC** | Earth system models of intermediate complexity |
| **EPS** | Encapsulated PostScript |
| **ERC** | emission reduction credit |

**ESE**            east-south-east

**fAPAR**          fraction of absorbed photosynthetically active radiation

**FDE**            finite difference equation

**FORTRAN**        FORmula TRANslation

**gawk**           GNU AWK

**GIS**            Geographic Information System

**GMT**            Greenwich Mean Time

**GNU**            GNU's Not Unix

**GPL**            General Public License

**GUI**            graphical user interface

**HFCs**           hydrofluorocarbons

**IDW**            inverse distance weighting

**ISO**            International Organization for Standardization

**JPEG**           Joint Photographic Experts Group

**LAI**            leaf area index

**LDD**            local drainage direction

**MODTRAN**        MODerate spectral resolution atmospheric TRANSsmittance
                   algorithm and computer model

**$N_2O$**         nitrous oxide

**NCM**            National Climate Message

**NERC**           Natural Environmental Research Council

**NIR**            near infra-red

**OAT**            one-at-a-time

**ODN**            Ordnance Datum Newlyn

**OS**             Ordnance Survey

**OS**             operating system

**PDF**            probability density function

**PDF**            Portable Document Format

| | |
|---|---|
| **PFCs** | perfluorocarbons |
| **PNG** | Portable Network Graphics |
| **PS** | PostScript |
| **RMSE** | root mean square error |
| **RPM** | RPM Package Manager |
| **RSPB** | Royal Society for the Protection of Birds |
| **SF$_6$** | sulfur hexafluoride |
| **SVG** | Scalable Vector Graphics |
| **TIN** | triangulated irregular network |
| **TSI** | total solar irradiance |
| **UNFCCC** | United Nations Framework Convention on Climate Change |
| **UK** | United Kingdom of Great Britain and Northern Ireland |
| **URL** | uniform resource locator |
| **UTC** | Coordinated Universal Time |
| **VDU** | visual display unit |
| **WECS** | wind energy conversion system |
| **WNW** | west-north-west |
| **WYSIWYG** | What You See Is What You Get |

# Appendix G

# List of Symbols

**Chapter 4 Wind Speed and Wind Power**

$A$          cross-sectional area of turbine rotors ($m^2$)

$c$          Weibull distribution scale parameter

$C_p$          Power coefficient

$k$          Weibull distribution shape parameter

$\Phi(u)$          Weibull distribution

$P$          wind power (W)

$N_g$          generator efficiency

$N_b$          mechanical efficiency

$\rho$          air density ($kg \cdot m^{-3}$)

$r$          radius (m)

$u$          wind speed ($m \cdot s^{-1}$)

$u_0$          wind speed at ground level ($m \cdot s^{-1}$)

$\bar{u}$          mean wind speed ($m \cdot s^{-1}$)

$u'$          threshold wind speed {$m \cdot s^{-1}$)

## Chapter 5  Solar Radiation at Earth's Surface

$c$            speed of light (approximately $3 \times 10^8$ m·s$^{-1}$)

$d$            distance between Earth and the sun (approximately $1.4 \times 10^{11}$ m)

$E_0$          total solar radiation incident at the top of Earth's atmosphere (W·m$^{-2}$)

$E_0(\lambda)$      spectral distribution of solar radiation incident at the top of Earth's atmosphere (W·m$^{-2}$·$\mu$m$^{-1}$)

$h$            Planck's constant ($6.626 \times 10^{-34}$ J·s)

$I_{\text{diffuse}}$      diffuse solar irradiance (W·m$^{-2}$)

$I_{\text{direct}}$      direct solar irradiance (W·m$^{-2}$)

$I_{\text{global}}$      global (total) solar irradiance (W·m$^{-2}$)

$k$            Boltzmann's constant ($1.3807 \times 10^{-23}$ J·K$^{-1}$)

$\lambda$            wavelength (m, $\mu$m or nm)

$\lambda_{\text{max}}$      wavelength of maximum emission ($\mu$m)

$m$           air mass number

$M$          emittance (W·m$^{-2}$)

$M(\lambda)$      spectral emittance (W·m$^{-2}$·$\mu$m)

$M_{\text{Earth}}$      spectral emittance of Earth (W·m$^{-2}$·$\mu$m)

$M_{\text{sun}}$      spectral emittance of the sun (W·m$^{-2}$·$\mu$m)

$\pi$            pi (3.14159265358979)

$\Psi$           solar zenith angle ($^{\circ}$)

$p_{\text{alt}}$      atmospheric pressure at a given altitude (mbar)

$p_{\text{sea}}$      atmospheric pressure at sea level (mbar)

$r_{\text{sun}}$      radius of the sun (approximately $6.96 \times 10^8$ m)

$\sigma$            Stefan-Boltzmann constant ($5.67 \times 10^{-8}$ W·m$^{-2}$·K$^{-4}$)

$\tau$            atmospheric transmittance

$T$           temperature (K or °C)

## Chapter 6  Light Interaction with a Plant Canopy

| | |
|---|---|
| $\alpha$ | absorptance |
| $\alpha(\lambda)$ | spectral absorptance |
| $\alpha_{\text{leaf}}(\lambda)$ | leaf spectral absorptance |
| $A$ | fractional area of ground covered by each leaf layer |
| $E$ | irradiance $(\text{W·m}^{-2})$ |
| $E(\lambda)$ | spectral irradiance $(\text{W·m}^{-2}\cdot\mu\text{m}^{-1})$ |
| $\lambda$ | wavelength $(\mu\text{m})$ |
| LAI | leaf area index |
| $M$ | radiant exitance $(\text{W·m}^{-2})$ |
| $M(\lambda)$ | spectral radiance exitance $(\text{W·m}^{-2}\cdot\mu\text{m}^{-1})$ |
| $M_{\text{leaf}}(\lambda)$ | spectral radiance exitance of leaves $(\text{W·m}^{-2}\cdot\mu\text{m}^{-1})$ |
| $M_{\text{soil}}(\lambda)$ | spectral radiance exitance of soil $(\text{W·m}^{-2}\cdot\mu\text{m}^{-1})$ |
| $M_{\text{surface}}(\lambda)$ | spectral radiance exitance of a mixed leaf and soil surface $(\text{W·m}^{-2}\cdot\mu\text{m}^{-1})$ |
| $N$ | number of leaf layers |
| $\rho$ | reflectance |
| $\rho(\lambda)$ | spectral reflectance |
| $\rho_{\text{canopy}}(\lambda)$ | canopy spectral reflectance |
| $\rho_{\text{leaf}}(\lambda)$ | leaf spectral reflectance |
| $\rho_{\text{soil}}(\lambda)$ | soil spectral reflectance |
| $\rho_{\text{surface}}(\lambda)$ | spectral reflectance of a mixed leaf and soil surface |
| $\tau$ | transmittance |
| $\tau(\lambda)$ | spectral transmittance |
| $\tau_{\text{leaf}}(\lambda)$ | leaf spectral transmittance |

## Chapter 7  Analytical and Numerical Solutions

| | |
|---|---|
| $dI^{\uparrow}$ | infinitesimal change in the flux traveling upward through a plant canopy (as a function of distance) |
| $dI^{\downarrow}$ | infinitesimal change in the flux traveling downward through a plant canopy (as a function of distance) |
| $dz$ | infinitesimal change in the distance traveled through a plant canopy |
| $I^{\uparrow}[z]$ | flux traveling upward from layer $z$ in a plant canopy |
| $I^{\uparrow}[z']$ | flux traveling upward from the soil substrate at the base of a plant canopy |
| $I^{\downarrow}[z]$ | flux traveling downward from layer $z$ in a plant canopy |
| $k$ | extinction (or attenuation) coefficient (Bouguer's Law) |
| $K$ | absorption coefficient (two-stream model) |
| LAI | leaf area index |
| $\Phi(0)$ | flux density at the top of a plant canopy |
| $\Phi(z)$ | flux density at distance/depth $z$ into a plant canopy |
| $R_C$ | canopy spectral reflectance |
| $R_L$ | leaf spectral reflectance |
| $R_S$ | soil spectral reflectance |
| $T_L$ | leaf spectral transmittance |
| $z$ | distance/depth into a plant canopy |

## Chapter 8  Population Dynamics

| | |
|---|---|
| $\alpha$ | competition on a member of species 1 caused by members of species 2 |
| $a$ | success rate with which predators capture prey |
| $\beta$ | competition on a member of species 2 caused by members of species 1 |
| $b$ | number of predator births related to number of prey consumed |
| $B$ | average number of offspring produced per individual per time step |
| $d$ | death rate of predators |
| $dN$ | change in population size (or density) over an infinitesimal interval of time |
| $dt$ | infinitesimal interval of time |

| | |
|---|---|
| $\frac{dN}{dt}$ | derivative of $N$ with respect to $t$ |
| $\Delta N$ | change in population size (or density) over a finite interval of time |
| $\Delta t$ | finite interval of time |
| $e$ | exponential function |
| $D$ | probability an individual will die during a given time step |
| $k_n$ | parameters of the Runge-Kutta method for numerical integration |
| $K$ | carrying capacity |
| $K_n$ | carrying capacity of species $n$ |
| $\lambda$ | discrete (or finite) rate of population growth |
| $N$ | population size (or density) |
| $N(t)$ | population size (or density) at time $t$ (continuous model) |
| $N_t$ | population size (or density) at time $t$ (discrete model) |
| $r$ | intrinsic (or instantaneous) rate of population growth |
| $R$ | difference between the per capita birth and death rates |
| $t$ | time |

## Chapter 9  Biospheric Feedback on Daisyworld

| | |
|---|---|
| $a_{black}$ | fractional area of planet covered by black daisies |
| $a_{soil}$ | fractional area of planet covered by exposed soil |
| $a_{suit}$ | fractional area of planet suitable for daisy growth |
| $a_{white}$ | fractional area of planet covered by white daisies |
| $A_{global}$ | global average albedo |
| $A_{black}$ | albedo of black daisies |
| $A_{soil}$ | albedo of soil |
| $A_{white}$ | albedo of white daisies |
| $\beta_{black}$ | intrinsic growth rate of black daisies |
| $\beta_{white}$ | intrinsic growth rate of white daisies |
| $\frac{da_x}{dt}$ | instantaneous rate of change in fractional area of planet covered by daisy species $x$ |

$\gamma$         instantaneous death rate of black and white daisies

$L$         solar luminosity

$M$         radiant exitance ($W \cdot m^{-2}$)

$q'$         factor describing thermal conduction

$\sigma$         Stefan-Boltzmann constant ($5.67 \times 10^{-8}\ W \cdot m^{-2} \cdot K^{-4}$)

$S$         solar constant ($917\ W \cdot m^{-2}$)

$T$         temperature ($°C$)

$T_{global}$         global average temperature ($°C$)

$T_{black}$         temperature of black daisies ($°C$)

$T_{white}$         temperature ($K$ or $°C$)

## Chapter 10 Modeling Incident Solar Radiation and Hydrological Networks over Natural Terrain

$\Delta$         distance between spatial samples in DEM (i.e., DEM cell size)

$\zeta$         angle between the surface normal (inclined) and the sun

$\theta$         gradient of the surface slope

$\Psi$         solar zenith angle

$\phi$         aspect (azimuth) of the surface slope

$\phi_s$         solar azimuth angle

$\delta$         solar declination angle

$\theta_L$         latitude of site

$E_0$         total exo-atmospheric solar irradiance

$\tau$         atmospheric transmittance

$m$         atmospheric air mass ratio

# REFERENCES

Abbott, M. B. and Refsgaard, J. C., 1996, *Distributed Hydrological Modelling* (Dordrecht: Kluwer Academic Publishers).

Aho, A. V., Kernighan, B. W. and Weinberger, P. J., 1988, *The AWK Programming Language* (Reading, MA: Addison-Wesley).

Alstad, D., 2001, *Basic Populus Models of Ecology* (Upper Saddle River, NJ, USA: Prentice Hall).

Atkinson, L., 2000, *Core PHP Programming: Using PHP to Build Dynamic Web Sites*, Second Edition (Upper Sadle River, NJ: Prentice Hall).

Barnsley, M. J., 1999, Digital remotely-sensed data and their characteristics, in *Geographical Information Systems: Principles, Techniques, Management and Applications*, edited by P. A. Longley, M. F. Goodchild, D. J. Maguire, and D. W. Rhind (New York: John Wiley), pp. 451–466.

Beer, J., Mende, W. and Stellmacher, R., 2000, The role of the sun in climate forcing. *Quaternary Science Reviews*, **19**, pp. 403–415.

Betz, A., 1926, *Windenergie und Ihre Ausnutzung durch Windmüllen* (Gottingen, Germany: Vandenhoeck and Ruprecht).

Beven, K. (Editor), 1998, *Distributed Hydrological Modelling: Applications of the TOPMODEL (Advances in Hydrological Processes)* (Chichester: John Wiley & Sons).

Beven, K., 2001, *Rainfall-Runoff Modelling: The Primer* (Chichester: John Wiley & Sons).

Biggs, N. L., 1989, *Discrete Mathematics* (Oxford: Clarendon Press).

Bonham-Carter, G. F., 1994, *Geographic Information Systems for Geoscientists: Modelling with GIS* (New York: Pergamon).

Borse, G. J., 1997, *Numerical Methods with MATLAB: A Resource for Scientists and Engineers* (Boston: PWS Publishing Company).

Bouman, B. A. M., 1992, Linking physical remote sensing models with crop growth simulation models, applied for sugar beet. *International Journal of Remote Sensing*, **13**, pp. 2565–2581.

Bowden, G. J., Barker, P. R., Shestopal, V. O. and Twidell, J. W., 1983, The Weibull distribution and wind power statistics. *Wind Energy*, **7**, pp. 85–98.

Brimicombe, A., 2003, *GIS, Environmental Modelling and Engineering* (London: Taylor & Francis).

British Computer Society, 1998, *A Glossary of Computing Terms*, Ninth Edition (Harlow, Essex: Addison Wesley Longman).

Buchanan, W., 1989, *Mastering Pascal and Delphi Programming* (Basingstoke, UK: Palgrave).

Burrough, P. A. and McDonnell, R. A., 1998, *Principles of Geographical Information Systems*, Second Edition (Oxford: Oxford University Press).

Buttenfield, B. and McMaster, R. B. (Editors), 1991, *Map Generalization: Making Rules for Knowledge Representation* (Harlow, Essex: Longman Scientific and Technical).

Camillo, P., 1987, A canopy reflectance model based on an analytical solution to the multiple scattering equation. *Remote Sensing of Environment*, **23**, pp. 453–477.

Campbell, G. S. and Norman, J. M., 1998, *An Introduction to Environmental Biophysics*, Second Edition (New York: Springer-Verlag).

Cao, M., Prince, S. D., Li, K., Tao, B., Small, J. and Shao, X., 2003, Response of terrestrial carbon uptake to climate interannual variability in China. *Global Change Biology*, **9**, pp. 536–546.

Chernoff, H., 1973, The use of faces to represent points in k-dimensional space graphically. *Journal of the American Statistical Association*, **68**, pp. 361–368.

Clarke, K. C., Parks, B. O. and Crane, M. P., 2002, *Geographic Information Systems and Environmental Modeling* (New York: Prentice Hall).

Cooper, K., Smith, J. A. and Pitts, D., 1982, Reflectance of a vegetation canopy using the Adding method. *Applied Optics*, **21**, pp. 4112–4118.

Cowlishaw, M., 1990, *The REXX Language: A Practical Approach to Programming*, Second Edition (Harlow, UK: Prentice Hall).

Cox, P. M., Huntingford, C. and Harding, R. J., 1998, A canopy conductance and photosynthesis model for use in a GCM land surface scheme. *Journal of Hydrology*, **212–213**, pp. 79–94.

Cramer, W., Bondeau, A., Woodward, F., Prentice, I., Betts, R., Brovkin, V., Cox, P., Fisher, V., Foley, J., Friend, A., Kucharik, C., Lomas, M., Rmankutty, N., Stitch, S., Smith, B., White, A. and Young-Molling, C., 2001, Global response of terrestrial ecosystem structure and function to $CO_2$ and climate change: Results from six dynamic global vegetation models. *Global Change Biology*, **7**, pp. 357–374.

Crommelynck, D. C. and Dewitte, S., 1997, Solar constant temporal and frequency characteristics. *Solar Physics*, **173**, pp. 177–191.

Dawkins, R., 1982, *The Extended Phenotype* (Oxford: W.H. Freeman).

Deaton, M. L. and Winebrake, J. J., 2000, *Dynamic Modeling of Environmental Systems* (New York: Springer-Verlag).

DiBona, C., Ockman, S. and Stone, M. (Editors), 1999, *Open Sources: Voices from the Open Source Revolution* (Sebastopol, CA: O'Reilly and Associates).

Donovan, T. M. and Welden, C. W., 2001, *Spreadsheet Exercises in Conservation Biology and Landscape Ecology* (Sunderland, MA: Sinauer Associates, Inc).

Donovan, T. M. and Welden, C. W., 2002, *Spreadsheet Exercises in Ecology and Evolution* (Sunderland, MA: Sinauer Associates, Inc).

Doolittle, W. F., 1981, Is nature really motherly? *Coevolution Quarterly*, **29**, pp. 58–65.

Dougherty, D., 1996, *sed and awk*, Second Edition (Sebastopol, CA: O'Reilly and Associates).

Edwards, D. and Hamson, M., 1989, *Guide to Mathematical Modelling* (London: Macmillan).

Fienberg, S. E., 1979, Graphical methods in statisics. *The American Statistician*, **33**, pp. 165–178.

Flanagan, D., 2001, *JavaScript: The Definitive Guide*, Fourth Edition (Sebastopol, CA: O'Reilly & Associates).

Fligge, M., Solanki, S. K., Pap, J. M., Fröhlich, C. and Wehrli, C., 2001, Variations of solar spectral irradiance from near UV to the infrared — measurements and results. *Journal of Atmospheric and Solar-Terrestrial Physics*, **63**, pp. 1479–1487.

Ford, A., 1999, *Modeling the Environment: An Introduction to Systems Dynamics Modeling of Environmental Systems* (Washington, D.C.: Island Press).

Forrester, J. W., 1961, *Industrial Dynamics* (Waltham, MA: Pegasus Communications).

Forrester, J. W., 1969, *Urban Dynamics* (Waltham, MA: Pegasus Communications).

Forrester, J. W., 1973, *World Dynamics*, Second Edition (Waltham, MA: Pegasus Communications).

Freris, L. L., 1990, *Wind Energy Conversion Systems* (London: Prentice Hall).

Friend, A. D., Stevens, A. K., Knox, R. G. and Cannell, M. R. G., 1997, A process-based, terrestrial biosphere model of ecosystem dynamics (Hybrid v3.0). *Ecological Modelling*, **95**, pp. 249–287.

Gao, W., 1993, A simple bidirectional-reflectance model applied to a tallgrass canopy. *Remote Sensing of Environment*, **45**, pp. 209–224.

Gao, W. and Lesht, B. M., 1997, Model inversion of satellite-measured reflectances for obtaining surface biophysical and bidirectional reflectance characteristics of grassland. *Remote Sensing of Environment*, **59**, pp. 461–471.

Gastellu-Etchegorry, J. P., Zagolski, F. and Romier, J., 1996, A simple anisotropic reflectance model for homogeneous multilayer canopies. *Remote Sensing of Environment*, **57**, pp. 22–38.

Gausman, H. W., 1977, Reflectance properties of leaf components. *Remote Sensing of Environment*, **6**, pp. 1–10.

Giordano, F. R., Weir, M. D. and Fox, W. P., 1997, *A First Course in Mathematical Modeling* (Pacific Grove, CA: Brooks/Cole).

Gipe, P., 1995, *Wind Energy Comes of Age* (New York: John Wiley & Sons).

Gipe, P., 1999, *Wind Energy Basics: A Guide to Small and Micro Wind Systems* (White River Junction, VT: Chelsea Green).

Gleick, J., 1987, *Chaos* (London: Heinemann).

Glickman, T. S. (Editor), 2000, *Glossary of Meteorology*, Second Edition (Boston, MA: American Meteorological Society).

Goel, N. S., 1987, Models of vegetation canopy reflectance and their use in the estimation of biophysical parameters from reflectance data. *Remote Sensing Reviews*, **3**, pp. 1–212.

Goel, N. S., 1989, Inversion of canopy reflectance models for estimation of biophysical parameters from reflectance data, in *Theory and Applications of Optical Remote Sensing*, edited by G. Asrar (New York: John Wiley & Sons), pp. 205–251.

Golding, E. W., 1977, *The Generation of Electricity by Wind Power* (London: Spon Press).

Goodchild, M. F., Steyaert, L. T., Parks, B. O., Johnston, C., Maidment, D., Crane, M. and Glendinning, S. (Editors), 1996, *GIS and Environmental Modeling: Progress and Research Issues* (New York: John Wiley & Sons).

Goossens, M., Mittelbach, F. and Samarin, A., 1994, *The LATEXCompanion* (Reading, MA: Addison-Wesley).

Goossens, M., Rahtz, S. and Mittelbach, F., 1997, *The LATEX Graphics Companion: Illustrating Documents with TEX and PostScript* (Reading, MA: Addison-Wesley).

Goudriaan, J. and van Laar, H. H., 1994, *Modelling Potential Crop Growth Processes* (Dordrecht: Kluwer).

Grant, L., 1987, Diffuse and specular characteristics of leaf reflectance. *Remote Sensing of Environment*, **22**, pp. 309–322.

Grossman, P., 1995, *Discrete Mathematics for Computing* (Basingstoke, Hampshire: MacMillan Press Ltd).

Gueymard, C. A., 2003, Direct solar transmittance and irradiance predictions with broadband models. Part I: Detailed theoretical performance assessment. *Solar Energy*, **74**, pp. 355–379.

Gueymard, C. A., Myers, D. and Emery, K., 2002, Proposed reference irradiance spectra for solar energy systems testing. *Solar Energy*, **73**, pp. 443–467.

Haefner, J. W., 1996, *Modeling Biological Systems* (New York: Chapman and Hall).

Hahn, B. D., 1994, *Fortran 90 for Scientists and Engineers* (London: Butterworth-Heinemann).

Hall, N. (Editor), 1991, *The New Scientist Guide to Chaos* (London: Penguin Books).

Hardisty, J., Taylor, D. M. and Metcalfe, S. E., 1993, *Computerised Environmental Modelling: A Practical Introduction Using Excel* (Chichester: John Wiley & Sons).

Harris, J. W. and Stocker, H., 1998, *Handbook of Mathematics and Computational Science* (New York: Springer-Verlag).

Harte, J., 1988, *Consider a Spherical Cow: A Course in Environmental Problem Solving* (Sausalito, CA: University Science Books).

Harte, J., 2001, *Consider a Cylindrical Cow: More Adventures in Environmental Problem Solving* (Sausalito, CA: University Science Books).

Huggett, R., 1980, *Systems Analysis in Geography* (Oxford: Clarendon Press).

Ichoku, C. and Karnieli, A., 1996, A review of mixture modelling techniques for sub-pixel land cover estimation. *Remote Sensing Reviews*, **13**, pp. 161–186.

Inselberg, A., 1985, The plane with parallel coordinates. *The Visual Computer*, **1**, pp. 69–92.

IPCC, 2001, *Climate Change 2001: The Scientific Basis* (Cambridge: Cambridge University Press).

Iqbal, M., 1983, *An Introduction to Solar Radiation* (New York: Academic Press).

Jacquemoud, S. and Baret, F., 1990, PROSPECT: A model of leaf optical properties spectra. *Remote Sensing of Environment*, **34**, pp. 75–92.

Jakeman, A. J., Beck, M. B. and McAleer, M. J. (Editors), 1993, *Modelling Change in Environmental Systems* (Chichester: John Wiley & Sons).

Jones, C. B., 1997, *Geographical Information Systems and Computer Cartography* (Harlow: Longman).

Jones, H. G., 1983, *Plants and Microclimate* (Cambridge: Cambridge University Press).

Kernighan, B. W. and Ritchie, D. M., 1988, *The C Programming Language*, Second Edition (Englewood Cliffs, NJ: Prentice-Hall).

Kirchner, J. W., 1989, The Gaia hypothesis: Can it be tested? *Reviews of Geophysics*, **27**, pp. 223–235.

Kirchner, J. W., 1990, Gaia metaphor unfalsifiable. *Nature*, **345**, pp. 470.

Kirchner, J. W., 2002, The Gaia hypothesis: Fact, theory, and wishful thinking. *Climate Change*, **52**, pp. 391–408.

Kirkby, M. J., Naden, P. S., Burt, T. P. and Butcher, D. P., 1993, *Computer Simulation in Physical Geography*, Second Edition (Chishester: John Wiley & Sons).

Kneizys, F. X., Robertson, D. C., Abreu, L. W., Acharya, P., Anderson, G. P., Rothman, L. S., Chetwynd, J. H., Selby, J. E. A., Shettle, E. P., Gallery, W. O., Berg, A., Clough, S. A. and Bernstein, L. S., 1996, The MODTRAN 2/3 Report and LOWTRAN 7 Model, Technical report, Ontar Corporation, Andover MA.

Kuusk, A., 1995, A fast, invertible canopy reflectance model. *Remote Sensing of Environment*, **51**, pp. 342–350.

Kuusk, A., 1996, A computer-efficient plant canopy reflectance model. *Computers and Geosciences*, **22**, pp. 149–163.

Lal, R., Kimble, J. M. and Stewart, B. A., 1999, *Global Climate Change and Tropical Ecosystems* (Boca Raton, FL: Lewis Publishers).

Lamport, L., 1994, *LaTeX: A Document Preparation System: User's Guide and Reference Manual*, Second Edition (Reading, MA: Addison-Wesley).

Lean, J., 1991, Variations in the sun's radiative output. *Reviews in Geophysics*, **29**, pp. 505–535.

Lee, R. B., Gibson, M. A., Wilson, R. S. and Thomas, S., 1995, Long-term total solar irradiance variability during sunspot cycle 22. *Journal of Geophysical Research*, **100**, pp. 1667–1675.

List, R. J. (Editor), 2000, *Smithsonian Meteorological Tables*, Sixth Edition (Washington, D.C.: Smithsonian Institution Press).

Liu, B. Y. and Jordan, R. C., 1960, The interrelationship and characteristic distribution of direct, diffuse, and total solar radiation. *Solar Energy*, **4**, pp. 1–19.

Lotka, A., 1925, *Elements of Physical Biology* (Baltimore, MD: Williams and Wilkins).

Lovelock, J., 1995a, *The Ages of Gaia*, Second Edition (Oxford: Oxford University Press).

Lovelock, J., 1995b, *Gaia: A New Look at Life on Earth*, Second Edition (Oxford: Oxford University Press).

Lutz, M., 1996, *Programming Python* (Sebastopol, CA: O'Reilly & Associates).

Manwell, J. F., McGowan, J. G. and Rogers, A. L., 2002, *Wind Energy Explained: Theory, Design and Application* (New York: John Wiley & Sons).

May, R. M., 1974, Biological populations with non-overlapping generations: Stable points, stable cycles and chaos. *Science*, **186**, pp. 645–647.

May, R. M., 1976, Simple mathematical models with very complicated dynamics. *Nature*, **261**, pp. 459.

May, R. M., 1991, The chaotic rhythms of life, in *The New Scientist Guide to Chaos*, edited by N. Hall (London: Penguin Books), pp. 82–95.

May, R. M., 2001, *Stability and Complexity in Model Ecosystems* (Princeton, NJ: Princeton University Press).

McGuffie, K. and Henderson-Sellers, A., 1997, *A Climate Modelling Primer*, Second Edition (Chichester: John Wiley & Sons).

McGuffie, K. and Henderson-Sellers, A., 2005, *A Climate Modelling Primer*, Third Edition (Chichester: John Wiley & Sons).

Merz, T., 1996, *PostScript and Acrobat/PDF* (Berlin: Springer-Verlag).

Monteith, J. L. and Unsworth, M., 1995, *Principles of Environmental Physics*, Second Edition (London: Arnold).

Moore, I. D., 1996, Hydrological Modeling and GIS, in *GIS and Environmental Modeling: Progress and Research Issues*, edited by M. F. Goodchild, L. T. Steyaert, B. O. Parks, D. Johnston, D. Maidment, M. Crane, and S. Glendinning (Fort Collins, CO: GIS World Books), pp. 143–148.

Mulligan, M., 2004, Modelling catchment hydrology, in *Environmental Modelling: Finding Simplicity in Complexity*, edited by J. Wainwright, and M. Mulligan (Chichester: John Wiley & Sons), pp. 106–121.

Myneni, R. B., Ross, J. and Asrar, G., 1990, A review of the theory of photon transport in leaf canopies. *Agricultural and Forest Meteorology*, **45**, pp. 1–153.

Niemeyer, P. and Knudsen, J., 2005, *Learning Java*, Third Edition (Sebastopol, CA: O'Reilly and Associates).

North, P. R. J., 1996, 3-dimensional forest light interaction-model using a Monte-Carlo method. *IEEE Transactions on Geoscience and Remote Sensing*, **34**, pp. 946–956.

Oualline, S., 1995, *Practical C++ Programming* (Sebastopol, CA: O'Reilly and Associates).

Oualline, S., 1997, *Practical C Programming*, Third Edition (Sebastopol, CA: O'Reilly and Associates).

Piff, M., 1992, *Discrete Mathematics: An Introduction for Software Engineers* (Cambridge: Cambridge University Press).

Pinty, B. and Verstraete, M. M., 1998, Modeling the scattering of light by homogeneous vegetation in optical remote sensing. *Journal of the Atmospheric Sciences*, **55**, pp. 137–150.

Rasool, S. I. and Schneider, S. H., 1971, Atmospheric carbon dioxide and aerosols: Effect of large increases on global climate. *Science*, **173**, pp. 138–141.

Raymond, E. S., 1999, *The Cathedral and the Bazaar: Musings on Linux and Open Source by an Accidental Revolutionary* (Sebastopol, CA: O'Reilly and Associates).

Robbins, A. D., 2001, *Effective AWK Programming*, Third Edition (Sebastopol, CA: O'Reilly and Associates).

Roughgarden, J., 1998, *Primer of Ecological Theory* (Upper Saddle River, NJ: Prentice Hall).

Russell, G., Jarvis, P. G. and Monteith, J. L., 1989, Absorption of radiation by canopies and stand growth, in *Plant Canopies: Their Growth, Form and Function*, edited by G. Russell, B. Marshall, and P. G. Jarvis (Cambridge: Cambridge University Press), pp. 20–39.

Saltelli, A., Chan, K. and Scott, E. M. (Editors), 2000, *Sensitivity Analysis* (Chichester: John Wiley & Sons).

Saltelli, A., Tarantola, S., Campologno, F. and Ratto, M., 2004, *Sensitivity Analysis in Practice: A Guide to Assessing Scientific Models* (Chichester: John Wiley & Sons).

Saunders, P. T., 1994, Evolution without natural selection: Further implications of the Daisyworld parable. *Journal of Theoretical Biology*, **166**, pp. 365–373.

Schott, R., 1997, *Remote Sensing: The Image Chain Approach* (New York: Oxford).

Schowengerdt, R. J., 1997, *Remote Sensing Models and Methods for Image Processing*, Second Edition (San Diego, CA: Academic Press).

Sellers, P. J., Randall, D., Collatz, C., Berry, J., Field, C., Dazlich, D., Zhang, C. and Collelo, G., 1996, A revised land surface parameterisation (SiB2) for atmospheric GCMs. Part I: Model formation. *Journal of Climate*, **9**, pp. 676–705.

Skidmore, A. K., 1989, A comparison of techniques for calculating gradient and aspect from a gridded digital elevation model. *International Journal of Geographical Information Systems*, **3**, pp. 323–334.

Taylor, B. N., 1995, Guide for the use of the International System of Units (SI), Technical Report Special Publication 811, National Institute of Standards and Technology, Gaitherburg, MD.

Thomas, D. and Hunt, A., 2000, *Programming Ruby: The Pragmatic Programmer's Guide* (New York: Addison-Wesley).

Tucker, C. J. and Garratt, M. W., 1977, Leaf optical system modelled as a stochastic process. *Applied Optics*, **16**, pp. 635–642.

Tufte, E., 2001, *The Visual Display of Quantitative Information*, Second Edition (Cheshire, CT: Graphics Press).

Twidell, J. W. and Weir, A. D., 2006, *Renewable Energy Resources*, Second Edition (Abingdon, UK: Taylor & Francis).

Verhoef, W., 1984, Light scattering by leaf layers with application to canopy reflectance modelling: The SAIL model. *Remote Sensing of Environment*, **16**, pp. 125–141.

Vermote, E. F., Tanré, D., Deuze, J. L., Herman, M. and Morcrette, J. J., 1997, Second simulation of the satellite signal in the solar spectrum, 6S: An overview. *IEEE Transactions on Geoscience and Remote Sensing*, **35**, pp. 675–686.

Verstraete, M. M., Pinty, B. and Dickinson, R. E., 1990, A physical model of the bidirectional reflectance of vegetation canopies. 1: Theory. *Journal of Geophysical Research-Atmospheres*, **95**, pp. 11755–11765.

Verstraete, M. M., Pinty, B. and Myneni, R. B., 1996, Potential and limitations of information extraction on the terrestrial biosphere from satellite remote-sensing. *Remote Sensing of Environment*, **58**, pp. 201–214.

Volterra, V., 1926, Variazioni e fluttuazioni del numero d'individui in specie animali conviventi. *Mem. R. Accad. Naz. dei Lincei*, **2**, pp. 31–113.

Wainwright, J. and Mulligan, M. (Editors), 2004, *Environmental Modelling: Finding Simplicity in Complexity* (Chichester: John Wiley & Sons).

Wall, L. and Schwartz, R., 1993, *Programming in Perl* (Sebastopol, CA: O'Reilly and Associates).

Ward, R. C. and Robinson, M., 1999, *Principles of Hydrology* (London: McGraw-Hill).

Watson, A. and Lovelock, J. E., 1983, Biological homeostasis of the global environment: The parable of the Daisyworld. *Tellus*, **35B**, pp. 284–289.

Watson, D. F., 1992, *Contouring: A Guide to the Analysis and Display of Spatial Data* (Oxford: Pergamon – Elsevier Science).

Welch, B., Jones, K. and Hobbs, J., 2003, *Practical Programming in Tcl and Tk* (New York: Prentice Hall).

Whittaker, R. H., 1975, *Communities and Ecosystems*, Second Edition (New York: Macmillan).

Whittaker, R. H., 1982, *Communities and Ecosystems*, Third Edition (New York: Macmillan).

Wiens, J. A., 1989, Spatial scaling in eclogy. *Functional Ecology*, **3**, pp. 385–397.

Williams, T. and Kelly, C., 1998, *gnuplot: An Interactive Plotting Program*, Dartmouth College, Hannover, NH.

Willis, T. and Newsome, B., 2005, *Beginning Visual Basic 2005* (Indianapolis, IN: Wrox Press).

Wilson, W., 2000, *Simulating Ecological and Evolutionary Systems in C* (Cambridge: Cambridge University Press).

Woodward, F. I., Smith, T. M. and Emanuel, W. R., 1995, A global land primary productivity and phytogeography model. *Global Biogeochemical Cycles*, **9**, pp. 471–490.

Zevenbergen, L. W. and Thorne, C. R., 1987, Quantitative analysis of land surface topography. *Earth Surface Processes and Landforms*, **12**, pp. 47–56.

# Index

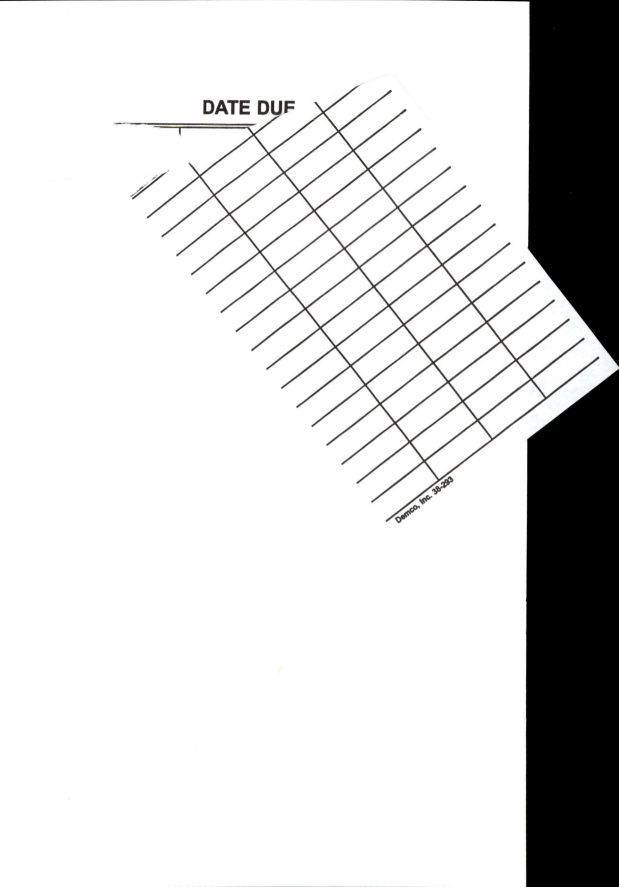

DATE DUE

Demco, Inc. 38-293

John M. Harris
Jeffry L. Hirst
Michael J. Mossinghoff

# Combinatorics and Graph Theory

With 124 Illustrations

 Springer

John M. Harris
Department of Mathematics
Furman University
Greenville, SC 29613
USA

Jeffry L. Hirst
Department of Mathematics
Appalachian State University
Boone, NC 28608
USA

Michael J. Mossinghoff
Department of Mathematics
UCLA
Los Angeles, CA 90095
USA

Mathematics Subject Classification (2000): 05-01, 05Cxx

Library of Congress Cataloging-in-Publication Data
Harris, John M. (John Michael), 1969–
    Combinatorics and graph theory / John M. Harris, Jeffry L. Hirst, Michael J. Mossinghoff.
       p.   cm. — (Undergraduate texts in mathematics)
    Includes bibliographical references and index.
    ISBN 0-387-98736-3 (alk. paper)
       1. Combinatorial analysis.   2. Graph theory.   I. Hirst, Jeffry L., 1957–
II. Mossinghoff, Michael J., 1964–   III. Title.   IV. Series.
QA165.H37   2000
511′.6—dc21                                                                                          99-049806

Printed on acid-free paper.

Production managed by A. Orrantia; manufacturing supervised by Jerome Basma.
Photocomposed pages prepared from the authors' LaTeX files by The Bartlett Press, Inc., Marietta, GA.
Printed and bound by R.R. Donnelley and Sons, Harrisonburg, VA.
Printed in the United States of America.

9 8 7 6 5 4 3 2 1

ISBN 0-387-98736-3 Springer-Verlag New York Berlin Heidelberg   SPIN 10708757

*To
Priscilla, Sophie,
Holly,
Kristine, and Amanda*

# Preface

*Three things should be considered: problems, theorems, and applications.*
— Gottfried Wilhelm Leibniz,
*Dissertatio de Arte Combinatoria*, 1666

This book grew out of several courses in combinatorics and graph theory given at Appalachian State University and UCLA in recent years. A one-semester course for juniors at Appalachian State University focusing on graph theory covered most of Chapter 1 and the first part of Chapter 2. A one-quarter course at UCLA on combinatorics for undergraduates concentrated on the topics in Chapter 2 and included some parts of Chapter 1. Another semester course at Appalachian State for advanced undergraduates and beginning graduate students covered most of the topics from all three chapters.

There are rather few prerequisites for this text. We assume some familiarity with basic proof techniques, like induction. A few topics in Chapter 1 assume some prior exposure to elementary linear algebra. Chapter 2 assumes some familiarity with sequences and series, especially Maclaurin series, at the level typically covered in a first-year calculus course. The text requires no prior experience with more advanced subjects, such as group theory.

While this book is primarily intended for upper-division undergraduate students, we believe that others will find it useful as well. Lower-division undergraduates with a penchant for proofs, and even talented high school students, will be able to follow much of the material, and graduate students looking for an introduction to topics in graph theory, combinatorics, and set theory may find several topics of interest.

Chapter 1 focuses on the theory of finite graphs. The first section serves as an introduction to basic terminology and concepts. Each of the following sections presents a specific branch of graph theory: trees, planarity, coloring, matchings, and Ramsey theory. These five topics were chosen for two reasons. First, they represent a broad range of the subfields of graph theory, and in turn they provide the reader with a sound introduction to the subject. Second, and just as important, these topics relate particularly well to topics in Chapters 2 and 3.

Chapter 2 develops the central techniques of enumerative combinatorics: the principle of inclusion and exclusion, the theory and application of generating functions, the solution of recurrence relations, Pólya's theory of counting arrangements in the presence of symmetry, and important classes of numbers, including the Fibonacci, Catalan, Stirling, Bell, and Eulerian numbers. The final section in the chapter continues the theme of matchings begun in Chapter 1 with a consideration of the stable marriage problem and the Gale–Shapley algorithm for solving it.

Chapter 3 presents infinite pigeonhole principles, König's Lemma, Ramsey's Theorem, and their connections to set theory. The systems of distinct representatives of Chapter 1 reappear in infinite form, linked to the axiom of choice. Counting is recast as cardinal arithmetic, and a pigeonhole property for cardinals leads to discussions of incompleteness and large cardinals. The last sections connect large cardinals to finite combinatorics and describe supplementary material on computability.

Following Leibniz's advice, we focus on problems, theorems, and applications throughout the text. We supply proofs of almost every theorem presented. We try to introduce each topic with an application or a concrete interpretation, and we often introduce more applications in the exercises at the end of each section. In addition, we believe that mathematics is a fun and lively subject, so we have tried to enliven our presentation with an occasional joke or (we hope) interesting quotation.

We would like to thank the Department of Mathematical Sciences at Appalachian State University and the Department of Mathematics at UCLA. We would especially like to thank our students (in particular, Jae-Il Shin at UCLA), whose questions and comments on preliminary versions of this text helped us to improve it. We would also like to thank the three anonymous reviewers, whose suggestions helped to shape this book into its present form. We also thank Sharon McPeake, a student at ASU, for her rendering of the Königsberg bridges.

In addition, the first author would like to thank Ron Gould, his graduate advisor at Emory University, for teaching him the methods and the joys of studying graphs, and for continuing to be his advisor even after graduation. He especially wants to thank his wife, Priscilla, for being his perfect match, and his daughter Sophie for adding color and brightness to each and every day. Their patience and support throughout this process have been immeasurable.

The second author would like to thank Judith Roitman, who introduced him to set theory and Ramsey's Theorem at the University of Kansas, using an early draft of her fine text. Also, he would like to thank his wife, Holly (the other Professor Hirst), for having the infinite tolerance that sets her apart from the norm.

The third author would like to thank Bob Blakley, from whom he first learned about combinatorics as an undergraduate at Texas A & M University, and Donald Knuth, whose class *Concrete Mathematics* at Stanford University taught him much more about the subject. Most of all, he would like to thank his wife, Kristine, for her constant support and infinite patience throughout the gestation of this project, and for being someone he can always, well, count on.

John M. Harris
Jeffry L. Hirst
Michael J. Mossinghoff

# Contents

# 1
# Graph Theory

> *"Begin at the beginning," the King said, gravely, "and go on till you come*
> *to the end; then stop."*
>
> — Lewis Carroll, *Alice in Wonderland*

The Pregolya River passes through a city once known as Königsberg. In the 1700s
seven bridges were situated across this river (Figure 1.1), and no resident of the
city was ever able to walk a route that crossed each of these bridges exactly once.
In 1736 the Swiss mathematician Leonhard Euler gave a proof that no such route
existed, and his work on the "Königsberg Bridge Problem" is considered by many
to be the beginning of the field of graph theory.

At first, the usefulness of Euler's ideas and of "graph theory" itself was found
only in solving puzzles and in analyzing games and other recreations. In the mid
1800s, however, people began to realize that graphs could be used to model many
things that were of interest in society. For instance, the "Four Color Map Conjec-

FIGURE 1.1. The bridges in Königsberg.

ture," introduced by DeMorgan in 1852, was a famous problem that was seemingly unrelated to graph theory. The conjecture stated that four is the maximum number of colors required to color any map where bordering regions are colored differently. This conjecture can easily be phrased in terms of graph theory, and Appel and Haken did just that when they proved the conjecture in 1976 [6].

The field of graph theory has blossomed throughout this century, as more and more modeling possibilities have been recognized. And as specific applications have increased in number and in scope, the theory itself has developed beautifully as well.

We will discuss five areas of graph theory in this chapter: trees, planarity, coloring, matchings, and Ramsey theory. While there are many topics that are worth investigating, we have chosen these because they relate particularly well to the other topics in this book. In Section 1.7 you can find pointers to many other sources on graph theory. We begin with some introductory graph theory concepts.

## 1.1   Introductory Concepts

*A definition is the enclosing a wilderness of idea within a wall of words.*
— Samuel Butler, *Higgledy-Piggledy*

### 1.1.1   Graphs and Their Relatives

A *graph* consists of two finite[1] sets, $V$ and $E$. Each element of $V$ is called a *vertex* (plural *vertices*). The elements of $E$, called *edges*, are unordered pairs of vertices. For instance, the set $V$ might be $\{1, 2, 3, 4, 5, 6, 7, 8\}$, and $E$ might be $\{\{1, 4\},$ $\{1, 5\}, \{2, 3\}, \{2, 5\}, \{2, 7\}, \{3, 6\}, \{4, 6\}, \{4, 7\}, \{7, 8\}\}$. Together, $V$ and $E$ are a graph $G$.

Graphs have natural visual representations. Look at the diagram in Figure 1.2. Notice that each element of $V$ is represented by a small circle and that each element of $E$ is represented by a line drawn between the corresponding two elements of $V$.

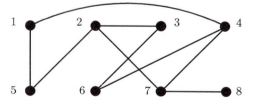

FIGURE 1.2. A visual representation of the graph $G$.

---

[1]We consider infinite graphs in Chapter 3.

FIGURE 1.3. A digraph.

FIGURE 1.4. A multigraph.

As a matter of fact, we can just as easily define a graph to be a diagram consisting of small circles, called vertices, and curves, called edges, where each curve connects two of the circles together. When we speak of a graph in this chapter, we will almost always refer to such a diagram.

We can obtain similar structures by altering our definition in various ways. Here are some examples.

1. By replacing our set $E$ with a set of *ordered* pairs of vertices, we obtain a *directed graph*, or *digraph* (Figure 1.3). Each edge of a digraph has a specific orientation.
2. If we allow repeated elements in our set of edges, technically replacing our set $E$ with a multiset, we obtain a *multigraph* (Figure 1.4).
3. By allowing edges to connect a vertex to itself ("loops"), we obtain a *pseudograph* (Figure 1.5).
4. Allowing our edges to be arbitrary subsets of vertices gives us *hypergraphs* (Figure 1.6).
5. By allowing $V$ or $E$ to be an infinite set, we obtain *infinite graphs*. Infinite graphs are studied in Chapter 3.

FIGURE 1.5. A pseudograph.

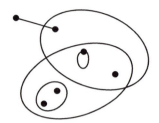

FIGURE 1.6. A hypergraph with 6 vertices and 5 edges.

In this chapter we will focus on finite, simple graphs: those without loops or multiple edges.

**Exercises**

1. Ten people are seated around a circular table. Each person shakes hands with everyone at the table except the person sitting directly across the table. Draw a graph that models this situation.

2. Six fraternity brothers (Adam, Bert, Chuck, Doug, Ernie, and Filthy Frank) need to pair off as roommates for the upcoming school year. Each person has compiled a list of the people with whom he would be willing to share a room.
   Adam's list: Doug
   Bert's list: Adam, Ernie
   Chuck's list: Doug, Ernie
   Doug's list: Chuck
   Ernie's list: Ernie
   Frank's list: Adam, Bert
   Draw a digraph that models this situation.

3. There are nine women's basketball teams in the Atlantic Coast Conference: Florida State (F), Georgia Tech (G), Clemson (C), Univ. of North Carolina (N), NC State (S), Duke (D), Univ. of Virginia (V), Univ. of Maryland (M), and Wake Forest Univ. (W). At a certain point in midseason,
   C has played D*, G
   D has played C*, S, W
   F has played N*, V
   G has played C, M
   M has played G, N
   N has played F*, M, W
   S has played D, V*
   V has played F, S*
   W has played D, N
   The asterisk(*) indicates that these teams have played each other twice. Draw a multigraph that models this situation.

4. Can you explain why no resident of Königsberg was ever able to walk a route that crossed each bridge exactly once? (For more background on this problem, see [12].)

## 1.1.2   The Basics

*When you read you begin with A, B, C;*
*When you sing you begin with Do Re Mi.*
> — Rodgers and Hammerstein, *The Sound of Music*

Getting It All Together

The vertex set of a graph $G$ is denoted by $V(G)$, and the edge set is denoted by $E(G)$. We may refer to these sets simply as $V$ and $E$ if the context makes the particular graph clear. For notational convenience, instead of representing an edge as $\{u, v\}$, we denote this simply by $uv$. The *order* of a graph $G$ is the cardinality of its vertex set, and the *size* of a graph is the cardinality of its edge set.

Given two vertices $u$ and $v$, if $uv \in E$, then $u$ and $v$ are said to be *adjacent*. If $uv \notin E$, $u$ and $v$ are *nonadjacent*. Furthermore, if an edge $e$ has a vertex $v$ as an endpoint, we say that $v$ and $e$ are *incident*.

The *neighborhood* of a vertex $v$, denoted by $N(v)$, is the set of vertices adjacent to $v$:

$$N(v) = \{x \in V \mid vx \in E\}.$$

Given a set $S$ of vertices, we define the neighborhood of $S$, denoted by $N(S)$, to be the union of the neighborhoods of the vertices in $S$.

The *degree* of $v$, denoted by $\deg(v)$, is the number of edges incident with $v$. In simple graphs, this is the same as the cardinality of the neighborhood of $v$. The *maximum degree* of a graph $G$, denoted by $\Delta(G)$, is defined to be

$$\Delta(G) = \max\{\deg(v) \mid v \in V(G)\}.$$

Similarly, the *minimum degree* of a graph $G$, denoted by $\delta(G)$, is defined to be

$$\delta(G) = \min\{\deg(v) \mid v \in V(G)\}.$$

For example, in Figure 1.2 vertices 1 and 5 are adjacent, while vertices 1 and 2 are nonadjacent. We also see that $\deg(3) = 2$, $\deg(7) = 3$, $N(4) = \{1, 6, 7\}$, $\Delta(G) = 3$, and $\delta(G) = 1$.

The following theorem is often referred to as the First Theorem of Graph Theory.

**Theorem 1.1.**  *In a graph $G$, the sum of the degrees of the vertices is equal to twice the number of edges. Alternatively, the number of vertices with odd degree is even.*

PROOF.   Let $S = \sum_{v \in V} \deg(v)$. Notice that in counting $S$, we count each edge exactly twice. Thus, $S = 2|E|$ (the sum of the degrees is twice the number of edges). Since $S$ is even, it must be that the number of vertices with odd degree is even. $\qquad\qquad\square$

A *path* in a graph is a sequence of distinct vertices $v_1, v_2, \ldots, v_k$ such that $v_i v_{i+1} \in E$ for $i = 1, 2, \ldots, k - 1$. The *length* of a path is the number of edges on the path. A *cycle* in a graph is a sequence of vertices $w_1, w_2, \ldots, w_{r-1}, w_r$

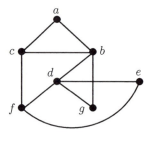

FIGURE 1.7.

such that $w_1, \ldots, w_{r-1}$ is a path, $w_1 = w_r$, and $w_{r-1}w_r \in E$. Essentially, a cycle is a closed path. The length of a cycle is defined to be the number of edges on the cycle. When we speak of an *odd* (*even*) *cycle*, we mean that the cycle has odd (even) length.

In Figure 1.7, $b, d, e, f, c$ is a path of length 4. The path $b, g$ has length 1, and the path $d$ has length 0 (how about that for a short path!). The sequence $c, b, d, g, b, a$ does not represent a path, since the vertex $b$ is repeated. Also in the figure, $a, b, c, a$ is a cycle of length 3 (an odd cycle), and $a, c, f, d, g, b, a$ is a cycle of length 6 (an even cycle). In fact, this graph contains cycles of length 3, 4, 5, 6, and 7.

### Taking It Apart

A graph is *connected* if every pair of vertices can be joined by a path. Informally, if one can pick up an entire graph by grabbing just one vertex, then the graph is connected. In Figure 1.8, $G_1$ is connected, and both $G_2$ and $G_3$ are not connected (or *disconnected*). Each maximal connected piece of a graph is called a *connected component*. In Figure 1.8, $G_1$ has one component, $G_2$ has three components, and $G_3$ has two components.

Given a graph $G$ and a vertex $v \in V(G)$, we let $G - v$ denote the graph obtained by removing $v$ and all edges incident with $v$ from $G$. If $S$ is a set of vertices, we let $G - S$ denote the graph obtained by removing each vertex of $S$ and all associated incident edges. If $e$ is an edge of $G$, then $G - e$ is the graph obtained by removing only the edge $e$ (its end vertices stay). Figure 1.9 gives examples of these operations.

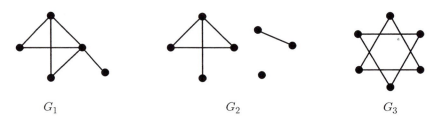

$G_1$                    $G_2$                    $G_3$

FIGURE 1.8. Connected and disconnected graphs.

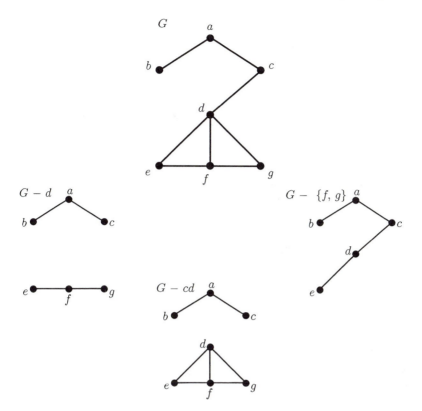

FIGURE 1.9. Deletion operations.

If the removal of a vertex $v$ from $G$ causes the number of components to increase, then $v$ is called a *cut vertex*. If the removal of an edge $e$ from $G$ causes the number of components to increase, then $e$ is called a *bridge*. In Figure 1.9, $d$ is a cut vertex and $cd$ is a bridge. There are two more cut vertices and two more bridges in this graph. Try to find them.

**Exercises**

1. If $G$ is a graph of order $n$, what is the maximum number of edges in $G$?
2. Prove that in any graph $G$, there exists a pair of vertices $u$ and $v$ such that $\deg(u) = \deg(v)$.
3. Let $G$ be a graph of order $n$ that is not connected. What is the maximum size of $G$?
4. Let $G$ be a graph of order $n$ and size strictly less than $n - 1$. Prove that $G$ is not connected.
5. Consider the graph shown in Figure 1.10. How many different paths have $c$ as an end vertex?
6. Is it true that a graph having exactly two vertices of odd degree must contain a path from one to the other? Give a proof or a counterexample.

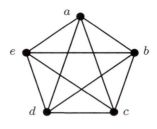

FIGURE 1.10.

**7.** Let $G$ be a graph such that $\deg(v) \geq 2$ for all $v \in V$. Prove that $G$ contains a cycle.

**8.** Let $P$ and $Q$ be two paths of maximum length in a connected graph $G$. Prove that $P$ and $Q$ have a common vertex.

**9.** Prove that an edge $e$ is a bridge of $G$ if and only if $e$ lies on no cycle of $G$.

**10.** Prove or disprove each of the following statements.

    **a.** If $G$ has no bridges, then $G$ has exactly one cycle.

    **b.** If $G$ has no cut vertices, then $G$ has no bridges.

    **c.** If $G$ has no bridges, then $G$ has no cut vertices.

### 1.1.3   Special Types of Graphs

*until we meet again ...*

— from *An Irish Blessing*

In this section we describe several types of graphs. We will run into many of them later in the chapter.

**1.** Complete Graphs
   The *complete graph* on $n$ vertices, denoted by $K_n$, is the graph of order $n$ where $uv \in E$ for all $u$ and $v \in V$ (Figure 1.11).

**2.** Empty Graphs
   The *empty graph* on $n$ vertices, denoted by $E_n$, is the graph of order $n$ where $E$ is the empty set (Figure 1.12).

**3.** Complements

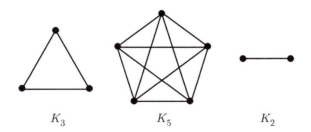

$K_3$        $K_5$        $K_2$

FIGURE 1.11. Examples of complete graphs.

$E_6$

FIGURE 1.12. An empty graph.

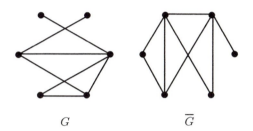

G                                $\overline{G}$

FIGURE 1.13. A graph and its complement.

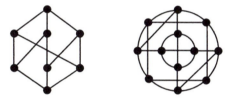

regular of degree 3            regular of degree 4

FIGURE 1.14. Examples of regular graphs.

Given a graph $G$, the *complement* of $G$, denoted by $\overline{G}$, is the graph whose vertex set is the same as $G$'s and whose edge set consists of all the edges that are *not* present in $G$ (Figure 1.13).

**4.** Regular Graphs

A graph $G$ is *regular* if every vertex has the same degree. $G$ is said to be *regular of degree r* (or *r-regular*) if $\deg(v) = r$ for all vertices $v$ in $G$. Complete graphs of order $n$ are regular of degree $n - 1$, and empty graphs are regular of degree 0. Two further examples are shown in Figure 1.14.

**5.** Cycles

The graph $C_n$ is simply a cycle on $n$ vertices (Figure 1.15).

**6.** Paths

The graph $P_n$ is simply a path on $n$ vertices (Figure 1.16).

**7.** Subgraphs

A graph $H$ is a *subgraph* of a graph $G$ if $V(H) \subseteq V(G)$ and $E(H) \subseteq E(G)$. In this case we write $H \subseteq G$, and we say that $G$ contains $H$. In Figure 1.17 $H_1$ and $H_2$ are both subgraphs of $G$, but $H_3$ is not.

FIGURE 1.15. The graph $C_7$.

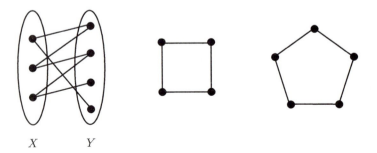

FIGURE 1.16. The graph $P_6$.

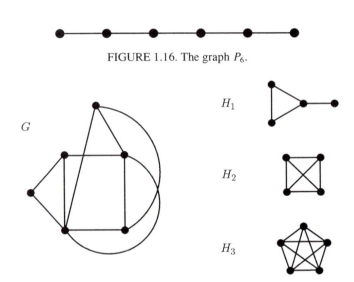

FIGURE 1.17. $H_1$ and $H_2$ are subgraphs of $G$.

**8.** Bipartite Graphs

A graph $G$ is *bipartite* if its vertex set can be partitioned into two sets $X$ and $Y$ in such a way that every edge of $G$ has one end vertex in $X$ and another in $Y$. In this case, $X$ and $Y$ are called the *partite sets*. The first two graphs in Figure 1.18 are bipartite. Since it is not possible to partition the vertices of the third graph into two such sets, the third graph is not bipartite.

FIGURE 1.18. Two bipartite graphs and one non-bipartite graph.

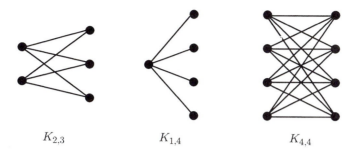

FIGURE 1.19. A few complete bipartite graphs.

A bipartite graph with partite sets $X$ and $Y$ is called a *complete bipartite graph* if its edge set is of the form $E = \{xy \mid x \in X, y \in Y\}$ (that is, if every possible connection of a vertex of $X$ with a vertex of $Y$ is present in the graph). Such a graph is denoted by $K_{|X|,|Y|}$. See Figure 1.19.

We conclude this section by stating and proving an interesting characterization of bipartite graphs.

**Theorem 1.2.** *A graph of order at least two is bipartite if and only if it contains no odd cycles.*

PROOF.  Let $G$ be a bipartite graph with partite sets $X$ and $Y$. Let $C$ be a cycle of $G$ and say that $C$ is $v_1, v_2, \ldots, v_k, v_1$. Assume without loss of generality that $v_1 \in X$. The nature of bipartite graphs implies then that $v_i \in X$ for all odd $i$, and $v_i \in Y$ for all even $i$. Since $v_k$ is adjacent to $v_1$, it must be that $k$ is even; and hence $C$ is an even cycle.

For the reverse direction of the theorem, let $G$ be a graph of order at least two such that $G$ contains no odd cycles. Without loss of generality, we can assume that $G$ is connected, for if not, we could treat each of its connected components separately. Let $v$ be a vertex of $G$, and define the set $X$ to be

$$X = \{x \in V(G) \mid \text{the shortest path from } x \text{ to } v \text{ has even length}\},$$

and let $Y = V(G) \setminus X$.

Now let $x$ and $x'$ be vertices of $X$, and suppose that $x$ and $x'$ are adjacent. If $x = v$, then the shortest path from $v$ to $x'$ has length one. But this implies that $x' \in Y$, a contradiction. So, it must be that $x \neq v$, and by a similar argument, $x' \neq v$. Let $P_1$ be a path from $v$ to $x$ of shortest length (a shortest $v$–$x$ path) and let $P_2$ be a shortest $v$–$x'$ path. Say that $P_1$ is $v = v_0, v_1, \ldots, v_{2k} = x$ and that $P_2$ is $v = w_0, w_1, \ldots, w_{2t} = x'$. The paths $P_1$ and $P_2$ certainly have $v$ in common. Let $v'$ be a vertex on both paths such that the $v'$–$x$ path, call it $P_1'$, and the $v'$–$x'$ path, call it $P_2'$, have only the vertex $v'$ in common. Essentially, $v'$ is the "last" vertex common to $P_1$ and $P_2$. It must be that $P_1'$ and $P_2'$ are shortest $v'$–$x$ and $v'$–$x'$ paths, respectively. Therefore (remembering that $x$ and $x'$ are adjacent), $v' = v_i = w_i$ for some $i$. But then, $v_i, v_{i+1}, \ldots, v_{2k}, w_{2t}, w_{2t-1}, \ldots, w_i$ is an odd cycle, and that is a contradiction.

FIGURE 1.20.

Thus, no two vertices in $X$ are adjacent to each other, and a similar argument shows that no two vertices in $Y$ are adjacent to each other. Therefore, $G$ is bipartite with partite sets $X$ and $Y$.    □

## Exercises

1. If $K_{r_1, r_2}$ is regular, prove that $r_1 = r_2$.
2. Determine whether $K_4$ is a subgraph of $K_{4,4}$. If yes, then exhibit it. If no, then explain why not.
3. List all of the connected subgraphs of $C_{34}$.
4. The concept of complete bipartite graphs can be generalized to define the complete multipartite graph $K_{r_1, r_2, \dots, r_k}$. This graph consists of $k$ sets of vertices $A_1$, $A_2$, ..., $A_k$ (with sizes $r_1, \dots, r_k$, respectively) where all possible "interset edges" are present and no "intraset edges" are present. Find expressions for the order and size of $K_{r_1, r_2, \dots, r_k}$.
5. The *line graph* $L(G)$ of a graph $G$ is defined in the following way: The vertices of $L(G)$ are the edges of $G$, $V(L(G)) = E(G)$; and two vertices in $L(G)$ are adjacent if and only if the corresponding edges in $G$ share a vertex.

   **a.** Let $G$ be the graph shown in Figure 1.20. Find $L(G)$.
   **b.** If a graph $G$ has $n$ vertices, find formulas for the order and size of $L(G)$.

## 1.2   Trees

> *"O look at the trees!"* they cried, *"O look at the trees!"*
> — Robert Bridges, *London Snow*

In this section we will look at the trees—but not the ones that sway in the wind or catch the falling snow. We will talk about graph-theoretic trees. Before moving on, glance ahead at Figure 1.21, and try to pick out which graphs are trees.

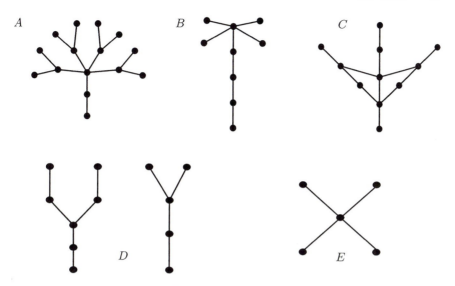

FIGURE 1.21. Which ones are trees?

## 1.2.1 Definitions and Examples

*Example, the surest method of instruction.*

— Pliny the Younger

In Figure 1.21 graphs *A*, *B*, and *E* are trees, while graphs *C* and *D* are not. Let us move on to some definitions.

A *tree* is a connected graph that contains no cycles. Graph-theoretic trees resemble the trees we see outside our windows. For example, graph-theoretic trees do not have cycles, just as the branches of trees in nature do not split and rejoin. The descriptive terminology does not stop here.

Graph *D* in Figure 1.21 is not a tree; rather, it is a *forest*. A forest is a collection of one or more trees. A vertex of degree 1 in a tree is called a *leaf*.

As in nature, graph-theoretic trees come in many shapes and sizes. They can be thin ($P_{10}$) or thick ($K_{1,1000}$), tall ($P_{1000}$) or short ($K_1$ and $K_2$). Yes, even the graphs $K_1$ and $K_2$ are considered trees (they are certainly connected and acyclic). In the spirit of our arboreal terminology, perhaps we should call $K_1$ a *stump* and $K_2$ a *twig*!

While we are on the subject of small trees, we should count a few of them. It is clear that $K_1$ and $K_2$ are the only trees of order 1 and 2, respectively. A moment's thought will reveal that $P_3$ is the only tree of order 3. Figure 1.22 shows the different trees of order 6 or less.

Trees sprout up as effective models in a wide variety of applications. We mention a few brief examples. A more thorough treatment of applications like these can be found in [124] and [52].

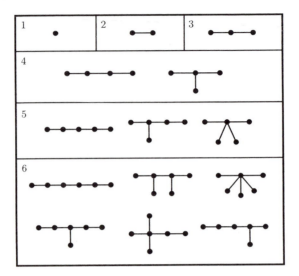

FIGURE 1.22. Trees of order 6 or less.

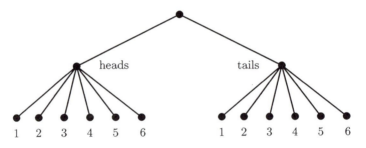

FIGURE 1.23. Outcomes of a coin/die experiment.

Examples

1. Trees are useful for modeling the possible outcomes of an experiment. For example, consider an experiment in which a coin is flipped and a 6-sided die is rolled. The leaves in the tree in Figure 1.23 correspond to the outcomes in the probability space for this experiment.

2. Programmers often use tree structures to facilitate searches and sorts and to model the logic of algorithms. For instance, the logic for a program that finds the maximum of four numbers $(w, x, y, z)$ can be represented by the tree shown in Figure 1.24. This type of tree is a *binary decision tree*.

3. Chemists can use trees to represent, among other things, saturated hydrocarbons—chemical compounds of the form $C_nH_{2n+2}$ (propane, for example). The bonds between the carbon and hydrogen atoms are depicted in the trees of Figure 1.25. The vertices of degree 4 are the carbon atoms, and the leaves represent the hydrogen atoms.

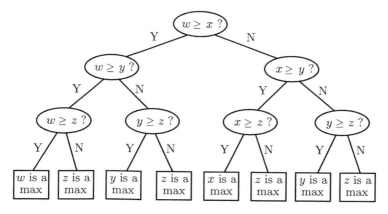

FIGURE 1.24. Logic of a program.

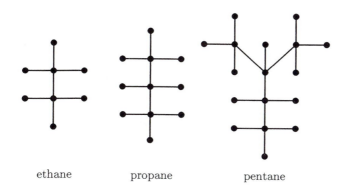

ethane             propane             pentane

FIGURE 1.25. A few saturated hydrocarbons.

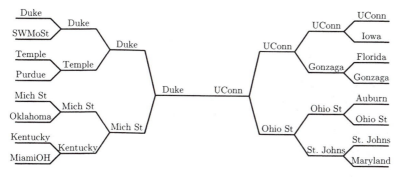

FIGURE 1.26. The 1999 Men's Sweet 16.

**4.** College basketball fans will recognize the tree in Figure 1.26. It displays final results for the "Sweet 16" portion of the 1999 NCAA men's basketball tournament.

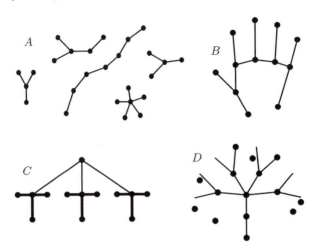

FIGURE 1.27. What would you name these graphs?

**Exercises**

1. Draw all trees of order 7. Hint: There are a prime number of them.
2. Draw all forests of order 6.
3. Let $T$ be a tree of order $n \geq 2$. Prove that $T$ is bipartite.
4. Graphs of the form $K_{1,n}$ are called *stars*. Prove that if $K_{r,s}$ is a tree, then it must be a star.
5. Match the graphs in Figure 1.27 with appropriate names: a palm tree, autumn, a path through a forest, tea leaves.

## 1.2.2    *Properties of Trees*

*And the tree was happy.*

— Shel Silverstein, *The Giving Tree*

Let us try an experiment. On a piece of scratch paper, draw a tree of order 16. Got one? Now count the number of edges in the tree. We are going to go out on a limb here and predict that there are 15. Since there are nearly 20,000 different trees of order 16, it may seem surprising that our prediction was correct. The next theorem gives away our secret.

**Theorem 1.3.** *If $T$ is a tree of order $n$, then $T$ has $n - 1$ edges.*

PROOF.    We induct on the order of $T$. For $n = 1$ the only tree is the stump ($K_1$), and it of course has 0 edges. Assume that the result is true for all trees of order less than $k$, and let $T$ be a tree of order $k$.

Choose some edge of $T$ and call it $e$. Since $T$ is a tree, it must be that $T - e$ is disconnected (see Exercise 7) and that it must consist of two connected components that are trees themselves (see Figure 1.28). Say that these two components of $T - e$

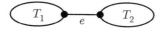

FIGURE 1.28.

are $T_1$ and $T_2$, with orders $k_1$ and $k_2$, respectively. Thus, $k_1$ and $k_2$ are less than $n$ and $k_1 + k_2 = k$.

Since $k_1 < k$, the theorem is true for $T_1$. Thus $T_1$ has $k_1 - 1$ edges. Similarly, $T_2$ has $k_2 - 1$ edges. Now, since $E(T)$ is the disjoint union of $E(T_1)$, $E(T_2)$, and $\{e\}$, we have $|E(T)| = (k_1 - 1) + (k_2 - 1) + 1 = k_1 + k_2 - 1 = k - 1$. This completes the induction. □

The next theorem extends the preceding result to forests. The proof is similar and appears as Exercise 4.

**Theorem 1.4.** *If F is a forest of order n containing k connected components, then F contains n − k edges.*

The next two theorems give alternative methods for defining trees. Two other methods are given in Exercises 5 and 6.

**Theorem 1.5.** *A graph of order n is a tree if and only if it is connected and contains n − 1 edges.*

PROOF. The forward direction of this theorem is immediate from the definition of trees and Theorem 1.3. For the reverse direction, suppose a graph $G$ of order $n$ is connected and contains $n - 1$ edges. We need to show that $G$ is acyclic. If $G$ did have a cycle, we could remove an edge from the cycle and the resulting graph would still be connected. In fact, we could keep removing edges (one at a time) from existing cycles, each time maintaining connectivity. The resulting graph would be connected and acyclic and would thus be a tree. But this tree would have fewer than $n - 1$ edges, and this is impossible by Theorem 1.3. Therefore, $G$ has no cycles, so $G$ is a tree. □

**Theorem 1.6.** *A graph of order n is a tree if and only if it is acyclic and contains n − 1 edges.*

PROOF. Again the forward direction of this theorem follows from the definition of trees and from Theorem 1.3. So suppose that $G$ is acyclic and has $n - 1$ edges. To show that $G$ is a tree we need to show only that it is connected. Let us say that the connected components of $G$ are $G_1, G_2, \ldots, G_k$. Since $G$ is acyclic, each of these components is a tree, and so $G$ is a forest. Theorem 1.4 tells us that $G$ has $n - k$ edges, implying that $k = 1$. Thus $G$ has only one connected component, implying that $G$ is a tree. □

It is not uncommon to look out a window and see leafless trees. Bare trees are not rare, especially in cooler climates. In graph theory, though, leafless trees are rare indeed. In fact, the stump ($K_1$) is the only such tree, and every other tree has

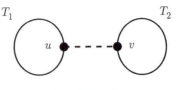

FIGURE 1.29.

at least two leaves. Take note of the proof technique of the following theorem. It is a standard graph theory induction argument.

**Theorem 1.7.**  *Let T be the tree of order n $\geq$ 2. Then T has at least two leaves.*

PROOF.    Again we induct on the order. The result is certainly true if $n = 2$, since $T = K_2$ in this case. Suppose the result is true for all orders from 2 to $n - 1$, and consider a tree $T$ of order $n \geq 3$. We know that $T$ has $n - 1$ edges, and since we can assume $n \geq 3$, $T$ has at least 2 edges. If every edge of $T$ is incident with a leaf, then $T$ has at least two leaves; and the proof is complete. So assume that there is some edge of $T$ that is not incident with a leaf, and let us say that this edge is $e = uv$. The graph $T - e$ is a pair of trees, $T_1$ and $T_2$, each of order less than $n$. Let us say, without loss of generality, that $u \in V(T_1)$, $v \in V(T_2)$, $|V(T_1)| = n_1$, and $|V(T_2)| = n_2$ (see Figure 1.29). Since $e$ is not incident with any leaves of $T$, we know that $n_1$ and $n_2$ are both at least 2, so the induction hypothesis applies to each of $T_1$ and $T_2$. Thus, each of $T_1$ and $T_2$ has two leaves. This means that each of $T_1$ and $T_2$ has at least one leaf that is not incident with the edge $e$. Thus the graph $(T - e) + e = T$ has at least two leaves.    □

We conclude this section with an interesting result about trees as subgraphs.

**Theorem 1.8.**  *Let T be be a tree with k edges. If G is a graph whose minimum degree satisfies $\delta(G) \geq k$, then G contains T as a subgraph. Alternatively, G contains every tree of order at most $\delta(G) + 1$ as a subgraph.*

PROOF.    We induct on $k$. If $k = 0$, then $T = K_1$, and it is clear that $K_1$ is a subgraph of any graph. Further, if $k = 1$, then $T = K_2$, and $K_2$ is a subgraph of any graph whose minimum degree is 1. Assume that the result is true for all trees with $k - 1$ edges ($k \geq 2$), and consider a tree $T$ with exactly $k$ edges. We know from Theorem 1.7 that $T$ contains at least two leaves. Let $v$ be one of them, and let $w$ be the vertex that is adjacent to $v$. Consider the graph $T - v$.

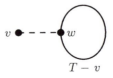

$T - v$

FIGURE 1.30.

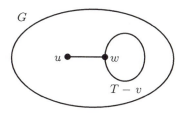

FIGURE 1.31. A copy of $T$ inside $G$.

Since $T - v$ has $k - 1$ edges, the induction hypothesis applies, so $T - v$ is a subgraph of $G$. We can think of $T - v$ as actually sitting inside $G$ (meaning $w$ is a vertex of $G$, too). Now, since $G$ contains at least $k + 1$ vertices and $T - v$ contains $k$ vertices, there exist vertices of $G$ that are not a part of the subgraph $T - v$. Further, since the degree in $G$ of $w$ is at least $k$, there must be a vertex $u$ not in $T - v$ that is adjacent to $w$ (Figure 1.31). The subgraph $T - v$ together with $u$ forms the tree $T$ as a subgraph of $G$. □

### Exercises

1. Draw each of the following, if you can. If you cannot, explain the reason.

   a. A 10-vertex forest with exactly 12 edges
   b. A 12-vertex forest with exactly 10 edges
   c. A 14-vertex forest with exactly 14 edges
   d. A 14-vertex forest with exactly 13 edges
   e. A 14-vertex forest with exactly 12 edges

2. Suppose a tree $T$ has an even number of edges. Show that at least one vertex must have even degree.
3. Let $T$ be a tree with max degree $\Delta$. Prove that $T$ has at least $\Delta$ leaves.
4. Let $F$ be a forest of order $n$ containing $k$ connected components. Prove that $F$ contains $n - k$ edges.
5. Prove that a graph $G$ is a tree if and only if for every pair of vertices $u$, $v$, there is exactly one path from $u$ to $v$.
6. Prove that $T$ is a tree if and only if $T$ contains no cycles, and for any new edge $e$, the graph $T + e$ has exactly one cycle.
7. Show that every edge in a tree is a bridge.
8. Show that every nonleaf in a tree is a cut vertex.
9. Find a shorter proof to Theorem 1.7. Hint: Start by considering a longest path in $T$.
10. Let $T$ be a tree of order $n > 1$. Show that the number of leaves is

$$2 + \sum_{\deg(v_i) \geq 3} (\deg(v_i) - 2),$$

where the sum is over all vertices of degree 3 or more.

**11.** For a graph $G$, define the average degree of $G$ to be

$$\text{avgdeg}(G) = \frac{\sum_{v \in V(G)} \deg(v)}{|V(G)|}.$$

If $T$ is a tree and $\text{avgdeg}(T) = a$, then find the number of vertices in $T$.

**12.** Let $T$ be a tree such that every vertex adjacent to a leaf has degree at least 3. Prove that some pair of leaves in $T$ has a common neighbor.

### 1.2.3  Spanning Trees

*Under the spreading chestnut tree ...*
— Henry W. Longfellow, *The Village Blacksmith*

The North Carolina Department of Transportation (NCDOT) has decided to implement a rapid rail system to serve eight cities in the western part of the state. Some of the cities are currently joined by roads or highways, and the state plans to lay the track right along these roads. Due to the mountainous terrain, some of the roads are steep and curvy; and so laying track along these roads would be difficult and expensive. The NCDOT hired a consultant to study the roads and to assign difficulty ratings to each one. The rating accounted for length, grade, and curviness of the roads; and higher ratings correspond to greater cost. The graph in Figure 1.32, call it the "city graph," shows the result of the consultant's investigation. The number on each edge represents the difficulty rating assigned to the existing road.

The state wants to be able to make each city accessible (but not necessarily directly accessible) from every other city. One obvious way to do this is to lay track along every one of the existing roads. But the state wants to minimize cost, so this solution is certainly not the best, since it would result in a large amount of unnecessary track. In fact, the best solution will not include a cycle of track anywhere, since a cycle would mean at least one segment of wasted track.

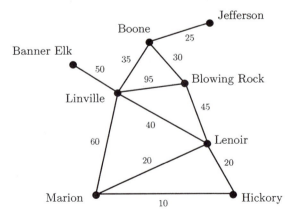

FIGURE 1.32. The city graph.

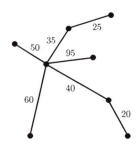

Total Weight = 325

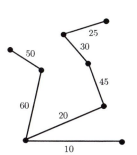

Total Weight = 240

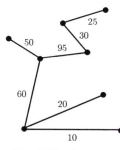

Total Weight = 290

FIGURE 1.33. Several spanning trees.

The situation above motivates a definition. Given a graph $G$ and a subgraph $T$, we say that $T$ is a *spanning tree* of $G$ if $T$ is a tree that contains every vertex of $G$.

So it looks as though the DOT just needs to find a spanning tree of the city graph, and they would like to find one whose overall rating is as small as possible. Figure 1.33 shows several attempts at a solution.

Of the solutions in the figure, the upper right solution has the least total weight— but is it the best one overall? Try to find a better one. We will come back to this problem soon.

Given a graph $G$, a *weight function* is a function $W$ that maps the edges of $G$ to the nonnegative real numbers. The graph $G$ together with a weight function is called a *weighted graph*. The graph in Figure 1.32 is a simple example of a weighted graph. Although one might encounter situations where negative valued weights would be appropriate, we will stick with nonnegative weights in our discussion.

It should be fairly clear that every connected graph has at least one spanning tree. In fact, it is not uncommon for a graph to have many different spanning trees. Figure 1.33 displays three different spanning trees of the city graph.

Given a connected weighted graph $G$, a spanning tree $T$ is called a *minimum weight spanning tree* if the sum of the weights of the edges of $T$ is no more than the sum for any other spanning tree of $G$.

There are a number of fairly simple algorithms for finding minimum weight spanning trees. Perhaps the best known is Kruskal's algorithm.

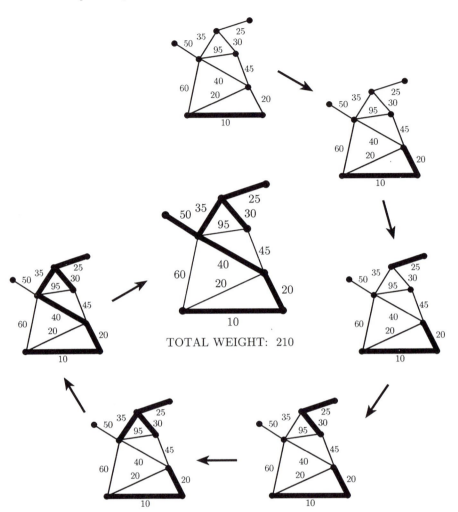

FIGURE 1.34. The stages of Kruskal's algorithm.

### Kruskal's Algorithm

Given a weighted graph $G$.

i. Find an edge of minimum weight and mark it.
ii. Among all of the unmarked edges that do not form a cycle with any of the marked edges, choose an edge of minimum weight and mark it.
iii. If the set of marked edges forms a spanning tree of $G$, then stop. If not, repeat step ii.

Figure 1.34 demonstrates Kruskal's algorithm applied to the city graph. The minimum weight is 210.

It is certainly possible for different trees to result from two different applications of Kruskal's algorithm. For instance, in the second step we could have chosen the edge between Marion and Lenoir instead of the one that was chosen. Even so, the total weight of resulting trees is the same, and each such tree is a minimum weight spanning tree.

It should be clear from the algorithm itself that the subgraph built is in fact a spanning tree of $G$. How can we be sure, though, that it has minimum total weight? The following theorem answers our question [97].

**Theorem 1.9.** *Kruskal's algorithm produces a spanning tree of minimum total weight.*

PROOF.   Let $G$ be a graph of order $n$, and let $T$ be a spanning tree obtained by applying Kruskal's algorithm to $G$. As we have seen, Kruskal's algorithm builds spanning trees by adding one edge at a time until a tree is formed. Let us say that the edges added for $T$ were (in order) $e_1, e_2, \ldots, e_{n-1}$. Suppose $T$ is *not* a minimum weight spanning tree. Among all minimum weight spanning trees of $G$, choose $T'$ to be a minimum weight spanning tree that agrees with the construction of $T$ for the longest time (i.e., for the most initial steps). This means that there exists some $k$ such that $T'$ contains $e_1, \ldots, e_k$, and no minimum weight spanning tree contains all of $e_1, \ldots, e_k, e_{k+1}$ (notice that since $T$ is not of minimum weight, $k < n - 1$).

Since $T'$ is a spanning tree, it must be that $T' + e_{k+1}$ contains a cycle $C$, and since $T$ contains no cycles, $C$ must contain some edge, say $e'$, that is *not* in $T$. If we remove the edge $e'$ from $T' + e_{k+1}$, then the cycle $C$ is broken and what remains is a spanning tree of $G$. Thus, $T' + e_{k+1} - e'$ is a spanning tree of $G$, and it contains edges $e_1, \ldots, e_k, e_{k+1}$. Furthermore, since the edge $e'$ must have been available to be chosen when $e_{k+1}$ was chosen by the algorithm, it must be that $w(e_{k+1}) \leq w(e')$. This means that $T' + e_{k+1} - e'$ is a spanning tree with weight no more than $T'$ that contains edges $e_1, \ldots, e_{k+1}$, contradicting our assumptions. Therefore, it must be that $T$ is a minimum weight spanning tree.   $\square$

**Exercises**

1. Prove that every connected graph contains at least one spanning tree.
2. Prove that a graph is a tree if and only if it is connected and has exactly one spanning tree.
3. Let $G$ be a connected graph with $n$ vertices and at least $n$ edges. Let $C$ be a cycle of $G$. Prove that if $T$ is a spanning tree of $G$, then $\overline{T}$, the complement of $T$, contains at least one edge of $C$.
4. Let $G$ be connected, and let $e$ be an edge of $G$. Prove that $e$ is a bridge if and only if it is in every spanning tree of $G$.
5. Using Kruskal's algorithm, find a minimum weight spanning tree of the graphs in Figure 1.35. In each case, determine (with proof) whether the minimum weight spanning tree is unique.
6. Give an example of a weighted graph that has a unique spanning tree *and* contains a cycle with two edges of identical weight.

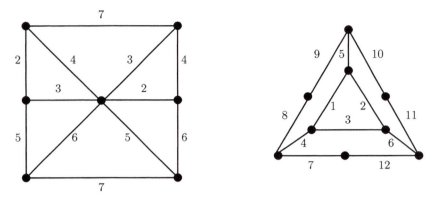

FIGURE 1.35. Two weighted graphs.

## 1.2.4   Counting Trees

*As for everything else, so for a mathematical theory: beauty can be perceived but not explained.*

— Arthur Cayley, [110]

In this section we discuss two beautiful results on counting the number of spanning trees in a graph. The next chapter studies general techniques for counting arrangements of objects, so these results are a sneak preview.

### Cayley's Tree Formula

Cayley's Tree Formula gives us a way to count the number of different labeled trees on $n$ vertices. In this problem we think of the vertices as being fixed, and we consider all the ways to draw a tree on those fixed vertices. Figure 1.36 shows three different labeled trees on three vertices, and in fact, these are the only three.

There are 16 different labeled trees on four vertices, and they are shown in Figure 1.37.

As an exercise, the ambitious student should try drawing all of the labeled trees on five vertices. The cautious ambitious student might wish to look ahead at Cayley's formula before embarking on such a task.

Cayley proved the following theorem in 1889 [24]. The proof technique that we will describe here is due to Prüfer [120] (with a name like that he was destined for

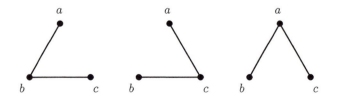

FIGURE 1.36. Labeled trees on three vertices.

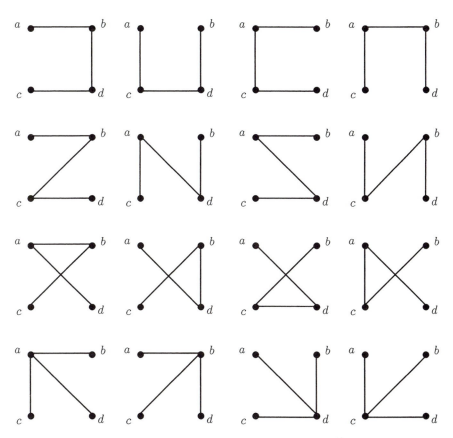

FIGURE 1.37. Labeled trees on four vertices.

mathematical greatness!). Prüfer's method is almost as noteworthy as the result itself. He counted the labeled trees by placing them in one-to-one correspondence with a set whose size is easy to determine—the set of all sequences of length $n - 2$ whose entries come from the set $\{1, \ldots, n\}$. There are $n^{n-2}$ such sequences.

**Theorem 1.10** (Cayley's Tree Formula). *There are $n^{n-2}$ distinct labeled trees of order $n$.*

The algorithm below gives the steps that Prüfer used to assign a particular sequence to a given tree, $T$, whose vertices are labeled $1, \ldots, n$. Each labeled tree is assigned a unique sequence.

**Prüfer's Method for Assigning a Sequence to a Labeled Tree**

Given: Labeled tree $T$.

1. Let $i = 0$, and let $T_0 = T$.
2. Find the leaf on $T_i$ with the smallest label and call it $v$.
3. Record in the sequence the label of $v$'s neighbor.

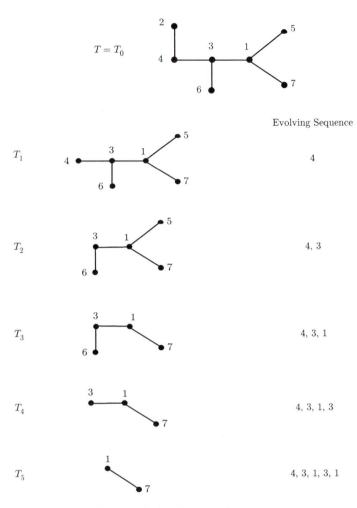

FIGURE 1.38. Creating a Prüfer sequence.

**4.** Remove $v$ from $T_i$ to create a new tree $T_{i+1}$.

**5.** If $T_{i+1} = K_2$, then stop. Otherwise, increment $i$ by 1 and go back to step 2.

Let us run through this algorithm with a particular graph. In Figure 1.38, tree $T = T_0$ has 7 vertices, labeled as shown. The first step is finding the leaf with smallest label: This would be 2. The neighbor of vertex 2 is the vertex labeled 4. Therefore, 4 is the first entry in the sequence. Removing vertex 2 produces tree $T_1$. The leaf with smallest label in $T_1$ is 4, and its neighbor is 3. Therefore, we put 3 in the sequence and delete 4 from $T_1$. Vertex 5 is the smallest leaf in tree $T_2 = T_1 - \{4\}$, and its neighbor is 1. So our sequence so far is 4, 3, 1. In $T_3 = T_2 - \{5\}$ the smallest leaf is vertex 6, whose neighbor is 3. In $T_4 = T_3 - \{6\}$, the smallest leaf is vertex

3, whose neighbor is 1. Since $T_5 = K_2$, we stop here. Our resulting sequence is 4, 3, 1, 3, 1.

Notice that in the previous example, none of the leaves of the original tree $T$ appears in the sequence. More generally, each vertex $v$ appears in the sequence exactly $\deg(v) - 1$ times. This is not a coincidence (see Exercise 1). We now present Prüfer's algorithm for assigning trees to sequences. Each sequence gets assigned a unique tree.

### Prüfer's Method for Assigning a Labeled Tree to a Sequence

Given: Sequence $\sigma = a_1, a_2, \ldots, a_k$.

1. Draw $k+2$ vertices and label them $v_1, v_2, \ldots, v_{k+2}$. Let $S = \{v_1, v_2, \ldots, v_{k+2}\}$.
2. Let $i = 0$, let $\sigma_0 = \sigma$, and let $S_0 = S$.
3. Let $j$ be the smallest number in $S_i$ that does not appear in the sequence $\sigma_i$.
4. Place an edge between vertex $v_j$ and the vertex whose subscript appears first in the sequence $\sigma_i$.
5. Remove the first number in the sequence $\sigma_i$ to create a new sequence $\sigma_{i+1}$. Remove the label $v_j$ from the set $S_i$ to create a new set $S_{i+1}$.
6. If the sequence $\sigma_{i+1}$ is empty, place an edge between the last two vertices in $S_{i+1}$, and stop. Otherwise, increment $i$ by 1 and return to step 3.

Let us apply this algorithm to a particular example. Let $\sigma = 4, 3, 1, 3, 1$ be our initial sequence to which we wish to assign a particular labeled tree. Since there are five terms in the sequence, our labels will come from the set $S = \{v_1, v_2, v_3, v_4, v_5, v_6, v_7\}$. After drawing the seven vertices, we look in the set $S = S_0$ to find the smallest subscript that does not appear in the sequence $\sigma = \sigma_0$. Subscript 2 is the one, and so we place an edge between vertices $v_2$ and $v_4$, the first subscript in the sequence. We now remove the first term from the sequence and the label $v_2$ from the set, forming a new sequence $\sigma_1 = 3, 1, 3, 1$ and a new set $S_1 = \{v_1, v_3, v_4, v_5, v_6, v_7\}$. The remaining steps in the process are shown in Figure 1.39.

You will notice that the tree that was created from the sequence $\sigma$ in the second example is the very same tree that created the sequence $\sigma$ in the first example. Score one for Prüfer!

### Matrix Tree Theorem

The second major result that we present in this section is the Matrix Tree Theorem, and like Cayley's Theorem, it provides a way of counting spanning trees of labeled graphs. While Cayley's Theorem in essence gives us a count on the number of spanning trees of complete labeled graphs, the Matrix Tree Theorem applies to labeled graphs in general. The theorem was proved in 1847 by Kirchhoff [91], and it demonstrates a wonderful connection between spanning trees and matrices. Before we state the theorem, we need to make a few definitions.

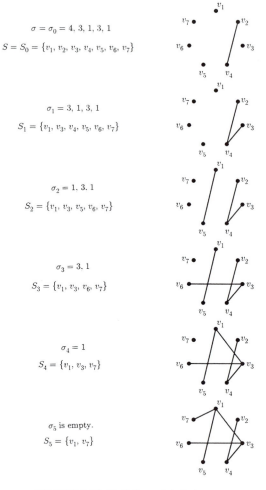

$\sigma = \sigma_0 = 4, 3, 1, 3, 1$

$S = S_0 = \{v_1, v_2, v_3, v_4, v_5, v_6, v_7\}$

$\sigma_1 = 3, 1, 3, 1$

$S_1 = \{v_1, v_3, v_4, v_5, v_6, v_7\}$

$\sigma_2 = 1, 3, 1$

$S_2 = \{v_1, v_3, v_5, v_6, v_7\}$

$\sigma_3 = 3, 1$

$S_3 = \{v_1, v_3, v_6, v_7\}$

$\sigma_4 = 1$

$S_4 = \{v_1, v_3, v_7\}$

$\sigma_5$ is empty.

$S_5 = \{v_1, v_7\}$

FIGURE 1.39. Building a labeled tree.

Let $G$ be a graph with vertices $v_1, v_2, \ldots, v_n$. The *adjacency matrix* of $G$ is the $n \times n$ matrix $A$ whose $(i, j)$ entry, denoted by $[A]_{i,j}$, is defined by

$$[A]_{i,j} = \begin{cases} 1 & \text{if } v_i \text{ and } v_j \text{ are adjacent,} \\ 0 & \text{otherwise.} \end{cases}$$

Let $G$ be a graph with vertices $v_1, v_2, \ldots v_n$. The *degree matrix* of $G$ is the $n \times n$ matrix $D$ whose $(i, j)$ entry, denoted by $[D]_{i,j}$, is defined by

$$[D]_{i,j} = \begin{cases} \deg(v_i) & \text{if } i = j, \\ 0 & \text{otherwise.} \end{cases}$$

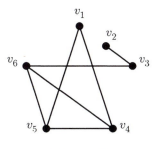

FIGURE 1.40.

The graph in Figure 1.40 has six vertices. Its degree matrix $D$ and adjacency matrix $A$ are

$$D = \begin{bmatrix} 2 & 0 & 0 & 0 & 0 & 0 \\ 0 & 1 & 0 & 0 & 0 & 0 \\ 0 & 0 & 2 & 0 & 0 & 0 \\ 0 & 0 & 0 & 3 & 0 & 0 \\ 0 & 0 & 0 & 0 & 3 & 0 \\ 0 & 0 & 0 & 0 & 0 & 3 \end{bmatrix}, \quad A = \begin{bmatrix} 0 & 0 & 0 & 1 & 1 & 0 \\ 0 & 0 & 1 & 0 & 0 & 0 \\ 0 & 1 & 0 & 0 & 0 & 1 \\ 1 & 0 & 0 & 0 & 1 & 1 \\ 1 & 0 & 0 & 1 & 0 & 1 \\ 0 & 0 & 1 & 1 & 1 & 0 \end{bmatrix}.$$

Let $M$ be an $n \times n$ matrix. The $i, j$ *cofactor* of $M$ is defined to be

$$(-1)^{i+j} \det(M(i \mid j)),$$

where $\det(M(i \mid j))$ represents the determinant of the $(n - 1) \times (n - 1)$ matrix formed by deleting row $i$ and column $j$ from $M$.

We are now ready to state the Matrix Tree Theorem, due to Kirchhoff. The proof that we give imitates those presented in [78] and [25].

**Theorem 1.11** (Matrix Tree Theorem). *If $G$ is a connected labeled graph with adjacency matrix $A$ and degree matrix $D$, then the number of unique spanning trees of $G$ is equal to the value of any cofactor of the matrix $D - A$.*

PROOF.    Suppose $G$ has $n$ vertices $(v_1, \ldots, v_n)$ and $k$ edges $(f_1, \ldots, f_k)$. Since $G$ is connected, we know that $k$ is at least $n - 1$. Let $N$ be the $n \times k$ matrix whose $i, j$ entry is defined by

$$[N]_{i,j} = \begin{cases} 1 & \text{if } v_i \text{ and } f_j \text{ are incident,} \\ 0 & \text{otherwise.} \end{cases}$$

$N$ is called the *incidence matrix* of $G$. Since every edge of $G$ is incident with exactly two vertices of $G$, each column of $N$ contains two 1's and $n - 2$ zeros. Let $M$ be the $n \times k$ matrix that results from changing the topmost 1 in each column to $-1$. To prove the result, we first need to establish two facts, which we call Claim A and Claim B.

Claim A. $MM^T = D - A$ (where $M^T$ denotes the transpose of $M$).

First, notice that the $(i, j)$ entry of $D - A$ is

$$[D - A]_{i,j} = \begin{cases} \deg(v_i) & \text{if } i = j, \\ -1 & \text{if } i \neq j \text{ and } v_i v_j \in E(G), \\ 0 & \text{if } i \neq j \text{ and } v_i v_j \notin E(G). \end{cases}$$

Now, what about the $(i, j)$ entry of $MM^T$? The rules of matrix multiplication tell us that this entry is the dot product of row $i$ of $M$ and column $j$ of $M^T$. That is,

$$\begin{aligned}[MM^T]_{i,j} &= \big([M]_{i,1}, [M]_{i,2}, \ldots, [M]_{i,k}\big) \cdot \big([M^T]_{1,j}, [M^T]_{2,j}, \ldots, [M^T]_{k,j}\big) \\ &= \big([M]_{i,1}, [M]_{i,2}, \ldots, [M]_{i,k}\big) \cdot \big([M]_{j,1}, [M]_{j,2}, \ldots, [M]_{j,k}\big) \\ &= \sum_{r=1}^{k} [M]_{i,r}[M]_{j,r}.\end{aligned}$$

If $i = j$, then this sum counts one for every nonzero entry in row $i$; that is, it counts the degree of $v_i$. If $i \neq j$ and $v_i v_j \notin E(G)$, then there is no column of $M$ in which both the row $i$ and row $j$ entries are nonzero. Hence the value of the sum in this case is 0. If $i \neq j$ and $v_i v_j \in E(G)$, then the only column in which both the row $i$ and the row $j$ entries are nonzero is the column that represents the edge $v_i v_j$. Since one of these entries is 1 and and the other is $-1$, the value of the sum is $-1$. We have shown that the $(i, j)$ entry of $MM^T$ is the same as the $(i, j)$ entry of $D - A$, and thus Claim A is proved.

Let $H$ be a subgraph of $G$ with $n$ vertices and $n - 1$ edges. Let $p$ be an arbitrary integer between 1 and $n$, and let $M'$ be the $(n-1) \times (n-1)$ submatrix of $M$ formed by all rows of $M$ except row $p$ and the columns that correspond to the edges in $H$.

Claim B. If $H$ is a tree, then $|\det(M')| = 1$. Otherwise, $\det(M') = 0$.

First suppose that $H$ is not a tree. Since $H$ has $n$ vertices and $n - 1$ edges, we know from earlier work that $H$ must be disconnected. Let $H_1$ be a connected component of $H$ that does not contain the vertex $v_p$. Let $M''$ be the $|V(H_1)| \times (n-1)$ submatrix of $M'$ formed by eliminating all rows other than the ones corresponding to vertices of $H_1$. Each column of $M''$ contains exactly two nonzero entries: 1 and $-1$. Therefore, the sum of all of the row vectors of $M''$ is the zero vector, so the rows of $M''$ are linearly dependent. Since these rows are also rows of $M'$, we see that $\det(M') = 0$.

Now suppose that $H$ is a tree. Choose some leaf of $H$ that is not $v_p$ (Theorem 1.7 lets us know that we can do this), and call it $u_1$. Let us also say that $e_1$ is the edge of $H$ that is incident with $u_1$. In the tree $H - u_1$, choose $u_2$ to be some leaf other than $v_p$. Let $e_2$ be the edge of $H - u_1$ incident with $u_2$. Keep removing leaves in this fashion until $v_p$ is the only vertex left. Having established the list of vertices $u_1, u_2, \ldots, u_{n-1}$, we now create a new $(n-1) \times (n-1)$ matrix $M^*$ by rearranging the rows of $M'$ in the following way: row $i$ of $M^*$ will be the row of $M'$ that corresponds to the vertex $u_i$.

An important (i.e., useful!) property of the matrix $M^*$ is that it is lower triangular (we know this because for each $i$, vertex $u_i$ is *not* incident with any of

$e_{i+1}, e_{i+2}, \ldots, e_{n-1}$). Thus, the determinant of $M^*$ is equal to the product of the main diagonal entries, which are either 1 or $-1$, since every $u_i$ is incident with $e_i$. Thus, $|\det(M^*)| = 1$, and so $|\det(M')| = 1$. This proves Claim B.

We are now ready to investigate the cofactors of $D - A = MM^T$. It is a fact from matrix theory that if the row sums and column sums of a matrix are all 0, then the cofactors all have the same value. (It would be a nice exercise—and a nice review of matrix skills—for you to try to prove this.) Since the matrix $MM^T$ satisfies this condition, we need to consider only one of its cofactors. We might as well choose $i$ and $j$ such that $i + j$ is even—let us choose $i = 1$ and $j = 1$. So, the $(1, 1)$ cofactor of $D - A$ is

$$\det\left((D - A)(1|1)\right) = \det\left(MM^T(1|1)\right)$$
$$= \det(M_1 M_1^T)$$

where $M_1$ is the matrix obtained by deleting the first row of $D - A$.

At this point we make use of the Cauchy–Binet Formula, which says that the determinant above is equal to the sum of the determinants of $(n - 1) \times (n - 1)$ submatrices of $M_1$ (for a more thorough discussion of the Cauchy–Binet Formula, see [16]). We have already seen (in Claim B) that any $(n - 1) \times (n - 1)$ submatrix that corresponds to a spanning tree of $G$ will contribute 1 to the sum, while all others contribute 0. This tells us that the value of $\det(D - A) = \det(MM^T)$ is precisely the number of spanning trees of $G$. $\qquad\square$

Figure 1.41 shows a labeled graph $G$ and each of its eight spanning trees. What does the Matrix Tree Theorem say about $G$? The degree matrix $D$ and adjacency

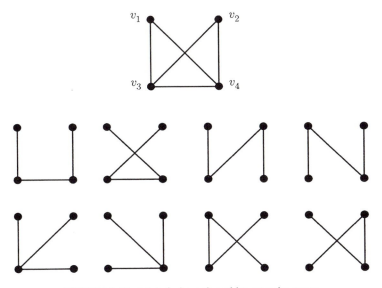

FIGURE 1.41. A labeled graph and its spanning trees.

matrix $A$ are

$$D = \begin{bmatrix} 2 & 0 & 0 & 0 \\ 0 & 2 & 0 & 0 \\ 0 & 0 & 3 & 0 \\ 0 & 0 & 0 & 3 \end{bmatrix}, \qquad A = \begin{bmatrix} 0 & 0 & 1 & 1 \\ 0 & 0 & 1 & 1 \\ 1 & 1 & 0 & 1 \\ 1 & 1 & 1 & 0 \end{bmatrix},$$

and so

$$D - A = \begin{bmatrix} 2 & 0 & -1 & -1 \\ 0 & 2 & -1 & -1 \\ -1 & -1 & 3 & -1 \\ -1 & -1 & -1 & 3 \end{bmatrix}.$$

The $(1, 1)$ cofactor of $D - A$ is

$$\det \begin{bmatrix} 2 & -1 & -1 \\ -1 & 3 & -1 \\ -1 & -1 & 3 \end{bmatrix} = 8.$$

Score one for Kirchhoff!

**Exercises**

1. Let $T$ be a labeled tree. Prove that the Prüfer sequence of $T$ will not contain any of the leaves' labels. Also prove that each vertex $v$ will appear in the sequence exactly $\deg(v) - 1$ times.
2. Determine the Prüfer sequence for the trees in Figure 1.42.
3. Draw and label a tree whose Prüfer sequence is

$$5, 4, 3, 5, 4, 3, 5, 4, 3.$$

4. Which trees have constant Prüfer sequences?
5. Which trees have Prüfer sequences with distinct terms?
6. Let $e$ be an edge of $K_n$. Use Cayley's Theorem to prove that $K_n - e$ has $(n - 2)n^{n-3}$ spanning trees.
7. Use the Matrix Tree Theorem to prove Cayley's Theorem. Hint: Look back at the discussion prior to the statement of the Matrix Tree Theorem.

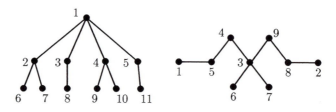

FIGURE 1.42. Two labeled trees.

# 1.3   Planarity

*Three civil brawls, bred of an airy word*
*By thee, old Capulet, and Montague,*
*Have thrice disturb'd the quiet of our streets . . .*
>  — William Shakespeare, *Romeo and Juliet*

The feud between the Montagues and the Capulets of Verona has been well doc-
umented, discussed, and studied. A fact that is lesser known, though, is that long
before Romeo and Juliet's time, the feud actually involved a third family—the
Hatfields.[2] The families' houses were fairly close together, and chance meetings
on the street were common and quite disruptive.

The townspeople of Verona became very annoyed at the feuding families. They
devised a plan to create separate, nonintersecting routes from each of the houses to
each of three popular places in town: the square, the tavern, and the amphitheater.
They hoped that if each family had its own route to each of these places, then the
fighting in the streets might stop.

Figure 1.43 shows the original layout of the routes. Try to rearrange them so
that no route crosses another route. We will come back to this shortly.

## 1.3.1   Definitions and Examples

*Define, define, well-educated infant.*
>  — William Shakespeare, *Love's Labour's Lost*

A graph $G$ is said to be *planar* if it can be drawn in the plane in such a way that
pairs of edges intersect only at vertices, if at all. If $G$ has no such representation,
$G$ is called *nonplanar*. A drawing of a planar graph $G$ in the plane in which edges
intersect only at vertices is called a *planar representation* (or a *planar embedding*)
of $G$.

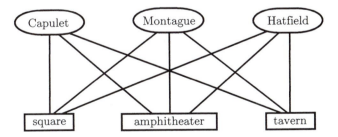

FIGURE 1.43. Original routes.

---

[2]The Hatfields eventually grew tired of feuding, and they left Verona in search of
friendlier territory. They found a nice spot in the mountains of West Virginia, right across
the river from a really nice family named McCoy.

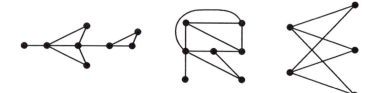

FIGURE 1.44. Examples of planar graphs.

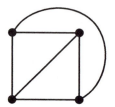

FIGURE 1.45. A planar representation of $K_4$.

Figure 1.44 shows examples of planar graphs. Notice that one of the drawings is not a planar representation—try to visualize untangling it.

Proving a graph to be planar is in some cases very simple—all that is required is to exhibit a planar representation of the graph. This is certainly quite easy to do with paths, cycles, and trees. What about complete graphs? $K_1$, $K_2$, and $K_3$ are clearly planar; Figure 1.45 shows a planar representation of $K_4$. We will consider $K_5$ shortly.

The Montague/Capulet/Hatfield problem essentially amounts to finding a planar representation of $K_{3,3}$. Unfortunately, the townspeople of Verona just had to learn to deal with the feuding families, for $K_{3,3}$ is nonplanar, and we will see an explanation shortly.

What is involved in showing that a graph $G$ is nonplanar? In theory, one would have to show that every possible drawing of $G$ is not a planar representation. Since considering every individual drawing is out of the question, we need some other tools.

Given a planar representation of a graph $G$, a *region* is a maximal section of the plane in which any two points can be joined by a curve that does not intersect any part of $G$.

Informally, if a cookie cutter has the shape of a planar representation of $G$, then the cookies are the regions (see Figure 1.46). The big region, $R_7$, is called the *exterior* (or outer) region.

It is quite natural to think of the regions as being *bounded by* the edges. For this reason we define the *bound degree* of $R$, denoted by $b(R)$, to be the number of edges that bound region $R$. For example, in Figure 1.46, $b(R_1) = b(R_4) = 4$, $b(R_2) = b(R_3) = b(R_5) = b(R_6) = 3$, and $b(R_7) = 12$.

Figure 1.47 displays six planar graphs along with the numbers of vertices, edges, and regions. Before continuing to the next section, study these numbers and try to

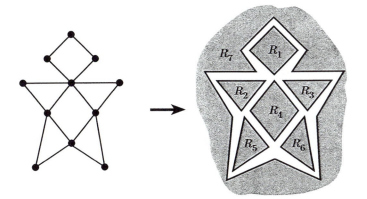

FIGURE 1.46. Six small cookies and one very large cookie.

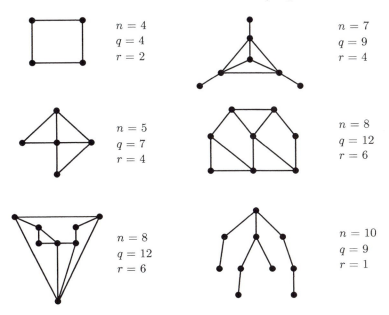

$n = 4$
$q = 4$
$r = 2$

$n = 7$
$q = 9$
$r = 4$

$n = 5$
$q = 7$
$r = 4$

$n = 8$
$q = 12$
$r = 6$

$n = 8$
$q = 12$
$r = 6$

$n = 10$
$q = 9$
$r = 1$

FIGURE 1.47. Is there a pattern?

find a pattern. You might also notice that two of drawings are actually the same graph. *This brings up an important point: The number of regions in a planar representation of a graph does not depend on the representation itself!*

**Exercises**

**1.** Find planar representations for each of the planar graphs in Figure 1.48.

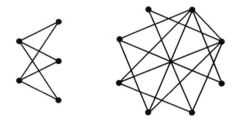

FIGURE 1.48.

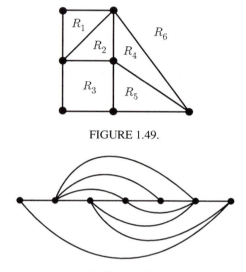

FIGURE 1.49.

FIGURE 1.50.

**2.** Give planar representations of the graph in Figure 1.49 such that each of the following is the exterior region.

   **a.** $R_1$
   **b.** $R_2$
   **c.** $R_3$
   **d.** $R_4$
   **e.** $R_5$

**3.** Explain why embedding a graph in a plane is essentially the same as embedding a graph on a sphere.

**4.** Write a nice proof of the fact that every tree is planar.

**5.** Draw a planar graph in which every vertex has degree exactly 5.

**6.** Let $G$ be a planar graph. Prove that $G$ is bipartite if and only if $b(R)$ is even for every region $R$.

**7.** In [50] and [145], Fáry and Wagner proved independently that every planar graph has a planar representation in which every edge is a straight line segment. Find such a representation for the graph in Figure 1.50.

### 1.3.2   Euler's Formula and Beyond

*Now I will have less distraction.*
                    — Leonard Euler, upon losing sight in his right eye [49]

Euler discovered a relationship between the numbers of vertices, edges, and regions of a graph, and his discovery is often called Euler's Formula [48].

**Theorem 1.12** (Euler's Formula).   *If $G$ is a connected planar graph with $n$ vertices, $q$ edges, and $r$ regions, then*

$$n - q + r = 2.$$

PROOF.   We induct on $q$, the number of edges. If $q = 0$, then $G$ must be $K_1$, a graph with 1 vertex and 1 region. The result holds in this case. Assume that the result is true for all connected planar graphs with fewer than $q$ edges, and assume that $G$ has $q$ edges.

*Case 1.* Suppose $G$ is a tree. We know from our work with trees that $q = n - 1$; and of course, $r = 1$, since a planar representation of a tree has only one region. Thus $n - q + r = n - (n - 1) + 1 = 2$, and the result holds.

*Case 2.* Suppose $G$ is not a tree. Let $C$ be a cycle in $G$, let $e$ be an edge of $C$, and consider the graph $G - e$. Compared to $G$, this graph has the same number of vertices, one edge fewer, and one region fewer, since removing $e$ coalesces two regions in $G$ into one in $G - e$. Thus the induction hypothesis applies, and in $G - e$,

$$n - (q - 1) + (r - 1) = 2,$$

implying that $n - q + r = 2$.

The result holds in both cases, and the induction is complete.    □

Euler's Formula is useful for establishing that a graph is nonplanar.

**Theorem 1.13.**   *$K_{3,3}$ is nonplanar.*

PROOF.   Suppose that $K_{3,3}$ were planar and that we had a planar representation. Since $n = 6$ and $q = 9$, Euler's Formula implies that such a planar representation of $K_{3,3}$ would have $r = 5$ regions. Now consider the sum

$$C = \sum_R b(R),$$

where the sum is over all regions $R$ in the representation of the graph. Since every edge of $G$ can be on the boundary of at most two regions, we get $C \leq 2q = 18$. On the other hand, since each region of $K_{3,3}$ has at least four edges on the boundary (there are no triangles in bipartite graphs), we see that $C \geq 4r = 20$. We have reached a contradiction. Therefore, $K_{3,3}$ is nonplanar.    □

**Theorem 1.14.**   *If $G$ is a planar graph with $n \geq 3$ vertices and $q$ edges, then $q \leq 3n - 6$. Furthermore, if equality holds, then every region is bounded by three edges.*

PROOF.    Again consider the sum

$$C = \sum_R b(R).$$

As previously mentioned, $C \le 2q$. Further, since each region is bounded by at least 3 edges, we have that $C \ge 3r$. Thus

$$3r \le 2q \quad \Rightarrow \quad 3(2 + q - n) \le 2q \quad \Rightarrow \quad q \le 3n - 6.$$

If equality holds, then $3r = 2q$, and it must be that every region is bounded by three edges.                                                                          □

We can use Theorem 1.14 to establish that $K_5$ is nonplanar.

**Theorem 1.15.** $K_5$ *is nonplanar.*

PROOF.    $K_5$ has 5 vertices and 10 edges. Thus $3n - 6 = 9 < 10 = q$, implying that $K_5$ is nonplanar.                                                              □

Exercise 5 in Section 1.3.1 asked for a planar graph in which every vertex has degree exactly 5. This next result says that such a graph is an extreme example.

**Theorem 1.16.** *If G is a planar graph, then G contains a vertex of degree at most five. That is, $\delta(G) \le 5$.*

PROOF.    Suppose $G$ has $n$ vertices and $q$ edges. If $n \le 6$, then the result is immediate, so we will suppose that $n > 6$. If we let $D$ be the sum of the degrees of the vertices of $G$, then we have

$$D = 2q \le 2(3n - 6) = 6n - 12.$$

If each vertex had degree 6 or more, then we would have $D \ge 6n$, which is impossible. Thus there must be some vertex with degree less than or equal to 5.                                                                          □

**Exercises**

1. $G$ is a planar graph of order 24, and it is regular of degree 3. How many regions are in a planar representation of $G$?
2. Let $G$ be a connected planar graph of order less than 12. Prove $\delta(G) \le 4$.
3. Prove that Euler's formula fails for disconnected graphs.
4. Let $G$ be a planar, triangle-free graph of order $n$ (that is, there are no $K_3$'s as subgraphs). Prove that $G$ has no more than $2n - 4$ edges.
5. Prove that there is no bipartite planar graph with minimum degree at least 4.
6. Let $G$ be a planar graph with $k$ components. Prove that

$$n - q + r = 1 + k.$$

7. Let $G$ be of order $n \ge 11$. Show that at least one of $G$ and $\overline{G}$ is nonplanar.
8. Show that the average degree (see Exercise 11 in Section 1.2.2) of a planar graph is less than six.

**9.** Prove that the converse of Theorem 1.14 is not true.

**10.** Find a 4-regular planar graph, and prove that it is unique.

**11.** A planar graph $G$ is called *maximal planar* if the addition of any edge to $G$ creates a nonplanar graph.

    **a.** Show that every region of a maximal planar graph is a triangle.

    **b.** If maximal planar graph has order $n$, how many edges and regions does it have?

### 1.3.3  Regular Polyhedra

*We are usually convinced more easily by reasons we have found ourselves than by those which have occurred to others.*

         — Blaise Pascal, *Pensées*

A polyhedron is a solid that is bounded by flat surfaces. Dice, bricks, pyramids, and the famous dome at Epcot Center are all examples of polyhedra. Polyhedra can be associated with graphs in a very natural way. Think of the polyhedra as having faces, edges, and corners (or vertices). The vertices and edges of the solid make up its skeleton, and the skeleton can be viewed as a graph. An interesting property of these skeleton graphs is that they are planar. One way to see this is to imagine taking hold of one of the faces and stretching it so that its edges form the boundary of the exterior region of the graph. The regions of these planar representations directly correspond to the faces of the polyhedra. Figure 1.51 shows a brick-shaped polyhedron, its associated graph, and a planar representation of the graph.

Because of the natural correspondence, we are able to apply some of what we know about planar graphs to polyhedra. The next theorem follows directly from Euler's Formula for planar graphs, Theorem 1.12.

**Theorem 1.17.** *If a polyhedron has V vertices, E edges, and F faces, then*

$$V - E + F = 2.$$

This next theorem is similar to Theorem 1.16. But first a definition.
Given a polyhedron $P$, define $\rho(P)$ to be

$$\rho(P) = \min\{b(R) \mid R \text{ is a region of } P\}.$$

**Theorem 1.18.** *For all polyhedra P, $3 \leq \rho(P) \leq 5$.*

FIGURE 1.51. A polyhedron and its graph.

PROOF.    Since one or two edges can never form a boundary, we know that $\rho(P) \geq 3$ for all polyhedra $P$. So we need to prove only the upper bound.

Let $P$ be a polyhedron and let $G$ be its associated graph. Suppose $P$ has $V$ vertices, $E$ edges, and $F$ faces. For each $k$, let $V_k$ be the number of vertices of degree $k$, and let $F_k$ be the number of faces of $P$ (or regions of $G$) of bound degree $k$. From our earlier remarks, if $k < 3$, then $V_k = F_k = 0$. Since every edge of $P$ touches exactly two vertices and exactly two faces, we find that

$$\sum_{k \geq 3} k V_k = 2E = \sum_{k \geq 3} k F_k.$$

If every face of $P$ were bounded by 6 or more edges, then we would have

$$2E = \sum_{k \geq 3} k F_k \geq \sum_{k \geq 6} 6 F_k = 6 \sum_{k \geq 6} F_k = 6F,$$

implying that $E \geq 3F$. Furthermore,

$$2E = \sum_{k \geq 3} k V_k \geq 3V,$$

implying that $V \leq \frac{2}{3} E$. Thus

$$E = V + F - 2 \leq \frac{2}{3} E + \frac{1}{3} E - 2 = E - 2,$$

and this, of course, is a contradiction. Therefore, some face of $P$ is bounded by fewer than 6 edges. Hence, $\rho(P) \leq 5$.    □

We may apply this result to derive a geometric fact known to the ancient Greeks.

A regular polygon is one that is equilateral and equiangular. We say a polyhedron is *regular* if its faces are mutually congruent, regular polygons and if the number of faces meeting at a vertex is the same for every vertex. The cube, whose faces are congruent squares, and the tetrahedron, whose faces are congruent equilateral triangles, are regular polyhedra. A fact that has been known for at least 2000 years is that there are only five regular polyhedra: the tetrahedron, the cube, the octahedron, the dodecahedron, and the icosahedron (see Figure 1.52). We can use a graph-theoretic argument to prove this.

**Theorem 1.19.**    *There are exactly five regular polyhedra.*

PROOF.    Let $P$ be a regular polyhedron, and let $G$ be its associated planar graph. Let $V$, $E$, and $F$ be the number of vertices, edges, and faces (regions) of $P$. Since the faces of $P$ are congruent, each is bordered by the same number of edges, say $k$. Theorem 1.18 tells us that $3 \leq k \leq 5$. Further, since the polyhedron $P$ is regular, it follows that the graph $G$ is also regular. Let us say that $G$ is regular of degree $r$ where $r \geq 3$. From Theorem 1.18, we obtain $rV = 2E = kF$. Now, Theorem 1.17 implies that

$$8 = 4V - 4E + 4F$$
$$= 4V - 2E + 4F - 2E$$

Tetrahedron

Cube

Octahedron

Dodecahedron

Icosahedron

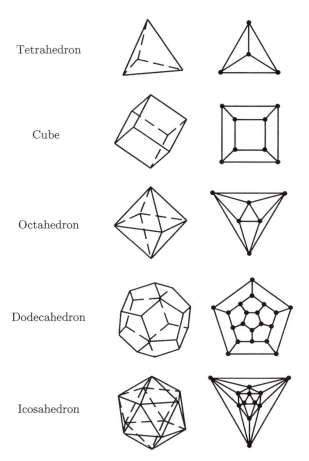

FIGURE 1.52. The five regular polyhedra.

$$= 4V - rV + 4F - kF$$
$$= (4 - r)V + (4 - k)F.$$

$V$ and $E$ are of course both positive, and since $3 \leq k \leq 5$ and $r \geq 3$, there are only five possible cases.

*Case* 1. Suppose $r = 3$ and $k = 3$. In this case, $V = F$ and $8 = V + F$, implying that $V = F = 4$. This is the tetrahedron. (The fact that the tetrahedron is the only regular polygon with $V = F = 4$ is based on a geometrical argument. This applies to the remaining four cases as well.)

*Case* 2. Suppose $r = 3$ and $k = 4$. Here we have $V = 8$ and $3V = 4F$. Thus $F = 6$, and $P$ is a cube.

*Case* 3. Suppose $r = 3$ and $k = 5$. In this case we have $8 = V - F$ and $3V = 5F$. Solving this system yields $V = 20$ and $F = 12$. This is a dodecahedron.

*Case* 4. Suppose $r = 4$ and $k = 3$. Here we have $F = 8$ and $4V = 3F$. Thus $V = 6$ and $P$ is an octahedron.

*Case 5.* Suppose $r = 5$ and $k = 3$. In this case we have $8 = -V + F$ and $5V = 3F$. Solving this system yields $V = 12$ and $F = 20$. This is an icosahedron.  □

**Exercises**

1. (From [25].) Show that the octahedron is $K_{r_1,\dots,r_n}$ for some $n$ and for some values $r_1, \dots, r_n$.
2. Find an example of a polyhedron different from the ones discussed in this section. Sketch the polyhedron, and draw the associated graph.
3. See whether you can find an alternative proof (not necessarily graph-theoretic) of the fact that there are only five regular polyhedra.

### 1.3.4   Kuratowski's Theorem

*...a pair so famous.*

— William Shakespeare, *Anthony and Cleopatra*

Our goal in this section is to compile a list of all nonplanar graphs. Since the list will be infinite (and since this book is not), we will make use of a clever characterization due to Kuratowski.

We have already established that both $K_{3,3}$ and $K_5$ are nonplanar, so we should put them at the top of our list. What other graphs should we include? Suppose $G$ is a graph that contains $K_{3,3}$ as a subgraph. This graph $G$ would have to be nonplanar, since a planar representation of it would have to contain a planar representation of $K_{3,3}$. So we can add to our list of nonplanar graphs all graphs that contain $K_{3,3}$ or $K_5$ as a subgraph. The graph in Figure 1.53 shows us that our list of nonplanar graphs is not yet complete. This graph is not planar, but it does not contain $K_5$ or $K_{3,3}$ as a subgraph. Of course, if we were to replace the two edges labeled $a$ and $b$ with a single edge $e$, then the graph would contain $K_5$ as a subgraph. This motivates the following definition.

Let $G$ be a graph. A *subdivision* of an edge $e$ in $G$ is a substitution of a path for $e$. We say that a graph $H$ is a *subdivision* of $G$ if $H$ can be obtained from $G$ by a finite sequence of subdivisions.

For example, the graph in Figure 1.53 contains a subdivision of $K_5$, and in Figure 1.54, $H$ is a subdivision of $G$.

We leave the proof of the following theorem to the exercises (see Exercise 1).

**Theorem 1.20.** *A graph $G$ is planar if and only if every subdivision of $G$ is planar.*

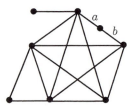

FIGURE 1.53.

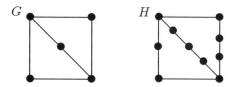

FIGURE 1.54. A graph and a subdivision.

Our list of nonplanar graphs now includes $K_{3,3}$, $K_5$, graphs containing $K_{3,3}$ or $K_5$ as subgraphs, *and* all graphs containing a subdivision of $K_{3,3}$ or $K_5$. The list so far stems from only two specific graphs: $K_{3,3}$ and $K_5$. A well-known theorem by Kuratowski [99] tells us that there are *no other* graphs on the list! The bottom line is that $K_{3,3}$ and $K_5$ are the only two real enemies of planarity.

Kuratowski proved this beautiful theorem in 1930, closing a long-open problem.[3] In 1954, Dirac and Schuster [38] found a proof that was slightly shorter than the original proof, and theirs is the proof that we will outline here. You will have a chance to prove some of the steps as exercises.

**Theorem 1.21** (Kuratowski's Theorem).   *A graph $G$ is planar if and only if it contains no subdivision of $K_{3,3}$ or $K_5$.*

SKETCH OF PROOF.   We have already discussed that if a graph $G$ is planar, it contains no subgraph that is a subdivision of $K_{3,3}$ or $K_5$. Thus we need to discuss only the reverse direction of the theorem.

Suppose $G$ is a graph that contains no subdivision of $K_{3,3}$ or $K_5$. Here are the steps that Dirac and Schuster used to prove the result.

1. Prove that $G$ is planar if and only if each block of $G$ is planar. (A *block of G* is a maximal connected subgraph of $G$ that has no cut vertex).
2. Explain why it suffices to show that a block is planar if and only if it contains no subdivision of $K_{3,3}$ or $K_5$. Assume that $G$ is a block itself (connected with no cut vertex).
3. Suppose that $G$ is a nonplanar block that contains no subdivision of $K_{3,3}$ or $K_5$ (and search for a contradiction).
4. Prove that $\delta(G) \geq 3$.
5. Establish the existence of an edge $e = uv$ such that the graph $G - e$ is also a block.
6. Explain why $G - e$ is a planar graph containing a cycle $C$ that includes both $u$ and $v$, and choose $C$ to have a maximum number of interior regions.
7. Establish several structural facts about the subgraphs inside and outside the cycle $C$.

---

[3]We should note here that Frink and Smith also discovered a proof of this fact in 1930, independently of Kuratowski. Since Kuratowski's result was published first, his name has traditionally been associated with the theorem (and the names Frink and Smith have traditionally been associated with footnotes like this one).

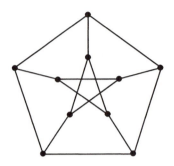

FIGURE 1.55. The Petersen Graph.

**8.** Use these structural facts to demonstrate the existence of subdivisions of $K_{3,3}$ or $K_5$, thus establishing the contradiction.

An application of Kuratowski's Theorem can be found in Exercise 2.    □

**Exercises**

**1.** Prove that a graph $G$ is planar if and only if every subdivision of $G$ is planar.
**2.** The Petersen graph, shown in Figure 1.55, is well known for the surprising connections it has with various areas of graph theory. Use Kuratowski's Theorem to prove that the Petersen graph is nonplanar.
**3.** Show that the Petersen graph is the line graph (see Exercise 5 Section 1.1.3) of $\overline{K_5}$.
**4.** Prove the first step of the proof of Kuratowski's Theorem.
**5.** Determine all graphs of the form $K_{r_1,\dots,r_n}$ that are planar.

## 1.4   Colorings

*One fish, two fish, red fish, blue fish.*

— Dr. Seuss

The senators in a particular state sit on various senate committees, and the committees need to schedule times for meetings. Since each senator must be present at each of his or her committee meetings, the meeting times need to be scheduled carefully. One could certainly assign a unique meeting time to each of the committees, but this plan may not be feasible, especially if the number of committees is large. We ask ourselves, given a particular committee structure, what is the fewest number of meeting times that are required? We can answer this question by studying graph coloring.

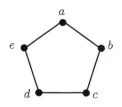

FIGURE 1.56. $C_5$ is 3-colorable.

## 1.4.1 Definitions

Given a graph $G$, a $k$-*coloring* of the vertices of $G$ is a partition of the vertex set $V$ into $k$ sets $C_1, C_2, \ldots, C_k$ such that for all $i$, no pair of vertices from $C_i$ are adjacent. If such a partition exists, $G$ is said to be $k$-*colorable*.

One can define an edge coloring in a similar manner, and indeed edge colorings are often useful in graph theory. In this section, however, we will use the term *coloring* strictly for vertex colorings. We will consider a particular type of edge coloring in the next section.

For example, the graph $C_5$ as shown in Figure 1.56 is 3-colorable, since we can choose $C_1 = \{a, c\}$, $C_2 = \{b, d\}$, and $C_3 = \{e\}$. Is $C_5$ 2-colorable?

Another way of viewing a $k$-coloring is as follows. Instead of assigning vertices to $k$ partitioning sets, assign each vertex one of $k$ colors. We need to make this assignment, though, in such a way that no pair of adjacent vertices gets the same color. Notice that each such coloring determines a specific partition of the vertex set: Vertices of the same color correspond to a single set in the partition.

It is natural to wonder how many colors are necessary to color a particular graph $G$. For instance, we know that three colors are enough for the graph in Figure 1.56, but is this the least required? A quick check of $C_5$ reveals that coloring with two colors is impossible. So three colors are necessary. This idea motivates a definition.

Given a graph $G$, the *chromatic number* of $G$, denoted by $\chi(G)$, is the smallest integer $k$ such that $G$ is $k$-colorable. In our example, we can say that $\chi(C_5) = 3$. What about odd cycles in general? (Try one!) What about even cycles? (Try one!) Here is a list of chromatic numbers for some common graphs. Verify them!

$$\chi(C_n) = \begin{cases} 2 & \text{if } n \text{ is even,} \\ 3 & \text{if } n \text{ is odd,} \end{cases}$$

$$\chi(P_n) = \begin{cases} 2 & \text{if } n \geq 2, \\ 1 & \text{if } n = 1, \end{cases}$$

$$\chi(K_n) = n,$$

$$\chi(E_n) = 1,$$

$$\chi(K_{m,n}) = 2.$$

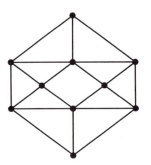

FIGURE 1.57. The Birkhoff Diamond.

**Exercises**

1. Find the chromatic number of each of the following graphs. Explain your answers completely.

   **a.** Each graph in Figure 1.86.
   **b.** Trees.
   **c.** Bipartite graphs.
   **d.** $K_{r_1, r_2, ..., r_t}$ (see Exercise 4 in Section 1.1.3).
   **e.** The Petersen Graph.
   **f.** The graph in Figure 1.57, called the Birkhoff Diamond.

2. Prove that a graph $G$ is $k$-colorable if and only if $\chi(G) \leq k$.
3. Senate committees $C_1$ through $C_7$ consist of the members shown: $C_1 =$ {Adams, Bradford, Charles}, $C_2 =$ {Charles, Davis, Eggers}, $C_3 =$ {Davis, Ford}, $C_4 =$ {Adams, Gardner}, $C_5 =$ {Eggers, Bradford, Gardner}, $C_6 =$ {Howe, Charles, Ford}, $C_7 =$ {Eggers, Howe}. Use the ideas of this section to determine the fewest number of meeting times that need to be scheduled for these committees.
4. (From [124].) In assigning frequencies to cellular phones, a zone gets a frequency to be used by all vehicles in the zone. Two zones that interfere (because of proximity or meteorological reasons) must get different frequencies. How many different frequencies are required if there are six zones, $a, b, c, d, e, f$, and zone $a$ interferes with zone $b$; $b$ interferes with $a$, $c$, and $d$; $c$ interferes with $b$, $d$, and $e$; $d$ interferes with $b$, $c$, and $e$; $e$ interferes with $c$, $d$, and $f$; and $f$ interferes with $e$ only?
5. When issuing seating assignments for third grade students, the teacher wants to be sure that if two students might interfere with one another, then they are assigned a different area of the room. There are six main troublemakers in the class: John, Jeff, Mike, Moe, Larry, and Curly. How many different areas are required in the room if John interferes with Moe and Curly; Jeff interferes with Larry and Curly; Mike interferes with Larry and Curly; Moe interferes with John, Larry, and Curly; Larry interferes with Jeff, Mike, Moe, and Curly; and Curly interferes with everyone?

6. Prove that adding an edge to a graph increases its chromatic number by at most one.

7. Prove that a graph $G$ is bipartite if and only if it is 2-colorable.

8. A graph $G$ is called *k-critical* if $\chi(G) = k$ and $\chi(G - v) < k$ for each vertex $v$ of $G$.

   **a.** Find all 1-critical and 2-critical graphs.
   **b.** Give an example of a 3-critical graph.
   **c.** If $G$ is $k$-critical, then show that $G$ is connected.
   **d.** If $G$ is $k$-critical, then show that $\delta(G) \geq k - 1$.
   **e.** Find all of the 3-critical graphs. Hint: Use part (d).

## 1.4.2    Bounds on Chromatic Number

*The point is, ladies and gentlemen, that greed, for lack of a better word, is good. Greed is right. Greed works.*

— Gordon Gekko, in *Wall Street*

In general, determining the chromatic number of a graph is hard. While small or well-known graphs (like the ones in the previous exercises) may be fairly easy, the number of possibilities in large graphs makes computing chromatic numbers difficult. We therefore often rely on bounds to give some sort of idea of what the chromatic number of a graph is, and in this section we consider some of these bounds.

If $G$ is a graph on $n$ vertices, then an obvious upper bound on $\chi(G)$ is $n$, since an $n$-coloring is always possible on a graph with $n$ vertices. This bound is exact for complete graphs, as it takes as many colors as there are vertices to color a complete graph. In fact, complete graphs are the only graphs for which this bound is sharp (see Exercise 5). We set this aside as Theorem 1.22.

**Theorem 1.22.** *For any graph $G$ of order $n$, $\chi(G) \leq n$.*

Let us now discuss a very basic graph coloring algorithm, the *greedy algorithm*. To color a graph having $n$ vertices using this algorithm, first label the vertices in some order—call them $v_1, v_2, \ldots, v_n$. Next, order the available colors in some way. We will denote them by the positive integers $1, 2, \ldots, n$. Then start coloring by assigning color 1 to vertex $v_1$. Next, if $v_1$ and $v_2$ are adjacent, assign color 2 to vertex $v_2$; otherwise, use color 1 again. In general, to color vertex $v_i$, use the first available color that has not been used for any of $v_i$'s previously colored neighbors. For example, the greedy algorithm produces the coloring on the right from the graph on the left in Figure 1.58. First, $v_1$ is assigned color 1; then $v_2$ is assigned color 1, since $v_2$ is not adjacent to $v_1$. Then $v_3$ is assigned color 1 since it is not adjacent to $v_1$ or $v_2$. Vertex $v_4$ is assigned color 2, then $v_5$ is assigned 2, and finally $v_6$ is assigned 2.

It is important to realize that the coloring obtained by the greedy algorithm depends heavily on the initial labeling of the vertices. Different labelings can (and often do) produce different colorings. Figure 1.59 displays the coloring obtained

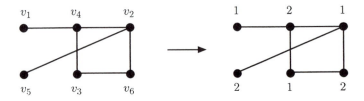

FIGURE 1.58. Applying the greedy algorithm.

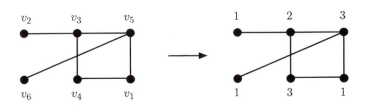

FIGURE 1.59. Applying it again.

from a different original labeling of the same graph. More colors are used in this second coloring. We see that while "greed works" in that the algorithm always gives a legal coloring, we cannot expect it to give us a coloring that uses the fewest possible colors.

The following bound improves Theorem 1.22.

**Theorem 1.23.** *For any graph $G$, $\chi(G) \leq \Delta(G)+1$, where $\Delta(G)$ is the maximum degree of $G$.*

PROOF.    Running the greedy algorithm on $G$ produces a legal coloring that uses at most $\Delta(G) + 1$ colors. This is because every vertex in the graph is adjacent to at most $\Delta(G)$ other vertices, and hence the largest color label used is at most $\Delta(G) + 1$. Thus, $\chi(G) \leq \Delta(G) + 1$.    □

Notice that we obtain equality in this bound for complete graphs and for cycles with an odd number of vertices. As it turns out, these are the only families of graphs for which the equality in Theorem 1.23 holds. This is stated in Brooks's Theorem [17]. The proof that we give is a modification of the one given by Lovász [102].

**Theorem 1.24** (Brooks's Theorem).    *If $G$ is a connected graph that is neither an odd cycle nor a complete graph, then $\chi(G) \leq \Delta(G)$.*

PROOF.    Let $G$ be a connected graph of order $n$ that is neither a complete graph nor an odd cycle. Let $k = \Delta(G)$. We know that $k \neq 0$ and $k \neq 1$, since otherwise $G$ is complete. If $k = 2$, then $G$ must be either an even cycle or a path. In either case, $\chi(G) = 2 = \Delta(G)$. So assume that $k = \Delta(G) \geq 3$.

We are now faced with three cases. In each case we will establish a labeling of the vertices of $G$ in the form $v_1, v_2, \ldots, v_n$. We will then use the greedy algorithm to color $G$ with no more than $k$ colors.

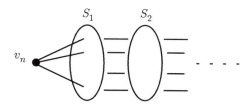

FIGURE 1.60. The sets $S_i$.

*Case* 1. Suppose that $G$ is not $k$-regular. Then there exists some vertex with degree less than $k$. Choose such a vertex and call it $v_n$. Let $S_0 = \{v_n\}$ and let $S_1 = N(v_n)$, the neighborhood of $v_n$. Further, let

$$S_2 = N(S_1) - \{v_n\} - S_1,$$
$$S_3 = N(S_2) - S_1 - S_2,$$
$$\vdots$$
$$S_i = N(S_{i-1}) - S_{i-2} - S_{i-1},$$

for each $i$ (Figure 1.60). Since $G$ is finite, there is some $t$ such that $S_t$ is not empty, and $S_r$ is empty for all $r > t$.

Next, label the vertices in $S_1$ with the labels $v_{n-1}, v_{n-2}, \ldots, v_{n-|S_1|}$. Label the vertices in $S_2$ with the labels $v_{n-|S_1|-1}, \ldots, v_{n-|S_1|-|S_2|}$. Continue labeling in this decreasing fashion until all vertices of $G$ have been labeled. The vertex with label $v_1$ is in the set $S_t$.

Let $u$ be a vertex in some $S_i$, $i \geq 1$. Since $u$ has at least one neighbor in $S_{i-1}$, it has at most $k - 1$ adjacencies with vertices whose label is less than its own. Thus, when the greedy algorithm gets to $u$, there will be at least one color from $\{1, 2, \ldots, k\}$ available. Further, since $\deg(v_n) < k$, there will be a color from $\{1, 2, \ldots, k\}$ available when the greedy algorithm reaches $v_n$. Thus, in this case the greedy algorithm uses at most $k$ colors to properly color $G$.

*Case* 2. Suppose that $G$ is $k$-regular and that $G$ has a cut vertex, say $v$. The removal of $v$ from $G$ will form at least two connected components. Say the components are $G_1, G_2, \ldots, G_t$. Consider the graph $H_1 = G_1 \cup \{v\}$ (the component $G_1$ with $v$ added back—see Figure 1.61). $H_1$ is a connected graph, and the degree of $v$ in $H$ is less than $k$. Using the method in Case 1, we can properly color $H_1$

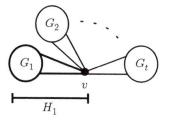

FIGURE 1.61. The graph $H_1$.

with at most $k$ colors. Similarly, we can properly color each $H_i = G_i - \{v\}$ with at most $k$ colors. Without loss of generality, we can assume that $v$ gets the same color in all of these colorings (if not, just permute the colors to make it so). These colorings together create a proper coloring of $G$ that uses at most $k$ colors. Case 2 is complete.

*Case* 3. Suppose that $G$ is $k$-regular and that it does not contain a cut vertex.

*Subcase* 3a. Suppose that for all $v$, the graph $G - v$ does not have a cut vertex.[4] Let $v$ be a vertex of $G$ with neighbors $v_1$ and $v_2$ such that $v_1 v_2 \notin E(G)$ (such vertices exist since $G$ is not complete). By the assumption in this subcase, the graph $G - \{v_1, v_2\}$ is connected.

*Subcase* 3b. Suppose there exists a vertex $v$ such that $G - v$ does have a cut vertex. Say this cut vertex is $w$. The graph $G - \{v, w\}$ must be disconnected. Let the components of $G - \{v, w\}$ be $G_1, G_2, \ldots, G_t$. Since $G$ has no cut vertex, it must be that $v$ has neighbors in each of the $G_i$'s. Let $v_1$ be a neighbor of $v$ in $G_1$, and let $v_2$ be a neighbor of $v$ in $G_2$. Since $v$ is not a cut vertex, the graph $G - \{v_1, v_2\}$ is connected.

In each subcase, we have identified vertices $v$, $v_1$, and $v_2$ such that $vv_1, vv_2 \in E(G)$, $v_1 v_2 \notin E(G)$, and $G - \{v_1, v_2\}$ is connected. We now proceed to label the vertices of $G$ in preparation for the greedy algorithm.

Let $v_1$ and $v_2$ be as labeled. Let $v$ be labeled $v_n$. Now choose a vertex adjacent to $v_n$ that is not $v_1$ or $v_2$ (such a vertex exists, since $\deg(v_n) \geq 3$). Label this vertex $v_{n-1}$. Next choose a vertex that is adjacent to either $v_n$ or $v_{n-1}$ and is not $v_1$, $v_2$, $v_n$, or $v_{n-1}$. Call this vertex $v_{n-2}$. We continue this process. Since $G - \{v_1, v_2\}$ is connected, then for each $i \in \{3, \ldots, n-1\}$, there is a vertex $v_i \in V(G) - \{v_1, v_2, v_n, v_{n-1}, \ldots, v_{i+1}\}$ that is adjacent to at least one of $v_{i+1}, \ldots, v_n$.

Now that the vertices are labeled, we can apply the greedy algorithm. Since $v_1 v_2 \notin E(G)$, the algorithm will give the color 1 to both $v_1$ and $v_2$. Since each $v_i$, $3 \leq i < n$, is adjacent to at most $k - 1$ predecessors, and since $v_n$ is adjacent to $v_1$ and $v_2$, the algorithm never requires more than $k = \Delta(G)$ colors. Case 3 is complete.                                                                                     □

The last two bounds involve some new concepts.

The *clique number* of a graph, denoted by $\omega(G)$, is defined as the order of the largest complete graph that is a subgraph of $G$. For example, in Figure 1.62, $\omega(G_1) = 3$ and $\omega(G_2) = 4$.

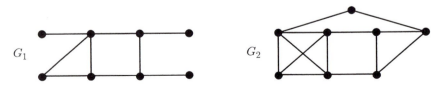

FIGURE 1.62. Graphs with clique numbers 3 and 4, respectively.

[4]This property is called *2-connected*. For more on connectivity, see [67], [25], or [146].

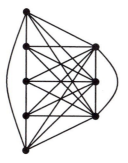

FIGURE 1.63. Is $\chi(G) = \omega(G)$?

A simple bound that involves clique number follows. We leave it to the reader to provide a (one or two line) explanation.

**Theorem 1.25.** *For any graph G, $\chi(G) \geq \omega(G)$.*

It is natural to wonder whether we might be able to strengthen this theorem and prove that $\chi(G) = \omega(G)$ for every graph $G$. Unfortunately, this is false. Consider the graph $G$ shown in Figure 1.63. The clique number of this graph is 5, and the chromatic number is 6 (see Exercise 2). Thus, we are stuck with a lower bound in Theorem 1.25.

A set of vertices in a graph is said to be an *independent* set of vertices if they are pairwise nonadjacent. The *independence number* of a graph $G$, denoted by $\beta(G)$, is defined to be the largest size of an independent set of vertices from $G$. As an example, consider the graphs in Figure 1.64. A largest independent set in $G_1$ is $\{c, d\}$, so $\beta(G_1) = 2$. A largest independent set in $G_2$ is $\{a, c, e\}$, so $\beta(G_2) = 3$.

The upper and lower bounds given in Theorem 1.26 concern this independence number, and the proofs are left as an exercise (see Exercise 6).

**Theorem 1.26.** *For any graph G of order n,*

$$\frac{n}{\beta(G)} \leq \chi(G) \leq n + 1 - \beta(G).$$

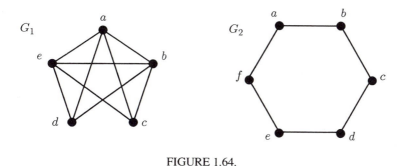

FIGURE 1.64.

**Exercises**

1. Recall that avgdeg($G$) denotes the average degree of vertices in $G$. Prove or give a counterexample to the following statement:

$$\chi(G) \leq 1 + \text{avgdeg}(G).$$

2. If $G$ is the graph in Figure 1.63, prove that $\chi(G) = 6$ and $\omega(G) = 5$.
3. Determine a necessary and sufficient condition for a graph to have a 2-colorable line graph.
4. If $m(G)$ denotes the number of edges in a longest path of $G$, prove that $\chi(G) \leq 1 + m(G)$.
5. Prove that the only graph $G$ of order $n$ for which $\chi(G) = n$ is $K_n$.
6. Prove that for any graph $G$ of order $n$,

$$\frac{n}{\beta(G)} \leq \chi(G) \leq n + 1 - \beta(G).$$

7. If $G$ is bipartite, prove that $\omega(\overline{G}) = \chi(\overline{G})$.
8. Let $G$ be a graph of order $n$. Prove that

   **a.** $n \leq \chi(G)\chi(\overline{G})$;
   **b.** $2\sqrt{n} \leq \chi(G) + \chi(\overline{G})$.

## 1.4.3    The Four Color Problem

*That doesn't sound too hard.*

— Star Wars

**The Four Color Problem**    *Is it true that the countries on any given map can be colored with four or fewer colors in such a way that adjacent countries are colored differently?*

The seemingly simple Four Color Problem was introduced in 1852 by Francis Guthrie, a student of Augustus DeMorgan. The first written reference to the problem is a letter from DeMorgan to Sir William Rowan Hamilton. Despite Hamilton's indifference, DeMorgan continued to talk about the problem with other mathematicians. In the years that followed, many of the world's top mathematical minds attempted either to prove or disprove the conjecture, and in 1879 Alfred Kempe announced that he had found a proof. In 1890, however, P.J. Heawood discovered an error in Kempe's proof. Kempe's work did have some positive features, though, for Heawood made use of Kempe's ideas to prove that five colors always suffice. In this section, we translate the Four Color Problem into a graph theory problem, and we prove the Five Color Theorem.

Any map can be represented by a planar graph in the following way: Represent each country on the map by a vertex, and connect two vertices with an edge whenever the corresponding countries share a nontrivial border (more than just a point). Some examples are shown in Figure 1.65.

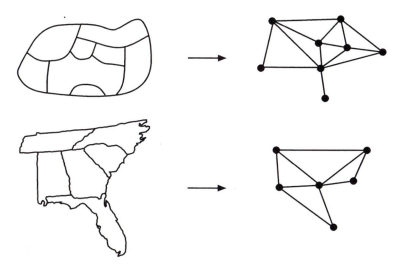

FIGURE 1.65. Graph representations of maps.

The regions on the map correspond to vertices on the graph, so a graph coloring yields a map coloring with no bordering regions colored the same. This natural representation allows us to see that a map is 4-colorable if and only if its associated graph is 4-colorable.

The Four Color Conjecture is equivalent to the following statement. A thorough discussion of this equivalence can be found in [25].

**Theorem 1.27** (Four Color Theorem). *Every planar graph is 4-colorable.*

When Heawood pointed out the error in Kempe's proof, it was back to the drawing board for many interested parties. People worked on the Four Color Problem for years and years trying numerous strategies. Finally, in 1976, Kenneth Appel and Wolfgang Haken, with the help of John Koch, announced that they had found a proof [6]. To complete their proof, they verified thousands of cases with computers, using over 1000 hours of computer time. As you might imagine, people were skeptical of this at first. Was this really a proof? How could an argument with so many cases be verified?

While the Appel–Haken proof *is* accepted as being valid, mathematicians still search for alternative proofs. Robertson, Sanders, Seymour, and Thomas [125] have probably come the closest to finding a short and clever proof, but theirs still requires a number of computer calculations.

In a 1998 article [139], Robin Thomas said the following.

For the purposes of this survey, let me telescope the difficulties with the A&H proof into two points: (1) part of the proof uses a computer and cannot be verified by hand, and (2) even the part that is supposedly hand-checkable has not, as far as I know, been independently verified in its entirety. . . . Neil Robertson, Daniel P. Sanders, Paul Seymour, and I tried to verify the Appel–

Haken proof, but soon gave up and decided that it would be more profitable to work out our own proof. . . . We were not able to eliminate reason (1), but we managed to make progress toward (2).

As mentioned earlier, Heawood [82] provided a proof of the Five Color Theorem in the late 1890s, and we present his proof here. Some of the ideas in his proof came from Kempe's attempt [90] to solve the Four Color Problem.

**Theorem 1.28** (Five Color Theorem).    *Every planar graph is 5-colorable.*

PROOF.    We induct on the order of $G$. Let $G$ be a planar graph of order $n$. If $n \leq 5$, then the result is clear. So suppose that $n \geq 6$ and that the result is true for all planar graphs of order $n - 1$. From Theorem 1.16, we know that $G$ contains a vertex, say $v$, having $\deg(v) \leq 5$.

Consider the graph $G'$ obtained by removing from $G$ the vertex $v$ and all edges incident with $v$. Since the order of $G'$ is $n - 1$ (and since $G'$ is of course planar), we can apply the induction hypothesis and conclude that $G'$ is 5-colorable. Now, we can assume that $G'$ has been colored using the five colors, named 1, 2, 3, 4, and 5. Consider now the neighbors of $v$ in $G$. As noted earlier, $v$ has at most five neighbors in $G$, and all of these neighbors are vertices in (the already colored) $G'$.

If in $G'$ fewer than five colors were used to color these neighbors, then we can properly color $G$ by using the coloring for $G'$ on all vertices other than $v$, and by coloring $v$ with one of the colors that is not used on the neighbors of $v$. In doing this, we have produced a 5-coloring for $G$.

So, assume that in $G'$ exactly five of the colors were used to color the neighbors of $v$. This implies that there are exactly five neighbors, call them $w_1$, $w_2$, $w_3$, $w_4$, $w_5$, and assume without loss of generality that each $w_i$ is colored with color $i$ (see Figure 1.66).

We wish to rearrange the colors of $G'$ so that we make a color available for $v$. Consider all of the vertices of $G'$ that have been colored with color 1 or with color 3.

*Case* 1. Suppose that in $G'$ there does not exist a path from $w_1$ to $w_3$ where all of the colors on the path are 1 or 3. Define a subgraph $H$ of $G'$ to be the union of all paths that start at $w_1$ and that are colored with either 1 or 3. Note that $w_3$ is not a vertex of $H$ and that none of the neighbors of $w_3$ are in $H$ (see Figure 1.67).

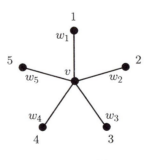

FIGURE 1.66.

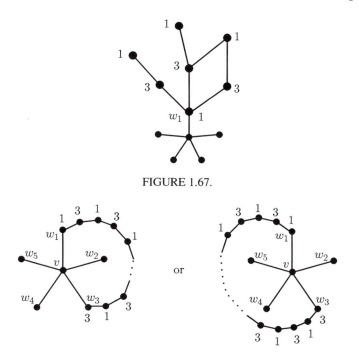

FIGURE 1.67.

FIGURE 1.68. Two possibilities.

Now, interchange the colors in $H$. That is, change all of the 1's into 3's and all of the 3's into 1's. The resulting coloring of the vertices of $G'$ is a proper coloring, because no problems could have possibly arisen in this interchange. We now see that $w_1$ is colored 3, and thus color 1 is available to use for $v$. Thus, $G$ is 5-colorable.

*Case* 2. Suppose that in $G'$ there does exist a path from $w_1$ to $w_3$ where all of the colors on the path are 1 or 3. Call this path $P$. Note now that $P$ along with $v$ forms a cycle that encloses either $w_2$ or $w_4$ (Figure 1.68).

So there does not exist a path from $w_2$ to $w_4$ where all of the colors on the path are 2 or 4. Thus, the reasoning in Case 1 applies! We conclude that $G$ is 5-colorable.                                                    □

## Exercises

1. Determine the chromatic number of the graph of the map of the United States.
2. Determine the chromatic number of the graph of the map of the countries of South America.
3. Determine the chromatic number of the graph of the map of the countries of Africa.
4. Determine the chromatic number of the graph of the map of the countries of Australia. Hint: This graph will be quite small!
5. Where does the proof of the Five Color Theorem go wrong for four colors?

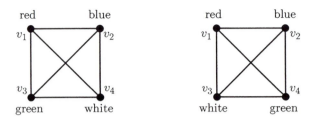

FIGURE 1.69. Two different colorings.

### 1.4.4   Chromatic Polynomials

*Everything should be made as simple as possible, but not simpler.*
— Albert Einstein

Chromatic polynomials, developed by Birkhoff in the early 1900s as he studied the Four Color Problem, provide us with a method of counting the number of different colorings of a graph.

Before we introduce the polynomials, we should clarify what we mean by different colorings. Given a graph $G$, suppose that its vertices are labeled $v_1$, $v_2$, $\ldots v_n$. A coloring of $G$ is an assignment of colors to these vertices, and we call two colorings $C_1$ and $C_2$ *different* if at least one $v_i$ receives a different color in $C_1$ than it does in $C_2$. For instance, the two colorings of $K_4$ shown in Figure 1.69 are considered different, since $v_3$ and $v_4$ receive different colorings.

If we restrict ourselves to four colors, how many different colorings are there of $K_4$? Since there are four choices for $v_1$, then three for $v_2$, etc., we see that there are $4 \cdot 3 \cdot 2 \cdot 1$ different colorings of $K_4$ using four colors. If six colors were available, there would be $6 \cdot 5 \cdot 4 \cdot 3$ different colorings. If only two were available, there would be no proper colorings of $K_4$.

In general, define $c_G(k)$ to be the number of different colorings of a graph $G$ using at most $k$ colors. So we have $c_{K_4}(4) = 24$, $c_{K_4}(6) = 360$, and $c_{K_4}(2) = 0$. In fact, if $k$ and $n$ are positive integers where $k \geq n$, then

$$c_{K_n}(k) = k(k-1)(k-2)\cdots(k-n+1).$$

Further, if $k < n$, then $c_{K_n}(k) = 0$. We also note that $c_{E_n}(k) = k^n$ for all positive integers $k$ and $n$.

A simple but important property of $c_G(k)$ is that $G$ is $k$-colorable if and only if $c_G(k) > 0$. Equivalently, $c_G(k) > 0$ if and only if $\chi(G) \leq k$.

Finding values of $c_G(k)$ is relatively easy for some well-known graphs. Computing this function in general, though, can be hard. Birkhoff and Lewis [13] developed a way to reduce this hard problem to an easier one. Before we see their method, we need a definition.

Let $G$ be a graph and let $e$ be an edge of $G$. Recall that $G - e$ denotes the graph where $e$ is removed from $G$. Define the graph $G/e$ to be the graph obtained from $G$ by removing $e$, identifying the end vertices of $e$, and leaving only one copy of any resulting multiple edges.

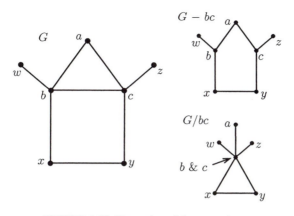

FIGURE 1.70. Examples of the operations.

As an example, a graph $G$ and the graphs $G - bc$ and $G/bc$ are shown in Figure 1.70.

**Theorem 1.29.**  *Let $G$ be a graph and $e$ be any edge of $G$. Then*

$$c_G(k) = c_{G-e}(k) - c_{G/e}(k).$$

PROOF.  Suppose that the end vertices of $e$ are $u$ and $v$, and consider the graph $G - e$.

How many $k$-colorings are there of $G - e$ where $u$ and $v$ are assigned the same color? If $C$ is a such a coloring of $G - e$, then $C$ can be thought of as a coloring of $G/e$, since $u$ and $v$ are colored the same. Similarly, any coloring of $G/e$ can also be thought of as a coloring of $G - e$ where $u$ and $v$ are colored the same. Thus, the answer to this question is $c_{G/e}(k)$.

Now, how many $k$-colorings are there of $G - e$ where $u$ and $v$ are assigned different colors? If $C$ is a such a coloring of $G - e$, then $C$ can be considered as a coloring of $G$, since $u$ and $v$ are colored differently. Similarly, any coloring of $G$ can also be thought of as a coloring of $G - e$ where $u$ and $v$ are colored differently. Thus, the answer to this second question is $c_G(k)$.

Thus, the total number of $k$-colorings of $G - e$ is

$$c_{G-e}(k) = c_{G/e}(k) + c_G(k),$$

and the result follows.                                                    □

For example, suppose we want to find $c_{P_4}(k)$. That is, how many ways are there to color the vertices of $P_4$ with $k$ colors available? We label the edges of $P_4$ as

FIGURE 1.71. The labeled edges of $P_4$.

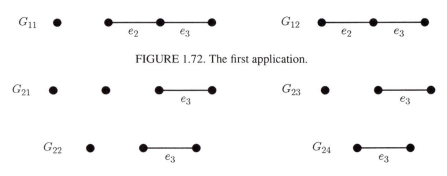

FIGURE 1.72. The first application.

FIGURE 1.73. The second application.

shown in Figure 1.71. The theorem implies that

$$c_{P_4}(k) = c_{P_4-e_1}(k) - c_{P_4/e_1}(k).$$

For convenience, let us denote $P_4 - e_1$ and $P_4/e_1$ by $G_{11}$ and $G_{12}$, respectively (see Figure 1.72). Applying the theorem again, we obtain

$$c_{P_4}(k) = c_{G_{11}-e_2}(k) - c_{G_{11}/e_2}(k) - c_{G_{12}-e_2}(k) + c_{G_{12}/e_2}(k).$$

Denote the graphs $G_{11} - e_2$, $G_{11}/e_2$, $G_{12} - e_2$, and $G_{12}/e_2$ by $G_{21}$, $G_{22}$, $G_{23}$, and $G_{24}$, respectively (see Figure 1.73).

Applying the theorem once more yields

$$\begin{aligned}
c_{P_4}(k) = {} & c_{G_{21}-e_3}(k) - c_{G_{21}/e_3}(k) - c_{G_{22}-e_3}(k) + c_{G_{22}/e_3}(k) \\
& - c_{G_{23}-e_3}(k) + c_{G_{23}/e_3}(k) + c_{G_{24}-e_3}(k) - c_{G_{24}/e_3}(k).
\end{aligned}$$

That is,

$$c_{P_4}(k) = c_{E_4}(k) - c_{E_3}(k) - c_{E_3}(k) + c_{E_2}(k) - c_{E_3}(k) + c_{E_2}(k) + c_{E_2}(k) - c_{E_1}(k).$$

Thus,

$$\begin{aligned}
c_{P_4}(k) &= k^4 - k^3 - k^3 + k^2 - k^3 + k^2 + k^2 - k \\
&= k^4 - 3k^3 + 3k^2 - k.
\end{aligned}$$

We should check a couple of examples. How many colorings of $P_4$ are there with one color?

$$c_{P_4}(1) = 1^4 - 3(1)^3 + 3(1)^2 - 1 = 0.$$

This, of course, makes sense. And how many colorings are there with two colors?

$$c_{P_4}(2) = 2^4 - 3(2)^3 + 3(2)^2 - 2 = 2.$$

Figure 1.74 shows these two colorings. Score one for Birkhoff!

As you can see, chromatic polynomials provide a way to count colorings, and the Birkhoff–Lewis theorem allows you to reduce a problem to a slightly simpler one. We should note that it is not always necessary to work all the way down to

red     blue     red     blue         blue     red     blue     red

FIGURE 1.74. Two 2-colorings of $P_4$.

empty graphs, as we did in the previous example. Once a graph $G$ is obtained for which the value of $c_G(k)$ is known, there is no need to reduce that one further.

We now present some properties of $c_G(k)$.

**Theorem 1.30.** *Let G be a graph of order n. Then*

1. *$c_G(k)$ is a polynomial in k of degree n,*
2. *the leading coefficient of $c_G(k)$ is 1,*
3. *the constant term of $c_G(k)$ is 0,*
4. *the coefficients of $c_G(k)$ alternate in sign, and*
5. *the absolute value of the coefficient of the $k^{n-1}$ term is the number of edges in G.*

We leave the proof of this theorem as an exercise (Exercise 3). One proof strategy is to induct on the number of edges in $G$ and use the Birkhoff–Lewis reduction theorem (Theorem 1.29).

Before leaving this section, we should note that Birkhoff considered chromatic polynomials of planar graphs, and he hoped to find one of them that had 4 as a root. If he had found one, then the corresponding planar graph would not be 4-colorable, and hence would be a counterexample to the Four Color Conjecture. Although he was unsuccessful in proving the Four Color Theorem, he still deserves credit for producing a very nice counting technique.

**Exercises**

1. Find chromatic polynomials for each of the following graphs. For each one, determine how many 5-colorings exist.

   **a.** $K_{1,3}$
   **b.** $K_{1,5}$
   **c.** $C_4$
   **d.** $C_5$
   **e.** $K_4 - e$
   **f.** $K_5 - e$

2. Show that $k^4 - 4k^3 + 3k^2$ is not a chromatic polynomial for any graph.
3. Prove Theorem 1.30.

# 1.5 Matchings

*Pardon me, do you have a match?*

            — James Bond, in *From Russia with Love*

The Senate committees that we discussed earlier need to form an executive council. Each committee needs to designate one of its members as an official representative to sit on the council, and council policy states that no senator can be the official representative for more than one committee. For example, let us suppose ·there are four committees: Senators $A$, $B$, $C$, and $D$ are on Committee 1; Senators $A$, $E$, and $F$ are members of Committee 2; Committee 3 consists of $E$, $F$, and $D$; and Senator $A$ is the only member of Committee 4. In this example, the executive council could consist of $A$, $E$, $F$, and $C$—representing Committees 4, 3, 2, and 1, respectively.

As another example, suppose Committee 1 consists of $W$, $X$ and $Y$; Committee 2 of $W$, $X$, and $Z$; Committee 3 of $W$, $Y$, and $Z$; Committee 4 of $X$, $Y$, and $Z$; and Committee 5 of $W$ and $Z$. It does not take long to see that it is impossible in this case to select official representatives according to the policy.

So a natural question arises: Under what circumstances can the executive council be formed successfully? In the sections that follow, we will see how graphs can be used to help answer this question.

### 1.5.1   Definitions

*And as to the meaning . . .*

— C.S. Calverly, *Ballad*

A *matching* in a graph is a set of independent edges. That is, it is a set of edges in which no pair shares a vertex. Given a matching $M$ in a graph $G$, the vertices belonging to the edges of $M$ are said to be *saturated* by $M$ (or *M-saturated*). The other vertices are *M-unsaturated*.

Consider the graph $G$ shown in Figure 1.75. An example of a matching in $G$ is $M_1 = \{ab, ce, df\}$. $M_2 = \{cd, ab\}$ is also a matching, and so is $M_3 = \{df\}$. We can see that $a$, $b$, $c$, $d$ are $M_2$-saturated and $e$, $f$, and $g$ are $M_2$-unsaturated. The only $M_1$-unsaturated vertex is $g$.

If a matching $M$ saturates every vertex of $G$, then $M$ is said to be a *perfect matching*. We will focus on perfect matchings in Section 1.5.4. For now, look at Figure 1.76. $G_1$ has a perfect matching, namely $\{ab, ch, de, fg\}$. None of $G_2$, $G_3$, and $G_4$ has a perfect matching. Why is this?

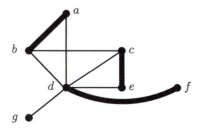

FIGURE 1.75. The matching $M_1$.

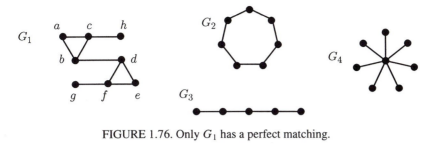

FIGURE 1.76. Only $G_1$ has a perfect matching.

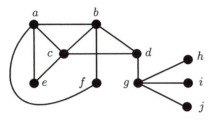

FIGURE 1.77.

A *maximum matching* in a graph is a matching that has the largest possible cardinality. A *maximal matching* is a matching that cannot be enlarged. In Figure 1.77, $M_1 = \{ae, bf, cd, gh\}$ is a maximum matching (since at most one of $gh$, $gi$, and $gj$ can be in any matching). The matching $M_2 = \{dg, af, bc\}$ is maximal, but not maximum.

**Exercises**

1. Determine whether the graph of Figure 1.78 has a perfect matching. If so, then exhibit it. If not, explain why.

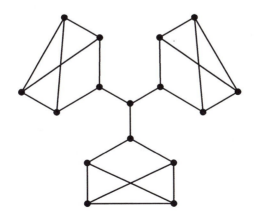

FIGURE 1.78. Is there a perfect matching?

**2.** Find the minimum size of a maximal matching in each of the following graphs.

   **a.** $C_{10}$
   **b.** $C_{11}$
   **c.** $C_n$

**3.** (From [25].) The *matching graph* $M(G)$ of a graph $G$ has the maximum matchings of $G$ as its vertices, and two vertices $M_1$ and $M_2$ of $M(G)$ are adjacent if $M_1$ and $M_2$ differ in only one edge. Show that each cycle $C_n$, $n = 3, 4, 5,$ or $6$, is the matching graph of some graph.

## 1.5.2 Hall's Theorem and SDRs

*I'll match that!*

— Monty Hall

In this section we consider several classic results concerning matchings. We begin with a few more definitions.

Given a graph $G$ and a matching $M$, an *M-alternating path* is a path in $G$ where the edges alternate between $M$-edges and non-$M$-edges. An *M-augmenting path* is an $M$-alternating path where both end vertices are $M$-unsaturated.

As an example, consider the graph $G$ and the matching $M$ indicated in Figure 1.79. An example of an $M$-alternating path is $c, a, d, e, i$. An example of an $M$-augmenting path is $j, g, f, a, c, b$. The reason for calling such a path "$M$-augmenting" will become apparent soon.

The following result is due to Berge [11].

**Theorem 1.31** (Berge's Theorem). *Let $M$ be a matching in a graph $G$. $M$ is maximum if and only if $G$ contains no $M$-augmenting paths.*

PROOF.   First, assume that $M$ is a maximum matching, and suppose that $P$ : $v_1, v_2, \ldots, v_k$ is an $M$-augmenting path. Due to the alternating nature of $M$-augmenting paths, it must be that $k$ is even and that the edges $v_2 v_3$, $v_4 v_5$, $\ldots$, $v_{k-2} v_{k-1}$ are all edges of $M$. We also see that the edges $v_1 v_2$, $v_3 v_4$, $\ldots$, $v_{k-1} v_k$ are not edges of $M$ (Figure 1.80).

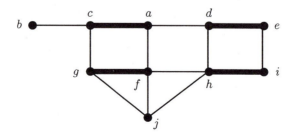

FIGURE 1.79. The graph $G$ and matching $M$.

FIGURE 1.80. An $M$-augmenting path.

But then if we define the set of edges $M_1$ to be

$$M_1 = (M \setminus \{v_2v_3, \ldots, v_{k-2}v_{k-1}\}) \cup \{v_1v_2, \ldots, v_{k-1}v_k\},$$

then $M_1$ is a matching that contains one more edge than $M$, a matching that we assumed to be maximum. This is a contradiction, and we can conclude that $G$ contains no $M$-augmenting paths.

For the converse, assume that $G$ has no $M$-augmenting paths, and suppose that $M'$ is a matching that is larger than $M$. Define a subgraph $H$ of $G$ as follows: Let $V(H) = V(G)$ and let $E(H)$ be the set of edges of $G$ that appear in exactly one of $M$ and $M'$. Now consider some properties of this subgraph $H$. Since each of the vertices of $G$ lies on at most one edge from $M$ and at most one edge from $M'$, it must be that the degree (in $H$) of each vertex of $H$ is at most 2. This implies that each connected component of $H$ is either a single vertex, a path, or a cycle. If a component is a cycle, then it must be an even cycle, since the edges alternate between $M$-edges and $M'$-edges. So, since $|M'| > |M|$, there must be at least one component of $H$ that is a path that begins and ends with edges from $M'$. But this path is an $M$-augmenting path, contradicting our assumption. Therefore, no such matching $M'$ can exist—implying that $M$ is maximum.                                    $\square$

Before we see Hall's classic matching theorem, we need to define one more term. If $G$ is a bipartite graph with partite sets $X$ and $Y$, we say that $X$ can be *matched into* $Y$ if there exists a matching in $G$ that saturates the vertices of $X$. Consider the two examples in Figure 1.81.

In the bipartite graph on the left, we see that $X$ can be matched into $Y$. In the graph on the right, though, it is impossible to match $X$ into $Y$ (why is this?). What conditions on a bipartite graph must exist if we want to match one partite set into the other? The answer to this question is found in the following result of Hall [77] (Philip, not Monty).

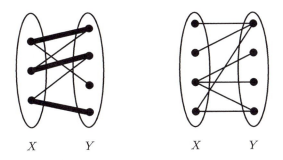

FIGURE 1.81.

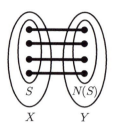

FIGURE 1.82.

Recall that the neighborhood of a set of vertices $S$, denoted by $N(S)$, is the union of the neighborhoods of the vertices of $S$.

**Theorem 1.32** (Hall's Theorem).    *Let $G$ be a bipartite graph with partite sets $X$ and $Y$. $X$ can be matched into $Y$ if and only if $|N(S)| \geq |S|$ for all subsets $S$ of $X$.*

PROOF.    First suppose that $X$ can be matched into $Y$, and let $S$ be some subset of $X$. Since $S$ itself is also matched into $Y$, we see immediately that $|S| \leq |N(S)|$ (see Figure 1.82). Now suppose that $|N(S)| \geq |S|$ for all subsets $S$ of $X$, and let $M$ be a maximum matching. Suppose that $u \in X$ is not saturated by $M$ (see Figure 1.83). Define the set $A$ to be the set of vertices of $G$ that can be joined to $u$ by an $M$-alternating path. Let $S = A \cap X$, and let $T = A \cap Y$ (see Figure 1.84). Notice now that Berge's Theorem implies that every vertex of $T$ is saturated by $M$ and that $u$ is the only unsaturated vertex of $S$. That is, every vertex of $T$ is saturated, and every vertex of $S \setminus \{u\}$ is saturated. This implies that $|T| = |S| - 1$. It follows from Berge's Theorem and the definition of $T$ that $N(S) = T$. But then we have that $|N(S)| = |S| - 1 < |S|$, and this is a contradiction. We conclude that such a vertex $u$ cannot exist in $X$ and that $M$ saturates all of $X$.    □

Given some family of sets $X$, a *system of distinct representatives*, or SDR, for the sets in $X$ can be thought of as a "representative" collection of distinct elements from the sets of $X$. For instance, let $S_1, S_2, S_3, S_4,$ and $S_5$ be defined as follows:

$$S_1 = \{2, 8\},$$
$$S_2 = \{8\},$$
$$S_3 = \{5, 7\},$$

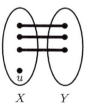

FIGURE 1.83.

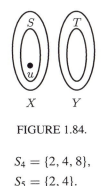

FIGURE 1.84.

$$S_4 = \{2, 4, 8\},$$
$$S_5 = \{2, 4\}.$$

The family $X_1 = \{S_1, S_2, S_3, S_4\}$ does have an SDR, namely $\{2, 8, 7, 4\}$. The family $X_2 = \{S_1, S_2, S_4, S_5\}$ does not have an SDR. So under what conditions will a finite family of sets have an SDR? We answer this question with the following theorem.

**Theorem 1.33.** *Let $S_1$, $S_2$, ..., $S_k$ be a collection of finite, nonempty sets. This collection has an SDR if and only if for every $t \in \{1, \ldots, k\}$, the union of any $t$ of these sets contains at least $t$ elements.*

PROOF. Since each of the sets is finite, then of course $S = S_1 \cup S_2 \cup \cdots \cup S_k$ is finite. Let us say that the elements of $S$ are $a_1, \ldots, a_n$.

We now construct a bipartite graph with partite sets $X = \{S_1, \ldots, S_k\}$ and $Y = \{a_1, \ldots, a_n\}$ (Figure 1.85). We place an edge between $S_i$ and $a_j$ if and only if $a_j \in S_i$.

Hall's Theorem now implies that $X$ can be matched into $Y$ if and only if $|A| \leq |N(A)|$ for all subsets $A$ of $X$. In other words, the collection of sets has an SDR if and only if for every $t \in \{1, \ldots, k\}$, the union of any $t$ of these sets contains at least $t$ elements. $\square$

Hall's Theorem is often referred to as Hall's Marriage Theorem. We will see more about this in Section 2.7.

**Exercises**

**1.** (From [28].) For the graphs of Figure 1.86, with matchings $M$ as shaded, find

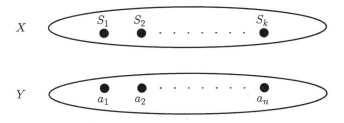

FIGURE 1.85. Constructing a bipartite graph.

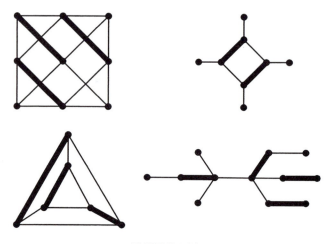

FIGURE 1.86.

**a.** an $M$-alternating path that is not $M$-augmenting;

**b.** an $M$-augmenting path if one exists; and, if so, use it to obtain a bigger matching.

2. For each of the following families of sets, determine whether the condition of Theorem 1.33 is met. If so, then find an SDR. If not, then show how the condition is violated.

   **a.** $\{1, 2, 3\}, \{2, 3, 4\}, \{3, 4, 5\}, \{4, 5\}, \{1, 2, 5\}$
   **b.** $\{1, 2, 4\}, \{2, 4\}, \{2, 3\}, \{1, 2, 3\}$
   **c.** $\{1, 2\}, \{2, 3\}, \{1, 2, 3\}, \{2, 3, 4\}, \{1, 3\}, \{3, 4\}$
   **d.** $\{1, 2, 5\}, \{1, 5\}, \{1, 2\}, \{2, 5\}$
   **e.** $\{1, 2, 3\}, \{1, 2, 4\}, \{1, 3, 4\}, \{1, 2, 3, 4\}, \{2, 3, 4\}$

3. Let $G$ be a bipartite graph. Show that $G$ has a matching of size at least $|E(G)|/\Delta(G)$.

4. Let $\Theta = \{S_1, S_2, \ldots, S_r\}$ be a family of distinct nonempty subsets of the set $\{1, 2, \ldots, n\}$. If the $S_i$ are all of the same cardinality, then prove that there exists an SDR of $\Theta$.

5. Let $M_1$ and $M_2$ be matchings in a bipartite graph $G$ with partite sets $X$ and $Y$. If $S \subseteq X$ is saturated by $M_1$ and $T \subseteq Y$ is saturated by $M_2$, show that there exists a matching in $G$ that saturates $S \cup T$.

6. (From [74].) Let $G$ be a bipartite graph with partite sets $X$ and $Y$. Let $\delta_X$ denote the minimum degree of the vertices in $X$, and let $\Delta_Y$ denote the maximum degree of the vertices in $Y$. Prove that if $\delta_X \geq \Delta_Y$, then there exists a matching in $G$ that saturates $X$.

## 1.5.3   The König–Egerváry Theorem

*What I tell you three times is true.*

              — Lewis Carroll, *The Hunting of the Snark*

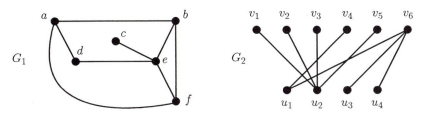

FIGURE 1.87.

The main theorem that we present in this section is very important, for it is closely related to several results from other areas of graph theory. We will discuss a few of these areas after we have proven the theorem.

A set $C$ of vertices in a graph $G$ is said to *cover* the edges of $G$ if every edge of $G$ is incident with at least one vertex of $C$. Such a set $C$ is called an *edge cover* of $G$.

Consider the graphs $G_1$ and $G_2$ in Figure 1.87. In $G_1$, the set $\{b, d, e, a\}$ is an edge cover, as is the set $\{a, e, f\}$. In fact, you can see by a little examination that there is no edge cover $G_1$ with fewer than three vertices. So we can say that $\{a, e, f\}$ is a minimum edge cover of $G_1$. In $G_2$, each of the following sets is an edge cover: $\{v_1, v_2, v_3, v_4, v_5, v_6\}$ (obviously) and $\{u_2, v_6, u_1\}$. What is the size of a minimum edge cover here?

We are now ready to prove the following result of König [95] and Egerváry [40].

**Theorem 1.34** (König–Egerváry Theorem). *Let $G$ be a bipartite graph. The maximum number of edges in a matching in $G$ equals the minimum number of vertices in an edge cover of $G$.*

PROOF.    Let $M$ be a maximum matching of $G$. Let $X$ and $Y$ be the partite sets of $G$, and let $W$ be the set of all $M$-unsaturated vertices of $X$ (see Figure 1.88). Note that $|M| = |X| - |W|$.

Now let $A$ be the set of vertices of $G$ that can be reached via an $M$-alternating path from some vertex of $W$. Let $S = A \cap X$, and let $T = A \cap Y$. We can note two things now: first, $S \setminus W$ is matched to $T$ (implying that $|W| = |S| - |T|$), and second, $N(S) = T$.

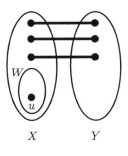

FIGURE 1.88.

If we let $C = (X \setminus S) \cup T$, we see that $C$ covers the edges of $G$. So $C$ is an edge cover of $G$, and $|C| = |X| - |S| + |T| = |X| - |W| = |M|$. Now suppose that $C'$ is any edge cover. Since each vertex of $C'$ can cover at most one edge of $M$, it must be that $|C'| \geq |M|$. We conclude then that $C$ is a minimum edge cover.    □

The König–Egerváry Theorem is one of several theorems in graph theory that relate the minimum of one thing to the maximum of something else. What follows are some examples of theorems that are very closely related to the König–Egerváry Theorem.

### Menger's Theorem

Let $G$ be a connected graph, and let $u$ and $v$ be vertices of $G$. If $S$ is a subset of vertices that does not include $u$ or $v$, and if the graph $G - S$ has $u$ and $v$ in different connected components, then we say that $S$ is a $u$, $v$-*separating set*.

The following result is known as Menger's Theorem [105].

**Theorem 1.35.** *Let G be a graph and let u and v be vertices of G. The maximum number of internally disjoint paths from u to v equals the minimum number of vertices in a u, v-separating set.*

### Max Flow Min Cut Theorem

A graph can be thought of as a flow network, where one vertex is specified to be the source of the flow and another is specified to be the receiver of the flow. As an amount of material flows from source to receiver, it passes through other intermediate vertices, each of which has a particular flow capacity. The *total flow* of a network is the amount of material that is able to make it from source to receiver. A *cut* in a network is a set of intermediate vertices whose removal completely cuts the flow from the source to the receiver. The *capacity of the cut* is simply the sum of the capacities of the vertices in the cut.

**Theorem 1.36.** *Let N be a flow network. The maximum value of total flow equals the minimum capacity of a cut.*

### Independent Zeros

If $A$ is an $m \times n$ matrix with real entries, a set of *independent zeros* in $A$ can be thought of as a set of ordered pairs $\{(i_1, j_1), (i_2, j_2), \ldots, (i_t, j_t)\}$ with the following properties:

**a.** the $(i_k, j_k)$ entry of $A$ is 0 for $k = 1, 2, \ldots, t$;
**b.** if $a \neq b$, then $i_a \neq i_b$ and $j_a \neq j_b$.

That is, none of the zeros in the set are in the same row or column.

Now, in this matrix $A$ one can draw lines through each row and column that contains a zero. Such a set of lines is said to cover the zeros of $A$.

**Theorem 1.37.** *The maximum number of independent zeros in A is equal to the minimum number of lines through rows or columns that together cover all the zeros of A.*

### Exercises

1. Use the König–Egerváry Theorem to prove Hall's Theorem.
2. Let $k$ be some fixed integer, $1 \le k \le n$, and let $G$ be some subgraph of $K_{n,n}$ with more than $(k-1)n$ edges. Prove that $G$ has a matching of size at least $k$.
3. Use the original statement of the König–Egerváry Theorem to prove Theorem 1.37.

## 1.5.4 Perfect Matchings

*It's a perfect ending.*

— Sophie, in *Anastasia*

We end this section on matchings by discussing perfect matchings. Recall that a perfect matching is a matching that saturates the entire vertex set of a graph. What kinds of graphs have perfect matchings? One thing that is clear is that a graph must be of even order in order to have a chance at having a perfect matching. But being of even order is certainly not enough to guarantee a perfect matching (look back at Figure 1.77).

We do know that $K_{2n}$, $C_{2n}$, and $P_{2n}$ have perfect matchings. The following result regarding perfect matchings in bipartite graphs is a corollary to Hall's Theorem. The proof is left as an exercise (Exercise 5).

**Corollary 1.38.** *If G is a bipartite graph that is regular of degree k, then G contains a perfect matching.*

It seems very natural to think that the more edges a graph has, the more likely it is that the graph will have a perfect matching. The following theorem verifies this thought, to a degree.

**Theorem 1.39.** *If G is a graph of order 2n such that $\delta(G) \ge n$, then G has a perfect matching.*

The simplest proof of Theorem 1.39 uses a theorem of Dirac that guarantees the existence of a cycle in $G$ that spans the vertex set.[5] The existence of such a cycle implies the existence of a perfect matching, since we could use every other edge on the cycle in the matching.

---

[5]This type of cycle is called a hamiltonian cycle. Hamiltonicity in graphs is a popular branch of graph theory, and it has been well studied. For more on hamiltonicity, see [67], [25], or [146].

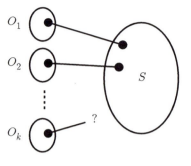

FIGURE 1.89.

In 1947 Tutte [141] provided perhaps the best known characterization of graphs with perfect matchings. A number of proofs of Tutte's Theorem have been published since then. The proof that we present is due to Anderson [3].

A definition first: Given a graph $G$, define $\Omega(G)$ to be the number of connected components of $G$ with odd order. Also, define $\Sigma(G)$ to be the number of connected components of $G$ with even order.

**Theorem 1.40** (Tutte's Theorem).    *Let $G$ be a graph of order $n \geq 2$. $G$ has a perfect matching if and only if $\Omega(G - S) \leq |S|$ for all subsets of $S$ of $V(G)$.*

PROOF.    We begin with the forward direction. Let $G$ be a graph that has a perfect matching. Suppose $S$ is a set of vertices and that $O_1, O_2, \ldots, O_k$ are the odd components of $G - S$. For each $i$, the vertices in $O_i$ can be adjacent only to other vertices in $O_i$ and to vertices in $S$. Since $G$ has a perfect matching, at least one vertex out of each of the $O_i$'s has to be matched with a different vertex in $S$. If $k > |S|$, then some $O_i$ would be left out (Figure 1.89). Thus, $k \leq |S|$.

For the reverse direction of the theorem, suppose that $|S| \geq \Omega(G - S)$ for all $S$. In particular, if $S = \emptyset$, then $\Omega(G - \emptyset) \leq 0$. This implies that there are no odd components of $G$—every component of $G$ is even. More generally, we make the following claim.

Claim A. For any proper subset $S$, $|S|$ and $\Omega(G - S)$ are either both even or both odd.

Let $C$ be some component of $G$. We know from earlier that $C$ has even order. If an even number of vertices is removed from $C$, then the number of odd components remaining must also be even. If an odd number of vertices is removed from $C$, then the number of odd components remaining must be odd. Since this is true for every component of $G$, it is true for all of $G$. Hence Claim A is proved.

We now proceed by induction on $n$, the order of the graph. If $n = 2$, then $G$ is $K_2$, which certainly has a perfect matching. Suppose now that the result is true for all graphs of even order up to $n$, and let $G$ be a graph of even order $n$. We now have two cases.

*Case* 1. Suppose that for every proper subset $S$, $\Omega(G - S) < |S|$. (That is, the strict inequality holds.) Claim A implies that $|S|$ and $\Omega(G - S)$ have the same parity, so we can say in this case that for all subsets $S$, $\Omega(G - S) \leq |S| - 2$. Let $uv \in E(G)$, and consider the graph $G - u - v$ (a graph with two fewer vertices than $G$). We would like to apply the induction hypothesis to $G - u - v$, so we need the following claim.

Claim B. For all subsets $S'$ of $V(G - u - v)$, $\Omega(G - u - v - S') \leq |S'|$.

If Claim B were not true, then $\Omega(G - u - v - S_1) > |S_1|$ for some subset $S_1$. But since $|S_1| = |S_1 \cup \{u, v\}| - 2$, we get $\Omega(G - u - v - S_1) > |S_1 \cup \{u, v\}|$, and this contradicts the assumption in this case. Claim B is proved.

Since Claim B is true, we can apply the induction hypothesis to $G - u - v$. That is, we can conclude that $G - u - v$ has a perfect matching. This matching, together with the edge $uv$, forms a perfect matching of $G$. Case 1 is complete.

*Case* 2. Suppose there exists a subset $S$ such that $\Omega(G - S) = |S|$. There may be a number of subsets $S$ that satisfy this condition—suppose without loss of generality that $S$ is a largest such set. Let $O_1, O_2, \ldots, O_k$ be the components of $G - S$ of odd order.

Claim C. $\Sigma(G - S) = 0$. That is, there are no even-ordered components of $G - S$.

Let $E$ be an even ordered component of $G - S$, and let $x$ be a vertex of $E$. The graph $G - S - x$ has exactly one more odd component than $G - S$. Thus, $|S \cup \{x\}| = \Omega(G - S - x)$. But this means that $S \cup \{x\}$ is a set larger than $S$ that satisfies the assumption of this case. Since we chose $S$ to be the largest, we have a contradiction. Therefore there are no even-ordered components of $G - S$. Claim C is proved.

Claim D. There exist vertices $s_1, s_2, \ldots, s_k \in S$ and vertices $v_1, v_2, \ldots, v_k$, where for each $i$ $v_i \in O_i$, such that $\{v_1 s_1, v_2 s_2, \ldots, v_k s_k\}$ is a matching.

For each $i \in \{1, \ldots, k\}$, define the set $S_i$ to be the set of vertices in $S$ that are adjacent to some vertex in $O_i$. Note that if $S_i = \emptyset$ for some $i$, then $O_i$ is completely disconnected from anything else in $G$, implying that $G$ itself has an odd component. Since this contradicts our assumption in this case, we can assume that each $S_i$ is nonempty. Furthermore, our initial assumptions imply that the union of any $r$ of the $S_i$'s has size at least $r$. Thus, the condition in Theorem 1.33 is satisfied, implying that there exists a system of distinct representatives for the family of sets $S_1, S_2, \ldots, S_k$. If we let these representatives be $s_1, s_2, \ldots, s_k$, and their adjacencies in the $O_i$'s be $v_1, v_2, \ldots, v_k$, then Claim D is proved.

The situation in $G$ is depicted in Figure 1.90, where $k = |S|$.

At this point, each vertex in $S$ has been matched to a vertex in an $O_i$. The goal at this point is to show that each $O_i - v_i$ has a perfect matching.

Let $W$ be some subset of vertices of (the even-ordered) $O_i - v_i$.

Claim E. $\Omega(O_i - v_i - W) \leq |W|$.

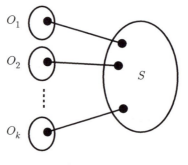

FIGURE 1.90.

If $\Omega(O_i - v_i - W) > |W|$, then, by Claim A, $\Omega(O_i - v_i - W) \geq |W| + 2$. But then,

$$\Omega(G - S - v_i - W) \geq |S| - 1 + |W| + 2$$
$$= |S| + |W| + 1$$
$$= |S \cup W \cup \{v_i\}|.$$

This contradicts our assumption, and thus Claim E is proved.

Since Claim E is true, each $O_i - v_i$ satisfies the induction hypothesis, and thus has a perfect matching. These perfect matchings together with the perfect matching shown in Figure 1.90 form a perfect matching of $G$, and so Case 2 is complete.                                                                                              □

We conclude this section by considering perfect matchings in regular graphs. If a graph $G$ is 1-regular, then $G$ itself *is* a perfect matching. If $G$ is 2-regular, then $G$ is a collection of disjoint cycles; as long as each cycle is even, $G$ will have a perfect matching.

What about 3-regular graphs? A graph that is 3-regular must be of even order, so is it possible that every 3-regular graph contains a perfect matching? In a word, no. The graph in Figure 1.91 is a connected 3-regular graph that does not have a perfect matching. Thanks to Petersen [114], though, we do know of a special class of 3-regular graphs that do have perfect matchings. Recall that a bridge in a graph is an edge whose removal would disconnect the graph. The graph in Figure 1.91 has three bridges.

**Theorem 1.41** (Petersen's Theorem).    *Every bridgeless, 3-regular graph contains a perfect matching.*

PROOF.    Let $G$ be a bridgeless, 3-regular graph, and suppose that it does not contain a perfect matching. By Tutte's Theorem, there must exist a subset $S$ of vertices where the number of odd components of $G - S$ is greater than $|S|$. Denote the odd-ordered components of $G - S$ by $O_1, O_2, \ldots, O_k$.

First, each $O_i$ must have at least one edge into $S$. Otherwise, there would exist an odd-ordered, 3-regular subgraph of $G$, and this is not possible, by Theorem 1.1. Second, since $G$ is bridgeless, there must be at least two edges joining each $O_i$ to

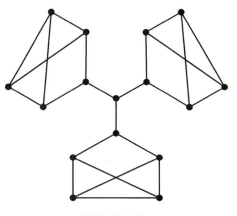

FIGURE 1.91.

$S$. Moreover, if there were only two edges joining some $O_i$ to $S$, then $O_i$ would contain an odd number of vertices with odd degree, and this cannot happen.

We can therefore conclude that there are at least three edges joining each $O_i$ to $S$. This implies that there are at least $3k$ edges coming into $S$ from the $O_i$'s. But since every vertex of $S$ has degree 3, the greatest number of edges incident with vertices in $S$ is $3|S|$, and since $3k > 3|S|$, we have a contradiction. Therefore, $G$ must have a perfect matching.    $\square$

It is probably not surprising that the Petersen of Theorem 1.41 is the same person for whom the Petersen graph (Figure 1.55) is named.

Petersen used this special graph as an example of a 3-regular, bridgeless graph whose edges cannot be partitioned into three separate, disjoint matchings.

## Exercises

1. Find a maximum matching of the graph shown in Figure 1.91.
2. Use Tutte's Theorem to prove that the graph in Figure 1.91 does not have a perfect matching.
3. Draw a connected, 3-regular graph that has both a cut vertex and a perfect matching.
4. Determine how many different perfect matchings there are in $K_{n,n}$.
5. Prove Corollary 1.38.
6. Characterize when $K_{r_1,r_2,\ldots,r_k}$ has a perfect matching.
7. Prove that every tree has at most one perfect matching.
8. Let $G$ be a subgraph of $K_{20,20}$. If $G$ has a perfect matching, prove that $G$ has at most 190 edges that belong to no perfect matching.
9. Use Tutte's Theorem to prove Hall's Theorem.

# 1.6   Ramsey Theory

*I have to go and see some friends of mine, some that I don't know, and some who aren't familiar with my name.*

— John Denver, *Goodbye Again*

We begin this section with a simple question: How many people are required at a gathering so that there must exist either three mutual acquaintances *or* three mutual strangers? We will answer this question soon.

Ramsey theory is named for Frank Ramsey, a young man who was especially interested in logic and philosophy. Ramsey died at the age of 27 in 1930—the same year that his paper "On a Problem of Formal Logic" was published. His paper catalyzed the development of the mathematical field now known as Ramsey theory. The study of Ramsey theory has burgeoned since that time. While many results in the subject are published each year, there are many questions whose answers remain elusive. In the words of Graham, Rothschild, and Spencer [71], "the field is alive and exciting."

## 1.6.1   Classical Ramsey Numbers

*An innocent looking problem often gives no hint as to its true nature.*

— Paul Erdős, [43]

A 2-*coloring* of the *edges* of a graph $G$ is any assignment of one of two colors to each of the edges of $G$. Figure 1.92 shows a 2-coloring of the edges of $K_5$ using red (thick) and blue (thin).

Let $p$ and $q$ be positive integers. The *(classical) Ramsey number* associated with these integers, denoted by $R(p, q)$, is defined to be the smallest integer $n$ such that every 2-coloring of the edges of $K_n$ either contains a red $K_p$ or a blue $K_q$ as a subgraph.

Read through that definition at least one more time, and then consider this simple example. We would like to find the value of $R(1, 3)$. According to the definition, this is the least value of $n$ such that every 2-coloring of the edges of $K_n$ either contains as a subgraph a $K_1$ all of whose edges are red, or a $K_3$ all of whose edges are blue. How many vertices are required before we know that we will have one

FIGURE 1.92. A 2-coloring of $K_5$.

of these objects in every coloring? If you have just one vertex, then no matter how you color the edges (ha-ha) of $K_1$, you will always end up with a red $K_1$. Thus, $R(1, 3) = 1$. We have found our first Ramsey number!

We should note here that the definition given for Ramsey number is in fact a good definition. That is, given positive integers $p$ and $q$, $R(p, q)$ does in fact exist. Ramsey himself proved this fact, and we will learn more about the proof of "Ramsey's Theorem" in Chapter 3.

Back to examples. We just showed that $R(1, 3) = 1$. Similar reasoning shows that $R(1, k) = 1$ for all positive integers $k$ (see Exercise 2).

How about $R(2, 4)$? We need to know the smallest integer $n$ such that every 2-coloring of the edges of $K_n$ contains either a red $K_2$ or a blue $K_4$. We claim that $R(2, 4) = 4$. To show this, we must demonstrate two things: first, that there exists a 2-coloring of $K_3$ that contains neither a red $K_2$ nor a blue $K_4$, and second, that any 2-coloring of the edges of $K_4$ contains at least one of these as a subgraph.

We demonstrate the first point. Consider the 2-coloring of $K_3$ given in Figure 1.93 (recall that red is thick and blue is thin—the edges in this coloring are all blue). This coloring of $K_3$ does not contain a red $K_2$, and it certainly does not contain a blue $K_4$. Thus $R(2, 4) > 3$.

For the second point, suppose that the edges of $K_4$ are 2-colored in some fashion. If any of the edges are red, then we have a red $K_2$. If none of the edges are red, then we have a blue $K_4$. So, no matter the coloring, we always get one of the two. This proves that $R(2, 4) = 4$.

What about $R(2, 5)$? $R(2, 8)$? $R(2, k)$? A similar argument reveals that $R(2, k) = k$ for all integers $k \geq 2$ (see Exercise 3).

### Exercises

1. How many different 2-colorings are there of $K_3$? of $K_4$? of $K_5$? of $K_{10}$?
2. Write a nice proof of the fact that $R(1, k) = 1$ for all positive integers $k$.
3. Write a nice proof of the fact that $R(2, k) = k$ for integers $k \geq 2$.
4. Prove that for positive integers $p$ and $q$, $R(p, q) = R(q, p)$.
5. If $2 \leq p' \leq p$ and $2 \leq q' \leq q$, then prove that $R(p', q') \leq R(p, q)$. Also, prove that equality holds if and only if $p' = p$ and $q' = q$.

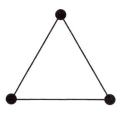

FIGURE 1.93. The edges of $K_3$ colored blue.

FIGURE 1.94. A 2-coloring of the edges of $K_5$.

## 1.6.2   Exact Ramsey Numbers and Bounds

*Take me to your leader.*

— proverbial alien

How many people are required at a gathering so that there must exist either three mutual acquaintances *or* three mutual strangers? We can rephrase this question as a problem in Ramsey theory: How many vertices do you need in an (edge) 2-colored complete graph for it to be necessary that there be either a red $K_3$ (people who know each other) or a blue $K_3$ (people who do not know each other)? As the next theorem states, the answer is 6.

**Theorem 1.42.**   $R(3, 3) = 6$.

PROOF.    We begin the proof by exhibiting (in Figure 1.94) a 2-coloring of the edges of $K_5$ that produces neither a red (thick) $K_3$ nor a blue (thin) $K_3$. This 2-coloring of $K_5$ demonstrates that $R(3, 3) > 5$. Now consider $K_6$, and suppose that each of its edges has been colored red or blue. Let $v$ be one of the vertices of $K_6$. There are five edges incident with $v$, and they are each colored red or blue, so it must be that $v$ is either incident with at least three red edges or at least three blue edges (think about this; it is called the Pigeonhole Principle—more on this in Chapter 3). Without loss of generality, let us assume that $v$ is incident with at least three red edges, and let us call them $vx$, $vy$, and $vz$ (see Figure 1.95).

Now, if none of the edges $xy$, $xz$, $yz$ is colored red, then we have a blue $K_3$ (Figure 1.96).

On the other hand, if at least one of $xy$, $xz$, $yz$ is colored red, we have a red $K_3$ (Figure 1.97).

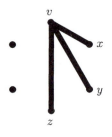

FIGURE 1.95.

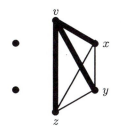

FIGURE 1.96.

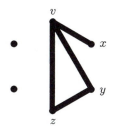

FIGURE 1.97.

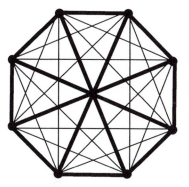

FIGURE 1.98. A 2-coloring of the edges in $K_8$.

Therefore, any 2-coloring of the edges of $K_6$ produces either a red $K_3$ or a blue $K_3$.    □

Let us determine another Ramsey number.

**Theorem 1.43.** $R(3, 4) = 9$.

PROOF.    Consider the 2-coloring of the edges of $K_8$ given in Figure 1.98.

A bit of examination reveals that this coloring produces no red (thick) $K_3$ and no blue (thin) $K_4$. Thus, $R(3, 4) \geq 9$. We now want to prove that $R(3, 4) \leq 9$, and we will use the facts that $R(2, 4) = 4$ and $R(3, 3) = 6$.

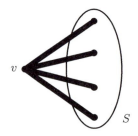

FIGURE 1.99.

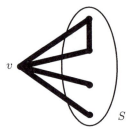

FIGURE 1.100.

Let $G$ be any complete graph of order at least 9, and suppose that the edges of $G$ have been 2-colored arbitrarily. Let $v$ be some vertex of $G$.

*Case* 1. Suppose that $v$ is incident with at least four red edges. Call the end vertices of these edges "red neighbors" of $v$, and let $S$ be the set of red neighbors of $v$ (see Figure 1.99).

Since $S$ contains at least four vertices, and since $R(2, 4) = 4$, the 2-coloring of the edges that are within $S$ must produce either a red $K_2$ or a blue $K_4$ *within* $S$ itself. If the former is the case, then we are guaranteed a red $K_3$ in $G$ (see Figure 1.100). If the latter is the case, then we are clearly guaranteed a blue $K_4$ in $G$.

*Case* 2. Suppose that $v$ is incident with at least six blue edges. Call the other end vertices of these edges "blue neighbors" of $v$, and let $T$ be the set of blue neighbors of $v$ (see Figure 1.101).

Since $T$ contains at least six vertices, and since $R(3, 3) = 6$, the 2-coloring of the edges that are within $T$ must produce either a red $K_3$ or a blue $K_3$ *within* $T$ itself. If the former is the case, then we are obviously guaranteed a red $K_3$ in $G$. If the latter is the case, then we are guaranteed a blue $K_4$ in $G$ (see Figure 1.102).

*Case* 3. Suppose that $v$ is incident with fewer than four red edges and fewer than six blue edges. In this case there must be at most nine vertices in $G$ altogether, and since we assumed at the beginning that the order of $G$ is at least 9, we can say that $G$ has order exactly 9. Further, we can say that $v$ is incident with exactly three red edges and exactly five blue edges. And since the vertex $v$ was chosen arbitrarily, we can assume that this holds true for every vertex of $G$.

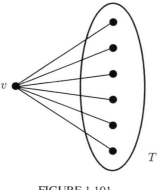

FIGURE 1.101.

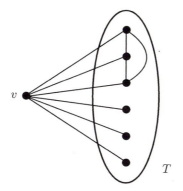

FIGURE 1.102.

Now if we consider the underlying "red" subgraph of $G$, we have a graph with nine vertices, each of which has degree 3. But this cannot be, since the number of vertices in $G$ with odd degree is even (the First Theorem of Graph Theory). Therefore, this case cannot occur.

We have therefore proved that any 2-coloring of the edges of a complete graph on 9 vertices (or more) produces either a red $K_3$ or a blue $K_4$. Hence, $R(3, 4) = 9$.                                                                                                         ☐

Some known Ramsey numbers are listed below.

$$R(1, k) = 1,$$
$$R(2, k) = k,$$
$$R(3, 3) = 6, \ R(3, 4) = 9, \ R(3, 5) = 14, \ R(3, 6) = 18,$$
$$R(3, 7) = 23, \ R(3, 8) = 28, \ R(3, 9) = 36,$$
$$R(4, 4) = 18, \ R(4, 5) = 25.$$

Bounds on Ramsey Numbers

Determining exact values of Ramsey numbers is extremely difficult in general. *In fact, the list given above is not only a list of some known Ramsey numbers, it is a list of all known Ramsey numbers.* Many people have attempted to determine other values, but to this day no other numbers are known.

However, there has been progress in finding bounds, and we state some important ones here. The first bound is due to Erdős and Szekeres [44], two major players in the development of Ramsey theory. Their result involves a quotient of factorials: Here, $n!$ denotes the product $1 \cdot 2 \cdots n$.

**Theorem 1.44.** *For positive integers p and q,*

$$R(p, q) \leq \frac{(p + q - 2)!}{(p - 1)!(q - 1)!}.$$

The next theorem gives a bound on $R(p, q)$ based on "previous" Ramsey numbers.

**Theorem 1.45.** *If $p \geq 2$ and $n \geq 2$, then*

$$R(p, q) \leq R(p - 1, q) + R(p, q - 1).$$

*Furthermore, if both terms on the right of this inequality are even, then the inequality is strict.*

The following bound is for the special case $p = 3$.

**Theorem 1.46.** *For every integer $q \geq 3$,*

$$R(3, q) \leq \frac{q^2 + 3}{2}.$$

The final bound that we present is due to Erdős [42]. It applies to the special case $p = q$. In the theorem, $\lfloor x \rfloor$ denotes the greatest integer less than or equal to $x$.

**Theorem 1.47.** *If $p \geq 3$, then*

$$R(p, p) > \lfloor 2^{n/2} \rfloor.$$

A number of other specific bounds are known:

$$34 \leq R(4, 6) \leq 53,$$
$$43 \leq R(5, 5) \leq 52,$$
$$58 \leq R(5, 6) \leq 94,$$
$$102 \leq R(6, 6) \leq 169.$$

Even with the sophisticated computing power that is available to us today, we are not able to compute values for more than a handful of Ramsey numbers. We have seen Paul Erdős' name a number of times now. He was a Hungarian mathematician who is certain to be counted as one of the greatest mathematicians of the twentieth

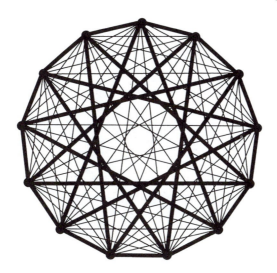

FIGURE 1.103. A 2-coloring of $K_{13}$.

century. Erdős once made the following comment regarding the difficulty in finding exact values of Ramsey numbers [34]:

> Suppose an evil alien would tell mankind "Either you tell me [the value of $R(5, 5)$] or I will exterminate the human race." ... It would be best in this case to try to compute it, both by mathematics and with a computer.
>
> If he would ask [for the value of $R(6, 6)$], the best thing would be to destroy him before he destroys us, because we couldn't [determine $R(6, 6)$].

**Exercises**

1. Prove that $R(3, 5) \geq 14$. The graph in Figure 1.103 will be very helpful.
2. Use Theorem 1.45 and the previous exercise to prove that $R(3, 5) = 14$.
3. Construct a graph and a 2-coloring that proves $R(4, 4) \geq 18$.
4. Use Theorem 1.45 and the previous exercise to prove that $R(4, 4) = 18$.
5. Use Theorem 1.45 to prove Theorem 1.46.

## 1.6.3   Graph Ramsey Theory

*All generalizations are dangerous, even this one.*

— Alexandre Dumas

Graph Ramsey theory is a generalization of classical Ramsey theory. Its development was due in part to the search for the elusive classical Ramsey numbers, for it was thought that the more general topic might shed some light on the search. The generalization blossomed and became an exciting field in itself. In this section we explain the concept of graph Ramsey theory, and we examine several results. These results, and more like them, can be found in [71].

Given two graphs $G$ and $H$, define the graph Ramsey number $R(G, H)$ to be the smallest value of $n$ such that any 2-coloring of the edges of $K_n$ contains either a red copy of $G$ or a blue copy of $H$. The classical Ramsey number $R(p, q)$ would in this context be written as $R(K_p, K_q)$.

The following simple result demonstrates the relationship between graph Ramsey numbers and classical Ramsey numbers.

**Theorem 1.48.** *If $G$ is a graph of order $p$ and $H$ is a graph of order $q$, then*

$$R(G, H) \leq R(p, q).$$

PROOF.   Ramsey proved that for every $p$ and $q$, $R(p, q)$ exists (we will prove this in Chapter 3). Let $n = R(p, q)$, and consider an arbitrary 2-coloring of $K_n$. By definition, $K_n$ contains either a red $K_p$ or a blue $K_q$. Since $G \subseteq K_p$ and $H \subseteq K_q$, there must either be a red $G$ or a blue $H$ in $K_n$. Hence, $R(G, H) \leq n = R(p, q)$.                                                                  □

Here is a result due to Chvátal and Harary [27] that relates $R(G, H)$ to the chromatic number of $G$, $\chi(G)$, and the size of the largest component of $H$, denoted by $C(H)$.

**Theorem 1.49.** $R(G, H) \geq (\chi(G) - 1)(C(H) - 1) + 1.$

PROOF.   Let $m = \chi(G) - 1$ and let $n = C(H) - 1$. Consider the graph $S$ formed by taking $m$ copies of $K_n$ and adding all of the edges in between each copy (Figure 1.104). Actually, $S = K_{mn}$. Now color all of the edges within each $K_n$ blue, and color all other edges red. From the way we have constructed the coloring, every red subgraph can be vertex colored with $m$ colors. Since $m < \chi(G)$, there can be no red $G$ present. Furthermore, any blue subgraph has at most $n = C(H) - 1$ vertices in its largest component. Hence, there can be no blue $H$ present. We have exhibited a 2-coloring of $K_{mn}$ that contains neither a red $G$ nor a blue $H$.                □

The next few theorems give exact graph Ramsey numbers for specific classes of graphs. The first is due to Chvátal [26], and the proof uses a few ideas from previous sections.

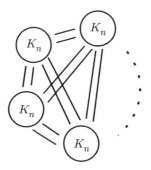

FIGURE 1.104. The graph $S$.

**Theorem 1.50.** *If $T_m$ is a tree with $m$ vertices, then*

$$R(T_m, K_n) = (m - 1)(n - 1) + 1.$$

PROOF.    If $m = 1$ or $n = 1$, then $R(T_m, K_n) = 1$ and the result holds. Assume then that $m$ and $n$ are both at least 2.

Claim A. $R(T_m, K_n) \geq (m - 1)(n - 1) + 1$.

Consider the graph that consists of $n - 1$ copies of $K_{m-1}$, with all possible edges between the copies of $K_{m-1}$. This graph is actually $K_{(m-1)(n-1)}$. Color the edges in each $K_{m-1}$ red, and color all of the other edges blue. Since each of the red subgraphs has order $m - 1$, no red $T_m$ can exist. Also, by this construction, no blue $K_n$ can exist. Since this 2-coloring contains no red $T_m$ and no blue $K_n$, it must be that $R(T_m, K_n) \geq (m - 1)(n - 1) + 1$.

Claim B. $R(T_m, K_n) \leq (m - 1)(n - 1) + 1$.

Let $G$ be $K_{(m-1)(n-1)+1}$, and suppose that its edges have been arbitrarily 2-colored. Let $G_r$ denote the subgraph of $G$ formed by the red edges, and let $G_b$ denote the subgraph of $G$ formed by the blue edges. If there is no blue $K_n$, then $\omega(G_b) \leq n-1$, and if so, then $\beta(G_r) \leq n - 1$, since $G_r$ is the complement of $G_b$. Thus by Theorem 1.26, $\chi(G_r) \geq m$. Let $H$ be a subgraph of $G_r$ that is $m$-critical. By part (d) of Exercise 8 in Section 1.4.1, $\delta(H) \geq m - 1$. By Theorem 1.8, $H$ contains $T_m$ as a subgraph, and therefore $G$ has a red $T_m$.    □

The next theorem is due to Burr [20].

**Theorem 1.51.** *If $T_m$ is a tree of order $m$ and if $m - 1$ divides $n - 1$, then*

$$R(T_m, K_{1,n}) = m + n - 1.$$

In the following theorem, $m K_2$ denotes the graph consisting of $m$ copies of $K_2$, and $n K_2$ has a similar meaning.

**Theorem 1.52.** *If $m \geq n \geq 1$, then*

$$R(m K_2, n K_2) = 2m + n - 1.$$

As we mentioned earlier, these results apply to specific classes of graphs. In general, determining values of $R(G, H)$ is quite difficult. So the generalization that was intended to solve hard classical Ramsey problems has produced hard problems of its own!

**Exercises**

1. Find $R(P_3, P_3)$.
2. Find $R(P_3, C_4)$.
3. Find $R(C_4, C_4)$.
4. Prove that $R(K_{1,3}, K_{1,3}) = 6$.
5. Prove that $R(2K_3, K_3) = 8$.

## 1.7   References

*Prince John: Are you finished?*
*Sir Robin of Locksley: I'm only just beginning.*

— *Robin Hood*

We have only just begun our walk through the field of graph theory. In this section we will provide references for those who are interested in further study.

The books by Chartrand and Lesniak [25] and West [146] provide very thorough introductions to a large number of topics in graph theory. Gould's book [67] covers a number of graph algorithms, from finding maximum matchings to testing planarity. Many interesting applications of graph theory can be found in texts by Gross and Yellen [74] and Foulds [52]. A good source for connections between graph theory and other mathematical topics is [10], edited by Beineke and Wilson. The text [78] by Harary is a thorough discussion of counting various types of graphs. For more on the development of the Four Color Problem, see the book by Aigner [2]. For a discussion on recent attempts to solve the Four Color Problem, see the article by Thomas [139]. Much more information regarding Ramsey theory can be found in the book [71] by Graham, Rothschild, and Spencer. And finally, a wonderful source for the history of graph theory and some of its famous problems is the book [12] by Biggs, Lloyd, and Wilson.

# 2

# Combinatorics

*[Combinatorics] has emerged as a new subject standing at the crossroads between pure and applied mathematics, the center of bustling activity, a simmering pot of new problems and exciting speculations.*
— Gian-Carlo Rota, [129, p. vii]

The formal study of combinatorics dates at least to Gottfried Wilhelm Leibniz's *Dissertatio de Arte Combinatoria* in the seventeenth century. The last half-century, however, has seen a huge growth in the subject, fueled by problems and applications from many fields of study. Applications of combinatorics arise, for example, in chemistry, in studying arrangements of atoms in molecules and crystals; biology, in questions about the structure of genes and proteins; physics, in problems in statistical mechanics; communications, in the design of codes for encryption, compression, and correction of errors; and especially computer science, for instance in problems of scheduling and allocating resources, and in analyzing the efficiency of algorithms.

Combinatorics is, in essence, the study of arrangements: pairings and groupings, rankings and orderings, selections and allocations. There are three principal branches in the subject. *Enumerative combinatorics* is the science of counting. Problems in this subject deal with determining the number of possible arrangements of a set of objects under some particular constraints. *Existential combinatorics* studies problems concerning the existence of arrangements that possess some specified property. *Constructive combinatorics* is the design and study of algorithms for creating arrangements with special properties.

Combinatorics is closely related to the theory of graphs. Many problems in graph theory concern arrangements of objects and so may be considered as combinatorial problems. For example, the theory of matchings and Ramsey theory, both studied in the previous chapter, have the flavor of existential combinatorics. Also, combinatorial techniques are often employed to address problems in graph theory. For example, in Section 2.3 we determine another method for finding the chromatic polynomial of a graph.

We focus on topics in enumerative combinatorics through most of this chapter, then turn to some questions in existential and constructive combinatorics in Section 2.7. Throughout this chapter we study arrangements of finite sets. Chapter 3 deals with arrangements and combinatorial problems involving infinite sets. Our study in this chapter includes the investigation of the following questions.

- Should a straight beat a flush in the game of poker? What about a full house?
- Suppose a lazy professor collects a quiz from each student in a class, then shuffles the papers and redistributes them randomly to the class for grading. How likely is it that no one receives his or her own quiz to grade?
- How many ways are there to make change for a dollar?
- How many different necklaces with twenty beads can be made using rhodonite, rose quartz, and lapis lazuli beads, if a necklace can be worn in any orientation?
- How many seating arrangements are possible for $n$ guests attending a wedding reception in a banquet room with $k$ round tables?
- Suppose 100 medical students rank 100 positions for residencies at hospitals in order of preference, and the hospitals rank the students in order of preference. Is there a way to assign the students to the hospitals in such a way that no student and hospital prefer each other to their assignment? Is there an efficient algorithm for finding such a matching?

## 2.1   Three Basic Problems

*The mere formulation of a problem is far more essential than its solution...*

— Albert Einstein

We begin our study of combinatorics by considering three basic counting problems.

**Problem 1.** How many ways are there to order a collection of $n$ different objects?

For example, how many ways are there to arrange the cards in a standard deck of 52 playing cards by shuffling? How many different batting orders are possible among the nine players on a baseball team? How many ways are there to arrange ten books on a shelf?

To order a collection of $n$ objects, we need to pick one object to be first, then another one to be second, and another one third, and so on. There are $n$ different choices for the first object, then $n - 1$ remaining choices for the second, and $n - 2$ for the third, and so forth, until just one choice remains for the last object. The total number of ways to order the $n$ objects is therefore the product of the integers between 1 and $n$. This number, called *n factorial*, is written $n!$. An ordering, or rearrangement, of $n$ objects is often called a *permutation* of the objects. Thus, the number of permutations of $n$ items is $n!$.

This calculation exemplifies a basic principle of counting often called the *product rule*. In general, suppose that constructing an arrangement of some sort requires making $m$ selections in sequence, and there are $n_i$ different choices at the $i$th step. Then the number of different possible arrangements is the product $n_1 n_2 \cdots n_m$.

Our second problem generalizes the first one.

**Problem 2.** How many ways are there to make an ordered list of $k$ objects from a collection of $n$ different objects?

For example, how many ways can a poll rank the top 20 teams in a college sport if there are 100 teams in the division? How many ways can a band arrange a play list of twelve songs if they know only 25 different songs?

Applying the same reasoning used in the first problem, we find that the answer to Problem 2 is the product $n(n - 1)(n - 2) \cdots (n - k + 1)$, or $n!/(n - k)!$. This number is sometimes denoted by $P(n, k)$, but products like this occur frequently in combinatorics, and a more descriptive notation is often used to designate them.

We define the *falling factorial power* $x^{\underline{k}}$ as a product of $k$ terms beginning with $x$, with each successive term one less than its predecessor:

$$x^{\underline{k}} = x(x - 1)(x - 2) \cdots (x - k + 1) = \prod_{i=0}^{k-1} (x - i). \qquad (1)$$

The expression $x^{\underline{k}}$ is pronounced "$x$ to the $k$ falling." Similarly, we define the *rising factorial power* $x^{\overline{k}}$ ("$x$ to the $k$ rising") by

$$x^{\overline{k}} = x(x + 1)(x + 2) \cdots (x + k - 1) = \prod_{i=0}^{k-1} (x + i). \qquad (2)$$

Thus, we see that $P(n, k) = n^{\underline{k}} = (n - k + 1)^{\overline{k}}$, and $n! = n^{\underline{n}} = 1^{\overline{n}}$. Also, the expressions $n^{\underline{0}}$, $n^{\overline{0}}$, and $0!$ all represent products having no terms at all. Multiplying any expression by such an empty product should not disturb the value of the expression, so the value of each of these degenerate products is taken to be 1.

Our third problem concerns unordered selections.

**Problem 3.** How many ways are there to select $k$ objects from a collection of $n$ objects, if the order of selection is irrelevant?

For example, how many different hands are possible in the game of poker? A poker hand consists of five cards drawn from a standard deck of 52 different cards. The order of the cards in a hand is unimportant, since players can rearrange their cards freely.

The solution to Problem 3 is usually denoted by $\binom{n}{k}$, or sometimes $C(n, k)$. The expression $\binom{n}{k}$ is pronounced "$n$ choose $k$."

We can find a formula for $\binom{n}{k}$ by using our solutions to Problems 1 and 2. Since there are $k!$ different ways to order a collection of $k$ objects, it follows that the product $\binom{n}{k}k!$ is the number of possible ordered lists of $k$ objects selected from the same collection of $n$ objects. Therefore,

$$\binom{n}{k} = \frac{n^{\underline{k}}}{k!} = \frac{n!}{k!(n-k)!}. \tag{3}$$

The numbers $\binom{n}{k}$ are called *binomial coefficients*, for reasons discussed in the next section. The binomial coefficients are ubiquitous in combinatorics, and we close this section with a few applications of these numbers.

1. The number of different hands in poker is $\binom{52}{5} = 52^{\underline{5}}/5! = 2\,598\,960$. The number of different thirteen-card hands in the game of bridge is $\binom{52}{13} = 635\,013\,559\,600$.

2. To play the Texas lottery, a gambler selects six different numbers between 1 and 50. The order of selection is unimportant. The number of possible lottery tickets is therefore $\binom{50}{6} = 15\,890\,700$.

3. Suppose we need to travel $m$ blocks east and $n$ blocks south in a regular grid of city streets. How many paths are there to our destination if we travel only east and south?

   We can represent a path to our destination as a sequence $b_1, b_2, \ldots, b_{n+m}$, where $b_i$ represents the direction we are traveling during the $i$th block of our route. Exactly $m$ of the terms in this sequence must be "east," and there are precisely $\binom{m+n}{m}$ ways to select $m$ positions in the sequence to have this value. The remaining $n$ positions in the sequence must all be "south," so the number of possible paths is $\binom{m+n}{m} = \frac{(m+n)!}{m!n!}$.

4. A standard deck of playing cards consists of four suits (spades, hearts, clubs, and diamonds), each with thirteen cards. Each of the cards in a suit has a different face value: a number between 2 and 10, or a jack, queen, king, or ace. How many poker hands have exactly three cards with the same face value?

   We can answer this question by considering how to construct such a hand through a sequence of simple steps. First, select one of the thirteen different face values. Second, choose three of the four cards in the deck having this value. Third, pick two cards from the 48 cards having a different face value. By the product rule, the number of possibilities is

$$\binom{13}{1}\binom{4}{3}\binom{48}{2} = 58656.$$

Poker aficionados will recognize that this strategy counts the number of ways to deal either of two different hands in the game: the "three of a kind" and the stronger "full house." A full house consists of a matched triple together with a matched pair, for example, three jacks and two aces; a three of a kind has only a matched triple. The number of ways to deal a full house is

$$\binom{13}{1}\binom{4}{3}\binom{12}{1}\binom{4}{2} = 3744,$$

since choosing a matched pair involves first selecting one of twelve different remaining face values, then picking two of the four cards having this value. The number of three of a kind hands is therefore $58656 - 3744 = 54912$.

We can also compute this number directly by modifying our first strategy. To avoid the possibility of selecting a matched pair in the last step, we can replace the term $\binom{48}{2} = 48 \cdot 47/2$ by $48 \cdot 44/2$, since the face value of the last card should not match any other card selected. Indeed, we calculate $13 \cdot 4 \cdot 48 \cdot 44/2 = 54912$. Notice that dividing by 2 is required in the last step, since the last two cards may be selected in any order.

## Exercises

1. In the C++ programming language, a variable name must start with a letter or the underscore character (_), and succeeding characters must be letters, digits, or the underscore character. Uppercase and lowercase letters are considered to be different. How many variable names having at most four letters can be formed in C++?

2. There are 29 teams in the National Basketball Association: 14 in the Western Conference, and 15 in the Eastern Conference.

   a. Suppose each of the teams in the league has one pick in the first round of the NBA draft. How many ways are there to arrange the order of the teams selecting in the draft?

   b. Suppose that each of the first three positions in the draft must be awarded to one of the thirteen teams that did not advance to the playoffs that year. How many ways are there to assign the first three positions in the draft?

   c. How many ways are there for eight teams from each conference to advance to the playoffs, if order is unimportant?

   d. Suppose that every team has three centers, four guards, and five forwards. How many ways are there to select an all-star team with the same composition from the Western Conference?

3. Compute the number of ways to deal each of the following five-card hands in poker.

   a. Straight: the values of the cards form a sequence of consecutive integers. A jack has value 11, a queen 12, and a king 13. An ace may have a value of 1 or 14, so A 2 3 4 5 and 10 J Q K A are both straights, but K A 2 3 4 is not. Furthermore, the cards in a straight cannot all be of the same suit (a flush).

   b. Flush: all five cards have the same suit (but not in addition a straight).

   c. Straight flush: both a straight and a flush. Make sure that your counts for straights and flushes do not include the straight flushes.

   d. Four of a kind.

**e.** Two distinct matching pairs.

**f.** Exactly one matching pair.

4. A superstitious resident of Amarillo, Texas, always picks three even numbers and three odd numbers when playing the lottery. What fraction of all possible lottery tickets have this property?

5. A ballot lists ten candidates for city council, eight candidates for the school board, and five bond issues. The ballot instructs voters to choose up to four people running for city council, rank up to three candidates for the school board, and approve or reject each bond issue. How many different ballots may be cast, if partially completed ballots are allowed?

6. Let $\Delta$ be the *difference operator*: $\Delta(f(x)) = f(x+1) - f(x)$. Show that

$$\Delta(x^{\underline{n}}) = nx^{\underline{n-1}},$$

and use this to prove that

$$\sum_{k=0}^{m-1} k^{\underline{n}} = \frac{m^{\underline{n+1}}}{n+1}.$$

## 2.2 Binomial Coefficients

*About binomial theorem I'm teeming with a lot o' news,*
*With many cheerful facts about the square of the hypotenuse.*
                                        — Gilbert and Sullivan, *The Pirates of Penzance*

The binomial coefficients possess a number of interesting arithmetic properties. In this section we study some of the most important identities associated with these numbers. Because binomial coefficients occur so frequently in this subject, knowing these essential identities will be helpful in our later studies.

The first identity generalizes our formula (3).

**Expansion.** *If n is a nonnegative integer and k is an integer, then*

$$\binom{n}{k} = \begin{cases} \dfrac{n!}{k!(n-k)!} & \text{if } 0 \leq k \leq n, \\ 0 & \text{otherwise.} \end{cases} \tag{4}$$

Designating the value of $\binom{n}{k}$ to be 0 when $k < 0$ or $k > n$ is sensible, for there are no ways to select fewer than zero or more than $n$ items from a collection of $n$ objects.

Notice that every subset of $k$ objects selected from a set of $n$ objects leaves a complementary collection of $n - k$ objects that are not selected. Counting the number of subsets with $k$ objects is therefore the same as counting the number of subsets with $n - k$ objects. This observation leads us to our second identity, which is easy to verify using the expansion formula.

**Symmetry.** *If n is a nonnegative integer and k is an integer, then*

$$\binom{n}{k} = \binom{n}{n-k}.$$ (5)

Before presenting the next identity, let us consider again the problem of counting poker hands. Suppose the ace of spades is the most desirable card in the deck (it certainly is in American Western movies), and we would like to know the number of five-card hands that include this card. The answer is the number of ways to select four cards from the other 51 cards in the deck, namely, $\binom{51}{4}$. We can also count the number of hands that do not include the ace of spades. This is the number of ways to pick five cards from the other 51, that is, $\binom{51}{5}$. But every poker hand either includes the ace of spades or does not, so

$$\binom{52}{5} = \binom{51}{5} + \binom{51}{4}.$$

More generally, suppose we distinguish one particular object in a collection of $n$ objects. The number of unordered collections of $k$ of the objects that include the distinguished object is $\binom{n-1}{k-1}$; the number of collections that do not include this special object is $\binom{n-1}{k}$. We therefore obtain the following identity.

**Addition.** *If n is a positive integer and k is any integer, then*

$$\binom{n}{k} = \binom{n-1}{k} + \binom{n-1}{k-1}.$$ (6)

We can prove this identity more formally using the expansion identity. It is easy to check that the identity holds for $k \leq 0$ or $k \geq n$. If $0 < k < n$, we have

$$
\begin{aligned}
\binom{n-1}{k} + \binom{n-1}{k-1} &= \frac{(n-1)!}{k!(n-1-k)!} + \frac{(n-1)!}{(k-1)!(n-k)!} \\
&= \frac{((n-k)+k)(n-1)!}{k!(n-k)!} \\
&= \frac{n!}{k!(n-k)!} \\
&= \binom{n}{k}.
\end{aligned}
$$

We can use this identity to create a table of binomial coefficients. Let $n \geq 0$ index the rows of the table, and let $k \geq 0$ index the columns. Begin by entering 1 in the first position of each row, since $\binom{n}{0} = 1$ for $n \geq 0$; then use (6) to compute the entries in successive rows of the table. The resulting pattern of numbers is called *Pascal's triangle*, after Blaise Pascal, who studied many of its properties in his *Traité du Triangle Arithmétique*, written in 1654 (see [39] for more information on its history). The first few rows of Pascal's triangle are shown in Table 2.1.

| $n$ | $k = 0$ | 1 | 2 | 3 | 4 | 5 | 6 | 7 | 8 | 9 | 10 | $2^n$ |
|-----|---------|---|---|---|---|---|---|---|---|---|----|-------|
| 0 | 1 | | | | | | | | | | | 1 |
| 1 | 1 | 1 | | | | | | | | | | 2 |
| 2 | 1 | 2 | 1 | | | | | | | | | 4 |
| 3 | 1 | 3 | 3 | 1 | | | | | | | | 8 |
| 4 | 1 | 4 | 6 | 4 | 1 | | | | | | | 16 |
| 5 | 1 | 5 | 10 | 10 | 5 | 1 | | | | | | 32 |
| 6 | 1 | 6 | 15 | 20 | 15 | 6 | 1 | | | | | 64 |
| 7 | 1 | 7 | 21 | 35 | 35 | 21 | 7 | 1 | | | | 128 |
| 8 | 1 | 8 | 28 | 56 | 70 | 56 | 28 | 8 | 1 | | | 256 |
| 9 | 1 | 9 | 36 | 84 | 126 | 126 | 84 | 36 | 9 | 1 | | 512 |
| 10 | 1 | 10 | 45 | 120 | 210 | 252 | 210 | 120 | 45 | 10 | 1 | 1024 |

TABLE 2.1. Pascal's triangle for binomial coefficients, $\binom{n}{k}$.

The next identity explains the origin of the name for these numbers: They appear as coefficients when expanding powers of the binomial expression $x + y$.

**The Binomial Theorem.** *If $n$ is a nonnegative integer, then*

$$(x + y)^n = \sum_{k} \binom{n}{k} x^k y^{n-k}. \tag{7}$$

The notation $\sum_{k}$ means that the sum is extended over all integers $k$. Thus, the right side of (7) is formally an infinite sum, but all terms with $k < 0$ or $k > n$ are zero by the expansion identity, so there are only $n + 1$ nonzero terms in this sum.

PROOF.    We prove this identity by induction on $n$. For $n = 0$, both sides evaluate to 1. Suppose then that the identity holds for a fixed nonnegative integer $n$. We need to verify that it holds for $n + 1$. Using our inductive hypothesis, then distributing the remaining factor of $(x + y)$, we obtain

$$(x + y)^{n+1} = (x + y) \sum_{k} \binom{n}{k} x^k y^{n-k}$$

$$= \sum_{k} \binom{n}{k} x^{k+1} y^{n-k} + \sum_{k} \binom{n}{k} x^k y^{n+1-k}.$$

Now we reindex the first sum, replacing each occurrence of $k$ by $k - 1$. Since the original sum extends over all values of $k$, the reindexed sum does, too. Thus

$$(x + y)^{n+1} = \sum_{k} \binom{n}{k - 1} x^k y^{n+1-k} + \sum_{k} \binom{n}{k} x^k y^{n+1-k}$$

$$= \sum_k \left( \binom{n}{k-1} + \binom{n}{k} \right) x^k y^{n+1-k}$$

$$= \sum_k \binom{n+1}{k} x^k y^{n+1-k},$$

by the addition identity. This completes the induction, and we conclude that the identity holds for all $n \geq 0$.    □

We note two important consequences of the binomial theorem. First, setting $x = y = 1$ in (7), we obtain

$$\sum_k \binom{n}{k} = 2^n. \tag{8}$$

Thus, summing across the $n$th row in Pascal's triangle yields $2^n$, and there are therefore exactly $2^n$ different subsets of a set of $n$ elements. These row sums are included in Table 2.1.

Second, setting $x = -1$ and $y = 1$ in (7), we find that the alternating sum across any row of Pascal's triangle is zero:

$$\sum_k (-1)^k \binom{n}{k} = 0. \tag{9}$$

This is obvious from the symmetry relation when $n$ is odd, but less clear when $n$ is even.

These two consequences of the binomial theorem concern sums over the lower index of binomial coefficients. The next identity tells the value of a sum over the upper index.

**Summing on the Upper Index.** *If $m$ and $n$ are nonnegative integers, then*

$$\sum_{k=0}^{n} \binom{k}{m} = \binom{n+1}{m+1}. \tag{10}$$

PROOF.    We use induction on $n$ to verify this identity. For $n = 0$, each side equals 1 if $m = 0$, and each side is 0 if $m > 0$. Suppose then that the identity holds for some fixed nonnegative integer $n$. We must show that it holds for the case $n + 1$. Let $m$ be a nonnegative integer. We obtain

$$\sum_{k=0}^{n+1} \binom{k}{m} = \binom{n+1}{m} + \sum_{k=0}^{n} \binom{k}{m}$$

$$= \binom{n+1}{m} + \binom{n+1}{m+1}$$

$$= \binom{n+2}{m+1}.$$

By induction, the identity holds for all $n \geq 0$.    □

To illustrate one last identity, we study the Texas lottery in more detail. Recall that a player selects six different numbers between 1 and 50 to enter the lottery. The largest prize is awarded to anyone matching all six numbers picked in a random drawing by lottery officials, but smaller prizes are given to players matching at least three of these numbers. To determine fair amounts for these smaller prizes, the state lottery commission needs to know the number of possible tickets that match exactly $k$ of the winning numbers, for every $k$.

Clearly, there is just one way to match all six winning numbers. There are $\binom{6}{5} = 6$ ways to pick five of the six winning numbers, and 44 ways to select one losing number, so there are $6 \cdot 44 = 264$ tickets that match five numbers. Selecting four of the winning numbers and two of the losing numbers makes $\binom{6}{4}\binom{44}{2} = 14190$ possible tickets, and in general we see that the number of tickets that match exactly $k$ of the winning numbers is $\binom{6}{k}\binom{44}{6-k}$. By summing over $k$, we count every possible ticket exactly once, so

$$\binom{50}{6} = \sum_k \binom{6}{k}\binom{44}{6-k}.$$

More generally, if a lottery game requires selecting $m$ numbers from a set of $m+n$ numbers, we obtain the identity

$$\binom{m+n}{m} = \sum_k \binom{m}{k}\binom{n}{m-k}.$$

That is, the number of possible tickets equals the sum over $k$ of the number of ways to match exactly $k$ of the $m$ winning numbers and $m-k$ of the $n$ losing numbers. More generally still, suppose a lottery game requires a player to select $\ell$ numbers on a ticket, and each drawing selects $m$ winning numbers. Using the same reasoning, we find that

$$\binom{m+n}{\ell} = \sum_k \binom{m}{k}\binom{n}{\ell-k}.$$

Now replace $\ell$ by $\ell + p$ and reindex the sum, replacing $k$ by $k + p$, to obtain the following identity.

**Vandermonde's Convolution.** *If $m$ and $n$ are nonnegative integers and $\ell$ and $p$ are integers, then*

$$\binom{m+n}{\ell+p} = \sum_k \binom{m}{p+k}\binom{n}{\ell-k}. \tag{11}$$

Notice that the lower indices in the binomial coefficients on the right side sum to a constant.

**Exercises**

1. Prove that there are exactly $2^n$ different subsets of a set of $n$ elements without using the binomial theorem.

2. Prove the absorption/extraction identity: If $n$ is a positive integer and $k$ is a nonzero integer, then

$$\binom{n}{k} = \frac{n}{k}\binom{n-1}{k-1}. \tag{12}$$

3. Prove the cancellation identity: If $n$ and $k$ are nonnegative integers and $m$ is an integer with $m \le n$, then

$$\binom{n}{k}\binom{k}{m} = \binom{n}{m}\binom{n-m}{k-m}. \tag{13}$$

This identity is very useful when the left side appears in a sum over $k$, for the right side has only a single occurrence of $k$.

4. Prove the parallel summation identity: If $m$ and $n$ are nonnegative integers, then

$$\sum_{k=0}^{n}\binom{m+k}{k} = \binom{m+n+1}{n}. \tag{14}$$

5. Prove the hexagon identity:

$$\binom{n-1}{k-1}\binom{n}{k+1}\binom{n+1}{k} = \binom{n-1}{k}\binom{n}{k-1}\binom{n+1}{k+1}. \tag{15}$$

Why is it called the hexagon identity?

6. Compute the value of the following sums. Your answer should be an expression involving one or two binomial coefficients.

   a. $\displaystyle\sum_{k}\binom{80}{k}\binom{k+1}{31}.$

   b. $\displaystyle\sum_{k\ge 0}\frac{1}{k+1}\binom{99}{k}\binom{200}{120-k}.$

7. Prove the binomial theorem for falling factorial powers,

$$(x+y)^{\underline{n}} = \sum_{k}\binom{n}{k}x^{\underline{k}}y^{\underline{n-k}},$$

and for rising factorial powers,

$$(x+y)^{\overline{n}} = \sum_{k}\binom{n}{k}x^{\overline{k}}y^{\overline{n-k}}.$$

8. In the Texas lottery, a gambler picks six numbers between 1 and 50; in the California lottery, a player chooses six numbers between 1 and 51. In both states, an entry wins something if at least three of the six numbers on the ticket are among the six numbers drawn. In which state is a player more likely to win something?

## 2.3    The Principle of Inclusion and Exclusion

*Get yourselves organized up there!*

— Wallace, in *A Close Shave*

Suppose there are 50 beads in a drawer: 25 are glass, 30 are red, 20 are spherical, 18 are red glass, 12 are glass spheres, 15 are red spheres, and 8 are red glass spheres. How many beads are neither red, nor glass, nor spheres?

We can answer this question by organizing all of this information using a Venn diagram with three overlapping sets: $G$ for glass beads, $R$ for red beads, and $S$ for spherical beads. See Figure 2.1. We are given that there are eight red glass spheres, so start by labeling the common intersection of the sets $G$, $R$, and $S$ in the diagram with 8. Then the region just above this one must have ten elements, since there are 18 red glass beads, and exactly eight of these are spherical. Continuing in this way, we determine the size of each of the sets represented in the diagram, and we conclude that there are exactly twelve beads in the drawer that are neither red, nor glass, nor spheres.

Alternatively, we can answer this question by determining the size of the set $G \cup R \cup S$ (does this make us counting GURUS?). Summing the number of elements in the sets $G$, $R$, and $S$ produces a number that is too large, since this sum counts the beads that are in more than one of these sets at least twice. We can try to compensate by subtracting the number of elements in the sets $G \cap R$, $G \cap S$, and $R \cap S$ from the sum. This produces a total that is too small, since the beads that have all three attributes are counted three times in the first step, then subtracted three times in the second step. Thus, we must add the number of elements in $G \cap R \cap S$ to the sum, and we find that

$$|G \cup R \cup S| = |G| + |R| + |S| - |G \cap R| - |G \cap S| - |R \cap S| + |G \cap R \cap S|.$$

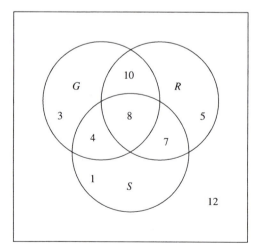

FIGURE 2.1. A solution using a Venn diagram.

Letting $N_0$ denote the number of beads with none of the three attributes, we then compute

$$N_0 = 50 - |G \cup R \cup S|$$
$$= 50 - |G| - |R| - |S| + |G \cap R| + |G \cap S| + |R \cap S| - |G \cap R \cap S|$$
$$= 50 - 25 - 30 - 20 + 18 + 12 + 15 - 8$$
$$= 12.$$

This suggests a general technique for solving some similar combinatorial problems. Suppose we have a collection of $N$ distinct objects, and each object may satisfy one or more properties that we label $a_1, a_2, \ldots, a_r$. Let $N(a_i)$ denote the number of objects having property $a_i$, $N(a_i a_j)$ the number having both property $a_i$ and property $a_j$, and in general let $N(a_{i_1} a_{i_2} \ldots a_{i_m})$ represent the number satisfying the $m$ properties $a_{i_1}, \ldots, a_{i_m}$. Let $N_0$ denote the number of objects having none of the properties. We prove the following theorem.

**Theorem 2.1** (Principle of Inclusion and Exclusion). *Using the notation above,*

$$N_0 = N - \sum_i N(a_i) + \sum_{i<j} N(a_i a_j) - \sum_{i<j<k} N(a_i a_j a_k) + \cdots$$
$$+ (-1)^m \sum_{i_1 < \cdots < i_m} N(a_{i_1} \ldots a_{i_m}) + \cdots + (-1)^r N(a_1 a_2 \ldots a_r).$$
$$(16)$$

PROOF.    Suppose an object satisfies none of the properties. Then the expression on the right side counts it precisely once, in the $N$ term. On the other hand, suppose an object satisfies precisely $m$ of the properties, with $m$ a positive number. Then it is counted once in the $N$ term, $m$ times in the $\sum N(a_i)$ term, $\binom{m}{2}$ times in the second sum, and in general $\binom{m}{k}$ times in the $k$th sum. Therefore, the total contribution on the right side from this object is

$$\sum_k (-1)^k \binom{m}{k} = 0$$

by (9). This completes the proof.    □

### The Euler $\varphi$ Function

Two integers are said to be *relatively prime* if their greatest common divisor is 1. If $n$ is a positive integer, let $\varphi(n)$ be the number of positive integers $m \leq n$ that are relatively prime to $n$. This function, called the Euler $\varphi$ function or the Euler totient function, is important in number theory. We can derive a formula for this function using the principle of inclusion and exclusion.

We must name a set and list a number of properties such that the number of elements in the set satisfying none of the properties is $\varphi(n)$. Suppose $n$ is divisible by precisely $r$ different primes, which we label $p_1$ through $p_r$. Select $\{1, 2, \ldots, n\}$ as the set, and let $a_i$ be the property "is divisible by $p_i$." Then $N_0 = \varphi(n)$, and it is easy to compute the terms on the right side of the equation in Theorem 2.1:

$N = n$, $N(a_i) = n/p_i$, $N(a_i a_j) = n/(p_i p_j)$, and so on. Therefore,

$$\varphi(n) = n - \sum_i \frac{n}{p_i} + \sum_{i<j} \frac{n}{p_i p_j} - \sum_{i<j<k} \frac{n}{p_i p_j p_k} + \cdots + (-1)^r \frac{n}{p_1 p_2 \cdots p_r}$$

$$= n \prod_{i=1}^{r} \left(1 - \frac{1}{p_i}\right).$$

Exercise 1 asks you to verify the last step.

For example, the primes dividing 24 are 2 and 3, so $\varphi(24) = 24(1-\frac{1}{2})(1-\frac{1}{3}) = 8$. The eight numbers between 1 and 24 that are relatively prime to 24 are 1, 5, 7, 11, 13, 17, 19, and 23.

## Counting Prime Numbers

Suppose $m$ is a composite positive integer, so $m$ can be written as a product of two integers that are both greater than 1: $m = ab$ with $1 < a \leq b$. Then $a^2 \leq m$, so $a \leq \sqrt{m}$, and so $m$ must be divisible by a prime number $p$ with $p \leq \sqrt{m}$.

We can use this observation, together with Theorem 2.1, to count the prime numbers between 1 and a given positive integer $n$. We start with the set $\{1, 2, \ldots, n\}$ and use the theorem to count the number of elements that remain when multiples of prime numbers $p \leq \sqrt{n}$ are excluded from the set. Since every composite number $m \leq n$ has a prime factor $p \leq \sqrt{m}$, excluding all of these numbers removes all the composite numbers from the set.

For example, for $n = 120$, the largest prime less than or equal to $\sqrt{n}$ is the fourth prime number, 7, so we require just four properties in Theorem 2.1 to exclude all the composite numbers in the set $\{1, 2, \ldots, 120\}$. The four properties are $a_1 =$ "is even," $a_2 =$ "is divisible by 3," $a_3 =$ "is divisible by 5," and $a_4 =$ "is divisible by 7." We compute $N(a_1) = 120/2 = 60$, $N(a_2) = 120/3 = 40$, $N(a_3) = 120/5 = 24$, and $N(a_4) = \lfloor 120/7 \rfloor = 24$. Here, $\lfloor x \rfloor$ denotes the *floor* of $x$, or *integer part* of $x$. It is defined to be the largest integer $m$ satisfying $m \leq x$. Similarly, the *ceiling* of $x$, denoted by $\lceil x \rceil$, is the smallest integer $m$ satisfying $x \leq m$.

Continuing our calculation, we compute $N(a_1 a_2) = \lfloor 120/6 \rfloor = 20$, $N(a_1 a_3) = \lfloor 120/10 \rfloor = 12$, etc., and find that $N_0 = 120 - (60 + 40 + 24 + 17) + (20 + 12 + 8 + 8 + 5 + 3) - (4 + 2 + 1 + 1) + 0 = 27$. But this is not the number of prime numbers between 1 and 120, for our method excludes the primes 2, 3, 5, and 7, and includes the nonprime 1. Accounting for these exceptions, we find that the number of primes between 1 and 120 is $N_0 + 4 - 1 = 30$.

## Chromatic Polynomials

Let $G$ be a graph. Recall that its chromatic polynomial $c_G(x)$ measures the number of ways to color the vertices of $G$ using at most $x$ colors in such a way that no two vertices connected by an edge have the same color. We can use Theorem 2.1 to compute chromatic polynomials.

Suppose $G$ has $n$ vertices, and consider the set of colorings of the vertices of $G$ using at most $x$ colors, so the number of colorings in this set is $N = x^n$. To find

$c_G(x)$, we must exclude all of the inadmissible colorings from this set. For each edge $e_i$ in the graph, select property $a_i$ to be "edge $e_i$ connects two vertices that have the same color." In this way, the colorings in the set that satisfy none of the properties are precisely the admissible colorings, so $N_0 = c_G(x)$.

For example, we compute the chromatic polynomial for the complete graph $K_3$ using this strategy. This graph has three edges, so we take $r = 3$ in the theorem. We compute $N(a_1) = N(a_2) = N(a_3) = x^2$, since every coloring satisfying one of the properties has two vertices with the same color, and the third vertex may be any color. Also, $N(a_1 a_2) = N(a_2 a_3) = N(a_1 a_3) = N(a_1 a_2 a_3) = x$, as every coloring satisfying more than one property must have all vertices colored identically. Thus, $c_{K_3}(x) = N_0 = x^3 - 3x^2 + 3x - x = x(x-1)(x-2) = x^{\underline{3}}$.

## Derangements

Suppose a lazy professor gives a quiz to a class of $n$ students, then collects the papers, shuffles them, and redistributes them randomly to the class for grading. The professor would prefer that no student receives his or her own paper to grade. What is the probability that this occurs? Is this probability substantially different for different class sizes? What do you think the limiting probability is as $n \to \infty$? Notice that as $n$ grows larger, there are more ways for at least one person to receive his or her own quiz back, but perhaps this increase is swamped by the growth of the total number of permutations possible.

Suppose we have $n$ objects in an initial configuration. A permutation of these objects in which the position of each object differs from its initial position is called a *derangement* of the objects. Since $n!$ denotes the number of permutations of $n$ objects, following [69] we denote the number of derangements of $n$ objects by $n_i$ (and since $n!$ is often pronounced "$n$ bang," perhaps $n_i$ should be pronounced "$n$ gnab").

We compute $n_i$ for some small values of $n$. For $n = 0$, there is just one permutation, and it vacuously satisfies the derangement condition, so $0_i = 1$. There is only one permutation of a single object, and it is not a derangement, so $1_i = 0$. Only one of the two permutations of two objects is a derangement, so $2_i = 1$, and exactly two of the six permutations of three objects satisfies the condition: If our original arrangement is $[1, 2, 3]$, then the derangements are $[2, 1, 3]$ and $[3, 1, 2]$. Thus $3_i = 2$. We find that $4_i = 9$: The derangements of $[1, 2, 3, 4]$ are $[2, 1, 4, 3]$, $[2, 3, 4, 1]$, $[2, 4, 1, 3]$, $[3, 1, 4, 2]$, $[3, 4, 1, 2]$, $[3, 4, 2, 1]$, $[4, 1, 3, 2]$, $[4, 3, 1, 2]$, and $[4, 3, 2, 1]$. Thus, the probability that a random permutation of a fixed number $n$ of objects is a derangement is respectively $1, 0, \frac{1}{2}, \frac{1}{3}$, and $\frac{3}{8}$ for $n = 0$ through 4.

We can use Theorem 2.1 to determine a formula for $n_i$. We select the original set to be the collection of all permutations of $n$ objects, and for $1 \le i \le n$ let $a_i$ denote the property that element $i$ remains in its original position in a permutation. Then $N_0$ is the number of permutations where no elements remain in their original position, so $N_0 = n_i$.

To compute $N(a_i)$, we see that element $i$ is fixed, but the other $n - 1$ elements may be arranged arbitrarily, so $N(a_i) = (n - 1)!$. Similarly, $N(a_i a_j) = (n - 2)!$

for $i < j$, $N(a_i a_j a_k) = (n - 3)!$ for $i < j < k$, and so on. Therefore,

$$n_i = n! - \sum_i (n - 1)! + \sum_{i<j} (n - 2)! - \cdots$$

$$+ (-1)^m \sum_{i_1 < \cdots < i_m} (n - m)! + \cdots + (-1)^n$$

$$= n! - \binom{n}{1}(n - 1)! + \binom{n}{2}(n - 2)! - \cdots$$

$$+ (-1)^m \binom{n}{m}(n - m)! + \cdots + (-1)^n$$

$$= \sum_m (-1)^m \binom{n}{m}(n - m)!$$

$$= n! \sum_{m=0}^{n} \frac{(-1)^m}{m!}.$$

Thus, the probability that a permutation of $n$ objects is a derangement is

$$\frac{n_i}{n!} = \sum_{m=0}^{n} \frac{(-1)^m}{m!},$$

and in the limit,

$$\lim_{n \to \infty} \frac{n_i}{n!} = \sum_{m \geq 0} \frac{(-1)^m}{m!} = e^{-1}. \tag{17}$$

Our lazy professor obtains a desired configuration about 36.8% of the time, for sizable classes.

### Exercises

1. Let $\alpha_1, \alpha_2, \ldots, \alpha_r$ be real numbers. Show that

$$\prod_{i=1}^{r} (1 - \alpha_i) = 1 - \sum_i \alpha_i + \sum_{i<j} \alpha_i \alpha_j - \sum_{i<j<k} \alpha_i \alpha_j \alpha_k + \cdots$$

$$+ (-1)^r \alpha_1 \alpha_2 \cdots \alpha_r.$$

2. **a.** Show that $\varphi(mn) = \varphi(m)\varphi(n)$ if $m$ and $n$ are relatively prime.
   **b.** Determine all integers $n$ satisfying $\varphi(n) = 12$, 13, or 14.
3. Use Theorem 2.1 to count the number of prime numbers less than 168.
4. Use Theorem 2.1 to determine the chromatic polynomial for each of the following graphs.

   **a.** The yield sign (add a single edge to the bipartite graph $K_{1,3}$).
   **b.** The bipartite graph $K_{2,3}$.

5. Is the probability that a permutation of $n$ objects is a derangement substantially different for $n = 12$ and $n = 120$? Quantify your answer.

**6.** Suppose our lazy professor collects a quiz and a homework assignment from a class of $n$ students one day, then distributes both the quizzes and the homework assignments back to the class in a random fashion for grading. Each student receives one quiz and one homework assignment to grade.

    **a.** What is the probability that every student receives someone else's quiz to grade, and someone else's homework to grade?

    **b.** What is the probability that no student receives both their own quiz and their own homework assignment to grade? In this case, some students may receive their own quiz, and others may receive their own homework assignment.

    **c.** Compute the limiting probability as $n \to \infty$ in each case.

**7.** Let $N_m$ denote the number of objects from a collection of $N$ objects that possess exactly $m$ of the properties $a_1, a_2, \ldots, a_r$. Generalize the principle of inclusion and exclusion by showing that

$$N_m = \sum_{k=m}^{r} (-1)^{k-m} \binom{k}{m} s_k, \tag{18}$$

where

$$s_k = \sum_{i_1 < \cdots < i_k} N(a_{i_1} \ldots a_{i_k}). \tag{19}$$

## 2.4    Generating Functions

> *And own no other function: each your doing,*
> *So singular in each particular,*
> *Crowns what you are doing in the present deed,*
> *That all your acts are queens.*
>     — William Shakespeare, *The Winter's Tale*, Act IV, Scene IV

Given a sequence $\{a_k\}$ with $k \geq 0$, the *generating function* for the sequence is defined as $G(x) = \sum_{k \geq 0} a_k x^k$. So $G(x)$ is a polynomial if $\{a_k\}$ is a finite sequence, and a power series if $\{a_k\}$ is infinite. For example, if $a_k = (-1)^k/k!$, then $G(x)$ is the Maclaurin series for $e^{-x}$, and for fixed $n$, if $a_k = \binom{n}{k}$, then $G(x) = (1+x)^n$, by the binomial theorem.

To illustrate how generating functions can be used to solve combinatorial problems, let us consider again the problem of determining the number of $k$-element subsets of an $n$-element set. Fix $n$, and let $a_k$ denote this number. Of course, we showed in Section 2.1 that $a_k = n^{\underline{k}}/k!$; here we derive this formula again using generating functions.

Suppose we wish to enumerate all subsets of the $n$-element set. To construct one subset, we must pick which elements to include in the subset and which to exclude. Let us denote the choice to omit an element by $x^0$, and the choice to include it by $x^1$. Using "+" to represent "or," the choice to include or exclude one element then is denoted by $x^0 + x^1$. We must make $n$ such choices to construct a subset, so using multiplication to denote "and," the expression $(x^0 + x^1)^n$ models the choices required to make a subset.

Since "and" distributes over "or" just as multiplication distributes over addition, we may expand this expression using standard rules of arithmetic to obtain representations for all $2^n$ subsets. For example, when $n = 3$, we obtain

$$\left(x^0 + x^1\right)^3 = x^0 x^0 x^0 + x^0 x^0 x^1 + x^0 x^1 x^0 + x^0 x^1 x^1$$
$$+ x^1 x^0 x^0 + x^1 x^0 x^1 + x^1 x^1 x^0 + x^1 x^1 x^1.$$

The first term represents the empty subset, the second the subset containing just the third item in the original set, etc. Writing 1 for $x^0$ and $x$ for $x^1$ and treating the expression as a polynomial, we find that $(1 + x)^3 = 1 + 3x + 3x^2 + x^3$, and the coefficient of $x^k$ is the number of subsets of a 3-element set having exactly $k$ items.

In general, we find that the generating function for the sequence $\{a_k\}$ is $(1+x)^n$, so $a_k = \binom{n}{k} = n^{\underline{k}}/k!$, by the binomial theorem. Since our proof of the binomial theorem relies only on basic facts of arithmetic, this argument gives an independent derivation for the number of $k$-element subsets of a set with $n$ elements.

This example illustrates the general strategy for using generating functions to solve combinatorial problems. First, express the problem in terms of determining one or more values of an unknown sequence $\{a_k\}$. Second, determine a generating function for this sequence, writing the monomial $x^k$ to represent selecting an object $k$ times, then using addition to represent alternative choices and multiplication to represent sequential choices. Third, use analytic methods to expand the generating function and determine the values of the encoded sequence.

In this section we show the power of this method by studying several combinatorial problems.

## 2.4.1 Double Decks

*I don't like the games you play, professor.*
— Roger Thornhill, in *North by Northwest*

How many five-card poker hands can be dealt from a double deck? Assume that the two decks are identical. More generally, how many ways are there to select $m$ items from $n$ different items, where each item can be selected at most twice? Let us denote this number by $t_{n,m}$, and let $G_n(x)$ be the generating function for the sequence $\{t_{n,m}\}$ for $m \geq 0$ and $n$ fixed.

We find that $G_n(x) = \left(1 + x + x^2\right)^n$, since each object may be selected zero times, one time, or two times. To find $t_{n,m}$, we must determine a formula for the coefficient of $x^m$ in $G_n(x)$. This is simply a matter of applying the binomial theorem twice:

$$G_n(x) = \left(1 + (x + x^2)\right)^n$$

$$= \sum_k \binom{n}{k} (x + x^2)^k$$

$$= \sum_k \binom{n}{k} x^k \sum_j \binom{k}{j} x^j$$

$$= \sum_k \sum_j \binom{n}{k}\binom{k}{j} x^{j+k}$$

$$= \sum_m \left(\sum_j \binom{n}{m-j}\binom{m-j}{j}\right) x^m,$$

where we obtained the last line by substituting $m$ for $j + k$. Therefore,

$$t_{n,m} = \sum_{j=0}^{\lfloor m/2 \rfloor} \binom{n}{m-j}\binom{m-j}{j}. \tag{20}$$

Thus, the number of five-card poker hands that can be dealt from a double deck is

$$\binom{52}{5}\binom{5}{0} + \binom{52}{4}\binom{4}{1} + \binom{52}{3}\binom{3}{2} = 3\,748\,160.$$

There is a simple combinatorial explanation for this expression. A five-card hand dealt from a double deck may have zero, one, or two cards repeated. There are $\binom{52}{5}$ hands with no cards repeated, $\binom{52}{4}\binom{4}{1}$ hands with exactly one card repeated, and $\binom{52}{3}\binom{3}{2}$ hands with exactly two cards repeated. A similar explanation applies for the general formula (20).

**Exercises**

**1.** How many five-card poker hands can be dealt from a triple deck?
**2.** How many five-card poker hands can be dealt from a quadruple deck?

### 2.4.2  Counting with Repetition

*Then, shalt thou count to three, no more, no less. Three shalt be the number thou shalt count, and the number of the counting shall be three. Four shalt thou not count, nor either count thou two, excepting that thou then proceed to three. Five is right out.*
*— Monty Python and the Holy Grail*

Suppose there is an inexhaustible supply of each of $n$ different objects. How many ways are there to select $m$ objects from the $n$ different objects, if you are allowed to select each object as many times as you like?

Let $a_{n,m}$ denote this number. Evidently, for fixed $n$, the generating function for $\{a_{n,m}\}_{m \geq 0}$ is

$$G_n(x) = \left(1 + x + x^2 + \cdots\right)^n = \left(\frac{1}{1-x}\right)^n,$$

since the sum is just a geometric series in $x$. This raises questions on convergence, for this formula is valid only for $|x| < 1$. We largely ignore these analytic issues, since we treat generating functions as formal series.

Thus, to find a formula for $a_{n,m}$, we must find the coefficient of $x^m$ in $G_n(x)$.

Let us consider a more general problem. Let $f(x) = (1 + x)^\alpha$, where $\alpha$ is a real number. Then $f'(0) = \alpha$, $f''(0) = \alpha(\alpha - 1)$, and in general, $f^{(k)}(0) = \alpha^{\underline{k}}$. Therefore, the Maclaurin series for $f(x)$ is

$$(1 + x)^\alpha = \sum_{k \geq 0} \frac{\alpha^{\underline{k}}}{k!} x^k.$$

Define the *generalized binomial coefficient* by

$$\binom{\alpha}{k} = \begin{cases} \alpha^{\underline{k}}/k! & \text{if } k \geq 0, \\ 0 & \text{if } k < 0. \end{cases} \tag{21}$$

Note that $\binom{\alpha}{k}$ equals the ordinary binomial coefficient whenever $\alpha$ is a nonnegative integer. We have the following theorem.

**Theorem 2.2** (Generalized Binomial Theorem). *If $|x| < 1$ or $\alpha$ is a nonnegative integer, then*

$$(1 + x)^\alpha = \sum_k \binom{\alpha}{k} x^k. \tag{22}$$

The proof of convergence may be found in many analysis texts, where it is often proved as a consequence of Bernstein's theorem on convergence of Taylor series (see for instance [5]). We do not supply the proof here.

Before solving our problem concerning selection with unlimited repetition, we note a useful identity for generalized binomial coefficients.

**Negating the Upper Index.** *If $\alpha$ is a real number and $k$ is an integer, then*

$$\binom{\alpha}{k} = (-1)^k \binom{k - \alpha - 1}{k}. \tag{23}$$

PROOF.  For $k < 0$, the identity is clear. For $k \geq 0$, we have

$$\binom{\alpha}{k} = \frac{1}{k!} \prod_{i=0}^{k-1} (\alpha - i).$$

We reindex this sum, replacing each $i$ by $k - 1 - i$, to obtain

$$\binom{\alpha}{k} = \frac{1}{k!} \prod_{i=0}^{k-1} (\alpha - (k - i - 1))$$

$$= \frac{(-1)^k}{k!} \prod_{i=0}^{k-1} (k - 1 - i - \alpha)$$

$$= (-1)^k \binom{k - \alpha - 1}{k}. \qquad \square$$

We may now solve our problem of determining $a_{n,m}$:

$$G_n(x) = (1 - x)^{-n}$$

$$= \sum_m \binom{-n}{m} (-x)^m$$

$$= \sum_m \binom{n + m - 1}{m} x^m.$$

Therefore, $a_{n,m} = \binom{n+m-1}{m}$.

For example, the number of five-card poker hands that can be dealt from a stack of five or more decks is $\binom{56}{5} = 3\,819\,816$.

Suppose we lay all 52 cards of a standard deck face up on a table. How many ways can we place five identical poker chips on the cards if we allow more than one chip to be placed on each card? To solve this, notice that each possible placement of chips corresponds to a hand of five cards, where repeated cards are allowed: If $k$ chips lie on a particular card, place that card into the hand $k$ times. Further, every such five-card hand can be represented by a judicious placement of chips. Therefore, the answer is the same as that of the previous example, $\binom{56}{5}$.

In general, the number of ways to place $m$ identical objects into $n$ distinguishable bins is the same as the number of ways to select $m$ objects from a set of $n$ objects with repetition allowed: The answer to both problems is $\binom{n+m-1}{m}$.

## Exercises

1. Prove the addition identity for generalized binomial coefficients: If $\alpha$ is a real number and $k$ is an integer, then

$$\binom{\alpha}{k} = \binom{\alpha - 1}{k} + \binom{\alpha - 1}{k - 1}.$$

2. Prove the absorption/extraction identity for generalized binomial coefficients: If $\alpha$ is a real number and $k$ is a nonzero integer, then

$$\binom{\alpha}{k} = \frac{\alpha}{k} \binom{\alpha - 1}{k - 1}.$$

3. Prove the cancellation identity for generalized binomial coefficients: If $\alpha$ is a real number and $k$ and $m$ are integers, then

$$\binom{\alpha}{k}\binom{k}{m} = \binom{\alpha}{m}\binom{\alpha - m}{k - m}.$$

4. Prove the parallel summation identity for generalized binomial coefficients: If $\alpha$ is a real number and $n$ is an integer, then

$$\sum_{k=0}^{n}\binom{\alpha + k}{k} = \binom{\alpha + n + 1}{n}.$$

5. Suppose that an unlimited number of jelly beans is available in each of five different colors: red, green, yellow, white, and black.

   a. How many ways are there to select twenty jelly beans?
   b. How many ways are there to select twenty jelly beans if we must select at least two jelly beans of each color?

6. A catering company brings fifty identical hamburgers to a party with twenty guests.

   a. How many ways can the hamburgers be divided among the guests?
   b. How many ways can the hamburgers be divided among the guests if every guest receives at least one hamburger?

## 2.4.3 Changing Money

*Jesus went into the temple, and began to cast out them that sold and bought in the temple, and overthrew the tables of the moneychangers ...*
                                                                     *— Mark 11:15*

We now turn to a problem popularized by the analyst and combinatorialist George Pólya [117]: How many ways are there to change a dollar? That is, how many combinations of pennies, nickels, dimes, quarters, half-dollars, and Susan B. Anthony dollars total $1? Our discussion of this problem follows the treatment of Graham, Knuth, and Patashnik [69].

Let us define $a_k$ to be the number of ways to make $k$ cents in change, and let $A(x)$ be a generating function for $a_k$: $A(x) = \sum_k a_k x^k$. Before analyzing this problem, pause a moment and make a guess. Do you think $a_{50}$ is more than 50 or less than 50? Is $a_{100}$ more than 100 or less than 100? How fast do you think $a_k$ grows as a function of $k$? Is it a polynomial in $k$? Exponential in $k$? Perhaps something between these?

To create a pile of change, we must make six choices, selecting a number of pennies, then nickels, then dimes, quarters, half-dollars, and dollars. We can model our choice of pennies by the sum

$$1 + x + x^2 + x^3 + \cdots = \frac{1}{1 - x}.$$

One might be tempted to use the same expression to model the different choices for each of the other coins, since we can pick any number of nickels, and any number of dimes, etc., but this would be incorrect. This would yield a generating function for the number of ways to select $k$ coins from a set of six different coins, not the number of ways to form $k$ cents. Instead, when choosing nickels, we select either zero cents, or five cents, or ten cents, and so on, so this selection is modeled as

$$1 + x^5 + x^{10} + x^{15} + \cdots = \frac{1}{1 - x^5}.$$

Therefore, the number of ways to make $k$ cents using either pennies or nickels is given by the generating function

$$\frac{1}{(1 - x)(1 - x^5)}.$$

Continuing in this way, we find that

$$A(x) = \frac{1}{(1 - x)(1 - x^5)(1 - x^{10})(1 - x^{25})(1 - x^{50})(1 - x^{100})}, \qquad (24)$$

so we merely need to find the coefficient of $a_k$ in the Maclaurin series for $A(x)$! This sounds rather daunting, so let us determine a few values of $a_k$ by hand first.

Let $\{p_k\}$ denote the number of ways to make $k$ cents using only pennies, so $p_k = 1$ for all $k$. Let $P(x)$ be the generating function for $\{p_k\}$, so $P(x) = 1/(1-x)$. Let $n_k$ be the number of ways to make $k$ cents using either pennies or nickels, so its generating function is

$$N(x) = \frac{P(x)}{1 - x^5}.$$

Thus $N(x) = P(x) + x^5 N(x)$, and equating coefficients we find that

$$n_k = \begin{cases} p_k & \text{if } 0 \le k \le 4, \\ p_k + n_{k-5} & \text{if } k \ge 5. \end{cases}$$

In the same way, let $d_k$ denote the number of ways to make $k$ cents using pennies, nickels, or dimes, and let $D(x)$ be its generating function. We then have $D(x) = N(x) + x^{10} D(x)$, and so

$$d_k = \begin{cases} n_k & \text{if } 0 \le k \le 9, \\ n_k + d_{k-10} & \text{if } k \ge 10. \end{cases}$$

There is a simple combinatorial interpretation for this equation. If $k < 10$, then we can choose only nickels and pennies to form $k$ cents, so $d_k = n_k$ in this case. If $k \ge 10$, we may form $k$ cents using only nickels and pennies, or we can choose one dime, then form the remaining $k - 10$ cents using dimes, nickels, and pennies. Thus $d_k = n_k + d_{k-10}$ in this case.

| k | 0 | 5 | 10 | 15 | 20 | 25 | 30 | 35 | 40 | 45 | 50 |
|---|---|---|----|----|----|----|----|----|----|----|----|
| $p_k$ | 1 | 1 | 1 | 1 | 1 | 1 | 1 | 1 | 1 | 1 | 1 |
| $n_k$ | 1 | 2 | 3 | 4 | 5 | 6 | 7 | 8 | 9 | 10 | 11 |
| $d_k$ | 1 | 2 | 4 | 6 | 9 | 12 | 16 | 20 | 25 | 30 | 36 |
| $q_k$ | 1 | | | | | 13 | | | | | 49 |
| $h_k$ | 1 | | | | | | | | | | 50 |
| $a_k$ | 1 | | | | | | | | | | |

| k | 55 | 60 | 65 | 70 | 75 | 80 | 85 | 90 | 95 | 100 |
|---|----|----|----|----|----|----|----|----|----|-----|
| $p_k$ | 1 | 1 | 1 | 1 | 1 | 1 | 1 | 1 | 1 | 1 |
| $n_k$ | 12 | 13 | 14 | 15 | 16 | 17 | 18 | 19 | 20 | 21 |
| $d_k$ | 42 | 49 | 56 | 64 | 72 | 81 | | 100 | | 121 |
| $q_k$ | | | | | 121 | | | | | 242 |
| $h_k$ | | | | | | | | | | 292 |
| $a_k$ | | | | | | | | | | 293 |

TABLE 2.2. Computing the number of ways to make $k$ cents in change.

Similarly, using $q_k$ for allowing quarters, $h_k$ for half dollars, and finally $a_k$ for Susan B. Anthony dollars, we have

$$q_k = \begin{cases} d_k & \text{if } 0 \le k \le 24, \\ d_k + q_{k-25} & \text{if } k \ge 25; \end{cases}$$

$$h_k = \begin{cases} q_k & \text{if } 0 \le k \le 49, \\ q_k + h_{k-50} & \text{if } k \ge 50; \end{cases}$$

$$a_k = \begin{cases} h_k & \text{if } 0 \le k \le 99, \\ h_k + a_{k-100} & \text{if } k \ge 100. \end{cases}$$

We may use these formulas to construct Table 2.2 below showing the number of ways to make $k$ cents with the different coin sets.

We find that there are precisely 50 ways to make 50 cents in change, and 293 ways to make one dollar in change.

This is a fairly efficient method to determine $a_k$, since apparently we can calculate this number using at most $5k$ arithmetic operations. But we can do much better! We can compute $a_k$ using at most a constant number of arithmetic operations, regardless of the value of $k$. To show this, let us first simplify $A(x)$ by exploiting the fact that all but one of the exponents in (24) is a multiple of 5. Let

$$B(x) = \frac{1}{(1-x)^2(1-x^2)(1-x^5)(1-x^{10})(1-x^{20})},$$

so that

$$A(x) = (1 + x + x^2 + x^3 + x^4)B(x^5).$$

Writing $b_k$ for the coefficient of $x^k$ in the Maclaurin series for $B(x)$ and equating coefficients, we find that

$$b_k = a_{5k} = a_{5k+1} = a_{5k+2} = a_{5k+3} = a_{5k+4}.$$

But this makes sense (or perhaps cents?), since the last few cents can be represented only using pennies. Now

$$B(x) = \frac{C(x)}{\left(1 - x^{20}\right)^6},$$

where

$$
\begin{aligned}
C(x) = {} & \left(1 + x + \cdots + x^{19}\right)^2 \left(1 + x^2 + \cdots + x^{18}\right)\left(1 + x^5 + x^{10} + x^{15}\right)\left(1 + x^{10}\right) \\
= {} & x^{81} + 2x^{80} + 4x^{79} + 6x^{78} + 9x^{77} + 13x^{76} + 18x^{75} + 24x^{74} + 31x^{73} \\
& + 39x^{72} + 50x^{71} + 62x^{70} + 77x^{69} + 93x^{68} + 112x^{67} + 134x^{66} + 159x^{65} \\
& + 187x^{64} + 218x^{63} + 252x^{62} + 287x^{61} + 325x^{60} + 364x^{59} + 406x^{58} \\
& + 449x^{57} + 493x^{56} + 538x^{55} + 584x^{54} + 631x^{53} + 679x^{52} + 722x^{51} \\
& + 766x^{50} + 805x^{49} + 845x^{48} + 880x^{47} + 910x^{46} + 935x^{45} + 955x^{44} \\
& + 970x^{43} + 980x^{42} + 985x^{41} + 985x^{40} + 980x^{39} + 970x^{38} + 955x^{37} \\
& + 935x^{36} + 910x^{35} + 880x^{34} + 845x^{33} + 805x^{32} + 766x^{31} + 722x^{30} \\
& + 679x^{29} + 631x^{28} + 584x^{27} + 538x^{26} + 493x^{25} + 449x^{24} + 406x^{23} \\
& + 364x^{22} + 325x^{21} + 287x^{20} + 252x^{19} + 218x^{18} + 187x^{17} + 159x^{16} \\
& + 134x^{15} + 112x^{14} + 93x^{13} + 77x^{12} + 62x^{11} + 50x^{10} + 39x^9 \\
& + 31x^8 + 24x^7 + 18x^6 + 13x^5 + 9x^4 + 6x^3 + 4x^2 + 2x + 1.
\end{aligned}
$$

We know from the previous section that

$$\frac{1}{(1 - z)^n} = \sum_k \binom{n + k - 1}{n - 1} z^k,$$

so

$$B(x) = C(x) \sum_k \binom{k + 5}{5} x^{20k}. \tag{25}$$

Therefore, writing $C(x) = \sum_k c_k x^k$, we have $a_{100} = b_{20} = c_0\binom{6}{5} + c_{20}\binom{5}{5} = 6 + 287 = 293$, and

$$
\begin{aligned}
a_{1000} = b_{200} \\
= c_0\binom{15}{5} + c_{20}\binom{14}{5} + c_{40}\binom{13}{5} + c_{60}\binom{12}{5} + c_{80}\binom{11}{5} \\
= 2\,103\,596.
\end{aligned}
$$

Our expression for computing $a_k$ is a sum having at most five terms, so this method allows us to compute $a_k$ using only a constant number of operations. Exercise 4 asks you to determine a bound on this number.

Finally, consider the crazy system of coinage where there is a coin minted worth $k$ cents for every $k \geq 1$. Let $p_k$ denote the number of ways to make $k$ cents in change in this system. Evidently, the generating function for this sequence is

$$\sum_{k \geq 0} p_k x^k = \prod_{m \geq 1} \frac{1}{1 - x^m}. \tag{26}$$

For example, $p_4 = 5$, since we can make four cents by using four pennies, or two pennies and one two-cent piece, or one penny and one three-cent piece, or two two-cent pieces, or one four-cent piece. By representing these five possibilities as the sums $1 + 1 + 1 + 1$, $1 + 1 + 2$, $1 + 3$, $2 + 2$, and $4$, we see that $p_k$ is the number of ways to write $k$ as a sum of one of more positive integers, disregarding the order of the summands. Such a representation is called a *partition* of $k$.

Suppose we have $k$ identical objects that we wish to split into a number of (indistinguishable) piles. How many ways can we do this? With four objects, there are five ways: We can place each object in a separate pile, or make three piles with one of them containing two of the objects, or make one pile with three objects and another with one, or form two piles with two objects in each, or place all four objects in a single pile. In general, each way of splitting $k$ objects into piles corresponds to a representation of $k$ as a sum of one or more positive integers, so the answer is simply the number of partitions of $k$.

Table 2.3 lists the first few values of the sequence $\{p_k\}$. There is a large body of research that studies properties of this sequence; see, for instance, [4]. Hardy and Ramanujan obtained a complicated exact formula for $p_k$ as a certain convergent series, and Rademacher later improved their formulation. We refer the reader to [4] for details. We mention only that this sequence grows quite rapidly. Asymptotically,

$$p_k \sim \frac{e^{\pi \sqrt{2k/3}}}{4k\sqrt{3}},$$

where $f(k) \sim g(k)$ means that $\lim_{k \to \infty} f(k)/g(k) = 1$.

**Exercises**

1. Use (25) to compute $a_{1999}$, the number of ways to make \$19.99 in change.
2. How many ways are there to select 100 coins from an inexhaustible supply of pennies, nickels, dimes, quarters, half-dollars, and Susan B. Anthony dollars?
3. Show that the number of ways to make $10m$ cents in change using only pennies, nickels, and dimes is $(m + 1)^2$.
4. Show that $a_k$ can be computed using equation (25) using at most 60 arithmetic operations. Optimize your method to show that $a_k$ can be computed using at most 31 arithmetic operations.
5. Prove that $a_k$ grows like $k^5$ by showing that there exist positive constants $c$ and $C$ such that $ck^5 < a_k < Ck^5$ for sufficiently large $k$.

| k | $p_k$ | k | $p_k$ | k | $p_k$ |
|---|-------|----|-------|----|-------|
| 0 | 1 | 11 | 56 | 22 | 1002 |
| 1 | 1 | 12 | 77 | 23 | 1255 |
| 2 | 2 | 13 | 101 | 24 | 1575 |
| 3 | 3 | 14 | 135 | 25 | 1958 |
| 4 | 5 | 15 | 176 | 26 | 2436 |
| 5 | 7 | 16 | 231 | 27 | 3010 |
| 6 | 11 | 17 | 297 | 28 | 3718 |
| 7 | 15 | 18 | 385 | 29 | 4565 |
| 8 | 22 | 19 | 490 | 30 | 5604 |
| 9 | 30 | 20 | 627 | 31 | 6842 |
| 10 | 42 | 21 | 792 | 32 | 8349 |

TABLE 2.3. The number of partitions of $k$.

6. The following coins were in circulation in the United States in 1875: the Indian-head penny, a bronze two-cent piece (last minted in 1873), a silver three-cent piece (also last minted in 1873), a nickel three-cent piece, the shield nickel (worth five cents), the seated liberty half-dime, dime, twenty-cent piece (produced for only four years beginning in 1875), quarter, half-dollar, and silver dollar, and the Indian-head gold dollar. (We ignore the trade dollar, minted for circulation between 1873 and 1878, as it was issued for overseas trade. This coin holds the distinction of being the only U.S. coin to be demonetarized.)

   a. How many ways were there to make twenty cents in change in 1875? How about twenty-five cents? Compute these values using the tabular method of this section.
   b. Write down a generating function in the form of a rational function for the number of ways to make $k$ cents in change in 1875; then use a computer algebra system to find the number of ways to make one dollar in change in 1875.

7. (Inspired in part by [69, Exercise 7.21].) A ransom note demands:

   (i) $10000 in unmarked fifty- and hundred-dollar bills, and
   (ii) the number of ways to award the cash.

   You realize that both old-fashioned and redesigned anticounterfeit bills are available in both denominations. Unfortunately, you have no access to modern computer algebra systems, although there is some simple accounting software on hand, should you want to use that.

   a. Answer the second demand of the ransom note. For extra credit, answer the first demand ⌣.
   b. Find a closed form for the number of ways to make $50m$ dollars using the two kinds of fifty- and hundred-dollar bills. Here, $m$ is a nonnegative integer.

**8.** How many ways are there to place ten identical glazed donuts into boxes if a single box can hold up to twelve donuts and there is an unlimited supply of identical boxes available?

**9.** A hungry math major visits the school's cafeteria and wants to know the number of ways $s_k$ to take $k$ servings of food, including at least one main course, an even number (possibly zero) of side vegetables, an odd number of rolls, and at least two desserts. The cafeteria's food can be distinguished only in the coarsest way: Every dish is either a main course, a side vegetable, a roll, or a dessert. There is an unlimited supply of each kind of dish available.

    **a.** Determine a closed form for the generating function $\sum_k s_k x^k$.

    **b.** Show that

$$s_k = \binom{\lfloor \frac{k+1}{2} \rfloor}{3} + \binom{\lceil \frac{k+1}{2} \rceil}{3}.$$

The quantities $\lfloor x \rfloor$ and $\lceil x \rceil$ are defined on page 98.

## 2.4.4  Fibonacci Numbers

*Attention! Attention! Ladies and gentlemen, attention! There is a herd of killer rabbits headed this way and we desperately need your help!*

                            *— Night of the Lepus*

Hey, shouldn't that be a *colony* of killer rabbits?

    Leonardo of Pisa, better known as Fibonacci, proposed the following harey problem in 1202. Assume that the rabbit population grows according to the following rules.

**1.** Every pair of adult rabbits produces a pair of baby rabbits, one of each gender, every month.

**2.** Baby rabbits become adult rabbits at age one month and produce their first offspring at age two months.

**3.** Rabbits are immortal.

Starting with a single pair of baby rabbits at the start of the first month, how many pairs of rabbits are there after $k$ months?

    Let $F_k$ denote this number. In the first month, there is one pair of baby rabbits, so $F_1 = 1$. Likewise, $F_2 = 1$, as there is one pair of adult rabbits in the second month. In the third month, we have one baby pair and one adult pair, so $F_3 = 2$, and in the fourth month, the babies become adults and the adults produce another pair of offspring, so there is one pair of babies and two pairs of adults: $F_4 = 3$. Continuing in this way, we record the population in the following table.

| $k$ | Baby pairs | Adult pairs | $F_k$ |
|---|---|---|---|
| 0 | 0 | 0 | 0 |
| 1 | 1 | 0 | 1 |
| 2 | 0 | 1 | 1 |
| 3 | 1 | 1 | 2 |
| 4 | 1 | 2 | 3 |
| 5 | 2 | 3 | 5 |
| 6 | 3 | 5 | 8 |
| 7 | 5 | 8 | 13 |

Notice that the number of pairs of adults in month $k$ equals the total number of pairs of rabbits in month $k - 1$. This is $F_{k-1}$. Also, the number of pairs of baby rabbits in month $k$ equals the number of adult pairs in month $k - 1$, which is the total number of pairs in month $k - 2$. This is $F_{k-2}$. Therefore;

$$F_k = F_{k-1} + F_{k-2}, \quad k \geq 2. \tag{27}$$

This recurrence, together with the initial conditions $F_0 = 0$ and $F_1 = 1$, determines the Fibonacci sequence $\{F_k\} = \{0, 1, 1, 2, 3, 5, 8, 13, 21, 34, 55, 89, \ldots\}$. This sequence appears frequently in combinatorial problems.

In this section we determine a closed form for $F_k$ by analyzing its generating function. We will adapt this technique to solve other recurrences later in this chapter.

Let $G(x)$ be the generating function for $\{F_k\}$. Then

$$\begin{aligned}
G(x) &= \sum_{k \geq 0} F_k x^k \\
&= F_0 + F_1 x + \sum_{k \geq 2} F_k x^k \\
&= x + \sum_{k \geq 2} (F_{k-1} + F_{k-2}) x^k \\
&= x + x \sum_{k \geq 2} F_{k-1} x^{k-1} + x^2 \sum_{k \geq 2} F_{k-2} x^{k-2} \\
&= x + x \sum_{k \geq 1} F_k x^k + x^2 \sum_{k \geq 0} F_k x^k \\
&= x + x G(x) + x^2 G(x),
\end{aligned}$$

and so

$$G(x) = \frac{-x}{x^2 + x - 1}.$$

Thus $F_k$ is the coefficient of $x^k$ in the Maclaurin series for this rational function. How can we determine this series without all the messy differentiation?

The trick is using partial fractions to write $G(x)$ as a sum of simpler rational functions. Write $x^2 + x - 1 = (x + \varphi)(x + \hat{\varphi})$, where $\varphi$ is the golden ratio, $\varphi = (1 + \sqrt{5})/2$, and $\hat{\varphi} = (1 - \sqrt{5})/2$. Write

$$\frac{-x}{x^2 + x - 1} = \frac{A}{x + \varphi} + \frac{B}{x + \hat{\varphi}}$$

and solve to find that $A = -\varphi/\sqrt{5}$ and $B = \hat{\varphi}/\sqrt{5}$. Thus

$$
\begin{aligned}
G(x) &= \frac{1}{\sqrt{5}}\left(\frac{\hat{\varphi}}{x+\hat{\varphi}} - \frac{\varphi}{x+\varphi}\right) \\
&= \frac{1}{\sqrt{5}}\left(\frac{1}{1+x/\hat{\varphi}} - \frac{1}{1+x/\varphi}\right) \\
&= \frac{1}{\sqrt{5}}\left(\frac{1}{1-\varphi x} - \frac{1}{1-\hat{\varphi}x}\right),
\end{aligned}
$$

since $\varphi\hat{\varphi} = -1$. Now, the two terms above are closed forms for simple geometric series, so

$$G(x) = \frac{1}{\sqrt{5}}\sum_{k\geq 0}(\varphi^k - \hat{\varphi}^k)x^k,$$

and therefore

$$F_k = \frac{\varphi^k - \hat{\varphi}^k}{\sqrt{5}}. \tag{28}$$

Notice that $|\hat{\varphi}| < 1$, so $F_k \sim \varphi^k/\sqrt{5}$: a large number of rabbits indeed.

### Exercises

1. Determine a closed form for the sum $F_0 + F_2 + F_4 + \cdots + F_{2n}$.
2. Prove that $F_n$ divides $F_{nm}$ for any nonnegative integers $n$ and $m$.
3. Solve the recurrence $a_k = 2a_{k-1} + 3a_{k-2}$, with $a_0 = 0$ and $a_1 = 8$.
4. The *Lucas numbers* are defined by $L_0 = 2$, $L_1 = 1$, and $L_k = L_{k-1} + L_{k-2}$ for $k \geq 2$. Find a formula for $L_k$ in terms of $\varphi$ and $\hat{\varphi}$.
5. Prove the following identities involving Fibonacci numbers and Lucas numbers.

   a. $L_n = F_{n+1} + F_{n-1}$, $n \geq 1$.

   b. $\displaystyle\sum_{k=0}^{n} F_k^2 = F_n F_{n+1}$.

   c. $\displaystyle\sum_{k=0}^{n} L_k^2 = L_n L_{n+1} + 2$.

   d. $\displaystyle\sum_{k=0}^{n}(-1)^k F_{n-k} = F_{n-1} - (-1)^n$, $n \geq 1$.

   e. $\displaystyle\sum_{k=0}^{n}(-1)^k L_{n-k} = L_{n-1} + 3(-1)^n$, $n \geq 1$.

   f. $F_{2n} = F_n L_n$.

   g. $L_{2n} = L_n^2 - 2(-1)^n$.

6. The Perrin sequence is defined by $a_0 = 3, a_1 = 0, a_2 = 2$, and $a_k = a_{k-2} + a_{k-3}$ for $k \geq 3$. The Padovan sequence is defined by $b_0 = 0$, $b_1 = 1$, $b_2 = 1$, and $b_k = b_{k-2} + b_{k-3}$ for $k \geq 3$.

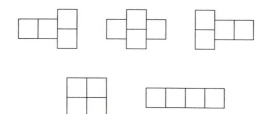

FIGURE 2.2. Hopscotch games with four squares.

**a.** Find generating functions in the form of rational functions for the Perrin sequence and the Padovan sequence.

**b.** Prove that $a_k = r^k + \alpha^k + \overline{\alpha}^k$, where $r$, $\alpha$, and $\overline{\alpha}$ are the three complex roots of $x^3 - x - 1$. Conclude that $a_k \sim r^k$.

The Perrin sequence has an interesting property: If $p$ is a prime number, then $p$ divides the $p$th term in the Perrin sequence, $p \mid a_p$. This was first noted by Lucas in 1878 [103] (perhaps Lucas would have been interested in Exercise 6 of Section 2.4.3). Thus we obtain a test for composite numbers: If $n$ does not divide $a_n$, then $n$ is not prime. Unfortunately, the converse is false: There are infinitely many composite $n$ with the property that $n \mid a_n$. This was proved by Grantham in 1997 [72].

**7.** In the children's game of hopscotch, a player hops across an array of squares drawn on the ground, landing on only one foot whenever there is just one square at a position, and landing on both feet when there are two. If every position has either one or two squares, how many different hopscotch games have exactly $n$ squares? Figure 2.2 shows the five different hopscotch games having four squares.

## 2.4.5    Recurrence Relations

*O me! O life!... of the questions of these recurring;*
                              — Walt Whitman, *Leaves of Grass*

In the "Tower of Hanoi" puzzle, one begins with a pyramid of $k$ disks stacked around a center pole, with the disks arranged from largest diameter on the bottom to smallest diameter on top. There are also two empty poles that can accept disks. The object of the puzzle is to move the entire stack of disks to one of the other poles, subject to three constraints:

**1.** Only one disk may be moved at a time.
**2.** Disks can be placed only on one of the three poles.
**3.** A larger disk cannot be placed on a smaller one.

How many moves are required to move the entire stack of $k$ disks onto another pole? Let $a_k$ denote this number. Clearly, $a_1 = 1$. To move $k$ disks, we must first move the $k - 1$ top disks to one of the other poles, then move the bottom disk to the third pole, then move the stack of $k - 1$ disks to that pole, so $a_k = 2a_{k-1} + 1$ for $k \geq 1$. Thus, $a_2 = 3$, $a_3 = 7$, $a_4 = 15$, and it appears that $a_k = 2^k - 1$.

We can certainly verify this formula by induction, but we wish to show how recurrences of this form can be solved using generating functions. Consider the more general recurrence

$$a_k = ba_{k-1} + c, \quad k \geq 1,$$

where $b$ and $c$ are constants. This is a linear recurrence relation, since $a_k$ is a linear function of the preceding values of the sequence. (The Fibonacci recurrence is also a linear recurrence relation.) If $c$ is zero, we call the recurrence *homogeneous*; otherwise, it is *inhomogeneous*.

Let $G(x)$ be the generating function for $\{a_k\}$. Then

$$G(x) = \sum_{k \geq 0} a_k x^k$$

$$= a_0 + \sum_{k \geq 1} \left( ba_{k-1} x^k + cx^k \right)$$

$$= a_0 + bx \sum_{k \geq 0} a_k x^k + cx \sum_{k \geq 0} x^k$$

$$= a_0 + bx G(x) + \frac{cx}{1 - x},$$

and so

$$G(x) = \frac{cx}{(1 - bx)(1 - x)} + \frac{a_0}{1 - bx}.$$

Assuming $b \neq 1$, we compute

$$\frac{cx}{(1 - bx)(1 - x)} = \frac{c}{b - 1} \left( \frac{1}{1 - bx} - \frac{1}{1 - x} \right),$$

so

$$G(x) = \left( a_0 + \frac{c}{b - 1} \right) \left( \frac{1}{1 - bx} \right) - \frac{c}{b - 1} \left( \frac{1}{1 - x} \right)$$

$$= \left( a_0 + \frac{c}{b - 1} \right) \sum_{k \geq 0} b^k x^k - \frac{c}{b - 1} \sum_{k \geq 0} x^k,$$

and therefore

$$a_k = \left( a_0 + \frac{c}{b - 1} \right) b^k - \frac{c}{b - 1}. \tag{29}$$

For example, to find the number of moves needed to solve the Tower of Hanoi puzzle, we set $a_0 = 0$, $b = 2$, and $c = 1$ to obtain $a_k = 2^k - 1$. Also, if we set $b = -\frac{1}{2}$ and $c = 2$, we find that $a_k = (-1)^k (a_0 - \frac{4}{3})/2^k + \frac{4}{3}$, so $a_k$ approaches $\frac{4}{3}$ as $k$ grows large, independent of the initial value $a_0$.

We conclude with a short list of useful generating functions. Since

$$\frac{1}{1-x} = \sum_{k \geq 0} x^k, \tag{30}$$

we differentiate both sides to find that

$$\frac{1}{(1-x)^2} = \sum_{k \geq 1} k x^{k-1},$$

and so

$$\frac{x}{(1-x)^2} = \sum_{k \geq 0} k x^k. \tag{31}$$

Thus we obtain a closed form for the generating function of the identity sequence $\{k\}$.

We integrate both sides of (30) to obtain the generating function for $\{1/k\}$:

$$-\ln(1-x) = \sum_{k \geq 1} \frac{x^k}{k}. \tag{32}$$

**Exercises**

1. Find a recurrence relation for the maximal number of regions of the plane separated by $k$ straight lines.
2. Solve the following recurrence relations.

   **a.** $a_k = a_{k-1} + c$.
   **b.** $a_k = b a_{k-1} + c r^k$.
   **c.** $a_k = b a_{k-1} + c k$.

3. Find a closed form for the generating function for the sequence $\{k^2\}_{k \geq 0}$.

## 2.4.6   Catalan Numbers

*zero, un, dos, tres, quatre, cinc, sis, set, vuit, nou, deu, onze, dotze, tretze, catorze, quinze, setze, disset, divuit, dinou, vint.*

How many ways are there to compute a product of $k + 1$ matrices? Matrix multiplication is associative but not commutative, so this is the number of ways to place $k - 1$ pairs of parentheses in the product $x_0 x_1 \ldots x_k$ in such a way that the order of multiplications is completely specified. Let $C_k$ denote this number.

Let us first compute a few values of $C_k$. There is only one way to compute the product of one or two matrices. There are two ways to group a product of three matrices, $(x_0 x_1) x_2$ and $x_0 (x_1 x_2)$, and there are five ways for a product of four matrices: $((x_0 x_1) x_2) x_3$, $(x_0 (x_1 x_2)) x_3$, $(x_0 x_1)(x_2 x_3)$, $x_0 ((x_1 x_2) x_3)$, and $x_0 (x_1 (x_2 x_3))$. A bit more work gives us 14 ways to compute a product of five matrices: There are five ways if one pair of parentheses is $x_0 (x_1 x_2 x_3 x_4)$, another five for $(x_0 x_1 x_2 x_3) x_4$, two for $(x_0 x_1)(x_2 x_3 x_4)$, and two more for $(x_0 x_1 x_2)(x_3 x_4)$. We record these numbers in the following table.

| $k$ | $C_k$ |
|-----|-------|
| 0   | 1     |
| 1   | 1     |
| 2   | 2     |
| 3   | 5     |
| 4   | 14    |

Can we determine a recurrence relation for $C_k$?

Suppose we group the terms so that the last multiplication occurs between $x_i$ and $x_{i+1}$:

$$(x_0 x_1 \ldots x_i)(x_{i+1} \ldots x_k).$$

Then there are $C_i$ ways to group the terms in the first part of the product, and $C_{k-1-i}$ ways for the second part, so there are $C_i C_{k-1-i}$ ways to group the remaining terms in this case. Summing over $i$, we obtain the following formula for the total number of ways to group the $k + 1$ terms:

$$C_k = \sum_{i=0}^{k-1} C_i C_{k-1-i}, \quad k \geq 1. \tag{33}$$

We compute

$$C_1 = C_0 C_0 = 1,$$
$$C_2 = C_0 C_1 + C_1 C_0 = 2,$$
$$C_3 = C_0 C_2 + C_1 C_1 + C_2 C_0 = 5,$$
$$C_4 = C_0 C_3 + C_1 C_2 + C_2 C_1 + C_3 C_0 = 14,$$
$$C_5 = C_0 C_4 + C_1 C_3 + C_2 C_2 + C_3 C_1 + C_4 C_0 = 42.$$

We would like to solve this recurrence to find a formula for $C_k$, so let us define the generating function for this sequence,

$$G(x) = \sum_{k \geq 0} C_k x^k.$$

Unlike other recurrences we have studied, this one is not linear, and has a variable number of terms. To solve it, we require one fact concerning products of generating functions.

If $A(x) = \sum_{k \geq 0} a_k x^k$ and $B(x) = \sum_{k \geq 0} b_k x^k$, then

$$A(x)B(x) = \sum_{k \geq 0} \left( \sum_{i=0}^{k} a_i b_{k-i} \right) x^k.$$

Let $c_k = \sum_{i=0}^{k} a_i b_{k-i}$. The sequence $\{c_k\}$ is called the *convolution* of the sequences $\{a_k\}$ and $\{b_k\}$. Thus, the generating function of the convolution of two sequences is the product of the generating functions of the sequences.

Using this fact, we find that

$$G(x) = \sum_{k \geq 0} C_k x^k$$

$$= C_0 + \sum_{k\geq 1} \left( \sum_{i=0}^{k-1} C_i C_{k-1-i} \right) x^k$$

$$= 1 + x \sum_{k\geq 0} \left( \sum_{i=0}^{k} C_i C_{k-i} \right) x^k$$

$$= 1 + x G(x)^2,$$

since $\{\sum_{i=0}^{k} C_i C_{k-i}\}$ is the convolution of $\{C_k\}$ with itself. Thus,

$$x G(x)^2 - G(x) + 1 = 0,$$

and so

$$G(x) = \frac{1 \pm \sqrt{1 - 4x}}{2x}.$$

Only one of these functions can be the generating function for $\{C_k\}$, and it must satisfy

$$\lim_{x\to 0} G(x) = C_0 = 1.$$

It is easy to check that the correct function is

$$G(x) = \frac{1 - \sqrt{1 - 4x}}{2x}.$$

We now expand $G(x)$ as a Maclaurin series to find a formula for $C_k$. Using the generalized binomial theorem and the identity for negating the upper index, we find that

$$(1 - 4x)^{1/2} = \sum_{k\geq 0} \binom{1/2}{k} (-4x)^k$$

$$= \sum_{k\geq 0} \binom{k - 3/2}{k} 4^k x^k$$

$$= 1 + \sum_{k\geq 1} \binom{k - 3/2}{k} 4^k x^k$$

$$= 1 + 4x \sum_{k\geq 0} \binom{k - 1/2}{k + 1} 4^k x^k.$$

Therefore,

$$G(x) = -2 \sum_{k\geq 0} \binom{k - 1/2}{k + 1} 4^k x^k,$$

and so

$$C_k = -2^{2k+1} \binom{k - 1/2}{k + 1}.$$

We determine a much simpler form for $C_k$. Expanding the generalized binomial coefficient and multiplying each term in the product by 2, we find that

$$C_k = -\frac{2^{2k+1}}{(k+1)!} \prod_{i=0}^{k} \left(k - \frac{1}{2} - i\right)$$

$$= -\frac{2^k}{(k+1)!} \prod_{i=0}^{k} (2k - 1 - 2i).$$

The product consists of all the odd numbers between $-1$ and $2k - 1$, so

$$C_k = \frac{2^k}{(k+1)!} \prod_{i=1}^{k} (2i - 1)$$

$$= \frac{2^k}{(k+1)!} \prod_{i=1}^{k} \frac{(2i-1)(2i)}{2i}$$

$$= \frac{1}{k!(k+1)!} \prod_{i=1}^{k} (2i - 1)(2i).$$

The remaining product is simply $(2k)!$, so

$$C_k = \frac{(2k)!}{k!(k+1)!} = \frac{1}{k+1} \binom{2k}{k}. \tag{34}$$

$C_k$ is called the $k$th *Catalan number*.

Incidentally, since $C_k$ is an integer, we have shown that $k + 1$ always divides the binomial coefficient $\binom{2k}{k}$. Can you find an independent arithmetic proof of this fact?

Sloane and Plouffe [136] remark that the Catalan numbers are perhaps the second most frequently occurring numbers in combinatorics, after the binomial coefficients. Indeed, Stanley [138, vol. 2] lists 66 different combinatorial interpretations of these numbers! We close with another problem whose solution involves the Catalan numbers.

A *rooted tree* is a tree with a distinguished vertex called the *root*. The vertices in a rooted tree form a hierarchy, with the root at the highest level, and the level of every other vertex determined by its distance from the root. Some familiar terms are often used to describe relationships between vertices in a rooted tree: If $v$ and $w$ are adjacent vertices and $v$ lies closer to the root than $w$, then $v$ is the *parent* of $w$, and $w$ is a *child* of $v$. Likewise, one may define siblings, grandparents, cousins, and other family relationships in a rooted tree.

We say that a rooted tree is *strictly binary* if every parent vertex has exactly two children. How many strictly binary trees are there with $k$ parent vertices? Do not take symmetry into account: If two trees are mirror images of one another, count both configurations. Figure 2.3 shows that there are five trees with three parent vertices.

It is easy to see that the number of strictly binary trees with $k$ parent vertices is $C_k$. By Exercise 2, every such tree has $k + 1$ leaves. Label these vertices

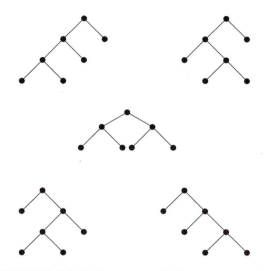

FIGURE 2.3. Strictly binary trees with three parent vertices.

with $x_0$ through $x_k$ from left to right in the tree. Then the tree determines an order of multiplication for the $x_i$. For example, the five trees in Figure 2.3 correspond to the multiplications $((x_0x_1)x_2)x_3$, $(x_0(x_1x_2))x_3$, $(x_0x_1)(x_2x_3)$, $x_0((x_1x_2)x_3)$, and $x_0(x_1(x_2x_3))$, respectively. Binary trees like these are often used in computer science to designate the order of evaluation of arithmetic expressions.

### Exercises

1. Show that every vertex in a rooted tree has at most one parent.
2. Show that a strictly binary tree having exactly $k$ parent vertices has exactly $k + 1$ leaves.
3. Suppose $2k$ people are seated around a table. How many ways are there for the $k$ pairs of people to shake hands simultaneously across the table in such a way that no arms cross?
4. How many ways are there to divide a convex polygon with $k + 2$ sides into triangles using exactly $k - 1$ noncrossing diagonals? (A diagonal is a line segment connecting two nonadjacent vertices.)
5. Show that the coefficient of $x^k$ in the Maclaurin series expansion of $(1 - (1 - 3x)^{1/3})/x$ is

$$\frac{1}{(k+1)!} \prod_{i=1}^{k} (3i - 1).$$

6. Use an arithmetic argument to show that $(2k)!$ is divisible by $k!(k + 1)!$. Hint: First compute the number of times a prime number $p$ divides $m!$.

# 2.5   Pólya's Theory of Counting

*Who are you who are so wise in the ways of science?*
— Sir Bedivere, in *Monty Python and the Holy Grail*

How many ways can King Arthur and his knights sit at the round table? How many different necklaces with $n$ beads can be formed using $m$ different kinds of beads?

Both these questions ask for a number of combinations in the presence of symmetry. Since there is no distinguished position at a round table, seating Arthur first, then Gawain, Percival, Bedivere, Tristram, and Galahad clockwise around the table yields the same configuration as seating Tristram first, then Galahad, Arthur, Gawain, Percival, and Bedivere in clockwise order. Similarly, we should consider two necklaces to be identical if we can transform one into the other by rotating the necklace or by turning it over.

Before answering these questions, let us first rephrase them in the language of group theory.

## 2.5.1   Permutation Groups

*I haven't fought just one person in a long time. I've been specializing in groups.*
— Fezzik, in *The Princess Bride*

A *group* consists of a set $G$ together with a binary operator $\circ$ defined on this set. The set and the operator must satisfy four properties.

- Closure. For every $a$ and $b$ in $G$, $a \circ b$ is in $G$.
- Associativity. For every $a$, $b$, and $c$ in $G$, $a \circ (b \circ c) = (a \circ b) \circ c$.
- Identity. There exists an element $e$ in $G$ that satisfies $e \circ a = a \circ e = a$ for every $a$ in $G$.
- Inverses. For every element $a$ in $G$, there exists an element $b$ in $G$ such that $a \circ b = b \circ a = e$. The element $b$ is called the *inverse* of $a$.

In addition, if $a \circ b = b \circ a$ for every $a$ and $b$ in $G$, we say that $G$ is an *abelian*, or *commutative*, group.

For example, the set of integers forms a group under addition. The identity element is 0, since $0 + i = i + 0 = i$ for every integer $i$, and the inverse of the integer $i$ is the integer $-i$. Similarly, the set of nonzero rational numbers forms a group under multiplication (with identity element 1), as does the set of nonzero real numbers.

We can also construct groups of permutations. A permutation of $n$ objects may be described by a function $\pi$ defined on the set $\{1, 2, \ldots, n\}$ by ordering the objects in some fashion, then taking $\pi(i) = j$ if the $i$th object in the ordering occupies the $j$th position in the permutation. For example, the permutation $[c, d, a, e, b]$ of the list $[a, b, c, d, e]$ is represented by the function $\pi$ defined on the set $\{1, 2, 3, 4, 5\}$, with $\pi(1) = 3$, $\pi(2) = 5$, $\pi(3) = 1$, $\pi(4) = 2$, and $\pi(5) = 4$. Notice that a function $\pi : \{1, \ldots, n\} \to \{1, \ldots, n\}$ arising from a permutation has the property that $\pi(i) \neq \pi(j)$ whenever $i \neq j$. Such a function is called an *injective*, or *one-to-one*, function. The map $\pi$ also has the property that for every $m$ with $1 \leq m \leq n$, there exists a number $i$ such that $\pi(i) = m$. A function like this that maps to every element in its range is called *surjective*, or *onto*, and a function that is both injective and surjective is said to be a *bijection*. Thus, every permutation of $n$ objects corresponds to a bijection $\pi$ on the set $\{1, 2, \ldots n\}$, and every such bijection corresponds to a permutation.

Let $S_n$ denote the set of all bijections on the set $\{1, 2, \ldots, n\}$. Exercise 3 asks you to verify that this set forms a group under the operation of composition of functions. For example, the identity element of the group is the identity map $\pi_0$, defined by $\pi_0(k) = k$ for each $k$, since $\pi \circ \pi_0 = \pi_0 \circ \pi = \pi$ for every $\pi$ in $S_n$. This group is called the *symmetric group* on $n$ elements.

The size of the group $S_n$ is the number of permutations of $n$ objects, so $|S_n| = n!$. Because of our correspondence, we normally refer to an element of $S_n$ as a permutation, rather than a bijection.

To specify a particular permutation $\pi$ in $S_n$, we need to name the value of $\pi(k)$ for each $k$. This is often written in two rows as follows:

$$\begin{pmatrix} 1 & 2 & 3 & \ldots & n \\ \pi(1) & \pi(2) & \pi(3) & \ldots & \pi(n) \end{pmatrix}.$$

For example,

$$\begin{pmatrix} 1 & 2 & 3 & 4 & 5 \\ 3 & 5 & 1 & 2 & 4 \end{pmatrix}.$$

denotes the permutation described earlier.

We can describe the permutations in a more succinct manner by using *cycle notation*. For example, in the permutation above, $\pi$ sends 1 to 3 and 3 to 1 and sends 2 to 5, 5 to 4, and 4 to 2. So we can think of $\pi$ as a combination of two *cycles*, $1 \to 3 \to 1$ and $2 \to 5 \to 4 \to 2$, and denote it by

$$(13)(254).$$

Of course, we could also denote this same permutation by the cycles $(542)(31)$, so to make our notation unique, we make two demands. First, the cycle containing 1 must appear first, followed by the cycle containing the smallest number not appearing in the first cycle, and so on. Second, we require the first number listed in each cycle to be the smallest number appearing in that cycle.

To simplify the notation, cycles of length 1 are usually omitted, so $(1253)(4)$ is written more simply as $(1253)$. The identity permutation is denoted by $(1)$.

The composition of two permutations is commonly written as a product in cycle notation. For example, if $\pi_1 = (13)(254)$ and $\pi_2 = (15423)$, then the composition $\pi_1 \circ \pi_2$ sends 1 to 4, 4 to 5, 5 to 2, 2 to 1, and 3 to 3, so $\pi_1 \circ \pi_2 = (1452)$. In cycle notation, we write

$$\pi_1 \circ \pi_2 = (13)(254)(15423) = (1452).$$

One can determine the cycle on the right easily by applying the two permutations in sequence. The right permutation sends 1 to 5, and the left one sends 5 to 4, so the composition sends 1 to 4. In the same way, we see that the composition sends 4 to 5, 5 to 2, 2 to 1, and 3 to itself.

Notice that $\pi_2 \circ \pi_1 = (15423)(13)(254) = (2435)$, so in general $S_n$ is not an abelian group.

A subset $H$ of a group $G$ is called a *subgroup* of $G$ if $H$ is itself a group under the same binary operation. The group $S_n$ contains many subgroups; for example, $\{(1),(12)\}$ is a subgroup of $S_n$ for every $n \geq 2$. We investigate three particularly important subgroups of $S_n$.

### The Cyclic Group

If $\pi$ is a permutation in $S_n$ and $m$ is a nonnegative integer, let $\pi^m$ denote the permutation obtained by composing $\pi$ with itself $m$ times, so $\pi^0 = (1)$, and $\pi^3 = \pi \circ \pi \circ \pi$. Let

$$\langle \pi \rangle = \{\pi^m \ : \ m \geq 0\}, \tag{35}$$

so that $\langle \pi \rangle$ is a subset of $S_n$. In fact (Exercise 4), $\langle \pi \rangle$ is a subgroup of $S_n$, and we call this group the cyclic subgroup generated by $\pi$ in $S_n$.

The *cyclic group* $C_n$ is the subgroup of $S_n$ generated by the permutation $(123\ldots n)$,

$$C_n = \langle (123\ldots n) \rangle. \tag{36}$$

Clearly, $C_n$ contains $n$ elements, since $n$ applications of the generating permutation are required to return to the identity permutation. For example, $(1234)^2 = (13)(24)$, $(1234)^3 = (1432)$, and $(1234)^4 = (1)$, so

$$C_4 = \{(1), \ (1234), \ (13)(24), \ (1432)\}. \tag{37}$$

The group $C_n$ may be realized as the group of rotational symmetries of a regular polygon having $n$ sides. For example, each of the permutations of (37) corresponds to a permutation of the vertices of Figure 2.4 obtained by rotating the square by 0, 90, 180, or 270 degrees.

### The Dihedral Group

The *dihedral group* $D_n$ is the group of symmetries of a regular polygon with $n$ sides, including reflections as well as rotations. Since $C_n$ consists of just the rotational symmetries of such a figure, evidently $C_n$ is a subgroup of $D_n$.

Referring to Figure 2.4, we see that $D_4$ consists of the four rotations of $C_4$, plus the four reflections $(12)(34)$, $(14)(23)$, $(13)$, and $(24)$. The first two permutations represent reflections about the vertical and horizontal axes of symmetry of the square; the last two represent flips about the diagonal axes of symmetry. In general, if $n$ is even, we obtain $n/2$ reflections through axes of symmetry that pass through opposite vertices, and $n/2$ reflections through axes that pass through midpoints of

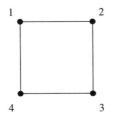

FIGURE 2.4. A square with labeled vertices.

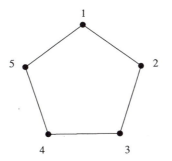

FIGURE 2.5. A regular pentagon with labeled vertices.

opposite edges. Combining these with the $n$ rotations of $C_n$, we find that $|D_n| = 2n$ in this case.

Using Figure 2.5, we find that $D_5$ consists of five rotations and five reflections,

$$D_5 = \{(1),\ (12345),\ (13524),\ (14253),\ (15432),\ (25)(34),$$
$$(13)(45),\ (15)(24),\ (12)(35),\ (14)(23)\}.$$

It is easy to see that we always obtain $n$ reflections if $n$ is odd, so $|D_n| = 2n$ for every $n \geq 1$.

### The Alternating Group

Every permutation can be expressed as a product of *transpositions*, cycles of length 2. For example, the cycle (123) can be written as the product (12)(23), and the permutation (1234)(567) equals the product (12)(23)(34)(56)(67). Such a decomposition is not unique; for instance, (123) may also be written as (23)(13), or (12)(23)(13)(13). However, the number of transpositions in any representation of one permutation is either always an even number, or always an odd number. Exercise 5 outlines a proof of this fact. If a permutation $\pi$ always decomposes into an even number of transpositions, we say that $\pi$ is an *even* permutation; otherwise, it is an *odd* permutation. Notice that the identity permutation is even, since it is represented by a product of zero transpositions.

The *alternating group* $A_n$ consists of the even permutations of $S_n$. For example, $A_3 = \{(1), (123), (132)\} = C_3$, and

$$A_4 = \{(1),\ (123),\ (132),\ (124),\ (142),\ (134),\ (143),$$
$$(234),\ (243),\ (12)(34),\ (13)(24),\ (14)(23)\}.$$

Exercises 6 and 7 ask you to verify that $A_n$ is a group of size $|A_n| = n!/2$ for $n \geq 2$, and that $A_n$ is not abelian for $n \geq 5$.

### Exercises

1. **a.** Suppose that $M$ is a finite set and $f : M \to M$ is an injective function. Show that $f$ is a bijection.
   **b.** Suppose that $M$ is a finite set and $f : M \to M$ is a surjective function. Show that $f$ is a bijection.
   **c.** Show that neither of these statements is necessarily true if $M$ is an infinite set.

2. In each part, determine all values of $n$ that satisfy the statement.

   **a.** $C_n$ is a subgroup of $A_n$.
   **b.** $D_n$ is a subgroup of $A_n$.
   **c.** $C_n$ is a subgroup of $D_{n+1}$.
   **d.** $C_n$ is a subgroup of $S_{n+1}$.

3. Verify that $S_n$ forms a group under composition of functions by checking that each of the required properties is satisfied.

   **a.** Closure. If $\pi_1$ and $\pi_2$ are bijections on $\{1, 2, \ldots, n\}$, show that $\pi_1 \circ \pi_2$ is also a bijection on $\{1, 2, \ldots, n\}$.
   **b.** Associativity. If $\pi_1$, $\pi_2$, and $\pi_3$ are in $S_n$, show that $\pi_1 \circ (\pi_2 \circ \pi_3)$ and $(\pi_1 \circ \pi_2) \circ \pi_3$ represent the same function in $S_n$.
   **c.** Identity. Check that $\pi_0 \circ \pi = \pi \circ \pi_0 = \pi$, for every $\pi$ in $S_n$. Here, $\pi_0$ is the identity map on $\{1, 2, \ldots, n\}$.
   **d.** Inverses. Given a bijection $\pi$ in $S_n$, construct a bijection $\pi^{-1}$ in $S_n$ satisfying $\pi \circ \pi^{-1} = \pi^{-1} \circ \pi = \pi_0$.

4. **a.** If $G$ is a group and $g$ is an element of $G$, show that $\langle g \rangle$ is a subgroup of $G$.
   **b.** Show that $\langle g \rangle$ is abelian.

5. Let $\mathbf{x}$ denote the vector of $n$ variables $(x_1, x_2, \ldots, x_n)$. Define

$$P(\mathbf{x}) = \prod_{1 \leq i < j \leq n} (x_i - x_j),$$

and if $\pi \in S_n$, let

$$P_\pi(\mathbf{x}) = \prod_{1 \leq i < j \leq n} (x_{\pi(i)} - x_{\pi(j)}).$$

   **a.** Show that $P_\pi(\mathbf{x}) = \pm P(\mathbf{x})$.
   **b.** Show that $P_\pi(\mathbf{x}) = -P(\mathbf{x})$ if $\pi$ is a transposition.

**c.** Conclude that no permutation $\pi$ in $S_n$ can be represented both as a product of an even number of transpositions and as a product of an odd number of transpositions.

**6. a.** Prove that $A_n$ is a group.
   **b.** Show that $A_n$ is not abelian for $n \geq 5$.
**7.** Let $n \geq 2$, let $B_n$ denote the set of odd permutations in $S_n$, and let $\tau$ be a transposition in $S_n$.

   **a.** Show that the map $T : S_n \rightarrow S_n$ defined by $T(\pi) = \tau \circ \pi$ is a bijection.
   **b.** Show that $T$ maps $A_n$ to $B_n$, and $B_n$ to $A_n$.
   **c.** Conclude that $|A_n| = n!/2$.

**8.** Determine the group of symmetries of each of the following objects.

   **a.** The vertices of a regular tetrahedron.
   **b.** The vertices of a cube.
   **c.** The vertices of a regular octahedron.

## 2.5.2 Burnside's Lemma

*Burnside had submitted the scheme to Meade and myself, and we both approved of it, as a means of keeping the men occupied.*
                              — *Personal Memoirs of U.S. Grant*

Armed with our knowledge of permutation groups, we now develop a general method for counting combinations in the presence of symmetry. In general, we are given a set of objects $S$, a set of colorings of these objects $C$, and a group of permutations $G$ representing symmetries possessed by configurations of the objects. We consider two colorings in $C$ to be equivalent if one of the permutations in $G$ transforms one coloring to the other, and we would like to determine the number of nonequivalent colorings in $C$.

For example, suppose $S = \{1, 2, 3, 4\}$ is the set of vertices of the square in Figure 2.4, and $C$ is the set of all possible colorings of these vertices using two colors, red and green. Let $rrgr$ denote the coloring where vertices 1, 2, and 4 are red and vertex 3 is green. Then

$$C = \{gggg, gggr, ggrg, ggrr, grgg, grgr, grrg, grrr, \\ rggg, rggr, rgrg, rgrr, rrgg, rrgr, rrrg, rrrr\}. \tag{38}$$

We consider two colorings in $C$ to be equivalent if one can be transformed to the other by a rotation of the square. For example, rotating the coloring $rrgr$ yields the equivalent colorings $rrrg$, $grrr$, and $rgrr$. So we choose $G$ to be the group of rotations, $C_4$. A permutation $\pi$ in $C_4$ is a function defined on the set $\{1, 2, 3, 4\}$, but $\pi$ induces a map $\pi^*$ defined on the set of colorings $C$ in a natural way. For example, if $\pi$ is the 180-degree rotation $(13)(24)$, then the induced map $\pi^*$ rotates a coloring by the same amount, so $\pi^*(rrgr) = grrr$, and $\pi^*(grgr) = grgr$.

If $c_1$ and $c_2$ are two equivalent colorings in $C$, so $\pi^*(c_1) = c_2$ for some $\pi \in G$, we write $c_1 \sim c_2$. Using the fact that $G$ is a group, it is easy to verify (Exercise 1) that the relation $\sim$ on the set of colorings is

- reflexive: $c \sim c$ for all colorings $c$,
- symmetric: $c_1 \sim c_2$ implies $c_2 \sim c_1$, and
- transitive: $c_1 \sim c_2$ and $c_2 \sim c_3$ implies $c_1 \sim c_3$.

A relation possessing these three properties is called an *equivalence relation*. By grouping together collections of mutually equivalent elements, an equivalence relation on a set partitions the set into a number of disjoint subsets, called *equivalence classes*. Our goal then is to determine the number of equivalence classes of $C$ under the relation $\sim$.

In our example, the group $C_4$ partitions our set of colorings (38) into six equivalence classes:

$$\{gggg\},$$
$$\{gggr, \, ggrg, \, grgg, \, rggg\},$$
$$\{ggrr, \, grrg, \, rggr, \, rrgg\},$$
$$\{grgr, \, rgrg\},$$
$$\{grrr, \, rgrr, \, rrgr, \, rrrg\},$$
$$\{rrrr\}.$$

Therefore, there are just six ways to color the vertices of a square using two colors, after discounting rotational symmetries.

We can now translate the problems from the introduction to this section into this more abstract setting. In the round table problem, $S$ is the set of $n$ places at the table, $G$ is $C_n$, and $C$ is the collection of the $n!$ seating assignments. In the necklace problem, $S$ is the set of $n$ bead positions, $G$ is $D_n$, and $C$ is the collection of the $m^n$ possible arrangements of the $m$ kinds of beads on the necklace.

Before presenting a general method to solve problems like these, we introduce three sets that will be useful in our analysis. Given a permutation $\pi$ in $G$, define $C_\pi$ to be the set of colorings that are invariant under action by the induced map $\pi^*$,

$$C_\pi = \{c \in C \, : \, \pi^*(c) = c\}. \tag{39}$$

This set is called the *invariant set* of $\pi$ in $C$. Similarly, given a coloring $c$ in $C$, define $G_c$ to be the set of permutations $\pi$ in $G$ for which $c$ is a fixed coloring,

$$G_c = \{\pi \in G \, : \, \pi^*(c) = c\}. \tag{40}$$

This set is called the *stabilizer* of $c$ in $G$. It is always a subgroup of $G$. Finally, let $\bar{c}$ be the set of colorings in $C$ that are equivalent to $c$ under the action of the group $G$,

$$\bar{c} = \{\pi^*(c) \, : \, \pi \in G\}. \tag{41}$$

The set $\overline{c}$ is thus the equivalence class of $c$ under the relation $\sim$. It is also called the *orbit* of $c$ under the action of $G$.

For example, if $C$ is given by (38) and $G$ is the dihedral group $D_4$, we have

$$\overline{gggr} = \{gggr, \; ggrg, \; grgg, \; rggg\}$$

and

$$G_{gggr} = \{(1), \; (13)\}.$$

Also,

$$\overline{grgr} = \{grgr, \; rgrg\}$$

and

$$G_{grgr} = \{(1), \; (13)(24), \; (13), \; (24)\}.$$

Notice that in both cases, the product of the size of the stabilizer of a coloring with the size of the equivalence class of the same coloring equals the number of elements in the group. The following lemma proves that this is always the case.

**Lemma 2.3.** *Suppose a group $G$ acts on a set of colorings $C$. For any coloring $c$ in $C$, we have $|G_c| \, |\overline{c}| = |G|$.*

PROOF. We prove this by showing that every permutation in $G$ may be represented in a unique way as a composition of a permutation in $G_c$ with a permutation in a particular set $P$, where $|P| = |\overline{c}|$. Suppose there are $m$ colorings in the equivalence class of $c$, $\overline{c} = \{c_1, c_2, \ldots, c_m\}$. For each $i$ between 1 and $m$, select a permutation $\pi_i \in G$ such that $\pi_i^*(c) = c_i$, and let $P = \{\pi_1, \pi_2, \ldots, \pi_m\}$.

Now let $\pi$ be an arbitrary permutation in $G$. Then $\pi^*(c) = c_i$ for some $i$, so $\pi^*(c) = \pi_i^*(c)$. Thus $(\pi_i^{-1} \circ \pi)^*(c) = c$, and so $\pi_i^{-1} \circ \pi \in G_c$. Since $\pi_i \circ (\pi_i^{-1} \circ \pi) = \pi$, we see that $\pi$ has at least one representation in the desired form. Suppose now that $\pi = \pi_i \circ \sigma = \pi_j \circ \tau$, for some $\pi_i$ and $\pi_j$ in $P$ and some $\sigma$ and $\tau$ in $G_c$. Then $\pi_i(\sigma(c)) = \pi_i(c) = c_i$ and $\pi_j(\tau(c)) = c_j$, so $c_i = c_j$, and hence $i = j$. Therefore, $\sigma = \tau$, so the representation of $\pi$ is unique. $\qquad\qquad \square$

The following formula for the number of equivalence classes of $C$ under the action of a group $G$ is usually named for Burnside (the English mathematician, not the American Civil War general), as it was popularized by his book [19]. This result was first proved by Frobenius [59], however, and Burnside even attributes the formula to Frobenius in the first edition of [19]. Further details on the history of this result appear in [109] and [149].

Briefly, Burnside's Lemma states that the number of equivalence classes of colorings is the average size of the invariant sets.

**Theorem 2.4** (Burnside's Lemma). *The number of equivalence classes $N$ of the set $C$ in the presence of the group of symmetries $G$ is given by*

$$N = \frac{1}{|G|} \sum_{\pi \in G} |C_\pi|. \tag{42}$$

PROOF.    If $P$ is a logical expression, let $[P]$ be 1 if $P$ is true and 0 if $P$ is false. Then

$$\frac{1}{|G|} \sum_{\pi \in G} |C_\pi| = \frac{1}{|G|} \sum_{\pi \in G} \sum_{c \in C} [\pi^*(c) = c]$$

$$= \frac{1}{|G|} \sum_{c \in C} \sum_{\pi \in G} [\pi^*(c) = c]$$

$$= \frac{1}{|G|} \sum_{c \in C} |G_c|$$

$$= \sum_{c \in C} \frac{1}{|\bar{c}|}$$

$$= \sum_{\bar{c}} \sum_{c \in \bar{c}} \frac{1}{|\bar{c}|}$$

$$= \sum_{\bar{c}} 1$$

$$= N.$$

We applied Lemma 2.3 to obtain the fourth line.                                                □

We may apply Burnside's Lemma to solve the problems we described earlier. In the round table problem, $|G| = n$. The invariant set of the identity permutation is the entire set of colorings, $C_{(1)} = C$, and the invariant set of any nontrivial rotation $\pi$ is empty, $C_\pi = \{\ \}$. Therefore, the number of nonequivalent seating arrangements is $|C|/n = (n-1)!$.

To determine the number of nonequivalent necklaces with four beads using two different kinds of beads, we calculate $|C_{(1)}| = 16$, $|C_{(13)}| = |C_{(24)}| = 8$, $|C_{(12)(34)}| = |C_{(13)(24)}| = |C_{(14)(23)}| = 4$, and $|C_{(1234)}| = |C_{(1432)}| = 2$. Therefore, $N = (16 + 2 \cdot 8 + 3 \cdot 4 + 2 \cdot 2)/8 = 6$. Last, we calculate the number of nonequivalent three-bead necklaces using three different kinds of beads. Here, $|C_{(1)}| = 27$, $|C_{(12)}| = |C_{(13)}| = |C_{(23)}| = 9$, and $|C_{(123)}| = |C_{(132)}| = 3$, so $N = 60/6 = 10$.

## Exercises

1. Show that $\sim$ is an equivalence relation on $C$.
2. Prove that $G_c$ is a subgroup of $G$.
3. How many different necklaces having five beads can be formed using three different kinds of beads if we discount:

   a. Both flips and rotations?
   b. Rotations only?
   c. Just one flip?

4. The commander of a space cruiser wishes to post four sentry ships arrayed around the cruiser at the vertices of a tetrahedron for defensive purposes, since an attack can come from any direction.

   **a.** How many ways are there to deploy the ships if there are two different kinds of sentry ships available, and we discount all symmetries of the tetrahedral formation?

   **b.** How many ways are there if there are three different kinds of sentry ships available?

**5. a.** How many ways are there to label the faces of a cube with the numbers 1 through 6 if each number may be used more than once?

   **b.** What if each number may only be used once?

## 2.5.3 The Cycle Index

   *1. Lance Armstrong.*
   *2. Alex Zuelle.*
   *3. Fernando Escartin.*

                     — Final standings, Tour de France, 1999

To use Burnside's Lemma to count the number of equivalence classes of a set of colorings $C$, we must compute the size of the invariant set $C_\pi$ associated with every permutation $\pi$ in a group of symmetries $G$. A simple observation allows us to compute the size of this set easily in many situations.

Suppose we wish to determine the number of ways to color $n$ objects using up to $m$ colors, discounting symmetries on the objects described by a group $G$. If a coloring is invariant under the action of a permutation $\pi$ in $G$, then every object permuted by one cycle of $\pi$ must have the same color. Therefore, if $\pi$ has $k$ disjoint cycles, the number of colorings invariant under the action of $\pi$ is $|C_\pi| = m^k$. For example, if $S$ is the set of vertices of a square and $G = D_4$, then $|C_{(1234)}| = m$, $|C_{(12)(34)}| = m^2$, $|C_{(13)(2)(4)}| = m^3$, and $|C_{(1)(2)(3)(4)}| = m^4$. Notice that it is essential to include the cycles of length 1 in these calculations.

With this in mind, we define the *cycle index* of a group $G$ of permutations on $n$ objects. For a permutation $\pi$ in $G$, define a monomial $M_\pi$ associated with $\pi$ in the following way. If $\pi$ is a product of $k$ cycles, and the $i$th cycle has length $\ell_i$, let

$$M_\pi = M_\pi(x_1, x_2, \ldots, x_n) = \prod_{i=1}^{k} x_{\ell_i}. \tag{43}$$

Here, $x_1, x_2, \ldots, x_n$ are indeterminates. The cycle index of $G$ is defined by

$$P_G(\mathbf{x}) = \frac{1}{|G|} \sum_{\pi \in G} M_\pi(\mathbf{x}), \tag{44}$$

where $\mathbf{x}$ denotes the vector $(x_1, x_2, \ldots, x_n)$.

For example, for $G = D_4$, we find that

$$M_{(1)(2)(3)(4)} = x_1^4,$$
$$M_{(13)(2)(4)} = M_{(1)(24)(3)} = x_1^2 x_2,$$
$$M_{(12)(34)} = M_{(13)(24)} = M_{(14)(23)} = x_2^2,$$
$$M_{(1234)} = M_{(1432)} = x_4.$$

Therefore,

$$P_{D_4}(x_1, x_2, x_3, x_4) = \frac{1}{8} \left( x_1^4 + 2x_1^2 x_2 + 3x_2^2 + 2x_4 \right), \tag{45}$$

and

$$P_{C_4}(x_1, x_2, x_3, x_4) = \frac{1}{4} \left( x_1^4 + x_2^2 + 2x_4 \right). \tag{46}$$

By Burnside's Lemma, the number of ways to color $n$ objects using up to $m$ colors, discounting the symmetries of $G$, is $P_G(m, m, \ldots, m)$. For example, the number of equivalence classes of four-bead necklaces composed using $m$ different kinds of beads is

$$P_{D_4}(m, m, m, m) = \frac{1}{8} \left( m^4 + 2m^3 + 3m^2 + 2m \right).$$

Substituting $m = 2$, we find there are six different colorings, as before.

Finally, let us compute the number of twenty-bead necklaces composed of rhodonite, rose quartz, and lapis lazuli beads. We must determine the cycle index for the group $D_{20}$. We find that eight of the rotations, those by $18k$ degrees with $k = 1, 3, 7, 9, 11, 13, 17$, or $19$, are a single cycle of length 20, yielding the term $8x_{20}$ in the cycle index. Four rotations, $k = 2, 6, 14$, and $18$, make two cycles of length 10, contributing $4x_{10}^2$. Rotations with $k = 4, 8, 12$, or $16$ make four cycles of length 5, adding $4x_5^4$, and $k = 5$ or $15$ contributes $2x_4^5$. The rotation with $k = 10$ yields $x_2^{10}$, and the identity adds $x_1^{20}$. Ten of the reflections, the ones about axes of symmetry that pass through midpoints of edges, are each represented by ten transpositions, contributing $10x_2^{10}$. The other ten reflections, flipping about opposite vertices, yield $10x_1^2 x_2^9$. Therefore,

$$P_{D_{20}}(x_1, x_2, \ldots, x_{20}) = \frac{1}{40} \left( x_1^{20} + 10x_1^2 x_2^9 + 11x_2^{10} + 2x_4^5 + 4x_5^4 + 4x_{10}^2 + 8x_{20} \right), \tag{47}$$

and the number of different twenty-bead necklaces that can be made using three kinds of beads is $P_{D_{20}}(3, 3, \ldots, 3) = 87\,230\,157$.

## Exercises

**1.** Show that the monomial $M_\pi$ defined in (44) has the property that $\sum_{i=1}^{k} \ell_i = n$.

**2. a.** Determine the cycle index for $S_4$ and $A_4$.

   **b.** Show that $P_{S_4}(m, m, m, m)$ may be written as a binomial coefficient.

**c.** Determine the smallest value of $m$ for which $P_{A_4}(m, m, m, m) > P_{S_4}(m, m, m, m)$.

3. Determine the number of different necklaces with 21 beads that can be made using four kinds of beads. Your equivalence classes should account for both rotations and flips.

4. Determine the number of eight-bead necklaces that can be made using red, green, blue, and white beads under each of the following groups of symmetries.

   **a.** $D_8$.

   **b.** A subgroup of $D_8$ having four elements. How does the answer depend on the subgroup you choose?

5. Determine the cycle index for the group of symmetries of the faces of a cube, and use this to determine the number of different six-sided dice that can be manufactured using $m$ different labels for the faces of the dice. Assume that each label may be used any number of times.

## 2.5.4  Pólya's Enumeration Formula

*I have yet to see any problem, however complicated, which, when looked at in the right way, did not become still more complicated.*

— Poul Anderson

We can use the cycle index to solve more complicated problems on arrangements in the presence of symmetry. Suppose we need to determine the number of equivalence classes of colorings of $n$ objects using the $m$ colors $y_1, y_2, \ldots, y_m$, where each color $y_i$ occurs a prescribed number of times. For example, how many different necklaces can be made using exactly two rhodonite, nine rose quartz, and nine lapis lazuli beads?

Let us define the *pattern inventory* of the colorings of $n$ objects using the $m$ colors as a generating function in $m$ variables,

$$H(y_1, y_2, \ldots, y_m) = \sum_{\mathbf{v}} a_{\mathbf{v}} y_1^{n_1} y_2^{n_2} \cdots y_m^{n_m}, \tag{48}$$

where the sum runs over all vectors $\mathbf{v} = (n_1, n_2, \ldots, n_m)$ of nonnegative integers satisfying $n_1 + n_2 + \cdots + n_m = n$, and $a_{\mathbf{v}}$ represents the number of nonequivalent colorings of the $n$ objects where the color $y_i$ occurs precisely $n_i$ times. For example, if we denote a rhodonite bead by $r$, a rose quartz bead by $q$, and a lapis lazuli bead by $l$, we see that the answer to our question above is the coefficient of $r^2 q^9 l^9$ in the generating function

$$H(r, q, l) = \sum_{\substack{i+j+k=20 \\ i,j,k \geq 0}} a_{(i,j,k)} r^i q^j l^k.$$

In his influential paper [116], Pólya found that the cycle index can be used to compute the pattern inventory in a simple way. Recall that each occurrence of $x_k$ in the cycle index arises from a permutation having a cycle of length $k$, and if a coloring is invariant under this permutation, then these $k$ elements must have the same color. So either each of the $k$ objects permuted by this cycle has color $y_1$, or each one has color $y_2$, etc. In the spirit of generating functions, this choice can be represented by the formal sum $y_1^k + y_2^k + \cdots + y_m^k$. Pólya found that substituting this expression for $x_k$ for each $k$ in the cycle index yields the pattern inventory for the coloring.

**Theorem 2.5** (Pólya's Enumeration Formula).    *Suppose $S$ is a set of $n$ objects and $G$ is a subgroup of the symmetric group $S_n$. Let $P_G(\mathbf{x})$ be the cycle index of $G$. Then the pattern inventory for the nonequivalent colorings of $S$ under the action of $G$ using colors $y_1, y_2, \ldots, y_m$ is*

$$P_G \left( \sum_{i=1}^{m} y_i, \sum_{i=1}^{m} y_i^2, \ldots, \sum_{i=1}^{m} y_i^n \right). \tag{49}$$

The proof we present follows Stanley [138, vol. 2].

PROOF.    Let $\mathbf{v} = (n_1, n_2, \ldots, n_m)$ be a vector of nonnegative integers of length $m$ whose components sum to $n$, and let $C_\mathbf{v}$ denote the set of colorings of $S$ where exactly $n_i$ of the objects have the color $y_i$, for each $i$. Let $C_{\mathbf{v},\pi}$ denote the invariant set of $C_\mathbf{v}$ under the action of a permutation $\pi$.

If a permutation $\pi$ in $G$ does not disturb a particular coloring, then every object permuted by one cycle of $\pi$ must have the same color. Therefore, $\left| C_{\mathbf{v},\pi} \right|$ is the coefficient of $y_1^{n_1} y_2^{n_2} \cdots y_m^{n_m}$ in $M_\pi(\sum y_i, \sum y_i^2, \ldots, \sum y_i^n)$, where $M_\pi$ is the monomial defined by (43). Let $\mathbf{y}^\mathbf{v}$ denote the term $y_1^{n_1} y_2^{n_2} \cdots y_m^{n_m}$. Then, summing over all permissible vectors $\mathbf{v}$, we obtain

$$\sum_\mathbf{v} \left| C_{\mathbf{v},\pi} \right| \mathbf{y}^\mathbf{v} = M_\pi \left( \sum_{i=1}^{m} y_i, \sum_{i=1}^{m} y_i^2, \ldots, \sum_{i=1}^{m} y_i^n \right).$$

Now we sum both expressions over all $\pi \in G$ and divide by $|G|$. On the left side, we have

$$\frac{1}{|G|} \sum_{\pi \in G} \sum_\mathbf{v} \left| C_{\mathbf{v},\pi} \right| \mathbf{y}^\mathbf{v} = \sum_\mathbf{v} \left( \frac{1}{|G|} \sum_{\pi \in G} \left| C_{\mathbf{v},\pi} \right| \right) \mathbf{y}^\mathbf{v}$$

$$= \sum_\mathbf{v} a_\mathbf{v} \mathbf{y}^\mathbf{v}$$

by Burnside's Lemma, and this is the pattern inventory (48). On the right side, using (44), we obtain (49), the cycle index of $G$ evaluated at $x_k = \sum_i y_i^k$:

$$\frac{1}{|G|} \sum_{\pi \in G} M_\pi \left( \sum_{i=1}^{m} y_i, \sum_{i=1}^{m} y_i^2, \ldots, \sum_{i=1}^{m} y_i^n \right)$$

$$= P_G \left( \sum_{i=1}^{m} y_i, \sum_{i=1}^{m} y_i^2, \ldots, \sum_{i=1}^{m} y_i^n \right). \qquad \square$$

For example, the pattern inventory for nonequivalent four-bead necklaces under $D_4$ using colors red $(r)$, green $(g)$, and blue $(b)$ is

$$P_{D_4}\left(r + g + b,\ r^2 + g^2 + b^2,\ r^3 + g^3 + b^3,\ r^4 + g^4 + b^4\right)$$
$$= r^4 + g^4 + b^4 + r^3g + rg^3 + r^3b + rb^3 + g^3b + gb^3$$
$$+ 2r^2g^2 + 2r^2b^2 + 2g^2b^2 + 2r^2gb + 2rg^2b + 2rgb^2.$$

The pattern inventory for nonequivalent four-bead necklaces under $C_4$ using the same three colors is

$$P_{C_4}\left(r + g + b,\ r^2 + g^2 + b^2,\ r^3 + g^3 + b^3,\ r^4 + g^4 + b^4\right)$$
$$= r^4 + g^4 + b^4 + r^3g + rg^3 + r^3b + rb^3 + g^3b + gb^3$$
$$+ 2r^2g^2 + 2r^2b^2 + 2g^2b^2 + 3r^2gb + 3rg^2b + 3rgb^2.$$

Notice that there are three nonequivalent necklaces with two red beads, one green bead, and one blue bead under $C_4$, but only two under $D_4$. Can you explain this?

Using (47) and Theorem 2.5, we may compute the pattern inventory for twenty-bead necklaces composed of rhodonite $(r)$, rose quartz $(q)$, and lapis lazuli $(l)$ beads. This pattern inventory is shown in Figure 2.6, where we see that there are exactly 231 260 different necklaces with two rhodonite, nine rose quartz, and nine lapis lazuli beads.

Pólya's enumeration formula has many applications in several fields, especially chemistry and physics. Pólya devotes a large portion of his paper [116] to applications involving enumeration of graphs, trees, and chemical isomers.

## Exercises

1. What is the pattern inventory for coloring $n$ objects using the $m$ colors $y_1, y_2, \ldots, y_m$ if the group of symmetries is $S_n$?
2. Use Pólya's enumeration formula to determine the number of six-sided dice that can be manufactured if each of three different labels must be placed on two of the faces.
3. The hydrocarbon benzene has six carbon atoms arranged at the vertices of a regular hexagon, and six hydrogen atoms, with one bonded to each carbon atom. Two molecules are said to be *isomers* if they are composed of the same number and types of atoms, but have different structure.

   a. Show that exactly three isomers (ortho-dichlorobenzene, meta-dichlorobenzene, and para-dichlorobenzene) may be constructed by replacing two of the hydrogen atoms of benzene with chlorine atoms.
   b. How many isomers may be obtained by replacing two of the hydrogen atoms with chlorine atoms, and two others with bromine atoms?

4. The hydrocarbon naphthalene has ten carbon atoms arranged in a double hexagon as in Figure 2.7, and eight hydrogen atoms attached at each of the positions labeled 1 through 8.

$r^{20} + r^{19} q + r^{19} l + 10 r^{18} q^2 + 10 r^{18} q l + 10 r^{18} l^2 + 33 r^{17} q^3 + 90 r^{17} q^2 l + 90 r^{17} q l^2 + 33 r^{17} l^3$
$\quad + 145 r^{16} q^4 + 489 r^{16} q^3 l + 774 r^{16} q^2 l^2 + 489 r^{16} q l^3 + 145 r^{16} l^4 + 406 r^{15} q^5 + 1956 r^{15} q^4 l$
$\quad + 3912 r^{15} q^3 l^2 + 3912 r^{15} q^2 l^3 + 1956 r^{15} q l^4 + 406 r^{15} l^5 + 1032 r^{14} q^6 + 5832 r^{14} q^5 l + 14724 r^{14} q^4 l^2$
$\quad + 19416 r^{14} q^3 l^3 + 14724 r^{14} q^2 l^4 + 5832 r^{14} q l^5 + 1032 r^{14} l^6 + 1980 r^{13} q^7 + 13608 r^{13} q^6 l$
$\quad + 40824 r^{13} q^5 l^2 + 67956 r^{13} q^4 l^3 + 67956 r^{13} q^3 l^4 + 40824 r^{13} q^2 l^5 + 13608 r^{13} q l^6 + 1980 r^{13} l^7$
$\quad + 3260 r^{12} q^8 + 25236 r^{12} q^7 l + 88620 r^{12} q^6 l^2 + 176484 r^{12} q^5 l^3 + 221110 r^{12} q^4 l^4 + 176484 r^{12} q^3 l^5$
$\quad + 88620 r^{12} q^2 l^6 + 25236 r^{12} q l^7 + 3260 r^{12} l^8 + 4262 r^{11} q^9 + 37854 r^{11} q^8 l + 151416 r^{11} q^7 l^2$
$\quad + 352968 r^{11} q^6 l^3 + 529452 r^{11} q^5 l^4 + 529452 r^{11} q^4 l^5 + 352968 r^{11} q^3 l^6 + 151416 r^{11} q^2 l^7$
$\quad + 37854 r^{11} q l^8 + 4262 r^{11} l^9 + 4752 r^{10} q^{10} + 46252 r^{10} q^9 l + 208512 r^{10} q^8 l^2 + 554520 r^{10} q^7 l^3$
$\quad + 971292 r^{10} q^6 l^4 + 1164342 r^{10} q^5 l^5 + 971292 r^{10} q^4 l^6 + 554520 r^{10} q^3 l^7 + 208512 r^{10} q^2 l^8$
$\quad + 46252 r^{10} q l^9 + 4752 r^{10} l^{10} + 4262 r^9 q^{11} + 46252 r^9 q^{10} l + 231260 r^9 q^9 l^2 + 693150 r^9 q^8 l^3$
$\quad + 1386300 r^9 q^7 l^4 + 1940568 r^9 q^6 l^5 + 1940568 r^9 q^5 l^6 + 1386300 r^9 q^4 l^7 + 693150 r^9 q^3 l^8$
$\quad + 231260 r^9 q^2 l^9 + 46252 r^9 q l^{10} + 4262 r^9 l^{11} + 3260 r^8 q^{12} + 37854 r^8 q^{11} l + 208512 r^8 q^{10} l^2$
$\quad + 693150 r^8 q^9 l^3 + 1560534 r^8 q^8 l^4 + 2494836 r^8 q^7 l^5 + 2912112 r^8 q^6 l^6 + 2494836 r^8 q^5 l^7$
$\quad + 1560534 r^8 q^4 l^8 + 693150 r^8 q^3 l^9 + 208512 r^8 q^2 l^{10} + 37854 r^8 q l^{11} + 3260 r^8 l^{12} + 1980 r^7 q^{13}$
$\quad + 25236 r^7 q^{12} l + 151416 r^7 q^{11} l^2 + 554520 r^7 q^{10} l^3 + 1386300 r^7 q^9 l^4 + 2494836 r^7 q^8 l^5$
$\quad + 3326448 r^7 q^7 l^6 + 3326448 r^7 q^6 l^7 + 2494836 r^7 q^5 l^8 + 1386300 r^7 q^4 l^9 + 554520 r^7 q^3 l^{10}$
$\quad + 151416 r^7 q^2 l^{11} + 25236 r^7 q l^{12} + 1980 r^7 l^{13} + 1032 r^6 q^{14} + 13608 r^6 q^{13} l + 88620 r^6 q^{12} l^2$
$\quad + 352968 r^6 q^{11} l^3 + 971292 r^6 q^{10} l^4 + 1940568 r^6 q^9 l^5 + 2912112 r^6 q^8 l^6 + 3326448 r^6 q^7 l^7$
$\quad + 2912112 r^6 q^6 l^8 + 1940568 r^6 q^5 l^9 + 971292 r^6 q^4 l^{10} + 352968 r^6 q^3 l^{11} + 88620 r^6 q^2 l^{12}$
$\quad + 13608 r^6 q l^{13} + 1032 r^6 l^{14} + 406 r^5 q^{15} + 5832 r^5 q^{14} l + 40824 r^5 q^{13} l^2 + 176484 r^5 q^{12} l^3$
$\quad + 529452 r^5 q^{11} l^4 + 1164342 r^5 q^{10} l^5 + 1940568 r^5 q^9 l^6 + 2494836 r^5 q^8 l^7 + 2494836 r^5 q^7 l^8$
$\quad + 1940568 r^5 q^6 l^9 + 1164342 r^5 q^5 l^{10} + 529452 r^5 q^4 l^{11} + 176484 r^5 q^3 l^{12} + 40824 r^5 q^2 l^{13}$
$\quad + 5832 r^5 q l^{14} + 406 r^5 l^{15} + 145 r^4 q^{16} + 1956 r^4 q^{15} l + 14724 r^4 q^{14} l^2 + 67956 r^4 q^{13} l^3$
$\quad + 221110 r^4 q^{12} l^4 + 529452 r^4 q^{11} l^5 + 971292 r^4 q^{10} l^6 + 1386300 r^4 q^9 l^7 + 1560534 r^4 q^8 l^8$
$\quad + 1386300 r^4 q^7 l^9 + 971292 r^4 q^6 l^{10} + 529452 r^4 q^5 l^{11} + 221110 r^4 q^4 l^{12} + 67956 r^4 q^3 l^{13}$
$\quad + 14724 r^4 q^2 l^{14} + 1956 r^4 q l^{15} + 145 r^4 l^{16} + 33 r^3 q^{17} + 489 r^3 q^{16} l + 3912 r^3 q^{15} l^2 + 19416 r^3 q^{14} l^3$
$\quad + 67956 r^3 q^{13} l^4 + 176484 r^3 q^{12} l^5 + 352968 r^3 q^{11} l^6 + 554520 r^3 q^{10} l^7 + 693150 r^3 q^9 l^8$
$\quad + 693150 r^3 q^8 l^9 + 554520 r^3 q^7 l^{10} + 352968 r^3 q^6 l^{11} + 176484 r^3 q^5 l^{12} + 67956 r^3 q^4 l^{13}$
$\quad + 19416 r^3 q^3 l^{14} + 3912 r^3 q^2 l^{15} + 489 r^3 q l^{16} + 33 r^3 l^{17} + 10 r^2 q^{18} + 90 r^2 q^{17} l + 774 r^2 q^{16} l^2$
$\quad + 3912 r^2 q^{15} l^3 + 14724 r^2 q^{14} l^4 + 40824 r^2 q^{13} l^5 + 88620 r^2 q^{12} l^6 + 151416 r^2 q^{11} l^7 + 208512 r^2 q^{10} l^8$
$\quad + \mathbf{231260\, r^2\, q^9\, l^9} + 208512 r^2 q^8 l^{10} + 151416 r^2 q^7 l^{11} + 88620 r^2 q^6 l^{12} + 40824 r^2 q^5 l^{13}$
$\quad + 14724 r^2 q^4 l^{14} + 3912 r^2 q^3 l^{15} + 774 r^2 q^2 l^{16} + 90 r^2 q l^{17} + 10 r^2 l^{18} + r q^{19} + 10 r q^{18} l$
$\quad + 90 r q^{17} l^2 + 489 r q^{16} l^3 + 1956 r q^{15} l^4 + 5832 r q^{14} l^5 + 13608 r q^{13} l^6 + 25236 r q^{12} l^7 + 37854 r q^{11} l^8$
$\quad + 46252 r q^{10} l^9 + 46252 r q^9 l^{10} + 37854 r q^8 l^{11} + 25236 r q^7 l^{12} + 13608 r q^6 l^{13} + 5832 r q^5 l^{14}$
$\quad + 1956 r q^4 l^{15} + 489 r q^3 l^{16} + 90 r q^2 l^{17} + 10 r q l^{18} + r l^{19} + q^{20} + q^{19} l + 10 q^{18} l^2 + 33 q^{17} l^3 + 145 q^{16} l^4$
$\quad + 406 q^{15} l^5 + 1032 q^{14} l^6 + 1980 q^{13} l^7 + 3260 q^{12} l^8 + 4262 q^{11} l^9 + 4752 q^{10} l^{10} + 4262 q^9 l^{11}$
$\quad + 3260 q^8 l^{12} + 1980 q^7 l^{13} + 1032 q^6 l^{14} + 406 q^5 l^{15} + 145 q^4 l^{16} + 33 q^3 l^{17} + 10 q^2 l^{18} + q l^{19} + l^{20}$

FIGURE 2.6. Pattern inventory for necklaces with twenty beads formed using three kinds of beads.

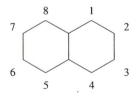

FIGURE 2.7. Naphthalene.

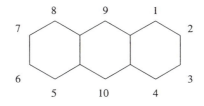

FIGURE 2.8. Anthracene.

**a.** Naphthol is obtained by replacing one of the hydrogen atoms of naphthalene with a hydroxyl group (OH). How many isomers of naphthol are there?

**b.** Tetramethylnaphthalene is obtained by replacing four of the hydrogen atoms of naphthalene with methyl groups (CH$_3$). How many isomers of tetramethylnaphthalene are there?

**c.** How many isomers may be constructed by replacing three of the hydrogen molecules of naphthalene with hydroxyl groups, and another three with methyl groups?

**d.** How many isomers may be constructed by replacing two of the hydrogen molecules of naphthalene with hydroxyl groups, two with methyl groups, and two with carboxyl groups (COOH)?

**5.** The hydrocarbon anthracene has fourteen carbon atoms arranged in a triple hexagon as in Figure 2.8, with ten hydrogen atoms bonded at the numbered positions.

**a.** How many isomers of trimethylanthracene can be formed by replacing three hydrogen atoms with methyl groups?

**b.** How many isomers can be formed by replacing four of the hydrogen atoms with chlorine, and two others with hydroxyl groups?

**6.** The molecule triphenylamine has three rings of six carbon atoms attached to a central nitrogen atom, as in Figure 2.9, and fifteen hydrogen atoms, with one attached to each carbon atom except the three carbons attached to the central nitrogen atom.

**a.** How many isomers can be formed by replacing six hydrogen atoms with hydroxyl groups?

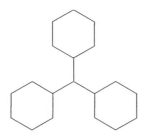

FIGURE 2.9. Triphenylamine.

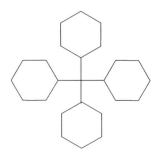

FIGURE 2.10. Tetraphenylmethane.

    **b.** How many isomers can be formed by replacing five hydrogen atoms with
    methyl groups, and five with fluorine atoms?

**7.** The hydrocarbon tetraphenylmethane consists of four rings of six carbon atoms,
each bonded to a central carbon atom, as in Figure 2.10, together with twenty
hydrogen atoms, with one hydrogen atom attached to each carbon atom in the
rings except for those attached to the carbon at the center.

    **a.** How many isomers can be formed by replacing five hydrogen atoms of
    tetraphenylmethane with chlorine?
    **b.** How many isomers can be formed by replacing five hydrogen atoms with
    bromine, and six others with hydroxyl groups?

**8.** Suppose a medical relief agency plans to design a symbol for their organization
in the shape of a regular cross, as in Figure 2.11. To symbolize the purpose
of the organization and emphasize its international constituency, its board of
directors decides that the cross should be white in color, with each of the twelve
line segments outlining the cross colored red, green, blue, or black, with an
equal number of lines of each color. If we discount rotations and flips, how
many different ways are there to design the symbol?

FIGURE 2.11. Symbol of a relief agency.

# 2.6    More Numbers

*Truly, I thought there had been one number more. . .*
— William Shakespeare, *The Merry Wives of Windsor*,
Act IV, Scene I

The combinatorial problems we introduce in this section lead us to four more important classes of numbers: Stirling numbers of the first and second kinds, Bell numbers, and Eulerian numbers. We study some important properties of these numbers, and introduce some different kinds of generating functions to assist us with our derivations. Analyzing these generating functions provides us with some interesting arithmetic insights linking ordinary powers, rising and falling factorial powers, and binomial coefficients.

## 2.6.1    Stirling Numbers of the First Kind

*The Round Table soon heard of the challenge, and of course it was a good deal discussed. . .*
— Mark Twain, *A Connecticut Yankee in King Arthur's Court*

Suppose King Arthur decides to divide his knights into committees in order to better govern Britain. How many ways are there to seat his $n$ knights at $k$ identical round tables if each table can seat any number of knights, and no table can be empty? Here, we must count all the different seating arrangements at each table: Recall that there are $(m - 1)!$ different ways to seat $m$ people at one round table.

Let us represent the $n$ knights by the integers 1 through $n$, and denote the seating of knights $K_1, K_2, \ldots, K_m$ in clockwise order around one table by $(K_1 K_2 \ldots K_m)$. Of course, $(K_2 K_3 \ldots K_m K_1)$ denotes the same arrangement of knights around the table, so to make our notation unique we demand that the knight represented by the smallest number appear first in the list. An arrangement of knights at the $k$ tables is then uniquely represented by a list of $k$ strings of integers in parentheses, where each integer between 1 and $n$ appears exactly once. For example, with six knights and three tables, we might seat knights 1, 3, and 5 in clockwise order around one table, knights 2 and 6 at another table, and knight 4 alone at the third table. This arrangement is denoted by (135)(26)(4). This is precisely the cycle notation we used to describe a permutation on six objects. We see that each seating arrangement of $n$ knights at $k$ tables corresponds to a unique permutation $\pi \in S_n$ having exactly $k$ cycles, and every such permutation corresponds to a unique seating arrangement.

We define the *Stirling number of the first kind*, denoted by $\left[ {n \atop k} \right]$ or $s(n, k)$, to be the number of ways to seat $n$ knights at $k$ tables, or, equivalently, the number of permutations $\pi \in S_n$ having exactly $k$ cycles. These are also called the *Stirling cycle numbers*. We derive a few properties of these numbers.

First, it is impossible to seat $n$ knights at zero tables, unless there are no knights, so

$$\begin{bmatrix} n \\ 0 \end{bmatrix} = \begin{cases} 1 & \text{if } n = 0, \\ 0 & \text{if } n > 0. \end{cases} \tag{50}$$

Second, if there is only one table, then

$$\begin{bmatrix} n \\ 1 \end{bmatrix} = (n-1)!, \quad n \ge 1. \tag{51}$$

Next, if there are $n$ tables, then each knight must sit at his own table, and if there are $n-1$ tables, then one pair of knights must sit at one table, and the others must each sit alone. Thus

$$\begin{bmatrix} n \\ n \end{bmatrix} = 1, \tag{52}$$

and

$$\begin{bmatrix} n \\ n-1 \end{bmatrix} = \binom{n}{2}. \tag{53}$$

There are no arrangements possible if there are more tables than knights, or a negative number of tables, so

$$\begin{bmatrix} n \\ k \end{bmatrix} = 0 \quad \text{if } k < 0 \text{ or } k > n. \tag{54}$$

Further, because of the correspondence between seating arrangements and permutations, we have

$$\sum_k \begin{bmatrix} n \\ k \end{bmatrix} = n!. \tag{55}$$

Consider now the case $n = 4$ and $k = 2$. To seat four knights at two tables, we may first seat three knights at one table, then seat the fourth knight at the second table. The number of arrangements in this case is $\begin{bmatrix} 3 \\ 1 \end{bmatrix} = 2$. Alternatively, we may first seat one of the first three knights at one table, and the other two at the second table. The fourth knight may then either join the single knight, or the table with two knights. There are two possibilities in the latter case, for the fourth knight may sit on either side of the knights already at the table. Thus, there are $3\begin{bmatrix} 3 \\ 2 \end{bmatrix} = 9$ possibilities in this case, and we find that $\begin{bmatrix} 4 \\ 2 \end{bmatrix} = 3\begin{bmatrix} 3 \\ 2 \end{bmatrix} + \begin{bmatrix} 3 \\ 1 \end{bmatrix} = 11$. Figure 2.12 shows these eleven arrangements when Tristram joins Bedivere, Lancelot, and Percival at two tables.

This technique generalizes to give us a recurrence relation for these numbers. To seat $n$ knights at $k$ tables, we may first seat $n-1$ knights at $k-1$ tables, then seat the last knight alone at the $k$th table. Alternatively, we may seat the first $n-1$ knights at $k$ tables, then insert the last knight at one of these tables. There are $m$ different places to insert the knight at a table that has $m$ knights already seated, so

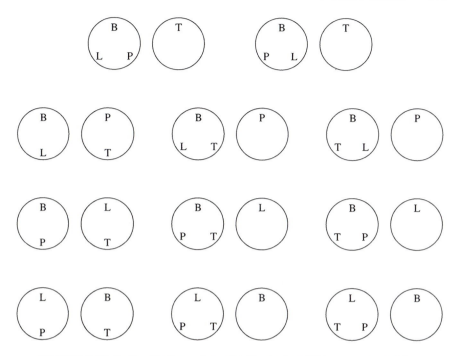

FIGURE 2.12. Seating Bedivere, Lancelot, Percival, and Tristram at two tables.

| $n$ | $k = 0$ | 1 | 2 | 3 | 4 | 5 | 6 | 7 | 8 | $n!$ |
|---|---|---|---|---|---|---|---|---|---|---|
| 0 | 1 | | | | | | | | | 1 |
| 1 | 0 | 1 | | | | | | | | 1 |
| 2 | 0 | 1 | 1 | | | | | | | 2 |
| 3 | 0 | 2 | 3 | 1 | | | | | | 6 |
| 4 | 0 | 6 | 11 | 6 | 1 | | | | | 24 |
| 5 | 0 | 24 | 50 | 35 | 10 | 1 | | | | 120 |
| 6 | 0 | 120 | 274 | 225 | 85 | 15 | 1 | | | 720 |
| 7 | 0 | 720 | 1764 | 1624 | 735 | 175 | 21 | 1 | | 5040 |
| 8 | 0 | 5040 | 13068 | 13132 | 6769 | 1960 | 322 | 28 | 1 | 40320 |

TABLE 2.4. Stirling numbers of the first kind, $\left[\begin{smallmatrix} n \\ k \end{smallmatrix}\right]$.

there are $n - 1$ different places to seat the last knight. Therefore,

$$\left[\begin{matrix} n \\ k \end{matrix}\right] = (n - 1)\left[\begin{matrix} n - 1 \\ k \end{matrix}\right] + \left[\begin{matrix} n - 1 \\ k - 1 \end{matrix}\right], \quad n \geq 1. \tag{56}$$

We can use this formula to compute the triangle of Stirling numbers of the first kind, just as we used the addition identity for binomial coefficients to obtain Pascal's triangle. Our computations appear in Table 2.4.

Recall that for fixed $n$ the generating function for the sequence of binomial coefficients has a particularly nice form: $\sum_k \binom{n}{k} x^k = (x + 1)^n$. We can use the identity (56) to obtain an analogous representation for the sequence of Stirling numbers of the first kind. Let $G_n(x) = \sum_k \left[ {n \atop k} \right] x^k$. Clearly, $G_0(x) = 1$, and for $n \geq 1$,

$$G_n(x) = \sum_k \left[ {n \atop k} \right] x^k$$

$$= (n-1) \sum_k \left[ {n-1 \atop k} \right] x^k + \sum_k \left[ {n-1 \atop k-1} \right] x^k$$

$$= (n-1) G_{n-1}(x) + x G_{n-1}(x),$$

so $G_n(x) = (x + n - 1) G_{n-1}(x)$. It is easy to verify by induction that this implies that $G_n(x) = x(x+1)(x+2) \cdots (x+n-1) = x^{\overline{n}}$. Thus,

$$x^{\overline{n}} = \sum_k \left[ {n \atop k} \right] x^k, \quad n \geq 0. \tag{57}$$

Therefore, the Stirling numbers of the first kind allow us to express rising factorial powers as linear combinations of ordinary powers.

**Exercises**

**1.** Use a combinatorial argument to show that

$$\left[ {n \atop 2} \right] = \frac{n!}{2} \sum_{m=1}^{n-1} \frac{1}{m(n-m)}.$$

**2.** Prove that

$$x^{\underline{n}} = \sum_k \left[ {n \atop k} \right] (-1)^{n-k} x^k. \tag{58}$$

## 2.6.2    Stirling Numbers of the Second Kind

*I don't understand these fractions.*

*— Close Encounters of the Third Kind*

How many ways are there to divide $n$ guests at a party into exactly $k$ groups, if we disregard the arrangement of people within each group? Rephrased, this problem asks for the number of ways to partition a set of $n$ objects into exactly $k$ nonempty subsets, so that each element in the original set appears exactly once among the $n$ subsets. For example, there are three ways to partition the set $\{a, b, c\}$ into two nonempty subsets: $\{a, b\}, \{c\}$; $\{a, c\}, \{b\}$; and $\{b, c\}, \{a\}$. There is just one way to partition $\{a, b, c\}$ into one subset, $\{a, b, c\}$; and just one way to partition $\{a, b, c\}$ into three subsets, $\{a\}, \{b\}, \{c\}$.

The number of ways to divide $n$ objects into exactly $k$ groups is denoted by $\left\{ {n \atop k} \right\}$, or $S(n, k)$. These numbers are called the *Stirling numbers of the second kind*, or the *Stirling set numbers*. Thus, $\left\{ {3 \atop 2} \right\} = 3$, and $\left\{ {3 \atop 1} \right\} = \left\{ {3 \atop 3} \right\} = 1$.

We begin by listing a few properties of these numbers. First, for $n \geq 1$ we have

$$\left\{ \begin{matrix} n \\ 1 \end{matrix} \right\} = \left\{ \begin{matrix} n \\ n \end{matrix} \right\} = 1, \tag{59}$$

since there is only one way to place $n$ people into a single group, and only one way to split them into $n$ groups. Second,

$$\left\{ \begin{matrix} n \\ 0 \end{matrix} \right\} = \begin{cases} 1 & \text{if } n = 0, \\ 0 & \text{if } n > 0, \end{cases} \tag{60}$$

for one cannot divide $n$ people into zero groups, unless there are no people. Third, to divide $n$ people into $n - 1$ groups, we must pick two people to be in one group, then place the rest of the people in groups by themselves, so

$$\left\{ \begin{matrix} n \\ n-1 \end{matrix} \right\} = \binom{n}{2}. \tag{61}$$

Next, we set

$$\left\{ \begin{matrix} n \\ k \end{matrix} \right\} = 0, \quad \text{if } k < 0 \text{ or } k > n. \tag{62}$$

Also, the Stirling number of the first kind distinguishes among the different ways to arrange $n$ people within $k$ groups, and the Stirling number of the second kind does not, so

$$\left\{ \begin{matrix} n \\ k \end{matrix} \right\} \leq \left[ \begin{matrix} n \\ k \end{matrix} \right] \tag{63}$$

for all $n \geq 0$ and all $k$.

Finally, we derive a recurrence relation for $\left\{ \begin{smallmatrix} n \\ k \end{smallmatrix} \right\}$. To divide $n$ people into $k - 1$ groups, we may divide $n - 1$ people into $k - 1$ groups, then place the last person in a group by herself. Alternatively, we can split $n - 1$ people into $k$ groups, then add the last person to one of these $k$ groups. There are $\left\{ \begin{smallmatrix} n-1 \\ k-1 \end{smallmatrix} \right\}$ different arrangements in the first case, and $k \left\{ \begin{smallmatrix} n-1 \\ k \end{smallmatrix} \right\}$ different groupings in the second. Therefore,

$$\left\{ \begin{matrix} n \\ k \end{matrix} \right\} = k \left\{ \begin{matrix} n-1 \\ k \end{matrix} \right\} + \left\{ \begin{matrix} n-1 \\ k-1 \end{matrix} \right\}, \quad n \geq 1. \tag{64}$$

For example, to partition the set $\{a, b, c, d\}$ into two subsets, we can place $d$ in its own set, yielding $\{a, b, c\}, \{d\}$, or we can split $\{a, b, c\}$ into two sets, then add $d$ to one of these sets. This yields the six different partitions

$$\{a, b, d\}, \{c\}; \quad \{a, c, d\}, \{b\}; \quad \{b, c, d\}, \{a\};$$
$$\{a, b\}, \{c, d\}; \quad \{a, c\}, \{b, d\}; \quad \{b, c\}, \{a, d\};$$

and $\left\{ \begin{smallmatrix} 4 \\ 2 \end{smallmatrix} \right\} = 2\left\{ \begin{smallmatrix} 3 \\ 2 \end{smallmatrix} \right\} + \left\{ \begin{smallmatrix} 3 \\ 1 \end{smallmatrix} \right\} = 7$.

We can use identity (64) to form a triangle of Stirling numbers of the second kind, as shown in Table 2.5. The sequence $\{b_n\}$ that appears in this table as the sum across the rows of the triangle is studied in the next section.

| $n$ | $k=0$ | 1 | 2 | 3 | 4 | 5 | 6 | 7 | 8 | $b_n$ |
|---|---|---|---|---|---|---|---|---|---|---|
| 0 | 1 | | | | | | | | | 1 |
| 1 | 0 | 1 | | | | | | | | 1 |
| 2 | 0 | 1 | 1 | | | | | | | 2 |
| 3 | 0 | 1 | 3 | 1 | | | | | | 5 |
| 4 | 0 | 1 | 7 | 6 | 1 | | | | | 15 |
| 5 | 0 | 1 | 15 | 25 | 10 | 1 | | | | 52 |
| 6 | 0 | 1 | 31 | 90 | 65 | 15 | 1 | | | 203 |
| 7 | 0 | 1 | 63 | 301 | 350 | 140 | 21 | 1 | | 877 |
| 8 | 0 | 1 | 127 | 966 | 1701 | 1050 | 266 | 28 | 1 | 4140 |

TABLE 2.5. Stirling numbers of the second kind, $\left\{ {n \atop k} \right\}$, and Bell numbers, $b_n$.

Exercise 1 analyzes the generating function for the sequence of Stirling numbers of the second kind with $n$ fixed. We obtain a more useful relation, however, if we replace the ordinary powers of $x$ in this generating function with falling factorial powers. For fixed $n$, let

$$F_n(x) = \sum_k \left\{ {n \atop k} \right\} x^{\underline{k}},$$

so $F_0(x) = 1$. If $n \geq 1$, then

$$F_n(x) = \sum_k \left( k \left\{ {n-1 \atop k} \right\} + \left\{ {n-1 \atop k-1} \right\} \right) x^{\underline{k}}$$

$$= \sum_k k \left\{ {n-1 \atop k} \right\} x^{\underline{k}} + \sum_k \left\{ {n-1 \atop k} \right\} x^{\underline{k+1}}$$

$$= \sum_k k \left\{ {n-1 \atop k} \right\} x^{\underline{k}} + \sum_k (x-k) \left\{ {n-1 \atop k} \right\} x^{\underline{k}}$$

$$= x F_{n-1}(x),$$

so by induction we obtain

$$x^n = \sum_k \left\{ {n \atop k} \right\} x^{\underline{k}}, \quad n \geq 0. \tag{65}$$

Therefore, Stirling numbers of the second kind allow us to express ordinary powers as combinations of falling factorial powers.

We can derive another useful formula by considering the generating function for the numbers $\left\{ {n \atop k} \right\}$ with $k$ fixed. Let

$$H_k(x) = \sum_{n \geq 0} \left\{ {n \atop k} \right\} x^n,$$

so $H_0(x) = 1$. For $k \geq 1$, we obtain

$$H_k(x) = \sum_{n \geq 1} \left\{ {n \atop k} \right\} x^n$$

$$= \sum_{n \geq 1} \left( k \left\{ \begin{matrix} n-1 \\ k \end{matrix} \right\} + \left\{ \begin{matrix} n-1 \\ k-1 \end{matrix} \right\} \right) x^n$$

$$= kx \sum_{n \geq 0} \left\{ \begin{matrix} n \\ k \end{matrix} \right\} x^n + x \sum_{n \geq 0} \left\{ \begin{matrix} n \\ k-1 \end{matrix} \right\} x^n$$

$$= kx H_k(x) + x H_{k-1}(x),$$

so

$$H_k(x) = \frac{x}{1 - kx} H_{k-1}(x),$$

and therefore

$$H_k(x) = \frac{x^k}{(1-x)(1-2x)\cdots(1-kx)}.$$

Next, we use partial fractions to expand this rational function. Our calculations are somewhat simpler if we multiply by $k!$ first, so we wish to find constants $A_1$, $A_2, \ldots, A_k$ such that

$$\frac{k! x^k}{\prod_{m=1}^{k}(1 - mx)} = \sum_{m=1}^{k} \frac{A_m}{1 - mx}.$$

Clearing denominators, we have

$$k! x^k = \sum_{m=1}^{k} A_m \prod_{j=1}^{m-1}(1 - jx) \prod_{j=m+1}^{k}(1 - jx),$$

and setting $x = 1/m$, we obtain

$$\frac{k!}{m^k} = A_m \prod_{j=1}^{m-1}\left(1 - \frac{j}{m}\right) \prod_{j=m+1}^{k}\left(1 - \frac{j}{m}\right),$$

so

$$k! = m A_m \prod_{j=1}^{m-1}(m - j) \prod_{j=m+1}^{k}(m - j)$$

$$= m A_m (m-1)! (-1)^{k-m} \prod_{j=m+1}^{k}(j - m)$$

$$= (-1)^{k-m} m! (k-m)! A_m,$$

and

$$A_m = (-1)^{k-m} \binom{k}{m}.$$

Thus

$$H_k(x) = \frac{1}{k!} \sum_{m=1}^{k} (-1)^{k-m} \frac{\binom{k}{m}}{1 - mx}$$

$$= \frac{1}{k!} \sum_{m=1}^{k} (-1)^{k-m} \binom{k}{m} \sum_{n \geq 0} (mx)^n$$

$$= \sum_{n \geq 0} \left( \frac{1}{k!} \sum_{m=1}^{k} (-1)^{k-m} \binom{k}{m} m^n \right) x^n,$$

and therefore

$$\left\{ \begin{matrix} n \\ k \end{matrix} \right\} = \frac{1}{k!} \sum_{m=0}^{k} (-1)^{k-m} \binom{k}{m} m^n, \tag{66}$$

for any nonnegative integers $n$ and $k$. This gives us a formula for the Stirling numbers of the second kind. For example, we may compute $\left\{ \begin{matrix} 6 \\ 3 \end{matrix} \right\} = \frac{1}{3!}(3 \cdot 1^6 - 3 \cdot 2^6 + 1 \cdot 3^6) = 90$.

### Exercises

1. Let $G_n(x) = \sum_k \left\{ \begin{matrix} n \\ k \end{matrix} \right\} x^k$, so $G_0(x) = 1$. Show that $G_n(x) = x(G_{n-1}(x) + G'_{n-1}(x))$ for $n \geq 1$, and use this recurrence to compute $G_4(x)$.

2. Show that

$$x^n = \sum_k \left\{ \begin{matrix} n \\ k \end{matrix} \right\} (-1)^{n-k} x^{\overline{k}}. \tag{67}$$

3. Use (58) and (65), or (57) and (67), to prove the following identities.

$$\sum_k \left[ \begin{matrix} n \\ k \end{matrix} \right] \left\{ \begin{matrix} k \\ m \end{matrix} \right\} (-1)^{(n-k)} = \begin{cases} 1 & \text{if } n = m, \\ 0 & \text{otherwise.} \end{cases}$$

$$\sum_k \left\{ \begin{matrix} n \\ k \end{matrix} \right\} \left[ \begin{matrix} k \\ m \end{matrix} \right] (-1)^{(n-k)} = \begin{cases} 1 & \text{if } n = m, \\ 0 & \text{otherwise.} \end{cases} \tag{68}$$

4. Suppose $\{r_1, \ldots, r_\ell\}$ and $\{s_1, \ldots, s_\ell\}$ are two sets of positive integers, $f(x) = \sum_{j=1}^{\ell} (x^{r_j} - x^{s_j})$, and $N$ is a positive integer. Prove that

$$\sum_{j=1}^{\ell} r_j^n = \sum_{j=1}^{\ell} s_j^n$$

for every $n$ with $1 \leq n \leq N$ if and only if $f^{(n)}(1) = 0$ for every $n$ with $1 \leq n \leq N$. Here, $f^{(n)}(x)$ denotes the $n$th derivative of $f(x)$.
For example, select $\{1, 5, 9, 17, 18\}$ and $\{2, 3, 11, 15, 19\}$ as the two sets, and select $N = 4$. Then $1 + 5 + 9 + 17 + 18 = 2 + 3 + 11 + 15 + 19 = 50$, $1^2 + 5^2 + 9^2 + 17^2 + 18^2 = 2^2 + 3^2 + 11^2 + 15^2 + 19^2 = 720$, $1^3 + 5^3 + 9^3 + 17^3 + 18^3 = 2^3 + 3^3 + 11^3 + 15^3 + 19^3 = 11600$, and $1^4 + 5^4 + 9^4 + 17^4 + 18^4 = 2^4 + 3^4 + 11^4 + 15^4 + 19^4 = 195684$; and $f(x) = x - x^2 + x^5 - x^3 + x^9 - x^{11} + x^{17} - x^{15} + x^{18} - x^{19}$ has $f^{(n)}(1) = 0$ for $1 \leq n \leq 4$.

### 2.6.3   Bell Numbers

*Silence that dreadful bell: it frights the isle. . .*
— William Shakespeare, *Othello*, Act II, Scene III

The *Bell number* $b_n$ is the number of ways to divide $n$ people into any number of groups. It is therefore a sum of Stirling numbers of the second kind,

$$b_n = \sum_k \left\{ {n \atop k} \right\}. \tag{69}$$

The first few values of the sequence appear in Table 2.5.

We can derive a recurrence relation for the Bell numbers. To divide $n$ people into groups, consider the different ways to form a group containing one particular person. We must choose some number $k$ of the other $n - 1$ people to join this person in one group, then divide the other $n - 1 - k$ people into groups. Therefore,

$$b_n = \sum_k \binom{n-1}{k} b_{n-1-k}.$$

Reindexing the sum by replacing $k$ with $n - 1 - k$ and applying the symmetry identity for binomial coefficients, we find the somewhat simpler relation

$$b_n = \sum_k \binom{n-1}{k} b_k, \quad n \geq 1. \tag{70}$$

Rather than analyze the ordinary generating function for the sequence of Bell numbers, we introduce another kind of generating function that is often useful in combinatorial analysis. The *exponential generating function* for the sequence $\{a_k\}$ is defined as the ordinary generating function for the sequence $\{a_k/k!\}$. For example, the exponential generating function for the constant sequence $a_k = c$ is $\sum_{k \geq 0} c x^k / k! = c e^x$, and for the sequence $a_k = (-1)^k k!$, it is $1/(1 + x)$. The exponential generating function for the sequence of Bell numbers is therefore

$$E(x) = \sum_{n \geq 0} \frac{b_n}{n!} x^n. \tag{71}$$

We determine a closed form for this series. Differentiating, we have

$$E'(x) = \sum_{n \geq 1} \frac{b_n}{(n-1)!} x^{n-1}$$

$$= \sum_{n \geq 1} \frac{1}{(n-1)!} \left( \sum_k \binom{n-1}{k} b_k \right) x^{n-1}$$

$$= \sum_{n \geq 1} \sum_{k=0}^{n-1} \frac{b_k}{k!(n-1-k)!} x^{n-1}$$

$$= \sum_{k \geq 0} \sum_{n \geq k+1} \frac{b_k}{k!(n-1-k)!} x^{n-1}$$

$$= \sum_{k \geq 0} \sum_{n \geq 0} \frac{b_k}{k!n!} x^{n+k}$$

$$= \left( \sum_{n \geq 0} \frac{x^n}{n!} \right) \left( \sum_{k \geq 0} \frac{b_k}{k!} x^k \right)$$

$$= e^x E(x).$$

Therefore,

$$(\ln E(x))' = e^x,$$

and so

$$\ln E(x) = e^x + c$$

for some constant $c$. Since $E(0) = b_0 = 1$, we must have $c = -1$. Thus

$$E(x) = e^{e^x - 1}. \tag{72}$$

We can use this closed form to determine a formula for $b_n$. Using the Maclaurin series for the exponential function twice, we find that

$$E(x) = \frac{1}{e} e^{e^x}$$

$$= \frac{1}{e} \sum_{k \geq 0} \frac{(e^x)^k}{k!}$$

$$= \frac{1}{e} \sum_{k \geq 0} \frac{1}{k!} \sum_{n \geq 0} \frac{(kx)^n}{n!}$$

$$= \frac{1}{e} \sum_{n \geq 0} \left( \sum_{k \geq 0} \frac{k^n}{k!} \right) \frac{x^n}{n!}.$$

Therefore,

$$b_n = \frac{1}{e} \sum_{k \geq 0} \frac{k^n}{k!}. \tag{73}$$

## Exercises

1. How many ways are there to put nine different dogs into pens if each pen can hold any number of dogs, and every pen is exactly the same?
2. Determine a closed form for the exponential generating function for each of the following sequences.

   a. $a_k = c^k$, with $c$ a constant.
   b. $a_k = 1$ if $k$ is even and 0 if $k$ is odd.
   c. $a_k = k$.

3. Verify that equation (73) for $b_n$ produces the correct value for $b_0$, $b_1$, and $b_2$.
4. Show that the series in equation (73) converges for every $n \geq 0$.

**5.** Use a combinatorial argument to show that

$$\left\{ {n \atop m} \right\} = \sum_k \binom{n-1}{k} \left\{ {n-k-1 \atop m-1} \right\}$$

$$= \sum_k \binom{n-1}{k} \left\{ {k \atop m-1} \right\},$$

for $n \geq 1$, and use this to derive the recurrence (70) for Bell numbers.

**6.** Let $E_k(x)$ denote the exponential generating function for the sequence of Stirling numbers of the first kind with $k$ fixed,

$$E_k(x) = \sum_{n\geq 0} \left[ {n \atop k} \right] \frac{x^n}{n!}.$$

Prove that

$$E_k'(x) = \frac{1}{1-x} E_{k-1}(x),$$

for $k \geq 1$, and use this to derive a closed form for $E_k(x)$,

$$\sum_{n\geq 0} \left[ {n \atop k} \right] \frac{x^n}{n!} = \frac{(-1)^k}{k!} (\ln(1-x))^k . \tag{74}$$

Comtet [32] uses this identity, together with (66) and (75), to derive a complicated formula due to Schlömilch for the Stirling numbers of the first kind. We include it here without proof:

$$\left[ {n \atop k} \right] = \sum_{m=0}^{n-k} (-1)^{n-k-m} \binom{n-1+m}{k-1} \binom{2n-k}{n-k-m} \left\{ {n-k+m \atop m} \right\}$$

$$= \sum_{m=0}^{n-k} \sum_{j=0}^{m} (-1)^{n-k-j} \binom{n-1+m}{k-1} \binom{2n-k}{n-k-m} \binom{m}{j} \frac{j^{n-k+m}}{m!}.$$

**7.** Use an argument similar to that of Exercise 6 to prove that

$$\sum_{n\geq 0} \left\{ {n \atop k} \right\} \frac{x^n}{n!} = \frac{1}{k!} (e^x - 1)^k \tag{75}$$

for every $k \geq 0$.

## 2.6.4  Eulerian Numbers

*Look out football, here we come! Houston Oilers, number one!*
                                  — *The Houston Oilers' Fight Song*

Suppose that a pipe organ having $n$ pipes needs to be installed at a concert hall. Each pipe has a different length, and the pipes must be arranged in a single row. Let us say that two adjacent pipes in an arrangement form an *ascent* if the one on the left is shorter than the one on the right, and a *descent* otherwise. Arranging the pipes from shortest to tallest yields an arrangement with $n - 1$ ascents and no descents; arranging from tallest to shortest results in no ascents and $n - 1$ descents.

Whether for aesthetic or acoustical reasons, the eccentric director of the concert hall demands that there be exactly $k$ ascents in the arrangement of the $n$ pipes. How many ways are there to install the organ? The answer is the *Eulerian number* $\left\langle {n \atop k} \right\rangle$. Stated in more abstract terms, $\left\langle {n \atop k} \right\rangle$ is the number of permutations $\pi$ of the integers $\{1, \ldots, n\}$ having $\pi(i) < \pi(i + 1)$ for exactly $k$ numbers $i$ between 1 and $n - 1$.

We list a few properties of these numbers. It is easy to see that there is only one arrangement of $n$ pipes with no ascents, and only one with $n - 1$ ascents, so

$$\left\langle {n \atop 0} \right\rangle = 1, \quad n \geq 0, \tag{76}$$

and

$$\left\langle {n \atop n-1} \right\rangle = 1, \quad n \geq 1. \tag{77}$$

The Eulerian numbers have a symmetry property similar to that of the binomial coefficients. An arrangement of $n$ pipes with $k$ ascents has $n - 1 - k$ descents, so reversing this arrangement yields a complementary configuration with $n - 1 - k$ ascents and $k$ descents. Thus,

$$\left\langle {n \atop k} \right\rangle = \left\langle {n \atop n-1-k} \right\rangle. \tag{78}$$

Next, by summing over $k$ we count every possible arrangement of pipes precisely once, so

$$\sum_k \left\langle {n \atop k} \right\rangle = n!. \tag{79}$$

We also note the degenerate cases

$$\left\langle {n \atop k} \right\rangle = 0, \quad \text{if } k < 0 \text{ or } k \geq n. \tag{80}$$

Let us consider a recurrence relation for the Eulerian numbers. To arrange $n$ pipes with exactly $k$ ascents, suppose we first place every pipe except the tallest into a configuration with exactly $k$ ascents. Then the tallest pipe can be inserted either in the first position, or between two pipes forming any ascent. Any other position would yield an additional ascent. There are therefore $k + 1$ different places to insert the tallest pipe in this case. Alternatively, we can line up the $n - 1$ shorter pipes so that there are $k - 1$ ascents, then insert the last pipe either at the end of the row or between two pipes forming any descent. There are $n - 2 - (k - 1) = n - k - 1$ descents, so there are $n - k$ different places to insert the tallest pipe in this case. It is impossible to create a permissible configuration by inserting the tallest pipe into any other arrangement of the $n - 1$ shorter pipes, so

$$\left\langle {n \atop k} \right\rangle = (k + 1) \left\langle {n-1 \atop k} \right\rangle + (n - k) \left\langle {n-1 \atop k-1} \right\rangle, \quad n \geq 1. \tag{81}$$

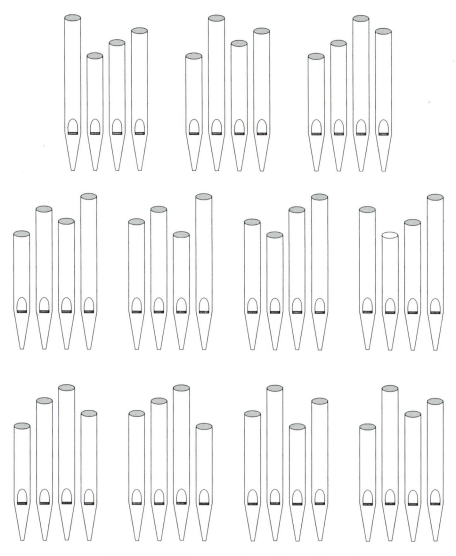

FIGURE 2.13. Four organ pipes with two ascents.

For example, $\left\langle {3 \atop 1} \right\rangle = 2\left\langle {2 \atop 1} \right\rangle + 2\left\langle {2 \atop 0} \right\rangle = 4$, and $\left\langle {4 \atop 2} \right\rangle = 3\left\langle {3 \atop 2} \right\rangle + 2\left\langle {3 \atop 1} \right\rangle = 3 + 8 = 11$. Figure 2.13 shows these eleven arrangements of four pipes with two ascents.

We can use the recurrence (81) to compute the triangle of Eulerian numbers, shown in Table 2.6.

We derive one interesting identity involving Eulerian numbers, binomial coefficients, and ordinary powers. Consider a sort of generating function for the sequence

| $n$ | $k=0$ | 1 | 2 | 3 | 4 | 5 | 6 | 7 | $n!$ |
|---|---|---|---|---|---|---|---|---|---|
| 0 | 1 | | | | | | | | 1 |
| 1 | 1 | | | | | | | | 1 |
| 2 | 1 | 1 | | | | | | | 2 |
| 3 | 1 | 4 | 1 | | | | | | 6 |
| 4 | 1 | 11 | 11 | 1 | | | | | 24 |
| 5 | 1 | 26 | 66 | 26 | 1 | | | | 120 |
| 6 | 1 | 57 | 302 | 302 | 57 | 1 | | | 720 |
| 7 | 1 | 120 | 1191 | 2416 | 1191 | 120 | 1 | | 5040 |
| 8 | 1 | 247 | 4293 | 15619 | 15619 | 4293 | 247 | 1 | 40320 |

TABLE 2.6. Eulerian numbers, $\left\langle {n \atop k} \right\rangle$.

$\{\left\langle {n \atop k} \right\rangle\}$ with $n$ fixed, where we use the binomial coefficient $\binom{x+k}{n}$ in place of $x^k$. Let

$$F_n(x) = \sum_k \left\langle {n \atop k} \right\rangle \binom{x+k}{n},$$

so that $F_0(x) = 1$. For $n \geq 1$,

$$
\begin{aligned}
F_n(x) &= \sum_k \left( (k+1)\left\langle {n-1 \atop k} \right\rangle + (n-k)\left\langle {n-1 \atop k-1} \right\rangle \right) \binom{x+k}{n} \\
&= \sum_k (k+1)\left\langle {n-1 \atop k} \right\rangle \binom{x+k}{n} + \sum_k (n-k)\left\langle {n-1 \atop k-1} \right\rangle \binom{x+k}{n} \\
&= \sum_k (k+1)\left\langle {n-1 \atop k} \right\rangle \binom{x+k}{n} + \sum_k (n-k-1)\left\langle {n-1 \atop k} \right\rangle \binom{x+k+1}{n} \\
&= \sum_k \left\langle {n-1 \atop k} \right\rangle \left( (k+1)\binom{x+k}{n} + (n-k-1)\left( \binom{x+k}{n} + \binom{x+k}{n-1} \right) \right) \\
&= \sum_k \left\langle {n-1 \atop k} \right\rangle \left( n\binom{x+k}{n} + (n-k-1)\binom{x+k}{n-1} \right) \\
&= \sum_k \left\langle {n-1 \atop k} \right\rangle \frac{(x+k)^{\underline{n-1}}}{(n-1)!} \left( (x+k-n+1) + (n-k-1) \right) \\
&= x \sum_k \left\langle {n-1 \atop k} \right\rangle \binom{x+k}{n-1} \\
&= x F_{n-1}(x).
\end{aligned}
$$

Therefore, $F_n(x) = x^n$, so

$$x^n = \sum_k \left\langle {n \atop k} \right\rangle \binom{x+k}{n}, \quad n \geq 0. \tag{82}$$

Thus, Eulerian numbers allow us to write ordinary powers as linear combinations of certain generalized binomial coefficients. For example, $x^4 = \binom{x}{4} + 11\binom{x+1}{4} + 11\binom{x+2}{4} + \binom{x+3}{4}$.

**Exercises**

1. Use an ordinary generating function to find a simple formula for $\left\langle \begin{array}{c} n \\ 1 \end{array} \right\rangle$.

2. A neurotic running back for an American football team will run between two offensive linemen only if the jersey number of the player on the left is less than the jersey number of the player on the right. The player will not run outside the last player on either end of the offensive line. The coach wants to be sure that the running back has at least three options on every play. If the coach always puts seven players on the offensive line, and there are fifteen players on the team capable of playing any position on the offensive line, each of whom has a different jersey number, how many formations of linemen are possible?

# 2.7 Stable Marriage

*How do I love thee? Let me count the ways.*
— Elizabeth Barrett Browning, Sonnet 43,
*Sonnets from the Portuguese*

Most of the problems we have considered in this chapter are questions in enumerative combinatorics, concerned with counting arrangements of objects subject to various constraints. In this section we consider a very different kind of combinatorial problem.

Suppose we must arrange $n$ marriages between $n$ men and $n$ women. Each man supplies us with a list of the women ranked according to his preference; each woman does the same for the men. Is there always a way to arrange the marriages so that no unmatched man and woman prefer each other to their assigned spouses? Such a pairing is called a *stable matching*.

Consider a simple example with $n = 2$. Suppose Aaron prefers Yvonne over Zoë and Björn prefers Zoë over Yvonne. We denote these preferences by

$$A : Y > Z,$$
$$B : Z > Y.$$

Suppose also that Yvonne and Zoë both prefer Aaron over Björn, so

$$Y : A > B,$$
$$Z : A > B.$$

Then the matching of Aaron with Zoë and Björn with Yvonne is unstable, for Aaron and Yvonne prefer each other over their partners. The preferences of Björn and Zoë are irrelevant: Indeed, Zoë would prefer to remain with Aaron in this case. On the other hand, the matching of Aaron with Yvonne and Björn with Zoë is stable, for no unmatched pair prefers to be together over their assigned partners.

The stable marriage problem has many applications in problems involving scheduling and assignments. We mention three examples.

1. Students and Universities.

   Suppose $n$ students have been accepted for $n$ positions at universities. Each student ranks the universities and each university ranks the students. Is there a way to assign the students to universities in such a way that no student and university prefer each other over their assignment?

   This problem is slightly different from the stable marriage problem, for one university might have more than one open position. We might call this variation the polygamous, or polyandrous, stable marriage problem.

2. Stable Roommates.

   Suppose $2n$ students must be paired off and assigned to $n$ dorm rooms. Each student ranks all of the others in order of preference. Is there always a stable pairing? This is another variation on the stable marriage problem, since we no longer have to pair elements from one set with elements from another. This variation is called the *stable roommates* problem.

3. Hospitals and Residents.

   The problem of assigning medical students to hospitals for residencies is analogous to problem 1 above concerning students and universities. In this case, however, a program has been used to make most of the assignments in the U.S. since 1952. The National Residents Matching Program was developed by a group of hospitals to try to ensure a fair method of hiring residents. We describe the algorithm it uses later in this section.

   It is important for this program to produce a stable matching, since no medical student is obligated to accept the position offered. (Since the program's inception, a large majority of the medical students have accepted their offer.)

The stable marriage problem is a question of existential combinatorics, since it asks whether a particular kind of arrangement exists. We might also consider it as a problem in constructive combinatorics, if we ask for an efficient algorithm for finding a stable matching whenever one does exist.

**Exercises**

1. Suppose that four fraternity brothers, Austin, Bryan, Conroe, and Dallas, need to pair off as roommates. Each of the four brothers ranks the other three brothers in order of preference. Prove that there is a set of rankings for which no stable matching of roommates exists.

2. Suppose $M_1$ and $M_2$ are two stable matchings between $n$ men and $n$ women, and we allow each woman to choose between the man she is paired with in $M_1$ and the partner she receives in $M_2$ (each woman always chooses the man she prefers). Show that the result is a stable matching between the men and the women.

3. Suppose that in the previous problem we assign each woman the man she likes less between her partners in the two matchings $M_1$ and $M_2$. Show that the result is again a stable matching.

## 2.7.1  The Gale–Shapley Algorithm

*Matchmaker, matchmaker, make me a match!*

*— Fiddler on the Roof*

In 1962, Gale and Shapley [60] proved that a stable matching between $n$ men and $n$ women always exists by describing an algorithm for finding such a matching. Their algorithm is essentially the same as the one used by the hospitals to select residents, although apparently no one realized this for several years [75, Chapter 1].

In the algorithm, we first choose either the men or the women to be the *proposers*. Suppose we select the men (the women will have their chance soon). Then the men take turns proposing to the women, and the women weigh the offers that they receive. More precisely, the Gale–Shapley algorithm has three principal steps.

*Step* 1. Label every man and woman as free.

*Step* 2. While some man $m$ is free, do the following.

Let $w$ be the highest-ranked woman on the preference list of $m$ to whom $m$ has not yet proposed. If $w$ is free, then label $m$ and $w$ as engaged to each other. If $w$ is engaged to $m'$ and $w$ prefers $m$ over $m'$, then label $m'$ as free and label $m$ and $w$ as engaged to one another. Otherwise, if $w$ prefers $m'$ over $m$, then $w$ remains engaged to $m'$ and $m$ remains free.

*Step* 3. Match all of the engaged couples.

For example, consider the problem of arranging marriages between five men, Mack, Mark, Marv, Milt, and Mort, and five women, Walda, Wanda, Wendy, Wilma, and Winny. The men's and women's preferences are listed in Table 2.7.

First, Mack proposes to Winny, who accepts, and Mark proposes to Wanda, who also accepts. Then Marv proposes to Winny. Winny likes Marv much better than her current fiancé, Mack, so Winny rejects Mack and becomes engaged to Marv. This leaves Mack without a partner, so he proceeds to the second name on his list,

|  | 1 | 2 | 3 | 4 | 5 |
|---|---|---|---|---|---|
| Mack | Winny | Wilma | Wanda | Walda | Wendy |
| Mark | Wanda | Winny | Wendy | Wilma | Walda |
| Marv | Winny | Walda | Wanda | Wilma | Wendy |
| Milt | Winny | Wilma | Wanda | Wendy | Walda |
| Mort | Wanda | Winny | Walda | Wilma | Wendy |
| | | | | | |
| Walda | Milt | Mort | Mack | Mark | Marv |
| Wanda | Milt | Marv | Mort | Mark | Mack |
| Wendy | Mort | Mack | Milt | Mark | Marv |
| Wilma | Mark | Mort | Milt | Mack | Marv |
| Winny | Marv | Mort | Mark | Milt | Mack |

TABLE 2.7. Preferences for five men and women.

Wilma. Wilma currently has no partner, so she accepts. Our engaged couples are now

(Mack, Wilma), (Mark, Wanda), and (Marv, Winny).

Next, Milt proposes to his first choice, Winny. Winny prefers her current partner, Marv, so she rejects Milt. Milt proceeds to his second choice, Wilma. Wilma rejects Mack in favor of Milt, and Mack proposes to his third choice, Wanda. Wanda prefers to remain with Mark, so Mack asks Walda, who accepts. Our engaged couples are now

(Mack, Walda), (Mark, Wanda), (Marv, Winny), and (Milt, Wilma).

Now our last unmatched man, Mort, asks his first choice, Wanda. Wanda accepts Mort over Mark, then Mark asks his second choice, Winny. Winny rejects Mark in favor of her current partner, Marv, so Mark proposes to his third choice, Wendy. Wendy is not engaged, so she accepts. Now all the men and women are engaged, so we have our matching:

(Mack, Walda), (Mark, Wendy), (Marv, Winny),

(Milt, Wilma), and (Mort, Wanda).

We prove that this is in fact a stable matching.

**Theorem 2.6.** *The Gale–Shapley algorithm produces a stable matching.*

PROOF.    First, each man proposes at most $n$ times, so the procedure must terminate after at most $n^2$ proposals. Thus, the procedure is an algorithm. Second, the algorithm always produces a matching. This follows from the observations that a woman, once engaged, is thereafter engaged to exactly one man, and every man ranks every woman, so the last unmatched man must eventually propose to the last unmatched woman. Third, we prove that the matching is stable. Suppose $m$ prefers $w$ to his partner in the matching. Then $m$ proposed to $w$, and was rejected in favor of another suitor. This suitor is ranked higher than $m$ by $w$, so $w$ must prefer her partner in the matching to $m$. Therefore, the matching is stable.    □

We remark that the Gale–Shapley algorithm is quite efficient: A stable matching is always found after at most $n^2$ proposals.

Suppose that we choose the women as the proposers. Does the algorithm produce the same stable matching? We test this by using the lists of preferences in Table 2.7. First, Walda proposes to Milt, who accepts. Next, Wanda proposes to Milt, and Milt prefers Wanda over Walda, so he accepts. Walda must ask her second choice, Mort, who accepts. Then Wendy proposes to Mort, who declines, so she asks Mack, and Mack accepts. Last, Wilma asks Mark, and Winny proposes to Marv, and both accept. We therefore obtain a different stable matching:

(Walda, Mort), (Wanda, Milt), (Wendy, Mack),

(Wilma, Mark), and (Winny, Marv).

|       | 1     | 2     | 3     | 4     | 5     |
|-------|-------|-------|-------|-------|-------|
| Mack  | Winny | Wilma | Wanda | **Walda** | Wendy |
| Mark  | Wanda | Winny | **Wendy** | Wilma | Walda |
| Marv  | **Winny** | Walda | Wanda | Wilma | Wendy |
| Milt  | Winny | **Wilma** | Wanda | Wendy | Walda |
| Mort  | **Wanda** | Winny | Walda | Wilma | Wendy |
|       |       |       |       |       |       |
| Walda | Milt  | Mort  | **Mack** | Mark  | Marv  |
| Wanda | Milt  | Marv  | **Mort** | Mark  | Mack  |
| Wendy | Mort  | Mack  | Milt  | **Mark** | Marv  |
| Wilma | Mark  | Mort  | **Milt** | Mack  | Marv  |
| Winny | **Marv** | Mort  | Mark  | Milt  | Mack  |

TABLE 2.8. Two stable matchings.

Only Winny and Marv are paired together in both matchings; everyone else receives a higher-ranked partner precisely when he or she is among the proposers. Table 2.8 illustrates this for the two different matchings. The pairing obtained with the men as proposers is in boldface; the matching resulting from the women as proposers is underlined.

The next theorem shows that this is no accident: The proposers always obtain the best possible stable matching, and those in the other group, which we call the proposees, always receive the worst possible stable matching. We define two terms before stating this theorem. We say a stable matching is *optimal* for a person $p$ if $p$ can do no better in any stable matching. So if $p$ is matched with $q$ in an optimal matching for $p$, and $p$ prefers $r$ over $q$, then there is no stable matching where $p$ is paired with $r$. Similarly, a stable matching is *pessimal* for $p$ if $p$ can do no worse in any stable matching. So if $p$ is matched with $q$ in a pessimal matching for $p$, and $p$ prefers $q$ over $r$, then there is no stable matching where $p$ is paired with $r$. Finally, a stable matching is optimal for a set of people $P$ if it is optimal for every person $p$ in $P$, and likewise for a pessimal matching.

**Theorem 2.7.** *The stable matching produced by the Gale–Shapley algorithm is independent of the order of proposers, optimal for the proposers, and pessimal for the proposees.*

PROOF. Suppose the men are the proposers. We first prove that the matching produced by the Gale–Shapley algorithm is optimal for the men, regardless of the order of the proposers. Order the men in an arbitrary manner, and suppose that a man $m$ and woman $w$ are matched by the algorithm. Suppose also that $m$ prefers a woman $w'$ over $w$, denoted by $m : w' > w$, and assume that there exists a stable matching $M$ with $m$ paired with $w'$. Then $m$ was rejected by $w'$ at some time during the execution of the algorithm. We may assume that this was the first time a potentially stable couple was rejected by the algorithm. Say $w'$ rejected $m$ in favor of another man $m'$, so $w' : m' > m$. Then $m'$ has no stable partner he prefers over $w'$, by our assumption. Let $w''$ be the partner of $m'$ in the matching $M$. Then

$w'' \neq w'$, since $m$ is matched with $w'$ in $M$, and so $m' : w' > w''$. But then $m'$ and $w'$ prefer each other to their partners in $M$, and this contradicts the stability of $M$.

The optimality of the matching for the proposers is independent of the order of the proposers, so the first statement in the theorem follows immediately.

Finally, we show that the algorithm is pessimal for the proposees. Suppose again that the men are the proposers. Assume that $m$ and $w$ are matched by the algorithm, and that there exists a stable matching $M$ where $w$ is matched with a man $m'$ and $w : m > m'$. Let $w'$ be the partner of $m$ in $M$. Since the Gale–Shapley algorithm produces a matching that is optimal for the men, we have $m : w > w'$. Therefore, $m$ and $w$ prefer each other over their partners in $M$, and this contradicts the stability of $M$.                                                                    $\square$

It is interesting to note that hospitals are the proposers in the National Residents Matching Program, so the matching produced by the program is optimal for the hospitals.

### Exercises

1. Our four fraternity brothers, Austin, Bryan, Conroe, and Dallas, plan to ask four women from the neighboring sorority, Willow, Xena, Yvette, and Zelda, to a dance on Friday night. Each person's preferences are listed in the following table.

|        | 1      | 2      | 3      | 4      |
|--------|--------|--------|--------|--------|
| Austin | Yvette | Xena   | Zelda  | Willow |
| Bryan  | Willow | Yvette | Xena   | Zelda  |
| Conroe | Yvette | Xena   | Zelda  | Willow |
| Dallas | Willow | Zelda  | Yvette | Xena   |
|        |        |        |        |        |
| Willow | Austin | Dallas | Conroe | Bryan  |
| Xena   | Dallas | Bryan  | Austin | Conroe |
| Yvette | Dallas | Bryan  | Conroe | Austin |
| Zelda  | Austin | Dallas | Conroe | Bryan  |

   **a.** What couples attend the dance, if each man asks the women in his order of preference, and each woman accepts the best offer she receives?

   **b.** Suppose the sorority hosts a "Sadie Hawkins" dance the following weekend, where the women ask the men out. Which couples attend this dance?

2. Determine a list of preferences for four men and four women where no one obtains his or her first choice, regardless of who proposes.

3. Determine a list of preferences for four men and four women where one proposer receives his or her lowest-ranked choice.

4. Determine a list of preferences for four men and four women where one proposer receives his or her lowest-ranked choice, and the rest of the proposers receive their penultimate choice.

5. (From [93].) Prove that the Gale–Shapley algorithm terminates after at most $n^2 - n + 1$ proposals by showing that at most one proposer receives his or her lowest-ranked choice.

6. Suppose that more than one woman receives her lowest-ranked choice when the men propose. Prove that there exist at least two stable matchings between the men and the women.

## 2.8    References

*You may talk too much on the best of subjects.*
                    — Benjamin Franklin, *Poor Richard's Almanack*

We list several references for the reader who wishes to embark on further study. The book [33] is an informal discussion of all types of numbers, including many common combinatorial sequences. The textbooks [124] and [140] are introductory treatments of combinatorics and graph theory, with an emphasis on applications. The text [119] is a set of notes on introductory topics in enumerative and constructive combinatorics, [122] is a classic introduction to combinatorics, and [143] is a thorough introduction to the subject, surveying many topics. Many topics in discrete mathematics and enumerative combinatorics are developed extensively in [69]. The texts [1], [32], [76], and [138] present more formal advanced treatments of many aspects of combinatorics. The collections [9] and [129] survey many topics in the theory and applications of combinatorics. Combinatorial identities are studied in [123], and automated techniques for deriving and proving identities involving binomial coefficients and certain other quantities are treated in [115]. The history of binomial coefficients and Pascal's triangle is studied in [39], and [73] observes some interesting patterns in the rows of Pascal's triangle. More details on generating functions and their applications can be found in [69] or [148]. The book [111] describes efficient algorithms for solving a number of problems in combinatorics and graph theory. The reference book [136] catalogs thousands of integer sequences, many of which arise in combinatorics and graph theory, and lists references to the literature for almost all of these sequences. Pólya's theory of counting, along with some generalizations, is described in the expository article [18]. The stable marriage problem is discussed in [119] as an introduction to constructive combinatorics, in [93] as motivation for the mathematical analysis of algorithms, and in detail in [75]. The collection [63] contains many influential papers in combinatorics and graph theory, including [117] and [121]. The two-volume set [70] is a collection of articles on the mathematics of Paul Erdős, including many contributions regarding his work in combinatorics and graph theory. The set [68] gives an overview of dozens of different areas of combinatorics and graph theory for the mathematician or computer scientist.

# 3

# Infinite Combinatorics and Graphs

*...the definitive clarification of the* nature of the infinite *has become necessary...*

— David Hilbert [84]

Infinite sets are very peculiar, and remarkably different from finite sets. This can be illustrated with a combinatorial example.

Suppose we have four pigeons and two pigeonholes. If we place the pigeons in the pigeonholes, one of the pigeonholes must contain at least two pigeons. This crowding will always occur, regardless of the arrangement we choose for the pigeons. Furthermore, the crowding will occur whenever there are more pigeons than holes. In general, if $P$ (*p*igeons) is a finite set, and $H$ (pigeon*h*oles) is a proper subset of $P$, then there is no matching between the elements of $P$ and $H$.

Now suppose that we have a pigeon for each real number in the interval $P = [0, 2]$. Put a leg tag on each pigeon with its real number. Also suppose that we have a pigeonhole for each real number in the interval $H = [0, 1]$. Put an address plate on each pigeonhole with its real number. Note that $H \subsetneq P$, so the set of address plate numbers is a proper subset of the set of leg tag numbers. For each $x \in [0, 2]$, place the pigeon tagged $x$ in the pigeonhole with address $x/2$. Using this arrangement, no two pigeons will be assigned to the same pigeonhole. Thus, if $P$ is infinite and $H$ is a proper subset of $P$, there may be a matching between the elements of $P$ and those of $H$.

Infinite sets behave differently from finite sets, and we have used ideas from graph theory and combinatorics to illustrate this difference. One of the justifications for studying infinite versions of combinatorial and graph-theoretic theorems is to gain more insight into the behavior of infinite sets and, by contrast, more insight into finite sets. Sections 3.1 and 3.2 follow this agenda, culminating in a proof of a finite combinatorial statement using infinite tools. We can also use combinatorial properties to distinguish between different sizes of infinite sets, as is done in Section 3.7. This requires the deeper understanding of the axioms for manipulating infinite sets provided by Sections 3.3 and 3.4, and a precise notion of size that appears in Section 3.5. Combinatorial and graph-theoretic properties can also illuminate the limitations of our axiom systems, as shown in Sections 3.6 and 3.8. The chapter concludes with a hint at the wealth of related topics and a list of references.

## 3.1    Pigeons and Trees

*I wonder about the trees.*

— Robert Frost, *The Sound of Trees*

The chapter introduction shows one way to use pigeons to distinguish between some finite and infinite sets. We could use this as a basis for defining finite sets, but this approach has some drawbacks that we will see in Section 3.4. It is more straightforward to say that a set is *infinite* if its not finite, and that a set is *finite* if its elements can be matched with a bounded initial segment of $\mathbb{N}$. For example, the set $\{A, B, C, D\}$ is finite, because the matching in Figure 3.1 exists. Note that the least integer not used in this matching is 4, which is also the size of the set $\{A, B, C, D\}$. This nifty trick, which results from using 0 in our matchings, reappears in Section 3.5.

Using the preceding notion of infinite and finite sets, we can propose another pigeonhole principle. Suppose we have an infinite number of pigeons that we stuff into a finite number of pigeonholes. Momentarily disregarding physical considerations, we must have at least one pigeonhole that contains an infinite number of pigeons. Letting $P$ be the set of pigeons, $H$ the set of holes, and $f$ the stuffing function, we obtain the following theorem.

**Theorem 3.1** (Infinite Pigeonhole Principle).    *Suppose $P$ is infinite, $H$ is finite, and $f : P \to H$. Then there is an element $h \in H$ such that the set $\{p \in P \mid f(p) = h\}$ is infinite.*

PROOF.    Let $P$, $H$, and $f$ be as in the hypothesis of the theorem. In particular, let $H = \{h_0, h_1, \ldots, h_n\}$. Suppose, by way of contradiction, that for each $h_i \in H$, the set $P_i = \{p \in P \mid f(p) = h_i\}$ has $s_i$ elements. Because $P$ can be written as

FIGURE 3.1. A matching.

$P = P_0 \cup P_1 \cup \ldots \cup P_n$, we see that $\sum_{i \le n} s_i$ is the size of $P$. Thus $P$ is finite, providing the desired contradiction.    □

A physicist might suggest that the density of matter resulting from cramming an unbounded number of pigeons into a bounded pigeonhole would yield a fusion explosion, obliterating any evidence that could be used by litigious animal rights advocates. Home experiments with actual pigeons are strongly discouraged. Despite the physical impracticality of our theorem, it is handy for proving a very nice theorem about trees.

As stated in Chapter 1, a tree is a connected acyclic graph. For big trees, it is handy to designate a root and label the vertices. Figure 3.2 is an example. As a convenient convention, we always think of the root $r$ as the bottom of the tree and vertices farther from $r$ as being higher in the tree. A *path* through a tree is a path leading up and away from the root. For example, $r, a_1, b_2$ and $r, a_1, b_1, c_0$ are paths in the tree above. The sequence $r, a_1, b_0$ is not a path, because $a_1 b_0$ is not an edge in the graph. If we add the edge $a_1 b_0$, the resulting graph is not a tree. (Find the cycle!)

A *level* in a tree is the collection of all vertices at a fixed distance from the root. The levels in our sample tree are $\{r\}$, $\{a_0, a_1\}$, $\{b_0, b_1, b_2\}$ and $\{c_0, c_1, c_2\}$. If $v$ is a vertex and $w$ is a neighboring vertex in the next higher level, then we call $w$ an immediate successor of $v$. In the sample, $b_1$ is an immediate successor of $a_1$, and $b_0$ is not. We can even say that $c_1$ is a successor of $a_1$, but not an immediate successor.

The vertex labels in the sample tree are arbitrary; if we want more than 26 levels, we could use a different labeling scheme. It is even possible to reuse labels in some circumstances, as shown in Exercise 2.

Now we are ready to state König's Lemma. The result concerns infinite trees, that is, trees with an infinite number of vertices. Essentially, König's Lemma says that big skinny trees are tall.

**Theorem 3.2** (König's Lemma).    *If $T$ is an infinite tree and each level of $T$ is finite, then $T$ contains an infinite path.*

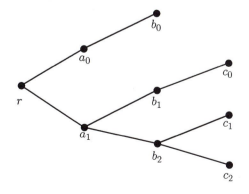

FIGURE 3.2. A tree with labels.

PROOF.    Let $T$ be an infinite tree in which every level is finite. Let $L_0 = \{r\}$, $L_1, L_2, \ldots$ be the levels of $T$. We will construct a path as follows. Let $r$ be the first element of the path. There are infinitely many vertices in $T$ above $r$. Each of these vertices is either in $L_1$ or above a unique vertex in $L_1$. Map each of the vertices above $r$ to the vertex of $L_1$ that it is equal to or above. We have mapped infinitely many vertices (pigeons) to the finitely many vertices of $L_1$ (pigeonholes). By Theorem 3.1, there is at least one vertex of $L_1$ that is above $r$ and has infinitely many vertices above it; pick one and call it $v_1$. The path so far is $r, v_1$. Since there are infinitely many vertices above $v_1$, we can replace $r$ by $v_1$ in the preceding argument and select $v_2$. Similarly, for each $n \in \mathbb{N}$, when we have found $v_n$ we can find $v_{n+1}$. Thus $T$ contains an infinite path.                          □

König's Lemma appears in the 1927 paper of Dénes König [94]. Some authors (e.g [112]) refer to the lemma as König's Infinity Theorem. The name König's Theorem is usually reserved for an unrelated result on cardinal numbers proved by Julius König, another (earlier) famous Hungarian mathematician.

### Exercises

1. Suppose we arrange finitely many pigeons in infinitely many pigeon holes. Use the Infinite Pigeonhole Principle to prove that there are infinitely many pigeonholes that contain no pigeons.
2. Reusing labels in trees.

    Here is an example of a tree where labels are reused.

    Note that in this tree, each vertex can be reached by a path corresponding to a unique sequence of labels. For example, there is exactly one vertex corresponding to $r, 0, 1$.

    a. Give an example of a tree with badly assigned labels, where two vertices have the same sequence of labels.
    b. Prove that if the immediate successors of each vertex in a tree have distinct labels, then no two vertices have matching sequences of labels.
    c. Prove the converse of part 2b.

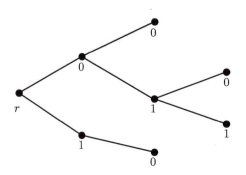

FIGURE 3.3. A tree with reused labels.

3. 2-coloring an infinite graph.
   Suppose $G$ is a graph with vertices $V = \{v_i \mid i \in \mathbb{N}\}$ and every subgraph of $G$ can be 2-colored. Use König's Lemma to prove that $G$ is 2-colorable. (Hint: Build a tree of partial colorings. Put the vertex $root, red, blue, blue$ in the tree if and only if assigning red to $v_0$, blue to $v_1$, and blue to $v_2$ yields a 2-coloring of the subgraph with vertices $\{v_0, v_1, v_2\}$. An infinite path through such a tree will be a coloring of $G$. You must prove that the tree is infinite and that each level is finite.)

4. Construct an infinite graph where each finite subgraph can be colored using a finite number of colors, but where infinitely many colors are needed to color the entire graph. (Hint: Use lots of edges.)

5. Heine–Borel Theorem on compactness of the real interval $[0, 1]$.
   Use König's Lemma to prove that if $(a_0, b_0), (a_1, b_1), \ldots$ are open intervals in $\mathbb{R}$ and $[0, 1] \subset (a_0, b_0) \cup (a_1, b_1) \cup \cdots$, then for some $n$, $[0, 1] \subset (a_0, b_0) \cup (a_1, b_1) \cup \ldots \cup (a_n, b_n)$. (Hint: Build a tree where the labels in the $i$th level are the closed intervals obtained by removing $(a_0, b_0) \cup (a_1, b_1) \cup \ldots \cup (a_i, b_i)$ from $[0, 1]$ and the successors of a vertex $v$ are labeled with subintervals of the interval for $v$. Use the fact that the intersection of any sequence of nested closed intervals is nonempty to show that the tree contains no infinite paths. Apply the contrapositive of König's Lemma.)

## 3.2   Ramsey Revisited

*Ah! the singing, fatal arrow,*
*Like a wasp it buzzed and stung him!*
                    — H.W. Longfellow, *The Song of Hiawatha*

Suppose that we 2-color the edges of $K_6$, the complete graph with six vertices, using the colors red and blue. As we proved in Chapter 1, the colored graph must contain a red $K_3$ or a blue $K_3$. Since we can 2-color $K_5$ in a way that prevents monochromatic triangles, $K_6$ is the smallest graph that must contain a monochromatic triangle. Thus, the Ramsey number $R(3, 3)$ is 6, as noted in Theorem 1.42. If we want to guarantee a monochromatic $K_4$ subgraph we must 2-color $K_{18}$, because $R(4, 4) = 18$. Exact values for $R(p, p)$ when $p \geq 5$ are not known, but by the Erdős–Szekeres bound (Theorem 1.44) we know that these Ramsey numbers exist.

Suppose that $G$ is the complete graph with vertices $V = \{v_i \mid i \in \mathbb{N}\}$. If we 2-color the edges of $G$, what can we say about monochromatic complete subgraphs? Since $G$ contains $K_6$, it must contain a monochromatic $K_3$. Similarly, since $G$ contains $K_{18}$, it must contain a monochromatic $K_4$. For $p \geq 5$, we know that $R(p, p)$ is finite and that $G$ contains $K_{R(p,p)}$ as a subgraph, so $G$ must contain a monochromatic $K_p$. So far we know that $G$ must contain arbitrarily large finite monochromatic complete subgraphs. As a matter of fact, $G$ contains an infinite complete monochromatic subgraph, though this requires some proof.

**Theorem 3.3.** *Let $G$ be the complete infinite graph with vertices $V = \{v_i \mid i \in \mathbb{N}\}$. Given any 2-coloring of the edges, $G$ will contain an infinite complete monochromatic subgraph.*

PROOF.    Suppose the edges of $G$ are colored using red and blue. We will build an infinite subsequence $\langle w_i \mid i \in \mathbb{N} \rangle$ of $V$ by repeatedly applying the pigeonhole principle (Theorem 3.1). Let $w_0 = v_0$. For each $i > 0$, the edge $v_0 v_i$ is either red or blue. Since this assigns $v_i$ to one of two colors for each $i > 0$, there is an infinite set of vertices $V_0$ such that all the edges $\{v_0 v \mid v \in V_0\}$ are the same color. Suppose we have selected $w_n$ and $V_n$. Let $w_{n+1}$ be the lowest-numbered vertex in $V_n$, and let $V_{n+1}$ be an infinite subset of $V_n$ such that the edges $\{w_{n+1} v \mid v \in V_{n+1}\}$ are the same color. This completes the construction of the sequence.

This sequence $\langle w_i \mid i \in \mathbb{N} \rangle$ has a very interesting property. If $i < j < k$, then $w_j$ and $w_k$ are both in $V_i$, and consequently $w_i w_j$ and $w_i w_k$ are the same color! We will say that a vertex $w_i$ is blue-based if $j > i$ implies $w_i w_j$ is blue, and red-based if $j > i$ implies $w_i w_j$ is red. Each vertex in the infinite sequence $\langle w_i \mid i \in \mathbb{N} \rangle$ is blue-based or red-based, so by the pigeonhole principle there must be an infinite subsequence $\langle w_{i_0}, w_{i_1}, \ldots \rangle$ where each element has the same color base. As a sample case, suppose the vertices in the subsequence are all blue-based. Then for each $j < k$, since $w_{i_j}$ is blue-based, the edge $w_{i_j} w_{i_k}$ is blue. Thus all the edges of the complete subgraph with vertices $\{w_{i_0}, w_{i_1}, \ldots\}$ are blue. If the subsequence vertices are red-based, then the edges of the associated infinite complete subgraph are red.    □

Using the preceding theorem, we can prove that the finite Ramsey numbers exist without relying on the Erdős–Szekeres bound.

**Theorem 3.4.** *For each $n \in \mathbb{N}$ there is an $m \in \mathbb{N}$ such that $R(n, n) = m$.*

PROOF.    By way of contradiction, suppose that there is an $n$ such that for every $m$ there is a 2-coloring of the edges of $K_m$ that contains no monochromatic $K_n$ subgraph. Let $G$ be the complete graph with vertices $V = \{v_i \mid i \in \mathbb{N}\}$. Let $E = \{e_i \mid i \in \mathbb{N}\}$ be an enumeration of the edges of $G$. Construct a tree $T$ of partial edge colorings of $G$ as follows. Include the sequence $root, c_0, c_1, c_2, \ldots, c_k$ in $T$ if and only if when edge $e_i$ is colored color $c_i$ for all $i \leq k$, the subgraph of $G$ containing $e_0, e_1, \ldots, e_k$ contains no monochromatic $K_n$. The $k$th level of $T$ contains at most $2^k$ vertices, so each level is finite. Since we have assumed that there is a way of coloring any $K_m$ so that no monochromatic $K_n$ appears, $T$ is infinite. By König's Lemma (Theorem 3.2), $T$ has an infinite path. This infinite path provides a 2-coloring of $G$ that contains no monochromatic $K_n$. Thus for this coloring, $G$ has no infinite complete monochromatic subgraph, contradicting the preceding theorem. Our initial supposition must be false, and so for each $n$, there is an $m$ such that $R(n, n) = m$.    □

We just used the infinite pigeonhole principle, infinite trees, and colorings of infinite graphs to prove a result about finite graphs! (In doing so, we are imitating Ramsey [121].) Besides being inherently fascinating, infinite constructions are very handy. Furthermore, the arguments are easily generalized. In order to take full advantage of our work, we need some new notation.

Here come the arrows! The notation $\kappa \rightarrow (\lambda)_c^2$ means that every $c$-colored complete graph on $\kappa$ vertices contains a monochromatic complete subgraph with $\lambda$ vertices. Most people pronounce $\kappa \rightarrow (\lambda)_c^2$ as "kappa arrows lambda 2 c." The statement that $R(3, 3) = 6$ combines the facts that $6 \rightarrow (3)_2^2$ ($K_6$ is big enough) and $5 \nrightarrow (3)_2^2$ ($K_5$ is not big enough). If we imitate set theorists and write $\omega$ for the size of the set $V = \{v_i \mid i \in \mathbb{N}\}$, we can rewrite Theorem 3.3 as $\omega \rightarrow \omega_2^2$. Abbreviating "for all $n$" by $\forall n$ and "there exists an $m$" by $\exists m$, Theorem 3.4 becomes $\forall n \exists m \; m \rightarrow (n)_2^2$.

Arrow notation is particularly useful if we want to use lots of colors. It is easy to check that if every use of two colors is replaced by some finite value $c$ in the proof of Theorem 3.3, the result still holds. The same can be said for Theorem 3.4. Consequently, for any $c \in \mathbb{N}$ we have

$$\omega \rightarrow (\omega)_c^2 \quad \text{and} \quad \forall n \exists m \; m \rightarrow (n)_c^2.$$

Note that when $c$ is largish, the arrow notation is particularly convenient. For example, the statement "$m$ is the least number such that $m \rightarrow (3)_9^2$" translates into Ramsey number notation as the cumbersome formula $R(3, 3, 3, 3, 3, 3, 3, 3, 3) = m$. Nobody would want to translate $m \rightarrow (3)_{1000}^2$. On the other hand, $R(3, 4) = 9$ does not translate into our arrow notation.

The 2 in $\kappa \rightarrow (\lambda)_c^2$ indicates that we are coloring *unordered pairs* of elements taken from a set of size $\kappa$. When we edge color a graph, we are indeed assigning colors to the pairs of vertices corresponding to the edges. However, we can extend Ramsey's theorem by coloring larger subsets. The resulting statements are still very combinatorial in flavor, though they no longer refer to edge colorings. For example, the notation $\kappa \rightarrow (\lambda)_c^n$ means that for any assignment of $c$ colors to the unordered $n$-tuples of $\kappa$, there is a particular color (say lime) and a subset $X \subset \kappa$ of size $\lambda$ such that no matter how we select $n$ elements from $X$, the resulting $n$-tuple will be lime colored. The proofs of Theorems 3.3 and 3.4 can be modified to prove the following theorems.

**Theorem 3.5** (Infinite Ramsey's Theorem).    *For all $n \in \mathbb{N}$ and $c \in \mathbb{N}$, $\omega \rightarrow (\omega)_c^n$.*

PROOF.    By induction on $n$. Exercise 2 gives hints.    □

**Theorem 3.6** (Finite Ramsey's Theorem).    *For all $k, n, c \in \mathbb{N}$, there is an $m \in \mathbb{N}$ such that $m \rightarrow (k)_c^n$.*

PROOF.    Follows from Theorem 3.5. Exercise 3 gives hints.    □

Throughout this section we have been very picky about our infinite sets. For example, $V = \{v_i \mid i \in \mathbb{N}\}$ has a built-in matching with $\mathbb{N}$. What happens if we look at graphs with a vertex for each real number? In Section 3.7 we will learn that the analog of Theorem 3.3 fails for an infinite graph of this sort. For what sorts of infinite graphs does Theorem 3.3 hold? To answer this question, we need a deeper understanding of the infinite.

**Exercises**

1. Let $X = \{x_i \mid i \in \mathbb{N}\}$ be a set. Suppose that the relation $\leq$ is a partial ordering on $X$. That is, for any $a, b, c \in X$, suppose that

   - $a \leq a$,
   - if $a \leq b$ and $b \leq a$, then $a = b$, and
   - if $a \leq b \leq c$, then $a \leq c$.

   Use Theorem 3.3 to prove that there is an infinite subset $Y \subset X$ such that either

   - for every $a, b \in Y$, either $a \leq b$ or $b \leq a$, or
   - for every $a, b \in Y$, both $a \not\leq b$ and $b \not\leq a$.

   A subset of the first type is called a *chain*, and a subset of the second type is called an *antichain*.

2. Prove Theorem 3.5. Begin by proving Theorem 3.5 for 2 colors. Proceed by induction on $n$. For $n = 1$, use the pigeonhole principle as a base case. For the induction step, assume that $\omega \to (\omega)_2^n$, and prove that $\omega \to (\omega)_2^{n+1}$ by imitating the proof of Theorem 3.3, substituting applications of $\omega \to (\omega)_2^n$ for the use of the pigeonhole principle.

   Given the theorem for 2 colors, there are many ways to prove it for other finite numbers of colors. You could replace 2 by $c$ everywhere in the proof you just did, or you could try proving the theorem for $c$ colors and $n$-tuples by using the theorem for 2 colors and $2n$-tuples.

3. Prove Theorem 3.6. Imitate the proof of Theorem 3.4, using Theorem 3.5 in place of Theorem 3.3.

4. One way to visualize coloring triples.

   We can represent a coloring of triples by attaching a claw to a triple that points in a particular direction. For example, the tripartite graph in Figure 3.4 represents coloring $\{0, 1, 2\}$ red and $\{1, 3, 4\}$ blue.

   Figure 3.5 represents a 2-coloring of all 10 triples that can be formed from the set $\{0, 1, 2, 3, 4\}$. You can check that every four-element subset of $\{0, 1, 2, 3, 4\}$

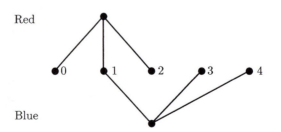

FIGURE 3.4. A tripartite graph representing a 2-coloring.

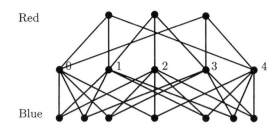

FIGURE 3.5. A 2-coloring of the triples from {0, 1, 2, 3, 4}.

contains a triple with a claw on the blue side and a triple with a claw on the red side. Thus, Figure 3.5 illustrates that $5 \nrightarrow (4)_2^3$.

**a.** Find a different coloring that shows that $5 \nrightarrow (4)_2^3$ and represent it as a tripartite graph. (How do you know that your coloring is significantly different?)

**b.** Find a tripartite graph that shows that $5 \nrightarrow (3)_2^2$.

**c.** Devise a way to draw a similar graph that shows that $6 \nrightarrow (3)_3^2$.

**d.** Find a tripartite graph that shows that $6 \nrightarrow (4)_2^3$. Since every triple gets a claw, make your life easier by drawing only the red claws.

## 3.3    ZFC

*No one shall be able to drive us from the paradise that Cantor created for us.*

— David Hilbert [84]

Paraphrasing Hilbert, in Cantor's paradise mathematicians can joyfully prove new and rich results by employing infinite sets. Since we have been living reasonably comfortably in this paradise since the beginning of the chapter, Hilbert's anxiety about eviction may seem misplaced. However, Russell and other mathematicians discovered some set-theoretic paradoxes that made the naïve use of infinite sets very questionable. Hilbert responded by calling for a careful investigation with the goal of completely clarifying the nature of the infinite.

One could argue that Hilbert's call (made in 1925) had already been answered by Zermelo in 1908. In the introduction to [152], Zermelo claimed to have reduced the entire theory created by Cantor and Dedekind to seven axioms and a few definitions. Although we now use formulations of the axioms of separation and replacement that more closely resemble those of Fraenkel and Skolem, the most commonly used axiomatization of set theory, ZFC, consists primarily of axioms proposed by Zermelo. The letters ZFC stand for **Z**ermelo, **F**raenkel, and Axiom of **C**hoice. Although Skolem does not get a letter, it would be hard to overestimate his influence in recasting ZFC as a first order theory.

ZFC succinctly axiomatizes what has become the de facto foundation for standard mathematical practice. With sufficient diligence, it would be possible to formalize every theorem appearing so far in this book and prove each of them from the axioms of ZFC. Since all these proofs can be carried out in a less formal setting, foundational concerns are insufficient motivation for adopting an axiomatic approach. However, many of the results in Sections 3.4 through 3.8 cannot even be stated without referring to ZFC. We will use ZFC as a base theory to explore the relative strength of some very interesting statements about sets. In particular, ZFC will be central to our discussion of large cardinals and infinite combinatorics.

## 3.3.1    Language and Logical Axioms

*The comfort of the typesetter is certainly not the summum bonum.*
— Gottlob Frege [57]

Before we discuss the axioms of ZFC, we need to list the symbols we will use. Although some of these symbols may be unfamiliar, they can be used as a very convenient shorthand.

Variables can be uppercase or lowercase letters with subscripts tacked on if we please. Good examples of variables include $A$, $B$, $x$, and $y_3$. The symbol $\emptyset$ denotes the empty set, and $\mathcal{P}$ and $\cup$ are function symbols for the power set and union. The exact meaning of $\emptyset$, $\mathcal{P}(x)$, and $\cup x$ are determined by the axioms in the next section. ($\cup x$ is *not* a typographical error; a discussion appears later.) A *term* is a variable, the symbol $\emptyset$, or the result of applying a function to a term. In ZFC, terms always denote sets. Consequently, all the objects discussed in ZFC are sets. Some early formalizations of set theory include distinct objects with no elements. These objects are usually called atoms or urelements. They do not show up in ZFC.

The atomic formulas of ZFC are $x \in y$ and $x = y$, where $x$ and $y$ could be any terms. As one would expect, the formula $x \in y$ means $x$ is an element of $y$. The connection between $\in$ and $=$ is partly determined by the axiom of extensionality (in the next section) and partly determined by the fact that $=$ really does denote the familiar equality relation.

All other formulas of ZFC are built up by repeatedly applying logical connectives and quantifiers to the atomic formulas. Table 3.1 lists typical formulas and their translations. The letters $\theta$ and $\psi$ denote formulas of ZFC.

| Formula | Translation |
|---------|-------------|
| $\neg\theta$ | not $\theta$ |
| $\theta \wedge \psi$ | $\theta$ and $\psi$ |
| $\theta \vee \psi$ | $\theta$ or $\psi$ |
| $\theta \rightarrow \psi$ | if $\theta$ then $\psi$ |
| $\theta \leftrightarrow \psi$ | $\theta$ if and only if $\psi$ |
| $\forall x\theta$ | for all sets $x$, $\theta$ holds |
| $\exists x\theta$ | there is a set $x$ such that $\theta$ holds |

TABLE 3.1. Translations of connectives and quantifiers.

Specifying that ZFC is a first order theory implicitly appends the axioms for predicate calculus with equality to the axioms for ZFC. In a nutshell, these logical axioms tell us that the connectives and quantifiers have the meanings shown in Table 3.1, and that $=$ is well behaved. In particular, we can substitute equal terms. Thus, if $x = y$ and $\theta(x)$ both hold, then $\theta(y)$ holds, too. As a consequence, we can prove the following theorem.

**Theorem 3.7.** *Equal sets have the same elements. Formally,*

$$x = y \rightarrow \forall t (t \in x \leftrightarrow t \in y).$$

PROOF.    Suppose $x = y$. Fix $t$. If $t \in x$, then by substitution, $t \in y$. Similarly, if $t \in y$, then $t \in x$. Our choice of $t$ was arbitrary, so $\forall t (t \in x \leftrightarrow t \in y)$.    □

We could completely formalize the preceding argument as a symbolic logic proof in any axiom system for predicate calculus with equality. Good formal axiom systems can be found in Mendelson [104] or Kleene [92] by readers with a frighteningly technical bent.

It is very convenient to write $x \subset y$ for the formula $\forall t (t \in x \rightarrow t \in y)$. Using this abbreviation and only the axioms of predicate calculus, we could prove that $\forall x (x \subset x)$, showing that every set is a subset of itself. We could also prove that containment is a transitive relation, which can be formalized as

$$\forall x \forall y \forall z ((x \subset y \wedge y \subset z) \rightarrow x \subset z).$$

The preceding results (which appear in the exercises) rely on logical axioms rather than on the actual nature of sets. To prove meaty theorems, we need more axioms.

### 3.3.2    Proper Axioms

> ...*I tasted the pleasures of Paradise, which produced these Hell torments*...
>
> — Pangloss, in *Candide*

The axiom system ZFC consists of nine basic axioms plus the axiom of choice. Typically, the nine basic axioms are referred to as ZF. In this section, we will examine the axioms of ZF, including their formalizations, some immediate applications, and a few random historical comments. This should be less painful than the affliction of Pangloss.

**1.** Axiom of extensionality: If $a$ and $b$ have the same elements, then $a = b$. Formally,

$$\forall x (x \in a \leftrightarrow x \in b) \rightarrow a = b.$$

This axiom is the converse of Theorem 3.7, so ZF can prove that $a = b$ if and only if $a$ and $b$ have exactly the same elements. Using this, we can prove the following familiar theorem about the connection between subsets and equality.

**Theorem 3.8.**   *For all sets a and b, a = b if and only if a ⊂ b and b ⊂ a.*
*Formally,*

$$\forall a \forall b (a = b \leftrightarrow (a \subset b \wedge b \subset a)).$$

PROOF.   First suppose that $a = b$. Since $a \subset a$ (see Exercise 1), by substitution
we have $a \subset b$ and $b \subset a$. Thus, $a = b \rightarrow (a \subset b \wedge b \subset a)$.

To prove the converse, suppose $a \subset b$ and $b \subset a$. Since $a \subset b$, for every
$x$ we have that if $x \in a$ then $x \in b$. Similarly, since $b \subset a$, $x \in b$ implies
$x \in a$. Summarizing, for all $x$, $x \in a \leftrightarrow x \in b$. By the axiom of extensionality,
$a = b$.    □

The axiom of extensionality and the preceding theorem give us strategies for
proving that sets are equal. Most proofs of set equality apply one of these two
approaches.

**2.**  Empty set axiom: Ø has no elements. Formally, $\forall x (x \notin \emptyset)$.

The empty set has some unusual containment properties. For example, it is a
subset of every set.

**Theorem 3.9.**   *Ø is a subset of every set. Formally,* $\forall t (\emptyset \subset t)$.

PROOF.   The proof relies on the mathematical meaning of implication. Suppose $t$
is a set. Pick any set $x$. By the empty set axiom, $x \notin \emptyset$, so $x \in \emptyset$ implies $x \in t$.
(When the hypothesis is false, the implication is automatically true. If I am the
king of the world, then you will send me all your money. The statement is true,
but no checks have arrived.) Formally, $\forall x (x \in \emptyset \rightarrow x \in t)$, so $\emptyset \subset t$.    □

The preceding proof also implies that $\emptyset \subset \emptyset$, although Exercise 1 provides a
more direct proof.

**3.**  Pairing axiom: For every $x$ and $y$, the pair set $\{x, y\}$ exists. Formally,

$$\forall x \forall y \exists z \forall t (t \in z \leftrightarrow (t = x \vee t = y)).$$

In the formal version of the axiom, the set $z$ has $x$ and $y$ as its only elements.
Thus, $z$ is $\{x, y\}$. The pair sets provided by the pairing axiom are unordered,
so $\{x, y\} = \{y, x\}$. The pairing axiom can be used to prove the existence of
single-element sets, which are often called singletons.

**Theorem 3.10.**   *For every x, the set $\{x\}$ exists. That is,* $\forall x \exists z \forall t (t \in z \leftrightarrow t = x)$.

PROOF.   Fix $x$. Substituting $x$ for $y$ in the pairing axiom yields a set $z$ such that
$\forall t (t \in z \leftrightarrow (t = x \vee t = x))$. By the axiom of extensionality, $z = \{x\}$.    □

The empty set axiom, the pairing axiom, and Theorem 3.10 on the existence of
singletons are all combined in Zermelo's original axiom of elementary sets [152].
As an immediate consequence he solves Exercise 4, showing that singleton sets
have no proper subsets.

The statement of the next axiom uses the union symbol in an unusual way. In
particular, we will write $\cup\{x, y\}$ to denote the familiar $x \cup y$. This prefix notation

is very convenient for writing unions of infinite collections of sets. For example, if $X = \{x_i \mid i \in \mathbb{N}\}$, then the infinite union $x_0 \cup x_1 \cup x_2 \cup \cdots$ can be written as $\cup X$, eliminating the use of pesky dots. The union axiom says that $\cup X$ contains the appropriate elements.

**4.** Union axiom: The elements of $\cup X$ are precisely the elements of the elements of $X$. Formally,

$$\forall t (t \in \cup X \leftrightarrow \exists y (t \in y \wedge y \in X)).$$

Exercise 5 is a verification that $\cup \{x, y\}$ is exactly the familiar set $x \cup y$. The notion of union extends naturally to collections of fewer than two sets, also. By the union axiom, $t \in \cup \{x\}$ if and only if there is a $y \in \{x\}$ such that $t \in y$, that is, if and only if $t \in x$. Thus, $\cup \{x\} = x$. For an exercise in wildly vacuous reasoning, try out Exercise 6, which shows that $\cup \emptyset = \emptyset$.

Like the union axiom, the power set axiom defines one of our built-in functions.

**5.** Power set axiom: The elements of $\mathcal{P}(X)$ are precisely the subsets of $X$. Formally,

$$\forall t (t \in \mathcal{P}(X) \leftrightarrow t \subset X).$$

This is the same power set operator that appears in the first chapter of dozens of mathematics texts. For example,

$$\mathcal{P}(\{a, b\}) = \{\emptyset, \{a\}, \{b\}, \{a, b\}\}.$$

If $X$ is a finite set of size $n$, then $\mathcal{P}(X)$ has $2^n$ elements. Thus for finite sets, the size of $\mathcal{P}(X)$ is always larger than the size of $X$. In Section 3.5 we will prove that this relation continues to hold when $X$ is infinite.

It may seem odd that we do not have other built-in functions, like intersection, set-theoretic difference, or Cartesian products. However, all these operations can be defined using the next axiom and are omitted in order to reduce redundancy in the axioms. Our version of the separation axiom is the invention of Skolem [134]. Both Skolem and Fraenkel [56] proposed emendations to Zermelo's version of the separation axiom.

**6.** Separation axiom: If $\psi(x)$ is a formula and $X$ is a set, then the set $\{x \in X \mid \psi(x)\}$ exists. More formally, given any set $X$ and any formula $\psi(x)$ in the language of ZFC, if $\psi(x)$ does not contain the variable $S$, then

$$\exists S \forall x (x \in S \leftrightarrow (x \in X \wedge \psi(x))).$$

Note that $\psi(x)$ may contain unquantified variables, which can be viewed as parameters. Thus $S$ can be defined in terms of $X$ and other given sets.

We can use the separation axiom to prove that intersections exist. It is nice to use intersection notation that is parallel to our union notation, so we write $\cap \{a, b\}$ for $a \cap b$. In general, an element should be in $\cap X$ precisely when it is in every element of $X$.

**Theorem 3.11.** *For any nonempty set X, ∩X exists. That is, for any set X there is a set Y such that*

$$\forall x(x \in Y \leftrightarrow \forall t(t \in X \rightarrow x \in t)).$$

PROOF.    Fix $X$. Let $Y = \{x \in \cup X \mid \forall t(t \in X \rightarrow x \in t)\}$. By the separation axiom, $Y$ exists. We still need to show that $Y$ is the desired set. By the definition of $Y$, if $x \in Y$, then $\forall t(t \in X \rightarrow x \in t)$. Conversely, if $\forall t(t \in X \rightarrow x \in t)$, then since $X$ is nonempty, $\exists t(t \in X \wedge x \in t)$. Thus $x \in \cup X$. Because $x \in \cup X$ and $\forall t(t \in X \rightarrow x \in t)$, we have $x \in Y$. Summarizing, $x \in Y$ if and only if $\forall t(t \in X \rightarrow x \in t)$. □

It is also possible to show that ∩$X$ is unique. (See Exercise 8.) Since we can show that for all $X$ the set ∩$X$ exists and is unique, we can add the function symbol ∩ to the language of ZFC. Of course, the symbol itself is subject to misinterpretation, so we need to add a defining axiom. The formula

$$\forall x(x \in \cap X \leftrightarrow \forall t(t \in X \rightarrow x \in t))$$

will do nicely. The resulting extended theory is more convenient to use, but proves exactly the same theorems, except for theorems actually containing the symbol ∩. Mathematical logicians would say that ZFC with ∩ is a *conservative extension* of ZFC.

Using the same process, we can introduce a variety of other set-theoretic functions. For example, we can specify a set that represents the ordered pair $(x, y)$, and define the Cartesian product $X \times Y = \{(x, y) \mid x \in X \wedge y \in Y\}$. Ordered $n$-tuples can be defined in a number of reasonable ways from ordered pairs. We could define the relative complement of $Y$ in $X$ by $X - Y = \{x \in X \mid x \notin Y\}$. See Exercises 9, 10, and 11 for more discussion of these operations.

There are some significant restrictions in the sorts of functions that can be conservatively added to ZFC. For example, it is acceptable to introduce the relative complement, but not a full-blown general complement. (Books usually use $\overline{X}$ or $X^c$ to denote a general complement.) Given a general complement, we could prove that $X \cup X^c$ existed. This would give us a set of all sets, but that is prohibited by the separation axiom.

**Theorem 3.12.** *There is no universal set. That is, there is no set U such that* $\forall x(x \in U)$.

PROOF.    Suppose by way of contradiction that $\forall x(x \in U)$. By the separation axiom, there is a set $X$ such that $X = \{z \in U \mid z \notin z\}$. Note that for all $z$, $z \in X$ if and only if $z \in U$ and $z \notin z$. Furthermore, $z \in U$ and $z \notin z$ if and only if $z \notin z$. Thus, $z \in X$ if and only if $z \notin z$ for any $z$ we care to choose. In particular, substituting $X$ for $z$ gives us $X \in X$ if and only if $X \notin X$, yielding the desired contradiction. □

The preceding proof contains the gist of Russell's paradox. Briefly, Russell's paradox says that the existence of $\{z \mid z \notin z\}$ leads inexorably to contradictions. Note that the existence of $\{z \mid z \notin z\}$ is **not** proved by the separation axiom, because the set is not bounded. For any bound $X$, we can prove that $\{z \in X \mid z \notin z\}$ exists; it is just a harmless subset of $X$. By requiring bounds on definable sets, we cleverly sidestep the paradoxes that ensnare the users of naïve set theory. For another experiment with Russell's style of argument, try Exercise 12.

Part of Hilbert's motivation for the rigorous study of set theory was to gain a deeper understanding of infinite sets. So far, our axioms do not guarantee the existence of a single infinite set. (Readers who love technical details may want to construct a model of axioms 1 through 6 in which every set is finite. The universe for this model will be infinite, but each element in the universe will be finite.)

One way to construct an infinite set is to start with $\emptyset$ and successively apply Theorem 3.10. If we let $x_0 = \emptyset$ and $x_{n+1} = \{x_n\}$ for each $n$, this yields a set for each natural number. In particular, $x_0 = \emptyset$, $x_1 = \{\emptyset\}$, $x_2 = \{\{\emptyset\}\}$, and so on. The next axiom affirms the existence of a set containing all these sets as elements.

7. Infinity axiom: There is a set $Z$ such that (i) $\emptyset \in Z$ and (ii) if $x \in Z$, then $\{x\} \in Z$. Formally,

$$\exists Z(\emptyset \in Z \wedge \forall x(x \in Z \to \exists y(y \in Z \wedge \forall t(t \in y \leftrightarrow t = x)))).$$

The axiom of infinity guarantees the existence of *some* set satisfying properties (i) and (ii). By applying the power set axiom, the separation axiom, and taking an intersection, we can find the smallest set with this property. For details, see Exercise 13.

Zermelo's axiomatization of set theory consists of axioms 1 through 7 plus the axiom of choice. We will discuss the axiom of choice shortly. In the meantime, there are two more axioms that have been appended to ZF that should be mentioned. The first of these is the axiom of replacement, versions of which were proposed by Fraenkel ([53], [54], and [55]), Skolem [134], and Lennes [100].

8. Replacement axiom: Ranges of functions restricted to sets exist. That is, if $f(x)$ is a function and $D$ is a set, then the set $R = \{f(x) \mid x \in D\}$ exists. More formally, if $\psi(x, y)$ is a formula of set theory such that

$$\forall x \forall y \forall z((\psi(x, y) \wedge \psi(x, z)) \to y = z),$$

then for every set $D$ there is a set $R$ such that

$$\forall y(y \in R \leftrightarrow \exists x(x \in D \wedge \psi(x, y))).$$

Note that the formula $\psi(x, y)$ in the formal statement of the axiom can be viewed as defining the relation $f(x) = y$. The replacement axiom is useful for proving the existence of large sets. In particular, if we assume that ZFC and the continuum hypothesis are consistent, in the absence of the replacement axiom it is impossible to prove that any sets of size greater than or equal to $\aleph_\omega$ exist. (To find out what $\aleph_\omega$ is, you have to stick around until Section 3.5.)

The final axiom of ZF is the regularity axiom. In a nutshell, it outlaws some rather bizarre behavior, for example having $x \in y \in x$. Attempts to avoid these strange constructs can be found in the work of Mirimanoff [107], but Skolem [134] and von Neumann [144] are usually given credit for proposing the actual axiom.

**9.** Regularity axiom: Every nonempty set $x$ contains an element $y$ such that $x \cap y = \emptyset$. Formally,

$$\forall x(x \neq \emptyset \to \exists y(y \in x \land x \cap y = \emptyset)).$$

The idea here is that $\in$ can be viewed as a partial ordering on any set by letting $x < y$ mean $x \in y$. The regularity axiom says that every set has a minimal element in this ordering. This rules out loops (like $x \in y \in x$) and infinite descending chains (like $\ldots \in x_3 \in x_2 \in x_1 \in x_0$). The following theorem shows that tight loops are outlawed.

**Theorem 3.13.** *For all $x$, $x \notin x$.*

PROOF. By way of contradiction, suppose $x \in x$. By Theorem 3.10, the set $X = \{x\}$ exists. The set $X$ is nonempty, so by the regularity axiom, there is an element $y \in X$ such that $X \cap y = \emptyset$. The only element of $X$ is $x$, so $y = x$ and $X \cap x = \emptyset$. However, $x \in X$ and $x \in x$, so $x \in X \cap x = \emptyset$, a contradiction.    $\square$

Summarizing this section, the proper axioms of ZF are:

**1.** Axiom of extensionality,
**2.** Empty set axiom,
**3.** Pairing axiom,
**4.** Union axiom,
**5.** Power set axiom,
**6.** Separation axiom,
**7.** Infinity axiom,
**8.** Replacement axiom, and
**9.** Regularity axiom.

We are still missing one axiom from Zermelo's list, the axiom of choice.

### 3.3.3  Axiom of Choice

> Vizzini: *...so I can clearly not choose the wine in front of me.*
> Man in black: *You've made your decision then?*
> Vizzini: [happily] *Not remotely!*
>
> — *The Princess Bride*

Suppose that we, like Vizzini, are faced with the task of selecting one glass from a set of two glasses. Since the set of glasses is nonempty, we can select one element and get on with our lives, which hopefully will be much longer than Vizzini's. To be very technical, the justification for our selection is the logical principle of existential instantiation. Similarly, using only axioms of ZF, we can select one element from any nonempty set, without regard for the size of the set. Furthermore, we could repeat this process any finite number of times, so we can choose one element from each set in any finite list of nonempty sets.

By contrast, making an infinite number of choices simultaneously can be problematic, depending on the circumstances. Suppose that we have an infinite collection of pairs of boots. We can pick one boot from each pair simply by specifying that we will select the left boot from each pair. Because each nonempty set (pair of boots) has a designated element (left boot), ZF suffices to prove the existence of the set of selected boots. Working in ZF, we cannot carry out the same process with an infinite collection of pairs of socks, because socks are not footed. We need a new axiom. In [130], Russell discusses this boot problem, though rather than selecting socks, he considers the case where "the left and right boots in each pair are indistinguishable." Cruel shoes indeed!

The axiom of choice guarantees the existence of a set of selected socks. The following version of the axiom is very close to that of Zermelo [152].

10. Axiom of choice (AC): If $T$ is a set whose elements are all sets that are nonempty and mutually disjoint, then $\cup T$ contains at least one subset having exactly one element in common with each element of $T$.

Most recent works use a formulation of the axiom of choice that asserts the existence of choice functions. In terms of socks, when a choice function is applied to a pair of socks, it outputs a designated sock. In the following statement, if $T$ is a set of pairs of socks, $t$ would be a pair of socks, and $f(t)$ would be a sock.

10'. Axiom of choice (AC2): If $T$ is a set of nonempty sets, then there is a function $f$ such that for every $t \in T$, $f(t) \in t$.

We use ZFC to denote ZF plus either version of AC. This is not imprecise, since we can prove that the two versions of the axiom of choice are interchangeable.

**Theorem 3.14.** *ZF proves that AC holds if and only if AC2 holds.*

PROOF.    First assume all the axioms of ZF plus AC. Let $T$ be a set of nonempty sets. Define the function $g$ with domain $T$ by setting

$$g(t) = \{(t, y) \mid y \in t\}$$

for each $t \in T$. Essentially, $g(t)$ looks like the set $t$ with a flag saying "I'm in $t$" attached to each element. By the replacement axiom, the set $Y = \{g(t) \mid t \in T\}$ exists. The elements of $Y$ are nonempty and disjoint, so by AC there is a set $S$ that contains exactly one element from each element of $Y$. Thus $S$ is a set of ordered pairs of the form $(t, y)$, where exactly one pair is included for each $t \in T$. Let $f(t)$ be the unique $y$ such that $(t, y) \in S$. Then $f$ is the desired choice function.

To prove the converse, assume ZF plus AC2. Let $T$ be a set whose elements are nonempty and disjoint. By AC2, there is a function $f$ such that for each $t \in T$, $f(t) \in t$. By the replacement axiom, $S = \{f(t) \mid t \in T\}$ exists. $S$ is the desired subset of $\cup T$.                                                                 □

Zermelo ([150], [151]) used AC to prove that every set can be well-ordered. Hartogs [81] extended Zermelo's result by proving that AC is actually equivalent to this well-ordering principle. Hartogs' result is identical in format to the equivalence result that we just proved. What really makes Hartogs' result and our equivalence theorem interesting is the fact that AC can neither be proved nor disproved in ZF.

(Technically, we just implicitly assumed that ZF is consistent. I assure you that many people make much more bizarre assumptions in their daily lives.) Gödel [65] proved that ZF cannot disprove AC, and Cohen ([30], [31]) showed that ZF cannot prove AC. Thus our equivalence theorem and the theorem of Hartogs list statements that we can add interchangeably to *strengthen* ZF. In later sections we will see more examples of equivalence theorems and more examples of statements that strengthen ZF and ZFC.

## Exercises

1. Prove that containment is reflexive. That is, prove $\forall x(x \subset x)$. (This requires only logical properties.)
2. Prove that containment is transitive. That is, prove

$$\forall x \forall y \forall z((x \subset y \wedge y \subset z) \to x \subset z).$$

   (This requires only logical properties.)
3. Prove that the empty set is unique. That is, if $\forall x(x \notin y)$, then $y = \emptyset$.
4. Prove that if $y \subset \{x\}$, then $y = \emptyset$ or $y = \{x\}$.
5. Prove that $\cup\{x, y\}$ is exactly the familiar set $x \cup y$. That is, prove that $t \in \cup\{x, y\}$ if and only if $t \in x$ or $t \in y$.
6. Prove that $\cup\emptyset = \emptyset$.
7. Find $\mathcal{P}(\emptyset)$, $\mathcal{P}(\mathcal{P}(\emptyset))$, and $\mathcal{P}(\mathcal{P}(\mathcal{P}(\emptyset)))$. (To make your answers look really bizarre and drive your instructor nuts, write $\{\ \}$ for $\emptyset$.)
8. Prove that $\cap X$ is unique. That is, show that if $Y$ is a set that satisfies $\forall x(x \in Y \leftrightarrow \forall t(t \in X \to x \in t))$ and $Z$ is a set that satisfies the formula $\forall x(x \in Z \leftrightarrow \forall t(t \in X \to x \in t))$, then $Y = Z$. (Proving the existence of $Y$ and $Z$ requires the separation axiom (see Theorem 3.11), but this problem uses the axiom of extensionality.)
9. Let $X - Y$ denote the set $\{x \in X \mid x \notin Y\}$.

   a. Prove that for every $X$ and $Y$, $X - Y$ exists.
   b. Prove that for every $X$ and $Y$, $X - Y$ is unique.

10. Representations of ordered pairs.

    a. Kuratowski [98] suggested that the ordered pair $(a, b)$ can be represented by the set $\{\{a, b\}, a\}$. (This encoding is still in use.) Using this definition, prove that $(a, b) = (c, d)$ if and only if $a = c$ and $b = d$.
    b. Using Kuratowski's encoding, show that if $X$ and $Y$ are sets, then the set $X \times Y$ defined by $X \times Y = \{(x, y) \mid x \in X \wedge y \in Y\}$ exists and is uniquely determined by $X$ and $Y$.
    c. Wiener [147] suggested that the ordered pair $(a, b)$ can be represented by the set $\{\{\{x\}, \emptyset\}, \{\{y\}\}\}$. If you dare, repeat parts 10a and 10b using this encoding.
    d. Show that encoding $(a, b)$ by $\{a, \{b\}\}$ leads to difficulties. (Find two distinct ordered pairs that have the same representation in this encoding.)

**11.** Representations of $n$-tuples.

    **a.** Usually, set theorists represent $(a, b, c)$ by $((a, b), c)$, where pairs are represented using the Kuratowski encoding from Exercise 10. Using this representation prove the following:

        (i) $(a, b, c) = (d, e, f) \leftrightarrow (a = d \wedge b = e \wedge c = f)$,

        (ii) $X \times Y \times Z$ exists, and

        (iii) $X \times Y \times Z$ is unique.

    **b.** To address type-theoretic concerns, Skolem [135] suggested representing $(a, b, c)$ by $((a, c), (b, c))$. Repeat part 11a with this encoding.

    **c.** Using parts 11a and 11b as the base cases in an induction argument, extend the statements in part 11a to $n$-tuples for each natural number $n$. (If you do this, you clearly have a great love for long technical arguments. You might as well repeat the whole mess with the Wiener encoding.)

    **d.** Show that encoding $(a, b, c)$ by $\{\{a, b, c\}, \{a, b\}, \{a\}\}$ leads to difficulties. (You can find distinct triples with the same representation, or you can find an ordered pair that has the same representation as an ordered triple.)

**12.** Prove that for all $X$, $\mathcal{P}(X) \not\subset X$. (Hint: Suppose that for some $X$, $\mathcal{P}(X) \subset X$. Define $Y = \{t \in X \mid t \notin t\}$. Show that $Y \in X$ and shop for a contradiction.)

**13.** Let $Z$ be the set provided by the infinity axiom. Let $T$ be the set of subsets of $Z$ that satisfy properties (i) and (ii) of the infinity axiom. Let $Z_0 = \cap T$.

    **a.** Prove that $T$ exists. (Hint: $T \subset \mathcal{P}(Z)$.)

    **b.** Prove that $Z_0$ exists.

    **c.** Prove that if $X$ satisfies properties (i) and (ii) of the infinity axiom, then $Z_0 \subset X$.

**14.** Use the regularity axiom to prove that for all $x$ and $y$ either $x \notin y$ or $y \notin x$.

## 3.4   The Return of der König

*And Aragorn planted the new tree in the court by the fountain.*
                  — J.R.R. Tolkien, *The Return of the King*

It may seem that the discussion of the last section strayed from our original topics in graph theory and combinatorics. However, AC is actually a statement about infinite systems of distinct representatives (SDR). As defined in Section 1.5.2, an SDR for a family of sets $T$ is a set that contains a distinct element from each set in $T$. For disjoint families, we have the following theorem.

**Theorem 3.15.** *ZF proves that the following are equivalent:*

**1.** *AC.*

**2.** *If $T$ is a family of disjoint nonempty sets, then there is a set $Y$ that is an SDR for $T$.*

PROOF.  First, assume ZF and AC and suppose that $T$ is a family of disjoint nonempty sets. By AC, there is a set $Y \subset \cup T$ that has exactly one element in common with each element of $T$. Since the elements of $T$ are disjoint, $Y$ is an SDR for $T$.

To prove the converse, suppose $T$ is a family of disjoint nonempty sets. Let $Y$ be an SDR for $T$. Then $Y \subset \cup T$, and $Y$ has exactly one element in common with each element of $T$, as required by AC.    □

What if $T$ is not disjoint? For finite families of sets, it is sufficient to know that every union of $k$ sets has at least $k$ elements. This is still necessary for infinite families, but no longer sufficient. Consider the family of sets $T = \{X_0, X_1, X_2, \ldots\}$ defined by $X_0 = \{1, 2, 3, \ldots\}$, $X_1 = \{1\}$, $X_2 = \{2\}$, and so on. The union of any $k$ sets from $T$ has at least $k$ elements. As a matter of fact, any collection of $k$ sets from $T$ has an SDR. However, the whole of $T$ has no SDR. To build an SDR for $T$, we must pick some $n$ as a representative for $X_0$. This immediately leaves us with no element to represent $X_n$. We are out of luck. Note that if we chuck $X_0$, we can find an SDR for the remaining sets. (There are not many options for the representatives; it is hard to go wrong.) The infinite set $X_0$ is the source of all our problems. If we allow only finite sets in the family, then we get a nice SDR existence theorem.

**Theorem 3.16.**  *Suppose $T = \{X_0, X_1, X_2, \ldots\}$ is a family of finite sets. $T$ has an SDR if and only if for every $k \in \mathbb{N}$ and every collection of $k$ sets from $T$, the union of these sets contains at least $k$ elements.*

PROOF.  Let $T = \{X_0, X_1, X_2, \ldots\}$ and suppose that each $X_i$ is finite. If $T$ has an SDR, then for any collection of $k$ sets, their representatives form a $k$ element subset of their union.

To prove the converse, assume that for every $k \in \mathbb{N}$, the union of any $k$ elements of $T$ contains at least $k$ elements. By Theorem 1.33, for each $k$ the subfamily $\{X_0, X_1, \ldots, X_k\}$ has an SDR. Let $Y$ be the tree whose paths are of the form $r, x_0, x_1, \ldots, x_k$, where $x_i \in X_i$ for $i \leq k$ and $\{x_0, x_1, \ldots, x_k\}$ is an SDR for $\{X_0, X_1, \ldots, X_k\}$. Since arbitrarily large finite subfamilies of $T$ have SDRs, the tree $Y$ is infinite. Furthermore, the size of the $k$th level of $Y$ is at most $|X_0| \cdot |X_1| \cdots |X_k|$, where $|X_i|$ denotes the size of $X_i$. Since these sets are all finite, each level is finite. By König's Lemma, $Y$ has an infinite path, and that path is an SDR for $T$.    □

In the preceding proof we made no immediately *obvious* use of AC. Here is a question: Have we actually avoided the use of AC, or did we merely disguise it? The answer is that we have used some of the strength of AC in a disguised form.

There are two very natural ways to restrict AC. Recall that AC considers a family of sets. We can either restrict the size of the sets or restrict the size of the family. If we require that each set is finite, we get the following statement.

Axiom of choice for finite sets (ACF): If $T$ is a family of finite, nonempty, mutually disjoint sets, then $\cup T$ contains at least one subset having exactly one element in common with each element of $T$.

If we specify that the family can be enumerated, we get the following statement. (We say that an infinite set is countable if it can be written in the form $\{x_0, x_1, x_2, \ldots\}$.)

Countable axiom of choice (CAC): If $T = \{X_0, X_1, X_2, \ldots\}$ is a family of nonempty, mutually disjoint sets, then $\cup T$ contains at least one subset having exactly one element in common with each element of $T$.

Combining both restrictions gives us CACF, the countable axiom of choice for finite sets. The statement of CACF looks like CAC with the added hypothesis that each $X_i$ is finite. This weak version of AC is exactly what we used in proving Theorem 3.16.

**Theorem 3.17.** *ZF proves that the following are equivalent:*

1. *König's Lemma.*
2. *Theorem* 3.16.
3. *CACF.*

PROOF.   The proof of Theorem 3.16 shows that 1 implies 2. The proofs that 2 implies 3 and 3 implies 1 are Exercises 1 and 2.   □

The relationships between our various versions of choice are very interesting. It is easy to see that ZF proves AC→CAC, AC→ACF, CAC→CACF, and ACF→CACF. It is not at all obvious, but can be shown, that not one of the converses of these implications is provable in ZF, and that CACF is not a theorem of ZF. To prove that ZF cannot prove these statements we would assume that ZF is consistent and build special models where each particular statement fails. The models are obtained by lifting results from permutation models or by forcing. Jech's book *The Axiom of Choice* [87] is an excellent reference.

Since ZF proves that König's Lemma is equivalent to CACF and CACF is not a theorem of ZF, we know that König's Lemma is not a theorem of ZF. Of course, ZFC can prove König's Lemma, so it is perfectly reasonable to think of it as a theorem of mathematics. Also, ZF can prove some restrictions of König's Lemma, for example if all the labels in the tree are natural numbers. Many applications of König's Lemma can be carried out with a restricted version.

Our proof closely ties König's Lemma to countable families of sets. As we will see in the next section, not all families are countable. We will see bigger sets where the lemma fails, and still bigger sets where it holds again. This is not the last return of König. (König means "king" in German.)

In the introduction to this chapter we noted that if $P$ is finite, then whenever $H$ is a proper subset of $P$ there is no matching between $P$ and $H$. A set $X$ is called *Dedekind finite* if no proper subset of $X$ can be matched with $X$. Thus, the introduction shows that if $X$ is finite, then $X$ is Dedekind finite. Exercise 4 shows that CAC implies the converse. Thus, in ZFC, the finite sets are exactly the Dedekind finite sets. This characterization of the finite sets requires use of a statement that is weaker than CAC, but not provable in ZF [87].

**Exercises**

1. Prove in ZF that Theorem 3.16 implies CACF. (Hint: You must use disjointness to show that the union of any $k$ sets contains at least $k$ elements.)
2. Challenging exercise. Prove König's Lemma using ZF and CACF. To do this, let $S$ be the set of nodes in the tree that have infinitely many successors. Find an enumeration for $S$. (It is easy to slip up and use full AC when finding the enumeration.) For each $s \in S$, let $X_s$ be the set of immediate successors of $s$ that have infinitely many successors. Apply CACF to the family $\{X_s \mid x \in S\}$. Use the selected vertices to construct a path through the tree.
3. Disjointification trick. Suppose that $\{X_n \mid n \in \mathbb{N}\}$ is a family of sets. For each $n \in \mathbb{N}$ let $\overline{X}_n = \{(n, x) \mid x \in X_n\}$. Show that $\{\overline{X}_n \mid n \in \mathbb{N}\}$ exists and is a disjoint family of sets.
4. Use CAC to prove that every infinite set has a countable subset. (Hint: Suppose that $W$ is infinite. For each $k \in \mathbb{N}$ let $W_k$ be the set of all subsets of $W$ of size $k$. Apply CAC to a disjointified version of the family $\{W_k \mid k \in \mathbb{N}\}$. Show that the union of the selected elements is a countable subset of $W$.)
5. Assume that every infinite set has a countable subset. Prove that if $X$ cannot be matched with any proper subset of itself, then $X$ is finite. (Hint: Suppose $X$ is infinite and use a countable subset of $X$ to find a matching between $X$ and a proper subset of $X$.)

## 3.5    Ordinals, Cardinals, and Many Pigeons

*Whenever Gutei Oshō was asked about Zen, he simply raised his finger. Once a visitor asked Gutei's boy attendant, "What does your master teach?" The boy too raised his finger. Hearing of this, Gutei cut off the boy's finger with a knife. The boy, screaming with pain, began to run away. Gutei called to him, and when he turned around, Gutei raised his finger. The boy suddenly became enlightened.*

— Mumon Ekai, *The Gateless Gate*

The previous section contains some references to infinite sets of different sizes. To make sense of this we need to know what it means for sets to be the same size. We can illustrate two approaches by considering some familiar sets. Thanks to the gentleness of my religious training, I have the same number of fingers on my left and right hands. This can be verified in two ways. I can count the fingers on my left hand, count the fingers on my right hand, and check that the results match. Note that in the process of counting, I am matching fingers with elements of a canonical ordered set, probably $\{1, 2, 3, 4, 5\}$. By emphasizing the matching process, I can verify the equinumerousness of my fingers without using any canonical set middleman. To do this, I match left thumb with right thumb, left forefinger with right forefinger, and so on. When my pinkies match, I know that I have the same number of fingers on my left and right hands. One advantage of this technique is that it works without modification for people with six or more fingers on each hand.

For infinite sets, either method works well. We will start by comparing sets directly, then study some canonical ordered sets, and finish the section off with some applications to pigeons and trees.

## 3.5.1  Cardinality

*The big one!*

— Connie Conehead

Suppose we write $X \precsim Y$ if there is a one-to-one function from $X$ into $Y$, and $X \sim Y$ if there is a one-to-one function from $X$ onto $Y$. Thus $X \sim Y$ means that there is a matching between $X$ and $Y$. If $X \precsim Y$ and $X \not\sim Y$, so $X$ can be embedded into but not onto $Y$, we will write $X \prec Y$. With this notation, we can describe relative sizes of some infinite sets.

First consider the sets $\mathbb{N} = \{0, 1, 2, \ldots\}$ and $\mathbb{Z} = \{\ldots, -2, -1, 0, 1, 2, \ldots\}$. Define the function $f : \mathbb{N} \to \mathbb{Z}$ by

$$f(n) = (-1)^{n+1} \cdot \left( \frac{2n + 1}{4} \right) + \frac{1}{4}.$$

It is not hard to verify that if $f(j) = f(k)$, then $j = k$, proving that $f$ is a one-to-one function. Additionally, if $m > 0$, then $f(2m - 1) = m$, and if $t \leq 0$, then $f(-2t) = t$, so $f$ maps the odd natural numbers onto the positive integers and the even natural numbers onto the negative integers and 0. Thus, $f$ witnesses that $\mathbb{N} \sim \mathbb{Z}$, and we now know that $\mathbb{N}$ and $\mathbb{Z}$ are the same size.

If $X$ is a set satisfying $\mathbb{N} \sim X$, then we say that $X$ is *countable* (or *countably infinite* if we are being very precise.) We just prove that $\mathbb{Z}$ is countable. Not every infinite set is countable, as shown by the following theorem of Cantor.

**Theorem 3.18** (Cantor's Theorem).  *For any set $X$, $X \prec \mathcal{P}(X)$. In particular, $\mathbb{N} \prec \mathcal{P}(\mathbb{N})$.*

PROOF.  Define $f : X \to \mathcal{P}(X)$ by setting $f(t) = \{t\}$ for each $t \in X$. Since $f$ is one-to-one, it witnesses that $X \precsim \mathcal{P}(X)$. It remains to show that $X \not\sim \mathcal{P}(X)$. Suppose $g : X \to \mathcal{P}(X)$ is any one-to-one function. We will show that $g$ is not onto. Let $y = \{t \in X \mid t \notin g(t)\}$. Suppose by way of contradiction that for some $x \in X$, $g(x) = y$. Because $g(x) = y$, $x \in g(x)$ if and only if $x \in y$, and by the definition of $y$, $x \in y$ if and only if $x \notin g(x)$. Concatenating, we get $x \in g(x)$ if and only if $x \notin g(x)$, a clear contradiction. Thus $y$ is not in the range of $g$, completing the proof.  □

One consequence of Cantor's Theorem is that any function from $\mathcal{P}(\mathbb{N})$ into $\mathbb{N}$ must not be one-to-one. More combinatorially stated, if we try to ram a pigeon for each element of $\mathcal{P}(\mathbb{N})$ into pigeonholes corresponding to the elements of $\mathbb{N}$, some pigeonhole must contain at least two pigeons. Another consequence is that by sequentially applying Cantor's Theorem to an infinite set, we get lots of infinite sets, including some very big ones.

**Corollary 3.19.** *There are infinitely many infinite sets of different sizes.*

PROOF.    $\mathbb{N}$ is infinite, and $\mathbb{N} \prec \mathcal{P}(\mathbb{N}) \prec \mathcal{P}(\mathcal{P}(\mathbb{N})) \cdots$ by Cantor's Theorem.    □

Using only the definition of $\sim$ and chasing some functions, we can prove that $\sim$ is an equivalence relation. (See Exercise 2.) In particular, for any sets $A$, $B$, and $C$, we have

- $A \sim A$,
- if $A \sim B$ then $B \sim A$, and
- if $A \sim B$ and $B \sim C$, then $A \sim C$.

These are handy shortcuts, and it would be nice to have analogous statements for the $\precsim$ relation. We can easily show that $A \precsim A$ and that if $A \precsim B$ and $B \precsim C$, then $A \precsim C$. (See Exercise 3.) Symmetry does not hold for the $\precsim$ relation, but we can prove that if $A \precsim B$ and $B \precsim A$ then $A \sim B$. This last statement is the Cantor–Bernstein Theorem, which is an incredibly handy shortcut for showing that sets are the same size. After discussing the proof and history of the theorem, we will look at some nice applications.

**Theorem 3.20** (Cantor–Bernstein Theorem).    *If $X \precsim Y$ and $Y \precsim X$, then $X \sim Y$.*

PROOF.    Suppose $f : A \to B$ and $g : B \to A$ are one-to-one functions. We will sketch the construction of a function $h : A \to B$ that is one-to-one and onto. Define a set of subsets of $A$ as follows. Let $A_0 = A$, $A_1 = g(B)$, and $A_n = g(f(A_{n-2}))$ for $n \geq 2$. In particular, writing $g \circ f(A)$ for $g(f(A))$, we have

$$A_0 = A,$$
$$A_1 = g(B),$$
$$A_2 = g \circ f(A),$$
$$A_3 = g \circ f \circ g(B),$$
$$A_4 = g \circ f \circ g \circ f(A), \text{ and}$$
$$A_5 = g \circ f \circ g \circ f \circ g(B).$$

Note that $A_n$ is defined with $n$ function applications. It goes "back and forth" $n$ times. Using induction as described in Exercise 4a, it is easy to prove the following claim.

Claim 1: For all $n$, $A_n \supset A_{n+1}$.

Given Claim 1, define the sets $A'_n = A_n - A_{n+1}$ for each $n$. Also define $A'_\omega = \bigcap_{n \in \mathbb{N}} A_n$. These sets form a partition of $A$ into disjoint pieces, as claimed below. Hints for the proof of this claim appear in Exercise 4b.

Claim 2: For every $x \in A$ there is a unique $n \in \{\omega, 0, 1, 2, \ldots\}$ such that $x \in A'_n$.

Define the function $h : A \to B$ by the following formula:

$$h(x) = \begin{cases} f(x) & \text{if } x \in \{A'_\omega, A'_0, A'_2, \ldots\}, \\ g^{-1}(x) & \text{if } x \in \{A'_1, A'_3, A'_5, \ldots\}. \end{cases}$$

By Claim 2, $h(x)$ is well-defined and has all of $A$ as its domain. It remains to show that $h(x)$ is one-to-one and onto. This can be accomplished by defining $B_0 = B$, and $B_{n+1} = f(A_n)$ for each $n \geq 0$. Imitating our work with the $A_n$, in Exercise 4c we prove the following.

Claim 3: For all $n$, $B_n \supseteq B_{n+1}$. Furthermore, if we define $B'_n = B_n - B_{n+1}$ and $B_\omega = \bigcap_{n \in \mathbb{N}} B_n$, then for every $y \in B$, there is a unique $n \in \{\omega, 0, 1, 2, \ldots\}$ such that $y \in B'_n$.

The partitions of $A$ and $B$ are closely related. In particular, Exercise 4d gives hints for proving the following claim.

Claim 4: For each $n \in \mathbb{N}$, $h(A'_{2n}) = B'_{2n+1}$ and $h(A'_{2n+1}) = B'_{2n}$. Also, $h(A'_\omega) = B'_\omega$.

Since $h$ matches the $A'$ pieces with the $B'$ pieces and is one-to-one and onto on these pieces, $h$ is the desired one-to-one and onto function.    □

One indication of the importance of the Cantor–Bernstein Theorem is the number of mathematicians who have produced proofs of it. Here is a partial listing. According to Levy [101], Dedekind proved the theorem in 1887. In 1895, Cantor [21] described the theorem as an easy consequence of a version of the axiom of choice. In the endnotes of [23], Jourdain refers to an 1896 proof by Schröder. Some texts, [108] for example, refer to the theorem as the Schröder–Bernstein Theorem. Bernstein proved the theorem without using the axiom of choice in 1898; this proof appears in Borel's book [15]. Later proofs were published by Peano [113], J. König [96], and Zermelo [151]. It is good to remember that the axioms for set theory were in flux during this period. These mathematicians were making sure that this very applicable theorem was supported by the axioms *du jour*.

Now we will examine a pair of applications of the Cantor–Bernstein Theorem. Note that we are freed of the tedium of constructing onto maps.

**Corollary 3.21.** $\mathbb{N} \sim \mathbb{Z}$.

PROOF.    Define $f : \mathbb{N} \to \mathbb{Z}$ by $f(n) = n$. Note that $f$ is one-to-one, so $\mathbb{N} \precsim \mathbb{Z}$. Define $g : \mathbb{Z} \to \mathbb{N}$ by

$$g(z) = \begin{cases} 2z & \text{if } z \geq 0, \\ 2|z| + 1 & \text{if } z < 0. \end{cases}$$

Note that $g$ is one-to-one, so $\mathbb{Z} \precsim \mathbb{N}$. By the Cantor–Bernstein Theorem, $\mathbb{N} \sim \mathbb{Z}$.    □

**Corollary 3.22.** $\mathbb{N} \sim \mathbb{N} \times \mathbb{N}$.

PROOF.    The function $f : \mathbb{N} \to \mathbb{N} \times \mathbb{N}$ defined by $f(n) = (0, n)$ is one-to-one, so $\mathbb{N} \precsim \mathbb{N} \times \mathbb{N}$. The function $g : \mathbb{N} \times \mathbb{N} \to \mathbb{N}$ defined by $g(m, n) = 2^{m+1} \cdot 3^{n+1}$ is also one-to-one, so $\mathbb{N} \times \mathbb{N} \precsim \mathbb{N}$. By the Cantor–Bernstein Theorem, $\mathbb{N} \sim \mathbb{N} \times \mathbb{N}$.    □

**Corollary 3.23.** $\mathcal{P}(\mathbb{N}) \sim \mathbb{R}$. *Consequently,* $\mathbb{R}$ *is uncountable.*

PROOF.    First we will prove that $\mathcal{P}(\mathbb{N}) \precsim \mathbb{R}$. Define $f : \mathcal{P}(\mathbb{N}) \to \mathbb{R}$ by setting $f(X) = \sum_{n \in X} 10^{-n}$ for each $X \in \mathcal{P}(\mathbb{N})$. As an example of how this map works, if $X$ consists of the odd natural numbers, then $f(X) = 0.10\overline{10}$. If $X$ and $Y$ are distinct subsets of $\mathbb{N}$, then they differ at some least natural number $n$, and $f(X)$ and $f(Y)$ will differ in the $n$th decimal place. Thus, $f$ is one-to-one, and so $\mathcal{P}(\mathbb{N}) \precsim \mathbb{R}$.

Now we must construct a one-to-one function $g : \mathbb{R} \to \mathcal{P}(\mathbb{N})$. Let $r$ be a real number. If we avoid decimal expansions that terminate in an infinite sequence of 9's, we can assume that $r$ has a unique decimal expansion of the form

$$(-1)^\epsilon \left( \sum_{i \in X_1} k_i 10^i + \sum_{j \in X_2} d_j 10^{-j} \right),$$

where $\epsilon \in \{0, 1\}$, each $k_i$ and $d_j$ is between 1 and 9, $X_1$ is a set of natural numbers, and $X_2$ is a set of nonzero natural numbers. In this representation, $(-1)^\epsilon$ is the sign of $r$, $\sum_{i \in X_1} k_i 10^i$ is the integer portion of $r$, and $\sum_{j \in X_2} d_j 10^{-j}$ is the fractional portion of $r$. Define the function $g$ by setting

$$g(r) = \{\epsilon\} \cup \{10^{2i+1} + k_i \mid i \in X_1\} \cup \{10^{2j} + d_j \mid j \in X_2\}$$

for each $r \in \mathbb{R}$. As an example of how this map works, $g(-12.305) = \{1, 1001, 12, 103, 1000005\}$. Since different reals differ in some decimal place, $g$ is one-to-one. By the Cantor–Bernstein Theorem, $\mathcal{P}(\mathbb{N}) \sim \mathbb{R}$.

By Theorem 3.18, $\mathbb{N} \prec \mathcal{P}(\mathbb{N})$. Together with $\mathcal{P}(\mathbb{N}) \sim \mathbb{R}$, this implies that $\mathbb{N} \prec \mathbb{R}$, so $\mathbb{R}$ is uncountable.    □

Note that we did not construct a one-to-one function from $\mathcal{P}(\mathbb{N})$ onto $\mathbb{R}$. The Cantor–Bernstein Theorem tells us that such a function exists, so we are not obligated to construct it. (If you are not already convinced that existence theorems are tremendously convenient, try doing a direct construction for the preceding corollary. This is intentionally *not* listed in the exercises.)

### 3.5.2    Ordinals and Cardinals

> *The aleph was heavy, like trying to carry a small engine block.*
> — William Gibson, *Mona Lisa Overdrive*

For Gibson, an aleph is a huge biochip of virtually infinite storage capacity. For a linguist, aleph is ℵ, the first letter of the Hebrew alphabet. For a set theorist, an aleph is a cardinal number. Saying that there are $\aleph_0$ natural numbers is like saying that there are five fingers on my right hand. Alephs are special sorts of ordinals, which are special sorts of well-ordered sets.

Suppose that $X$ is set and $\leq$ is an ordering relation on $X$. We write $x < y$ when $x \leq y$ and $x \neq y$. The relation $\leq$ is a linear ordering on $X$ if the following properties hold for all $x$, $y$, and $z$ in $X$.

Antisymmetry: $(x \leq y \wedge y \leq x) \rightarrow x = y$.
Transitivity: $x \leq y \rightarrow (y \leq z \rightarrow x \leq z)$.
Trichotomy: $x < y \vee x = y \vee y < x$.

Familiar examples of linear orderings include $\mathbb{N}$, $\mathbb{Z}$, $\mathbb{Q}$, and $\mathbb{R}$ with the usual orderings. We say that a linear ordering is a *well-ordering* if every nonempty subset has a least element. Since every subset of $\mathbb{N}$ has a least element, $\mathbb{N}$ is a well-ordering (using the usual ordering). Since the open interval $(0, 1)$ has no least element, the usual ordering does not well-order $\mathbb{R}$. An analyst would say that $0$ is the greatest lower bound of $(0, 1)$, but $0$ is not the least element of $(0, 1)$ because $0 \notin (0, 1)$. The following theorem gives a handy characterization of well-ordered sets.

**Theorem 3.24.** *Suppose $X$ with $\leq$ is a linearly ordered set. $X$ is well-ordered if and only if $X$ contains no infinite descending sequences.*

PROOF.   We will prove the contrapositive version, that is, $X$ is not well-ordered if and only if $X$ contains an infinite descending sequence.

First suppose $X$ is not well-ordered. Then $X$ has a nonempty subset $Y$ with no least element. Pick an element $x_0$ in $Y$. Since $x_0$ is not the least element of $Y$, there is an element $x_1$ in $Y$ such that $x_0 > x_1$. Continuing in this fashion, we obtain $x_0 > x_1 > x_2 > \cdots$, an infinite descending sequence.

To prove the converse, suppose $X$ contains $x_0 > x_1 > x_2 > \cdots$, an infinite descending sequence. Then the set $Y = \{x_i \mid i \in \mathbb{N}\}$ is a nonempty subset of $X$ with no least element. Thus, $X$ is not well-ordered.   $\square$

For any set $X$, the $\in$ relation defines an ordering on $X$. To see this, for $x, y \in X$, let $x \leq_\in y$ if $x \in y$ or $x = y$. In general, this is not a particularly pretty ordering. For example, assuming that $a \neq b$, the set $X = \{a, b, \{a, b\}\}$ is not linearly ordered by the $\leq_\in$ relation. On the other hand, $Y = \{a, \{a, b\}, \{a, \{a, b\}\}\}$ is well-ordered by the $\leq_\in$ relation. In a moment, we will use this property as part of the definition of an ordinal number.

A set $X$ is *transitive* if for all $y \in X$, if $x \in y$ then $x \in X$. A transitive set that is well-ordered by $\leq_\in$ is called an *ordinal*. The ordinals have some interesting properties.

**Theorem 3.25.** *Suppose $X$ is a set of ordinals and $\alpha$ and $\beta$ are ordinals. Then the following hold:*

1. $\cup X$ *is an ordinal.*
2. $\alpha \cup \{\alpha\}$ *is an ordinal.*
3. $\alpha \in \beta$ *or* $\alpha = \beta$ *or* $\beta \in \alpha$.

PROOF.   See Exercises 10, 13, 14, 15, and 16.                                    □

The first two properties in the preceding theorem give good ways to build new ordinals from old ones. For example, a little vacuous reasoning shows that $\emptyset$ is an ordinal. By the theorem, the sets $\emptyset \cup \{\emptyset\} = \{\emptyset\}$, $\{\emptyset\} \cup \{\{\emptyset\}\} = \{\emptyset, \{\emptyset\}\}$, and $\{\emptyset, \{\emptyset\}\} \cup \{\{\emptyset, \{\emptyset\}\}\} = \{\emptyset, \{\emptyset\}, \{\emptyset, \{\emptyset\}\}\}$ are all ordinals. Set theorists have special names for these finite ordinals. They write

$$\emptyset = 0,$$
$$\{\emptyset\} = \{0\} = 1,$$
$$\{\emptyset, \{\emptyset\}\} = \{0, 1\} = 2,$$
$$\{\emptyset, \{\emptyset\}, \{\emptyset, \{\emptyset\}\}\} = \{0, 1, 2\} = 3,$$

and so on for all $k \in \mathbb{N}$. Since each $k$ is a set, we can define $\omega$ by the formula $\omega = \cup_{k \in \mathbb{N}} k$, and use the first property in the theorem to see that $\omega$ is an ordinal. We do not have to stop here. Since $\omega$ is an ordinal, so is $\omega \cup \{\omega\} = \{\omega, 0, 1, 2, \ldots\}$, and we start all over. Sometimes, texts write $\alpha + 1$ for the ordinal $\alpha \cup \{\alpha\}$ and call ordinals like 1, 2, 3, and $\omega + 1$ successor ordinals. Ordinals that are not successors are called *limit ordinals*. The set $\omega$ is a good example of a limit ordinal.

Traditionally, greek letters are used to denote ordinals. Also, we usually write $\alpha \leq \beta$ rather than $\alpha \leq_\in \beta$. Consequently, for ordinals the formula $\alpha < \beta$ means the same thing as $\alpha \in \beta$. Because ordinals are transitive, $\alpha \in \beta$ implies $\alpha \subset \beta$, although the converse is not always true.

There are three ways to think about ordinals and well-orderings. First, every ordinal is a well-ordered set under the $\leq$ relation. Second, the class of all ordinals is well-ordered by the $\leq$ relation. Third, every well-ordered set looks just like an ordinal. The next theorem is a precise expression of the way that ordinals act as canonical well-orderings.

**Theorem 3.26.**   *Every nonempty well-ordered set is order isomorphic to an ordinal. That is, if $X$ is well-ordered by $\leq$, then there is an ordinal $\alpha$ and a function $h : X \to \alpha$ such that $h$ is one-to-one and onto, and for all $x$ and $y$ in $X$, $x \leq y$ implies $h(x) \leq h(y)$.*

PROOF.   Let $X$ be a well-ordered set. For each $x \in X$, define the initial segment for $x$ by $I_x = \{t \in X \mid t \leq x\}$. Let $W$ be the subset of $X$ consisting of all elements $x$ such that $I_x$ is order isomorphic to an ordinal. Note that for each $x \in W$, $I_x$ is order isomorphic to a unique ordinal. By the replacement axiom, we can construct a set $A$ of all the ordinals isomorphic to initial segments of $X$. Let $\alpha = \cup A$; by Theorem 3.25, $\alpha$ is an ordinal. If $x, y \in W$, $x \leq y$, $\gamma$ and $\delta$ are ordinals, and $h_x$ and $h_y$ are order isomorphisms such that $h_x : I_x \to \gamma$ and $h_y : I_y \to \delta$, then for all $t < x$, $h_y(t) = h_x(t)$. Using the replacement axiom, we can form the set of all the order isomorphisms corresponding to the initial segments, and concatenate them to build a new function $h$. This new function is an order isomorphism from $W$ onto $\alpha$. To complete the proof, we claim that $W = X$. Suppose not; since $X$ is well-ordered, we can find a least $t$ in $X$ such that $t \notin W$. If we extend $h$ by setting

$h(t) = \alpha$, then $h$ witnesses that $I_t$ is order isomorphic to $\alpha + 1$. Thus $t \in W$, yielding a contradiction and completing the proof. □

The next theorem shows that using AC we can well-order any set, broadening the applicability of the preceding theorem. Our proof of the "well-ordering principle" uses ideas from Zermelo's [150] original proof. The proof can also be viewed as a special case of Zorn's Lemma. See Exercise 18 for more about Zorn's Lemma.

**Theorem 3.27.** *Every set can be well-ordered.*

PROOF. Let $X$ be a set. We will construct a one-to-one map $h : \alpha \to X$ from an ordinal $\alpha$ onto $X$. This suffices to prove the theorem, since the elements of $\alpha$ are well-ordered and $h$ matches elements of $\alpha$ with elements of $X$.

By AC we can pick $x \in X - t$ for each nonempty $t \subset X$. There are two things to note here. First, $x_t$ is never an element of $t$. This is important later in the proof. Second, this is the only use of AC in this entire proof. This is handy for the exercises.

Suppose that $f : \alpha \to Y$ is a one-to-one map of an ordinal $\alpha$ onto a set $Y \subset X$. For each $\beta < \alpha$, let $f[\beta]$ denote $\{f(\delta) \mid \delta \in \beta\}$. (Remember, since $\beta$ and $\alpha$ are ordinals, $\beta < \alpha$ is the same thing as $\beta \in \alpha$.) We say that $f$ is a $\gamma$-function if $f(\beta) = x_{f[\beta]}$ for every $\beta \in \alpha$. Let $\Gamma$ be the set of all $\gamma$-functions. The $\gamma$-functions cohere nicely; if $f$ and $g$ are $\gamma$-functions and $\beta$ is in both of their domains, then $f(\beta) = g(\beta)$. (See Exercise 17.) If we view the functions in $\Gamma$ as sets of ordered pairs, $\cup\Gamma$ is also a set of ordered pairs. Since the functions cohere and are one-to-one, $\cup\Gamma$ is actually a one-to-one function; call it $h$. By Theorem 3.25, the union of the ordinals that are domains of the functions in $\Gamma$ is also an ordinal, so for some ordinal $\alpha$, $h : \alpha \to X$. Furthermore, $h$ is a $\gamma$-function.

It gets better. Suppose that $h$ does not map $\alpha$ onto $X$, so $h[\alpha] \subsetneq X$. Then we can define an extension $h'$ by setting $h'(\beta) = h(\beta)$ for $\beta < \alpha$ and $h'(\alpha) = x_{h[\alpha]}$. This extension $h'$ is also a $\gamma$-function, so $h' \in \Gamma$. Applying the definition of $h$, we find that $h'(\alpha)$ is in the range of $h$. But $h'(\alpha) = x_{h[\alpha]}$ and the range of $h$ is $h[\alpha]$, so we have $x_{h[\alpha]} \in h[\alpha]$, contradicting the statement two paragraphs back. Summarizing, $h$ is a one-to-one map of $\alpha$ onto $X$, so $X$ is well-ordered. □

Combining the last two theorems yields the following corollary.

**Corollary 3.28.** *For every set $X$ there is a unique least ordinal $\alpha$ such that $X \sim \alpha$.*

PROOF. Fix $X$. By Theorem 3.27, $X$ can be well-ordered. By Theorem 3.26, $X \sim \beta$ for some ordinal $\beta$. Let $A = \{\gamma \leq \beta \mid \gamma \sim X\}$ be the set of ordinals less than or equal to $\beta$ that are equinumerous with $X$. Since $A$ is a nonempty set of ordinals well-ordered by $\leq$, $A$ contains a least element, $\alpha$. Let $\delta$ be any ordinal such that $X \sim \delta$. By Theorem 3.25, $\delta < \alpha$ or $\alpha \leq \delta$. If $\delta < \alpha$, then $\delta \leq \beta$ and we have $\delta \in A$, contradicting the minimality of $\alpha$. Thus $\alpha \leq \delta$, and $\alpha$ is unique. □

Since every set has a unique least equinumerous ordinal, we can define $|X|$ as the least ordinal $\alpha$ such that $X \sim \alpha$. We say that an ordinal $\kappa$ is a *cardinal number* if $|\kappa| = \kappa$. In slogan form, a cardinal number is the least ordinal of its cardinality.

The finite pigeonhole principle asserts that every finite ordinal is a cardinal. Thus, 0, 1, and 17324 are all cardinals. The infinite pigeonhole principle shows that $\omega$ cannot be mapped one-to-one into any finite cardinal, so $\omega$ is a cardinal number; indeed, it is the least infinite cardinal. On the other hand, $\omega + 1 \sim \omega$ and $\omega + 2 \sim \omega$, so $\omega + 1$ and $\omega + 2$ are not cardinals. The elements of the next larger cardinal cannot be matched with the elements of $\omega$, so the next larger cardinal is uncountable.

Even though every cardinal number is an ordinal, we have special notation to distinguish the cardinals. When we are thinking of $\omega$ as a cardinal, we denote it with an aleph, so $\omega = \aleph_0$. The next larger (and therefore uncountable) cardinal is $\aleph_1$. Proceeding in this way, and using unions at limit ordinals, we can define $\aleph_\alpha$ for every ordinal number $\alpha$. For example, the least cardinal bigger than $\aleph_0, \aleph_1, \aleph_2, \ldots$ is $\aleph_\omega$. Assuming AC, for every infinite set $X$, there is an ordinal $\alpha$ such that $|X| = \aleph_\alpha$.

The ordinals are like a long string of beads. The infinite cardinals, which are the alephs, appear like infrequent pearls along the string. The ordinals are good for counting steps in order (like rosary beads), and the cardinals are good for summing up sizes (like abacus beads). For finite sets, cardinals and ordinals are identical. Thus $|\{A, B, C, D\}| = 4 = \{0, 1, 2, 3\}$ and $\{A, B, C, D\} \sim \{0, 1, 2, 3\}$. In general, the matching approach to measuring the sizes of sets agrees with the cardinality approach. This is formalized in the following theorem.

**Theorem 3.29.** *For all sets $X$ and $Y$, $|X| = |Y|$ if and only if $X \sim Y$.*

PROOF.    Suppose $|X| = |Y| = \kappa$. Then $X \sim \kappa$ and $Y \sim \kappa$, so $X \sim Y$. Conversely, if $X \sim Y$, let $\kappa_1 = |X|$ and $\kappa_2 = |Y|$. Since $\kappa_1 \sim X \sim Y \sim \kappa_2$, we have $\kappa_1 \sim \kappa_2$. Since $\kappa_1$ and $\kappa_2$ are cardinals, $\kappa_1 \sim \kappa_2$ implies $\kappa_1 = \kappa_2$. Thus $|X| = |Y|$.    □

### 3.5.3    Pigeons Finished Off

*Every Sunday you'll see*
*My sweetheart and me,*
*As we poison the pigeons in the park.*

— Tom Lehrer

At this point, we know quite a bit about stuffing pigeons into pigeonholes. For example, if $p$ and $h$ are finite cardinal numbers, we have the following finite pigeonhole principle.

- If we put $p$ pigeons into $h$ pigeonholes and $h < p$, then some pigeonhole contains at least two pigeons.

The idea here is that any function from the set of larger cardinality into the set of smaller cardinality must fail to be one-to-one. By the Cantor–Bernstein Theorem, this holds for infinite cardinals as well. Thus for any cardinals $\kappa$ and $\lambda$, we get the following analogue of the finite pigeonhole principle.

- If we put $\kappa$ pigeons into $\lambda$ pigeonholes and $\lambda < \kappa$, then some pigeonhole contains at least two pigeons.

The preceding infinite analogue of the finite pigeonhole principle is not the same as the infinite pigeonhole principle of Theorem 3.1. Here is a restatement of Theorem 3.1 using our notation for cardinals.

- If we put $\aleph_0$ pigeons into $h$ pigeonholes and $h < \aleph_0$, then some pigeonhole contains $\aleph_0$ pigeons.

The infinite pigeonhole principle says that some pigeonhole is infinitely crowded. This does not transfer directly to higher cardinalities. For example, we can put $\aleph_\omega$ pigeons into $\aleph_0$ pigeonholes is such a way that every pigeonhole has fewer than $\aleph_\omega$ pigeons in it. To do this, put $\aleph_0$ pigeons in the 0th hole, $\aleph_1$ pigeons in the 1st hole, $\aleph_2$ pigeons in the 2nd hole, and so on. The total number of pigeons is $\aleph_\omega = \cup_{n \in \omega} \aleph_n$, but each hole contains $\aleph_n$ pigeons for some $n < \omega$. This peculiar behavior stems from the singular nature of $\aleph_\omega$.

A cardinal $\kappa$ is *singular* if there is a cardinal $\lambda < \kappa$ and a function $f : \lambda \to \kappa$ such that $\cup_{\alpha < \lambda} f(\alpha) = \kappa$. (Remember, $\kappa$ is transitive, so if $f(\alpha) \in \kappa$, then $f(\alpha) \subset \kappa$.) As an example, if we define $f : \aleph_0 \to \aleph_\omega$ by $f(n) = \aleph_n$, then $\cup_{\alpha < \aleph_0} f(\alpha) = \aleph_\omega$, showing that $\aleph_\omega$ is singular. Any infinite cardinal number that is not singular is called *regular*. One good example of a regular cardinal is $\aleph_0$; it is not equal to any finite union of finite cardinals. We can generalize the infinite pigeonhole principle for regular cardinals, but to prove the new result, we will need the following theorem.

**Theorem 3.30.**  *For every infinite cardinal $\kappa$, $|\kappa \times \kappa| = \kappa$.*

We will postpone the proof of Theorem 3.30 and jump right to the avian corollary. Since this is the last pigeonhole principle in this book, we will call it ultimate.

**Corollary 3.31** (Ultimate Pigeonhole Principle).  *The following are equivalent:*

1. *$\kappa$ is a regular cardinal.*
2. *If we put $\kappa$ pigeons into $\lambda < \kappa$ pigeonholes, then some pigeonhole must contain $\kappa$ pigeons.*

PROOF.  First suppose that $\lambda < \kappa$ and $\kappa$ is regular. Suppose that $g : \kappa \to \lambda$ is an assignment of $\kappa$ pigeons to $\lambda$ pigeonholes. For each $\alpha < \lambda$, let $f(\alpha) = |\{x \in \kappa \mid g(x) = \alpha\}|$, so $f(\alpha)$ is the population of the $\alpha$th pigeonhole. Suppose, by way of contradiction, that $f(\alpha) < \kappa$ for each $\alpha$. Because each $f(\alpha)$ is a cardinal, $\mu = \cup_{\alpha < \lambda} f(\alpha)$ is a cardinal. Furthermore, $\mu < \kappa$ because $\kappa$ is regular. For each $\alpha$, $f(\alpha) \leq \mu$, so the population of the $\alpha$th pigeonhole can be matched with a subset of $\mu$. Since there are $\lambda$ pigeonholes, the entire pigeon population can be matched with a subset of $\mu \times \lambda$, so $\kappa \leq |\mu \times \lambda|$. Let $\nu = \max\{\mu, \lambda\}$. Since $\mu \leq \nu$ and $\lambda \leq \nu$, $|\mu \times \lambda| \leq |\nu \times \nu|$. By Theorem 3.30, $|\nu \times \nu| = \nu$. Since $\mu < \kappa$ and $\lambda < \kappa$, we have $\nu < \kappa$, and concatenating inequalities yields

$$\kappa \leq |\mu \times \lambda| \leq |\nu \times \nu| = \nu < \kappa,$$

a contradiction. Thus, for some $\alpha$, $f(\alpha) = \kappa$ and the $\alpha$th pigeonhole contains $\kappa$ birds.

To prove the converse, suppose that $\kappa$ is a singular cardinal. Then there is a cardinal $\lambda < \kappa$ and a function $f : \lambda \to \kappa$ such that $\cup_{\alpha<\lambda} f(\alpha) = \kappa$. Define $g : \kappa \to \lambda$ by letting $g(\beta)$ be the least $\alpha$ such that $\beta \in f(\alpha)$. Since $\cup_{\alpha<\lambda} f(\alpha) = \kappa$ and $\lambda$ is well-ordered, $g$ is well-defined and maps each element of $\kappa$ to an element of $\lambda$. Furthermore, for each $\alpha < \lambda$,

$$|\{\beta \in \kappa \mid g(\beta) = \alpha\}| \leq |f(\alpha)| < \kappa.$$

Thus $g$ can be viewed as an assignment of $\kappa$ pigeons to $\lambda$ pigeonholes in which the population of each pigeonhole is less than $\kappa$.    □

The first part of the preceding proof can be adapted to prove that lots of cardinals are regular. This is a nice fact, since it means that we can apply the pigeonhole principle in lots of situations.

**Corollary 3.32.** *For each ordinal $\alpha$, the cardinal $\aleph_{\alpha+1}$ is regular.*

PROOF.    We will sketch the argument. Suppose $f : \lambda \to \aleph_{\alpha+1}$, where $\lambda$ is a cardinal such that $\lambda < \aleph_{\alpha+1}$. Then $\lambda \leq \aleph_\alpha$, and for each $\beta < \lambda$, $|f(\beta)| \leq \aleph_\alpha$. Applying Theorem 3.30 yields $|\cup_{\alpha<\lambda} f(\alpha)| \leq |\aleph_\alpha \times \aleph_\alpha| = \aleph_\alpha < \aleph_{\alpha+1}$.    □

We should list the regular cardinals we have found. $\aleph_0$ is regular, and by the preceding corollary so are $\aleph_1(=\aleph_{0+1})$, $\aleph_2(=\aleph_{1+1})$, $\aleph_3$, $\aleph_4$, and so on. We have seen that the limit cardinal $\aleph_\omega$ is singular; the subscript cannot be written as $\alpha + 1$, so this does not contradict Corollary 3.32. However, $\aleph_{\omega+1}$ is regular, as are $\aleph_{\omega+2}$, $\aleph_{\omega+3}$, and so on. Our only good example of a regular limit cardinal is $\aleph_0$. We do not have an example of an uncountable regular limit cardinal. The reason for this is explained in Section 3.6.

It seems that Theorem 3.30 is a handy way to bound the sizes of unions. Here is a nice way to capsulize that.

**Corollary 3.33.** *If $\kappa$ is an infinite cardinal and $|X_\alpha| \leq \kappa$ for each $\alpha < \kappa$, then $|\cup_{\alpha<\kappa} X_\alpha| \leq \kappa$. In particular, a countable or finite union of at most countable sets is at most countable.*

PROOF.    Suppose $|X_\alpha| \leq \kappa$ for each $\alpha < \kappa$. For each $\alpha$, let $g_\alpha : X_\alpha \to \kappa$ be a one-to-one map. Define $f : \cup_{\alpha<\kappa} X_\alpha \to \kappa \times \kappa$ by $f(x) = (\alpha, g_\alpha(x))$, where $\alpha$ is the least ordinal such that $x \in X_\alpha$. The function $f$ is one-to-one, so $\cup_{\alpha<\kappa} X_\alpha \precsim \kappa \times \kappa$. Thus by Theorem 3.30, $|\cup_{\alpha<\kappa} X_\alpha| \leq |\kappa \times \kappa| = \kappa$. To prove the particular case, let $\kappa = \aleph_0$.    □

We have used Theorem 3.30 repeatedly, but still have not proved it. It is time to pay the piper.

PROOF OF THEOREM 3.30.    We will use induction to prove that if $\kappa$ is an infinite cardinal, $|\kappa \times \kappa| = \kappa$. For the base case, apply Corollary 3.22 to get $|\aleph_0 \times \aleph_0| = \aleph_0$. As the induction hypothesis, assume $|\lambda \times \lambda| = \lambda$ for every infinite cardinal $\lambda < \kappa$. Since $\kappa \precsim \kappa \times \kappa$, by the Cantor–Bernstein Theorem, it suffices to show that $\kappa \times \kappa \precsim \kappa$.

Define the ordering $<$ on $\kappa \times \kappa$ as follows. Let $(\alpha, \beta), (\alpha', \beta') \in \kappa \times \kappa$ and let $\mu = \max\{\alpha, \beta\}$ and $\mu' = \max\{\alpha', \beta'\}$. We say that $(\alpha, \beta) < (\alpha', \beta')$ if and only if

$\mu < \mu'$, or
$\mu = \mu'$ and $\alpha < \alpha'$, or
$\mu = \mu'$ and $\alpha = \alpha'$ and $\beta < \beta'$.

Informally, this relation sorts $\kappa \times \kappa$ by looking at maxima, then first elements, and then second elements. A routine but technical argument shows that $<$ is a well-ordering of $\kappa \times \kappa$. By Theorem 3.26, $\kappa \times \kappa$ under $<$ is order isomorphic to some ordinal. Let $\delta$ denote that ordinal, and let $f : \kappa \times \kappa \to \delta$ be the order isomorphism. If $\delta \leq \kappa$, then $\kappa \times \kappa \precsim \delta \precsim \kappa$ and the proof is complete.

Suppose by way of contradiction that $\kappa < \delta$. Since $\delta$ is an ordinal, we know that $\kappa \in \delta$, so there is an element $(\sigma, \tau) \in \kappa \times \kappa$ such that $f(\sigma, \tau) = \kappa$. Let $\mu = \max\{\sigma, \tau\}$ and note that $\sigma < \kappa, \tau < \kappa$, and consequently $\mu < \kappa$. Furthermore, by definition of the well-ordering on $\kappa \times \kappa$,

$$\{(\alpha, \beta) \in \kappa \times \kappa \mid f(\alpha, \beta) < \kappa\} \subset \mu \times \mu,$$

so $\kappa \precsim \mu \times \mu$ and $\kappa \leq |\mu \times \mu|$. Let $\lambda = |\mu|$. Since $\lambda \sim \mu$, we have $|\mu \times \mu| = |\lambda \times \lambda|$. Since $\mu < \kappa$, $\lambda$ is a cardinal less than $\kappa$, so by the induction hypothesis, $|\lambda \times \lambda| = \lambda < \kappa$. Concatenating inequalities yields $\kappa \leq |\mu \times \mu| = |\lambda \times \lambda| < \kappa$, which is a contradiction and completes the proof.    $\square$

### Exercises

1. Define $f : \mathbb{N} \to \mathbb{Z}$ by $f(n) = (-1)^{n+1}(\frac{1}{4})(2n + 1) + (\frac{1}{4})$.

   a. Show that $f$ is one-to-one. (Assume $f(j) = f(k)$ and prove $j = k$.)
   b. Show that $f$ is onto. (Show that if $m > 0$, then $f(2m - 1) = m$ and if $t \leq 0$ then $f(-2t) = t$.)

2. Prove that $\sim$ is an equivalence relation. That is, show that for all sets $A$, $B$ and $C$, the following hold:

   a. $A \sim A$,
   b. $A \sim B \to B \sim A$, and
   c. $(A \sim B \wedge B \sim C) \to A \sim C$.

3. Show that for all sets $A$, $B$, and $C$, the following hold:

   a. $A \precsim A$ (so $\precsim$ is reflexive), and
   b. $(A \precsim B \wedge B \precsim C) \to A \precsim C$ (so $\precsim$ is transitive).

4. Details of the Cantor–Bernstein proof. This problem uses notation from the proof of Theorem 3.20.

   a. Use induction to prove Claim 1. As a base case, show $A_0 \supset A_1 \supset A_2$. For the induction step, assume $A_n \supset A_{n+1} \supset A_{n+2}$ and show that $A_{n+2} \supset A_{n+3}$, using the fact that $A_n \supset A_{n+1}$ implies $g \circ f(A_n) \supset g \circ f(A_{n+1})$.

**b.** Prove Claim 2. Use the fact that either $x \in A'_\omega$ or there is a least $j$ such that $x \notin A_j$ to show that each $x$ is in some $A'_n$. (Here $n$ is a natural number or $\omega$.) To prove that $x$ is in a unique $A'_n$, suppose that $x$ is in two such sets, and seek a contradiction.

**c.** Prove Claim 3. Use Claim 1 to get a short proof that $B_n \supset B_{n+1}$. The remainder of the argument parallels the proof of Claim 2.

**d.** Prove Claim 4. To show that $h(A'_{2n}) = B'_{2n+1}$, note that because $f$ is one-to-one we have $f(A_{2n}) - f(A_{2n-1}) = f(A_{2n} - A_{2n-1})$, and so

$$B'_{2n+1} = B_{2n+1} - B_{2n} = f(A_{2n}) - f(A_{2n-1}) = f(A'_{2n}) = h(A'_{2n}).$$

The proof that $B'_{2n} = h(A'_{2n+1})$ is similar. The proof that $h(A'_\omega) = B'_\omega$ relies on the fact that since $f$ is one-to-one, we have $f(\cap_{n \in \omega} A_n) = \cap_{n \in \omega} f(A_n)$.

**5.** Let $\mathbb{Q}$ denote the set of rationals. Prove that $\mathbb{Q} \sim \mathbb{N}$.

**6.** Let *Seq* denote the set of all finite sequences of natural numbers. Prove that *Seq* $\sim \mathbb{N}$.

**7.** Without using Theorem 3.24, prove that $\mathbb{Z}$ with the usual ordering is not well-ordered.

**8.** Using Theorem 3.24, prove that $\mathbb{Z}$ with the usual ordering is not well-ordered.

**9.** Repeat Exercises 7 and 8 for the set $\mathbb{Q}^+ = \mathbb{Q} \cap [0, \infty)$ with the usual ordering.

**10.** Suppose that $X$ is a transitive set.

**a.** Prove that $X' = X \cup \{X\}$ is transitive.

**b.** Prove that $\cup X$ is transitive.

**11.** Give an example of a nontransitive set where $\leq_\in$ is a transitive relation. (Hint: The relation $\leq_\in$ is vacuously transitive on every two-element set.)

**12.** Give an example of a transitive set where $\leq_\in$ is not a transitive relation. (Hint: There is an example with three elements.)

**13.** Prove that if $\alpha$ is an ordinal, then so is $\alpha' = \alpha \cup \{\alpha\}$. (Hint: Exercise 10a shows that $\alpha'$ is transitive. Use the fact that $\alpha$ is well-ordered by $\leq_\in$ to show that $\alpha'$ is too.)

**14.** Prove that if $\alpha$ and $\beta$ are ordinals, then $\alpha \cap \beta = \alpha$ or $\alpha \cap \beta = \beta$. (Hint: Let $C = \alpha \cap \beta$ and suppose that $C \neq \alpha$ and $C \neq \beta$. Let $\gamma$ be the least element of $\alpha$ such that $\gamma \notin C$. Show that $\gamma = C$, so $C \in \alpha$. Similarly, $C \in \beta$, so $C \in \alpha \cap \beta = C$, contradicting Theorem 3.13.)

**15.** Prove that if $X$ is a set of ordinals, then $\cup X$ is an ordinal. (Hint: Exercise 10b shows that $\cup X$ is transitive. The fact that $\leq_\in$ well-orders each element of $X$ helps in proving antisymmetry and transitivity. Exercise 14 is useful in showing that trichotomy holds. Use the axiom of regularity to show that $\cup X$ has no infinite descending sequences.)

**16.** Prove that if $\alpha$ and $\beta$ are ordinals, then $\alpha \in \beta$, $\alpha = \beta$, or $\beta \in \alpha$. (Hint: Do Exercise 14 first.)

**17.** Details from the proof of Theorem 3.27.

Let $f : \beta_1 \to X$ and $g : \beta_2 \to X$ be $\gamma$-functions as defined in the proof of Theorem 3.27. Prove that if $\beta \in \beta_1 \cap \beta_2$, then $f(\beta) = g(\beta)$. (If $f$ and $g$

disagree, then there is a least $\beta$ such that $f(\beta) \neq g(\beta)$. For this $\beta$, $f[\beta] = g[\beta]$. Apply the definition of a $\gamma$-function.)

**18.** Zorn's Lemma and AC.

Prove in ZF that the following are equivalent:

1. AC.
2. Zorn's Lemma: If $P$ is a partial ordering (transitive and antisymmetric) such that every chain (linearly ordered subset) has an upper bound in $P$, then $P$ contains a maximal element (an element with no elements above it.)
3. Every set can be well-ordered.

a. Prove that 1 implies 2. (Hint: Adapt the proof of Theorem 3.27. Suppose $P$ has no maximal elements. For each chain $C$, let $x_C$ be an upper bound of $C$ that is not an element of $C$. Call $C$ a $\gamma$-chain if for every $p \in C$, $x_{\{y \in C \mid y < p\}} = p$. Use the union of the $\gamma$-chains to derive a contradiction.)
b. Prove that 2 implies 3. (Hint: Fix a set $X$ to well-order. Let $P$ be the set of all one-to-one maps from ordinals to subsets of $X$. $P$ is partially ordered by function extension. Show that every chain has an upper bound. Show that a maximal element maps an ordinal one-to-one onto $X$.)
c. Prove that 3 implies 1.

**19.** Describe a way to place $\aleph_0$ pigeons in $\aleph_0$ pigeonholes so that every pigeonhole contains at most one pigeon and $\aleph_0$ of the pigeonholes are empty. (Hint: $|\mathbb{Z}| = |\mathbb{N}| = \aleph_0$.)

**20.** Describe a way to place $\aleph_0$ pigeons in $\aleph_0$ pigeonholes so that every pigeonhole contains $\aleph_0$ pigeons. (Hint: $|\aleph_0 \times \aleph_0| = \aleph_0$.)

**21.** Describe a way to place $\aleph_0$ pigeons in $\aleph_0$ pigeonholes so that for each $\kappa \leq \aleph_0$ exactly one pigeonhole contains exactly $\kappa$ pigeons.

**22.** Describe a way to place $\aleph_0$ pigeons in $\aleph_0$ pigeonholes so that for each $\kappa \leq \aleph_0$ exactly $\aleph_0$ pigeonholes contain exactly $\kappa$ pigeons.

**23.** Show that if we put $\kappa$ pigeons into $\lambda$ pigeonholes and $\kappa < \lambda$, then $\lambda$ pigeonholes will remain empty. (Hint: The set of all pigeonholes is the union of the empty holes and the occupied holes.)

## 3.6   Incompleteness and Cardinals

> ...we never assumed that (ZFC) included all the "true" facts.
>
> — Levy [101]

In the last section we noted the absence of an example of an uncountable regular limit cardinal. We do not have an example because in ZFC we cannot prove that uncountable regular limit cardinals exist, assuming that ZFC is consistent. We have to *assume* that ZFC is consistent, because ZFC cannot prove that either. The last sentence is essentially Gödel's Second Incompleteness Theorem, which is where we will start our exploration of large cardinals.

## 3.6.1  Gödel's Theorems for PA and ZFC

*Gödel's 1931 (paper) was undoubtably the most exciting and the most cited article in mathematical logic and foundations to appear in the first eighty years of the (twentieth) century.*

— Kleene [66]

Gödel's First Incompleteness Theorem [64] is a statement about the provability of a statement in formal Peano Arithmetic (PA). In a nutshell, Gödel's first theorem says that if PA is $\omega$-consistent, then there is a formula $G$ such that PA does not prove $G$ and PA does not prove $\neg G$. In order to make this clear, we should discuss PA, $\omega$-consistency, and the formula $G$.

The axioms of PA are an attempt to describe the important properties of the natural numbers under the operations of successor (adding one), addition, and multiplication. PA includes predicate calculus and axioms that say that

- 0 is not the successor of any element,
- $x + 0 = x$ and $x + (n + 1) = (x + n) + 1$,
- $x \cdot 0 = 0$ and $x \cdot (n + 1) = x \cdot n + x$, and
- the distributive laws hold.

PA also includes an induction scheme that can be used to prove a wealth of facts about the natural numbers. At one point, it was thought that PA might be able to prove every true statement about $\mathbb{N}$, but then Gödel's work ruled that possibility out.

We say that a theory is *consistent* if there is no formula $A$ such that the theory proves both $A$ and $\neg A$. A theory that is inconsistent and includes predicate calculus can prove every formula. Thus a theory is consistent if and only if there is some formula the theory cannot prove. When we say PA is $\omega$-consistent, we mean PA cannot prove both $\exists x A(x)$ and every formula in the list $\neg A(0), \neg A(1), \neg A(2)$, and so on. Assuming $\omega$-consistency is very reasonable, but a little stronger than assuming regular old consistency. Rosser [128] devised a way to prove Gödel's first theorem assuming only the consistency of PA. His replacement for Gödel's sentence $G$ is a slightly more complicated formula.

Informally, Gödel's formula $G$ says "there is no proof in PA of the formula $G$." This is encoded in the language of arithmetic. Given our daily exposure to word processors and automated spelling and grammar checkers, we are used to the idea that formulas and lists of formulas (like proofs) can be represented as strings of zeros and ones, and that such strings can be viewed as integers and described by arithmetical formulas. It is very remarkable that Gödel devised and utilized an encoding scheme in 1931, long before the advent of electronic computers.

The method for making $G$ refer to $G$ is very entertaining. Let $G_0(x)$ be the formula that says "there is no number that encodes a proof in PA of the formula obtained by substituting the number $x$ for the free variable in the formula encoded by the number $x$." Suppose that $n$ is the number that encodes $G_0(x)$. Note that the formula obtained by substituting $n$ for the free variable in the formula encoded by $n$ is exactly $G_0(n)$. Informally, $G_0(n)$ says "there is no number that encodes a proof in PA of $G_0(n)$." Thus, $G_0(n)$ is the desired formula $G$.

Once we have the encoding procedures in hand and have created the formula $G$, the remainder of the proof of Gödel's First Incompleteness Theorem is straightforward. Suppose that PA is $\omega$-consistent. First, suppose that PA proves $G$. Then this proof is encoded by a number $n$ and PA proves that "there is a number that encodes a proof in PA of $G$." Thus, PA proves $\neg G$, contradicting the consistency of PA. Now suppose that PA proves $\neg G$. Then PA proves that "there is a number that encodes a proof of $G$." By the $\omega$-consistency of PA, we can find some number that actually does encode such a proof, and so PA proves $G$. Again, we have contradicted the consistency of PA.

Gödel's Second Incompleteness Theorem [64] says that if PA is consistent, then there is no proof in PA that PA is consistent. Much of the machinery used for the first theorem applies here also. The formula that asserts that PA is consistent, $Con_{PA}$, is an encoding of the sentence "there are no numbers $x$ and $y$ such that $x$ encodes a proof in PA of a formula and $y$ encodes a proof in PA of the negation of that formula." It is possible to prove in PA that $Con_{PA} \rightarrow G$. Thus, if PA proved $Con_{PA}$, then PA would prove $G$, contradicting Gödel's First Incompleteness Theorem.

Perhaps the most remarkable quality of the incompleteness theorems is the ubiquity of their applicability. The theorems utilize only a few important features of PA and therefore apply to a wide variety of formal theories. In particular, the proofs of the theorems rely heavily on the ability to carry out a modest amount of arithmetic and the ability to check proofs in a mechanical fashion. Consequently, if a theory has enough axioms (to prove facts about encoding) but not too many axioms (so proof checking is not incomprehensibly complicated), then both incompleteness theorems apply. For example, both incompleteness theorems hold for ZFC. Thus, assuming that ZFC is consistent, there is a formula $Z$ such that ZFC proves neither $Z$ nor $\neg Z$, and ZFC does not prove $Con_{ZFC}$. The incompleteness theorems also hold for ZF and for any extensions of ZF by a finite number of axioms.

### 3.6.2   Inaccessible Cardinals

*Better to reign in L, than serve in Heav'n.*
                                    — Milton, *Paradise Lost* (slightly misquoted)

If $\kappa$ is an uncountable regular limit cardinal, then we say that $\kappa$ is *weakly inaccessible*. Our goal is to prove that the existence of weakly inaccessible cardinals is not provable in ZFC. At the end of this section we will link this back to our study of pigeonhole principles. The plan for achieving the goal is straightforward. The first step is to prove in ZFC that if there is a weakly inaccessible cardinal, then ZFC is consistent. Then we apply Gödel's Second Incompleteness Theorem and get the desired result. The first step requires a journey to $L$, the constructible universe.

We will build the constructible universe in stages. Let $L_0 = \emptyset$. If $L_\alpha$ is defined, let $L_{\alpha+1}$ be the set of all subsets of $L_\alpha$ that are definable by restricted formulas with parameters from $L_\alpha$. To be precise, a set $X$ will be placed in $L_{\alpha+1}$ if all of the following conditions hold.

- $X \subset L_\alpha$,
- $u_1, u_2, \ldots, u_n \in L_\alpha$ is a finite list of parameters,

- $\psi$ is a formula in the language of set theory,
- $\psi$ does not contain the power set or union symbols,
- each quantifier in $\psi$ is of the form $\exists x \in L_\alpha$ or $\forall x \in L_\alpha$, and
- $X = \{y \in L_\alpha \mid \psi(y, u_1, u_2, \ldots, u_n)\}$.

If $\alpha$ is a limit ordinal, let $L_\alpha = \cup_{\beta < \alpha} L_\beta$. This suffices to define $L_\alpha$ for each ordinal number $\alpha$. Note that for each $\alpha$, $L_\alpha$ is a set. The constructible universe, $L$, is the class defined by $L = \cup L_\alpha$, where the union ranges over all ordinal numbers. Neither $\{L_\alpha \mid \alpha$ is an ordinal number$\}$ nor $L$ itself are sets, but it is convenient to refer to them using set-theoretic notation.

The finite levels of the constructible universe are simple in structure. For $k < \omega$, each $L_k$ is finite and $L_{k+1} = \mathcal{P}(L_k)$. By definition $L_0 = \emptyset$, and so $L_1 = \mathcal{P}(\emptyset) = \{\emptyset\}$ and $L_2 = \mathcal{P}(\{\emptyset\}) = \{\emptyset, \{\emptyset\}\}$. Using ordinal notation, $L_0 = 0$, $L_1 = 1$, and $L_2 = 2$. At the next level, $L_3$ breaks this pattern, since $L_3 = \mathcal{P}(L_2) = \{0, 1, \{1\}, 2\}$, which is not an ordinal. $L_\omega$ is defined by a union, so every element of $L_\omega$ is an element of $L_k$ for some $k < \omega$. Beyond $L_\omega$, the sets become vastly more complicated very rapidly. If $\kappa$ is weakly inaccessible, the $L_\kappa$ is an abridged version of the entire universe. This is stated more formally in the following theorem.

**Theorem 3.34.** *If $\kappa$ is weakly inaccessible, then $L_\kappa$ is a model of ZFC. That is, if we restrict the quantifiers in the axioms to $L_\kappa$, then the axioms of ZFC all hold.*

COMMENTS ON THE PROOF. For a detailed treatment, see [37] or [35]. We will provide only a hint of some of the main issues in the proof. The basic idea is to verify that the axioms of ZFC hold in $V_\kappa$ in much the same way that one would show that the axioms defining vector spaces hold in $\mathbb{R}^3$. Some axioms are very easy to manage. For example, $\emptyset \in L_1$, so the empty set axiom holds. If $a$ and $b$ are in $L_\alpha$, then $\{a, b\}$ is in $L_{\alpha+1}$, so some instances of the pairing axiom are easy to verify. Verification of the infinity axiom relies on $\kappa$ being larger than $\aleph_0$, since every set in $L_{\aleph_0}$ is finite. The verification of the power set axiom is particularly tricky and relies on the assumption that $\kappa$ is a regular limit cardinal.    □

Now we can finish our proof of the unprovability of the existence of a weakly inaccessible cardinal. The proof relies on the fact that any set of axioms with a model must be consistent. If $M$ is a model for a set of axioms $T$, then every theorem that can be proved from $T$ is true in $M$. (This is actually what makes proofs useful. If you prove a theorem from the axioms for vector spaces, then it has to be true in every vector space.) If $T$ is inconsistent, then it proves a contradiction, which would have to be true in $M$. But models are concrete (think of $\mathbb{R}^3$ as a model for the vector space axioms), so no contradiction can be true in a model. For a more technical discussion of models, truth, and provability, see [104].

**Theorem 3.35.** *If ZFC is consistent, then ZFC does not prove the existence of a weakly inaccessible cardinal.*

PROOF. Assume that ZFC is consistent. Suppose, by way of contradiction, that ZFC proves that there is a cardinal $\kappa$ that is weakly inaccessible. Then ZFC proves that the set $L_\kappa$ exists. By Theorem 3.34 we know that $L_\kappa$ is a model of ZFC, so

ZFC is consistent. Thus ZFC proves the consistency of ZFC, contradicting Gödel's Second Incompleteness Theorem and completing the proof. □

The unprovability of the existence of weakly inaccessible cardinals is not like the unprovability of AC in ZF. If we write $I$ for the statement "there is a weakly inaccessible cardinal" and write ZFC $\vdash I$ for "ZFC proves $I$," then the preceding theorem says that ZFC $\nvdash I$, provided that ZFC is consistent. To show that $I$ is *independent* of ZFC (like AC is for ZF) we would also need to prove ZFC $\nvdash \neg I$ assuming that ZFC is consistent. Thanks to Gödel, we know that this is an unattainable goal. To see this, suppose (for an eventual contradiction) that from $Con_{ZFC}$ we can prove in ZFC that ZFC $\nvdash \neg I$. If ZFC $\nvdash \neg I$, then $Con_{ZFC+I}$. Thus, our hypothesis boils down to ZFC $\vdash Con_{ZFC} \rightarrow Con_{ZFC+I}$. Since ZFC+$I$ is an extension of ZFC, we have ZFC+$I \vdash Con_{ZFC} \rightarrow Con_{ZFC+I}$. Theorem 3.34 shows that ZFC+$I \vdash Con_{ZFC}$. Concatenating the last two lines, we obtain ZFC+$I \vdash Con_{ZFC+I}$, contradicting Gödel's Second Incompleteness Theorem for ZFC+$I$.

Finally, we should summarize the combinatorial implications of this section. We know that if ZFC is consistent, then it cannot prove the existence of a weakly inaccessible cardinal. Also, $\kappa$ is weakly inaccessible if and only if it is an uncountable regular limit cardinal. By the Ultimate Pigeonhole Principle, $\kappa$ is regular if and only if whenever $\kappa$ pigeons are placed in fewer than $\kappa$ pigeonholes, then some hole contains $\kappa$ pigeons. So a weakly inaccessible cardinal is an uncountable limit cardinal with this pigeonhole property. In Section 3.5 we proved that $\aleph_0$ and $\aleph_{\alpha+1}$ for each ordinal $\alpha$ have this pigeonhole property. In this section we have proved that we cannot prove the existence of any more cardinals with this property.

### 3.6.3 A Small Collage of Large Cardinals

*All for one ... and more for me.*
— Cardinal Richelieu in *The Three Musketeers*

A cardinal number is said to be *large* if there is no proof in ZFC of its existence. We just met our first large cardinal, the weakly inaccessible cardinal. There are many other large cardinals related to combinatorial principles. This section describes the ones we need for the next two sections.

Recall that a weakly inaccessible cardinal is an uncountable regular limit cardinal. We say that a cardinal $\kappa$ is a *strong limit* if for every $\lambda < \kappa$ we have $|\mathcal{P}(\lambda)| < \kappa$. An uncountable regular strong limit cardinal is called *strongly inaccessible* (or just inaccessible). Every strongly inaccessible cardinal is weakly inaccessible. If we assume the generalized continuum hypothesis (GCH), then every weakly inaccessible cardinal is also strongly inaccessible. (See Exercise 6.) The hypothesis GCH is independent of ZFC, and has inspired a great deal of interesting work [89].

We say that $\kappa$ is *weakly compact* if $\kappa$ is uncountable and $\kappa \rightarrow (\kappa)_2^2$. This arrow notation is the same used in Section 3.2, so $\kappa \rightarrow (\kappa)_2^2$ means that if we color the edges of a complete graph with $\kappa$ vertices using two colors, then it must contain a monochromatic complete subgraph with $\kappa$ vertices. These cardinals reappear in the next section.

We will also look at some results concerning subtle cardinals, which are also defined in terms of colorings of unordered $n$-tuples. Suppose that $\kappa$ is a cardinal and let $[\kappa]^n$ denote the set of $n$-element subsets of $\kappa$. We say that a function $S : [\kappa]^n \rightarrow \mathcal{P}(\kappa)$ is an $(n, \kappa)$-*sequence* if for each element of $[\kappa]^n$ of the form $\alpha_1 < \alpha_2 < \ldots < \alpha_n < \kappa$, we have $S(\{\alpha_1, \alpha_2, \ldots, \alpha_n\}) \subset \alpha_1$. A subset $C \subset \kappa$ is *closed* if the limit of each sequence of elements in $C$ is either $\kappa$ or in $C$. The subset $C \subset \kappa$ is *unbounded* if for each $\alpha \in \kappa$, there is a $\beta \in C$ such that $\alpha < \beta$. We abbreviate closed and unbounded by writing *club*. The cardinal $\kappa$ is $n$-*subtle* if for every $(n, \kappa)$-sequence $S$ and every club set $C \subset \kappa$, there exist elements $\beta_1, \beta_2, \ldots, \beta_{n+1} \in C$ such that

$$S(\beta_1, \beta_2, \ldots, \beta_n) = \beta_1 \cap S(\beta_2, \beta_3, \ldots, \beta_{n+1}).$$

The basic idea is that given a coloring $S$ and a large set $C$, the large set must contain some elements that are monochromatic. The $n$-subtle cardinals are closely related to $n$-ineffable cardinals. More information on both these types of cardinals can be found in [7], [8], [37], and especially [83].

One variation on coloring $n$-tuples for fixed values of $n$ is to color $n$-tuples for all $n \in \omega$ simultaneously. Let $[\kappa]^{<\omega}$ denote the set of all finite subsets of $\kappa$. We say $\kappa$ is a *Ramsey cardinal* and write $\kappa \rightarrow (\kappa)_2^{<\omega}$ if for every function $f : [\kappa]^{<\omega} \rightarrow 2$ there is a set $X$ of size $\kappa$ such that for each $n$, $f$ is constant on $[X]^n$. Note that the same $X$ works for all $n$, though when $j \neq k$ the $j$-tuples may not be the same color as the $k$-tuples.

One type of cardinal frequently mentioned in the literature is bigger than anything we have listed so far. We say that $\kappa$ is a *measurable cardinal* if there is a $\kappa$-additive two-valued measure on $\kappa$. Roughly, this means that there is a way of assigning a value $\mu(X)$ to each $X \subset \kappa$ so that $\mu$ acts a lot like the measures that appear in analysis.

One way to organize all these cardinals is by comparing the sizes of the least example of each type of cardinal. Suppose we assign letters as follows:

$W$ :    Least weakly inaccessible cardinal,
$I$ :    Least strongly inaccessible cardinal,
$C$ :    Least weakly compact cardinal,
$S1$:    Least 1-subtle cardinal,
$S2$:    Least 2-subtle cardinal,

$\vdots$

$Sn$:    Least $n$-subtle cardinal,

$\vdots$

$R$ :    Least Ramsey cardinal,
$M$ :    Least measurable cardinal.

Then we have the following relationships:

$$W \leq I < C < S1 < S2 < \ldots < Sn < \ldots < R < M.$$

The proofs of these relationships are frequently nontrivial. Good references include [37] and [88].

**Exercises**

1. For each $0 \neq k \in \omega$ show that $|L_k| = 2^{k-1}$.
2. Using Exercise 1, prove that $L_\omega$ is countable.
3. Prove that for each ordinal $\alpha$, $L_\alpha$ is transitive. (Hint: $L_\alpha$ is transitive if $x \in y \in L_\alpha$ implies $x \in L_\alpha$. Use induction on the ordinals.)
4. Prove that if $\alpha < \beta$, then $L_\alpha \subset L_\beta$.
5. Prove that if $\kappa$ is a limit cardinal and $x \in L_\beta$ for some $\beta < \kappa$, then $\cup x \in L_\kappa$.
6. The generalized continuum hypothesis (GCH) asserts that for every $\alpha$, $|\mathcal{P}(\aleph_\alpha)| = \aleph_{\alpha+1}$. Assuming GCH, prove that every limit cardinal is a strong limit cardinal. As a corollary, show that GCH implies that every weakly inaccessible cardinal is strongly inaccessible.
7. Construct a 2-coloring $f$ of $[\omega]^{<\omega}$ such that $f$ is constant on $[\omega]^n$ for each $n$, but no pair has the same color as any triple.

## 3.7   Weakly Compact Cardinals

*Watch out for that tree!*

— *George of the Jungle* theme song

Theorem 3.3 says that if we 2-color a complete graph $G$ with $\aleph_0$ vertices, then it must contain a monochromatic subgraph with $\aleph_0$ vertices. In arrow notation, this is written as $\aleph_0 \rightarrow (\aleph_0)_2^2$, where the first $\aleph_0$ is the size of $G$ and the second $\aleph_0$ is the size of the desired monochromatic subgraph. It would be nice to know what other cardinals $\kappa$ satisfy $\kappa \rightarrow (\kappa)_2^2$. We call an uncountable cardinal with this property *weakly compact*. Judging from the next theorem, our list of weakly compact cardinals may be very short.

**Theorem 3.36.** $|\mathbb{R}| \nrightarrow (\aleph_1)_2^2$ *and consequently,* $\aleph_1 \nrightarrow (\aleph_1)_2^2$.

PROOF.   Let $\kappa = |\mathbb{R}|$ and let $g : \kappa \rightarrow \mathbb{R}$ be a matching between the ordinals less than $\kappa$ and the reals. Let $G$ be a complete graph with $\kappa$ vertices. We can think of each vertex of $G$ as having two labels, an ordinal $\alpha < \kappa$ and a real number $g(\alpha)$. Color the edges of $G$ using the scheme

$$\chi(\alpha\beta) = \begin{cases} \text{red} & \text{if } \alpha < \beta \leftrightarrow g(\alpha) < g(\beta), \\ \text{blue} & \text{if } \alpha < \beta \leftrightarrow g(\beta) < g(\alpha). \end{cases}$$

Informally, $\chi$ colors the edge $\alpha\beta$ red if the order on the ordinal labels agrees with the order on the real number labels, and colors the edge blue if the orders disagree.

Suppose that $S$ is a subgraph of $G$ and $|S| = \aleph_1$. We will show that $S$ is not monochromatic. Since the ordinal labels for the vertices of $S$ are a well-ordered subset of $\kappa$, we can list them in increasing order as $\langle \alpha_\gamma \mid \gamma < \aleph_1 \rangle$. We consider two cases.

First, suppose that $S$ is red. Then the ordering on the real labels of the vertices of $S$ agrees with the ordering on the ordinal labels. This gives us an uncountable well-ordered increasing sequence of reals, $\langle g(\alpha_\gamma) \mid \gamma < \aleph_1 \rangle$. Using the fact that $\mathbb{Q}$ is dense in $\mathbb{R}$, for each $\gamma \in \aleph_1$, choose a rational $q_\gamma$ such that $g(\alpha_\gamma) < q_\gamma < g(\alpha_{\gamma+1})$. Then $\langle q_\gamma \mid \gamma < \aleph_1 \rangle$ is an uncountable sequence of distinct rationals, contradicting the countability of $\mathbb{Q}$. (See Exercise 5 in Section 3.5.) Thus, $S$ is not red.

Second, suppose that $S$ is blue. This yields an uncountable decreasing sequence of reals. By choosing $q_\gamma$ such that $g(\alpha_\gamma) > q_\gamma > g(\alpha_{\gamma+1})$ we obtain a contradiction in the same fashion as in the preceding case.

Summarizing, $S$ is neither red nor blue. $G$ contains no monochromatic subgraph of size $\aleph_1$.

To prove the last statement in the theorem, we note that by Theorem 3.18 and Corollary 3.23, $\mathbb{N} \prec \mathbb{R}$. Thus $\aleph_0 < |\mathbb{R}|$, and so $|\mathbb{R}| \geq \aleph_1$. Since there is a way to color a graph with $|\mathbb{R}|$ vertices so that no $\aleph_1$-sized monochromatic subgraphs exist, we can certainly do the same for a (possibly smaller) graph with a mere $\aleph_1$ vertices. □

Summarizing, we know that $\aleph_0 \to (\aleph_0)_2^2$, but $\aleph_1 \not\to (\aleph_1)_2^2$. We can generalize Theorem 3.36 to show that $\aleph_{\alpha+1} \not\to (\aleph_{\alpha+1})_2^2$, eliminating all the successor cardinals from our hunt for weakly compact cardinals.

**Theorem 3.37.** $|\mathcal{P}(\aleph_\alpha)| \not\to (\aleph_{\alpha+1})_2^2$.

PROOF.    Imitate the preceding proof using $\mathcal{P}(\aleph_\alpha)$ in the role of $\mathbb{R}$. To do this, prove and use the fact that when $\mathcal{P}(\aleph_\alpha)$ is ordered by the relation

$$X < Y \quad \text{if and only if} \quad \min((X - Y) \cup (Y - X)) \in Y,$$

it contains no increasing or decreasing sequences of size $\aleph_{\alpha+1}$. □

At this point we know that any weakly compact cardinal must be an uncountable limit cardinal. In the following theorem, we emulate Erdős and Tarski [46] in showing that any weakly compact cardinal is inaccessible, and therefore large. Their studies of weakly compact cardinals were motivated by questions in infinite combinatorics stated in their 1943 paper [45].

**Theorem 3.38.**    *If $\kappa$ is weakly compact, then $\kappa$ is strongly inaccessible.*

PROOF.    Suppose that $\kappa \to (\kappa)_2^2$. We need to show that $\kappa$ is regular and a strong limit cardinal.

First suppose that $\kappa$ is not regular, so that for some $\lambda$ there is a function $f : \lambda \to \kappa$ such that $\cup_{\alpha<\lambda} f(\alpha) = \kappa$. We may assume that $f$ is increasing. We will use $f$ to construct a 2-coloring of a complete graph with $\kappa$ vertices. For each $\alpha < \beta < \kappa$, color the edge $\alpha\beta$ using the scheme

$$\chi(\alpha\beta) = \begin{cases} \text{red} & \text{if } \exists\gamma(\alpha < f(\gamma) \leq \beta), \\ \text{blue} & \text{otherwise.} \end{cases}$$

Informally, $f$ chops $\kappa$ into $\lambda$ intervals. An edge is red if it connects two intervals. Thus, no red subgraph can be larger than size $\lambda$. Also, each interval chopped out by $f$ is smaller than $\kappa$, so there is no blue subgraph of size $\kappa$. This contradicts $\kappa \to (\kappa)_2^2$, proving that no function like $f$ exists and that $\kappa$ is regular.

Now suppose that $\aleph_\alpha < \kappa$. If $\kappa \leq |\mathcal{P}(\aleph_\alpha)|$, then by Theorem 3.37, $\kappa \not\rightarrow (\aleph_{\alpha+1})^2_2$. But $\aleph_{\alpha+1} \leq \kappa$, so this implies that $\kappa \not\rightarrow (\kappa)^2_2$, contradicting weak compactness. Thus, if $\aleph_\alpha < \kappa$, then $|\mathcal{P}(\aleph_\alpha)| < \kappa$, proving that $\kappa$ is a strong limit cardinal.    □

We have seen that if $\kappa$ is uncountable and $\kappa \rightarrow (\kappa)^2_2$, then $\kappa$ is a large cardinal. By the results in Section 3.6, we know that ZFC cannot prove that such cardinals exist. Interestingly enough, increasing the number of colors (to any value less that $\kappa$) or increasing the size of the $n$-tuples (from 2 to any $n < \omega$) does not lead to larger cardinals. That is, if $\lambda < \kappa, n \in \omega$, and $\kappa$ is weakly compact, then $\kappa \rightarrow (\kappa)^n_\lambda$. (Details are left as exercises.) We could color $n$-tuples for all $n \in \omega$, and look for a single set that is monochromatic for $n$-tuples for each $n$. This leads to the Ramsey cardinals mentioned in Section 3.6, which are considerably larger than the weakly compact cardinals.

In Section 3.1 we saw a close relationship between Ramsey's Theorem and König's Lemma. This relationship reappears at higher cardinalities. We say that a cardinal $\kappa$ has the *tree property* if whenever $T$ is a tree with $\kappa$ many nodes and every level of $T$ has size less than $\kappa$, then $T$ must have a path of size $\kappa$. König's Lemma (Theorem 3.2) says that $\aleph_0$ has the tree property. The next theorem shows the connection between weakly compact cardinals and cardinals with the tree property.

**Theorem 3.39.** *$\kappa$ is weakly compact if and only if $\kappa$ is strongly inaccessible and has the tree property.*

POINTERS.    Excellent proofs of this result can be found in the books of Drake (Theorems 3.5 and 3.7 in Chapter 7 of [37]), Jech (Lemma 29.6 in Chapter 5 of [88]), and Roitman (Theorem 36 in Chapter 7 of [127]).    □

Theorem 3.39 does not characterize the smaller cardinals with the tree property. As we have already noted, $\aleph_0$ has the tree property. However, ZFC proves that $\aleph_1$ does not. There is a tree $T$ such that $|T| = \aleph_1$, the cardinality of each level of $T$ is less than $\aleph_1$, and $T$ has no paths of size $\aleph_1$. Such a tree is called an $\aleph_1$-Aronszajn tree, and stands as a counterexample to $\aleph_1$ having the tree property. The existence of an $\aleph_2$-Aronszajn tree is deducible from GCH in ZFC. On the other hand, Silver has shown that from the existence of a weakly compact cardinal we can prove the consistency of ZFC and "$\aleph_2$ has the tree property." Mitchell and Silver have a number of other results pertaining to the tree property. Finally, using the ordered list of cardinals that appears in Section 3.6, if $\kappa$ is the least strongly inaccessible cardinal, then $\kappa$ is strictly less than the least weakly compact cardinal, so by Theorem 3.39, $\kappa$ cannot have the tree property. Thus if $\kappa$ is the least strongly inaccessible cardinal, then there is a $\kappa$-Aronszajn tree.

### Exercises

1. Prove that if $\kappa \rightarrow (\kappa)^2_2$, then $\kappa \rightarrow (\kappa)^2_\lambda$ for every $\lambda < \kappa$.
2. Prove that if $\kappa \rightarrow (\kappa)^2_\lambda$ for each $\lambda < \kappa$, then $\kappa \rightarrow (\kappa)^3_2$.

3. Assuming GCH, show that if $\kappa$ is the least weakly inaccessible cardinal, then there is a $\kappa$-Aronszajn tree.
4. Find a proof that there is an $\aleph_1$-Aronszajn tree. (Hint: A library is a good place to look for proofs.)

## 3.8    Finite Combinatorics with Infinite Consequences

*Does mathematics need new axioms?*

— Solomon Feferman [51]

In this section we will state a remarkable result due to H. Friedman [58]. Friedman has concocted a finite combinatorial statement that he refers to as Proposition B. Under the appropriate consistency assumptions, he shows the following:

1. For each $k \in \omega$, the axiom system consisting of ZFC plus "there is a $k$-subtle cardinal" *cannot* prove Proposition B.
2. The axiom system consisting of ZFC plus "for every $k \in \omega$ there is a $k$-subtle cardinal" *can* prove Proposition B.

Thus, the use of large cardinals is *required* in the proof of the finite combinatorial statement of Proposition B.

Proposition B is not simple, but it can be understood with a modest effort. We will start with the full statement, and then gradually define any mysterious terminology.

**Proposition B**  *For every $k > 0$ and $p > 0$, there is an $n$ such that if $\{f_X \mid X \subset [n]^k\}$ is any #-decreasing function assignment for $[n]^k$, then we can find sets $A$ and $E$ satisfying*

- *$E$ is subset of $\{0, 1, 2, \ldots, n\}$ with $p$ elements,*
- *$[E]^k \subset A \subset [n]^k$, and*
- *$f_A$ has no more than $k^k$ regressive values on $[E]^k$.*

In the statement of the proposition, $p$, $n$, and $k$ represent natural numbers. Here, the notation $[n]^k$ denotes the set of all ordered $k$-tuples with coordinates in $\{0, 1, 2, \ldots, n-1\}$.

If $x$ is a $k$-tuple, then $|x|$ is the maximum coordinate of $x$ and $\min(x)$ is the minimum coordinate. For example, if $x = (1, 3, 0)$, then $|x| = 3$ and $\min(x) = 0$. If $f$ is a function mapping $k$-tuples to $k$-tuples, we say that $y$ is a *regressive value* for $f$ on $[E]^k$ if for some $x \in [E]^k$, $y = f(x)$ and $|y| < \min(x)$. For example, if $f(7, 6, 5) = (1, 3, 2)$, then $(1, 3, 2)$ is a regressive value for $f$ because $3 < 5$. A single regressive value may have many witnesses. The function $g : [8]^3 \to [8]^3$ defined by $g(x, y, z) = (0, 0, 0)$ for all $0 \le x, y, z < 8$ has only one regressive value, namely $(0, 0, 0)$.

A function assignment on $[n]^k$ assigns one function to each nonempty subset $X \subset [n]^k$. The function assigned to $X$ is required to map $X$ to $X$, so we always

have $f_X : X \to X$. We say that the function assignment $\{f_X \mid X \subset [n]^k\}$ is
#-decreasing if whenever $A \subset [n]^k$ and $x \in [n]^k$, either

- $f_{A \cup \{x\}}$ and $f_A$ agree on all elements of $A$, or
- there is a $y$ such that $|y| > |x|$ and $|f_A(y)| > |f_{A \cup \{x\}}(y)|$.

Some sense of the nature of #-decreasing function assignments can be gained
from looking at a very small example. Suppose we set $k = 1$ and consider the
possible function assignments for $[2]^1$. Since we are required to set $f_{\{0\}}(0) = 0$
and $f_{\{1\}}(1) = 1$, every function assignment on $[2]^1$ is completely determined by
the values of $f_{\{0,1\}}$. We will look at all four possible cases.

> Case 01: Suppose $f_{\{0,1\}}(0) = 0$ and $f_{\{0,1\}}(1) = 1$. Since $f_{\{0,1\}}$ extends both $f_{\{0\}}$
> and $f_{\{1\}}$, this is a #-decreasing function assignment.
> Case 00: Suppose $f_{\{0,1\}}(0) = 0$ and $f_{\{0,1\}}(1) = 0$. Since $f_{\{0,1\}}$ extends $f_{\{0\}}$ and
> $f_{\{1\}}(1) = 1 > 0 = f_{\{1\} \cup \{0\}}(1)$, this is also a #-decreasing function assignment.
> Case 10: Suppose $f_{\{0,1\}}(0) = 1$ and $f_{\{0,1\}}(1) = 0$. This is an acceptable function
> assignment, but it is not #-decreasing. If we let $A = \{0\}$ and set $x = 1$, then
> $f_{A \cup \{x\}}(0) = 1 \neq 0 = f_A(0)$, so the first clause of the definition of #-decreasing
> fails. Because $x = 1$, there is no $y \in A$ such that $|y| > |x|$, so the second clause
> fails also.
> Case 11: Suppose $f_{\{0,1\}}(0) = 1$ and $f_{\{0,1\}}(1) = 1$. This is an acceptable function
> assignment. Because $f_{\{0,1\}}(0) = 1$, imitating the argument in the preceding case
> will show that this function assignment is not #-decreasing.

In general, the second clause of the definition of #-decreasing function assignment forces the values of the functions to be pushed down. One might think that
this would always lead to a proliferation of regressive values. Proposition B asserts
that if we start with a large enough domain, then even when the function assignment is #-decreasing there will be a function on a large subset that is not regressive
at too many values.

Proposition B is a remarkable statement. Perhaps it is too remarkable to be true.
It is important to remember that proving the proposition requires large cardinal
assumptions that are well beyond the strength of the axioms of ZFC. Should we
automatically tack these large cardinal axioms onto our collection of everyday
set theory axioms? Feferman [51] would consider this rash. Does Friedman's result offer new insights into the consequences of assuming large cardinal axioms?
Absolutely.

## Exercises

1. Let $I$ denote the function assignment on $[n]^k$ defined by setting $f_A(x) = x$ for
   every $A$ and every $x \in A$. Show that $I$ is a #-decreasing function assignment.
2. Let $M$ denote the function assignment on $[n]^1$ defined by setting $f_A(x) = \min(A)$ for every $A$ and every $x \in A$.

   a. Show that $M$ is a #-decreasing function assignment.

**b.** If $f_A \in M$, what is the maximum number of regressive values that $f_A$ can have? What is the maximum number of witnesses that a regressive value for $f_A$ can have?

**3.** Let $S$ denote the function assignment on $[n]^1$ defined by setting

$$f_A(x) = \begin{cases} x & \text{if } x < |A|, \\ \min(A) & \text{if } x = |A|, \end{cases}$$

for every $A$ and every $x \in A$. Is $S$ a #-decreasing function assignment?

**4.** Find a function assignment on $[3]^1$ that is not #-decreasing. If you would like a challenge, find all of them.

## 3.9  Points of Departure

*Where do you want to go today?* ®

— Microsoft

This chapter can be summarized as a study of pigeonhole principles, König's Lemma, and Ramsey's Theorem and their connections to cardinal numbers. Our slightly obsessive focus on cardinals is not the only approach to studying infinite objects. In this section we take brief peeks at computability, reverse mathematics, and the analytical hierarchy. For each topic area, we outline an application to graph theory and combinatorics, and state an open question. The section closes with a list of Ramsey-style theorems and named trees.

### Computability

We say that a function $f : \mathbb{N} \to \mathbb{N}$ is *computable* if there is a program written in C that given an input of $n$ always halts with output $f(n)$. Any programming language can be substituted for C. A set $M \subset \mathbb{N}$ is *computable* if its characteristic function $\chi_M$ is a computable function. The function $\chi_M$ is defined by setting $\chi_M(n) = 1$ if $n \in M$ and $\chi_M(n) = 0$ if $n \notin M$. Our definition extends in the obvious way to handle functions on $\mathbb{N}^k$ and subsets of $\mathbb{N}^k$. The computable functions are sometimes called Turing computable functions or general recursive functions. The study of these functions is called computability theory or recursion theory. Good introductory texts include [14] and [137], and [62] is a survey of computability in graph theory and combinatorics.

Every computable function has a program. We can think of this program as an integer; it certainly would be stored in a computer as a string of zeros and ones that could be thought of as a single big integer. We use $\{m\}(n) = k$ as shorthand for saying that when the machine with code $m$ receives input $n$, it halts with output $k$. We frequently refer to these codes as *indices* for computable functions.

Not every subset of $\mathbb{N}$ is computable. For example, consider the set

$$H = \{m \in \mathbb{N} \mid \{m\}(m) = 0\}.$$

To show that $H$ is not computable, we consider $\chi_H$ and prove the following theorem.

**Theorem 3.40.** $\chi_H$ *is not computable.*

PROOF.  Suppose that $\chi_H$ is computable. Let $m$ be the code for a program that computes $\chi_H$. Then

$$\begin{aligned}
\chi_H(m) = 0 \quad &\leftrightarrow \quad \{m\}(m) = 0 \\
&\leftrightarrow \quad m \in H \\
&\leftrightarrow \quad \chi_H(m) = 1.
\end{aligned}$$

Since $0 \neq 1$, $\chi_H$ is not computable.    □

Because computer programs process in discrete steps, we can examine computations at various stages. We write $\{m\}(n) \downarrow_s$ if the program with index $m$ and input $n$ halts in fewer than $s$ steps. We also write $\{m\}(n) \downarrow$ if the program eventually halts. It is easy to mechanically check whether a program halts in $s$ steps, but impossible to mechanically determine which programs will eventually halt. Indeed, the set $H$ is referred to as the solution to the self-halting problem.

Using our extended shorthand, we can encode noncomputable sets like $H$ in graph theory problems. For example, suppose we define a coloring $F$ on $[\mathbb{N}]^3$ by the following rule. Let $\{i, j, k\}$ be any three-element subset of $\mathbb{N}$, assume that we have $i < j < k$, and let

$$F(i, j, k) = \begin{cases} \text{red} & (\forall m \leq i)\, \{m\}(m) \downarrow_j \leftrightarrow \{m\}(m) \downarrow_k, \\ \text{blue} & \text{otherwise.} \end{cases}$$

Thus $F(i, j, k)$ is red when for every machine with a code $m \leq i$, we get the same information about whether or not $\{m\}(m)$ halts by checking $j$ steps as we do by checking $k$ steps. If $F(i, j, k)$ is blue, then for some $m \leq i$ the machine $\{m\}(m)$ halts between step $j$ and step $k$. By Ramsey's Theorem, $F$ must have an infinite monochromatic set; call it $T = \{t_0, t_1, t_2, \ldots\}$ and assume that we have listed the elements in increasing order. No matter how we select three elements from $T$, applying $F$ should always give us the same color. Since there are only finitely many machines with codes less than $t_0$, we cannot have one of them stopping between $t_i$ steps and $t_{i+1}$ steps for every $i > 1$. (Smell any pigeons?) Thus $T$ must be a red monochromatic set. The elements of $T$ are very handy for computing the set $H$. Pick any machine $m$. We know that $m \leq t_m$, and if $\{m\}(m)$ ever halts, it must halt by step $t_{m+1}$. To see this, suppose that $\{m\}(m)$ halts at a later step $k$. Then $t_{k+1} > k$ and $F(t_m, t_{m+1}, t_{k+1})$ would be blue, contradicting the claim that $T$ is a monochromatic red set. Thus, using $\chi_T$ as a subprogram, we could easily write a program that computes $\chi_H$. Since $\chi_H$ is not computable, $\chi_T$ is not computable either, and we have proved the following theorem.

**Theorem 3.41.**  *There is a computable coloring of $[\mathbb{N}]^3$ that has no infinite computable monochromatic set.*

Many other types of coloring problems lead to noncomputable sets. For example, there is a computable 2-regular 2-colorable graph with no computable 2-coloring. However, bumping up the size of our color palette can lead to computable colorings. Schmerl [131] proved that if $G$ is a computable $d$-regular graph that is $n$-colorable, then $G$ has a computable $2n - 1$ coloring. Depending on the value of $d$, it may or may not be possible to find a computable coloring with fewer than $2n - 1$ colors. For $2 \leq n \leq m \leq 2n - 2$, let $D(n, m)$ be the least degree $d$ such that there is a $d$-regular graph with no recursive $m$ coloring. Beyond a few bounds (see [62] and [131]), little is known about the values of $D(n, m)$. For example, we know that $6 \leq D(5, 5) \leq 7$, but the exact value of $D(5, 5)$ is not known.

### Reverse Mathematics

Reverse mathematics is a program of mathematical logic that was founded by H. Friedman (as in Section 3.8) and S. Simpson. The goal of the program is to measure the logical strength of mathematical theorems by proving that each one is equivalent to some statement in a hierarchy of axioms for second order arithmetic. These proofs are carried out in a weak base system, $RCA_0$, which consists of PA with restricted induction plus a comprehension axiom that essentially asserts the existence of computable subsets of $\mathbb{N}$. We refer to these systems as *second order arithmetic* because the formulas include variables for numbers and variables for sets of numbers. Simpson's book [133] is a comprehensive source for information about reverse mathematics. A number of theorems about infinite graphs have been analyzed, including the following example.

**Theorem 3.42.** *$RCA_0$ can prove that the following are equivalent:*

1. *Every 2-regular graph with no cycles of odd length is bipartite.*
2. *Weak König's Lemma: Every infinite tree in which every node is labeled 0 or 1 contains an infinite path.*

Since a graph is bipartite if and only if it is 2-colorable, it is not hard to adapt the solution of Exercise 3 of Section 3.1 to prove that 2 implies 1. The proof that 1 implies 2 is less obvious, and the published proof uses an intermediate equivalent statement [86].

There are many open questions in reverse mathematics that are related to combinatorics and graph theory. For example, can $RCA_0$ prove that Ramsey's Theorem for pairs and two colors implies Weak König's Lemma? It is known that the converse is independent of $RCA_0$ and that Ramsey's Theorem for triples and two colors implies a much stronger version of König's Lemma.

### The Analytical Hierarchy

Subsets of $\mathbb{N}$ that are definable by formulas of second order arithmetic are called *analytical*, and can be organized by the complexity of their defining formulas. The resulting hierarchy of sets is analogous (but not identical) to the hierarchy of projective sets studied by descriptive set theorists. Both [126] and [85] provide useful background for the study of the analytical hierarchy.

Before we can state any results, we need some terminology. A set is $\Sigma_1^1$-*definable* if it is definable by a formula containing no universal set quantifiers. We say that a set is $\Sigma_1^1$-*complete* if it is $\Sigma_1^1$-definable and every other $\Sigma_1^1$-definable set is 1-reducible to it. Here, $B$ is 1-*reducible* to $A$ if there is a computable one-to-one function $f$ such that for all $n$, $n \in B$ if and only if $f(n) \in A$. Naïvely, each $\Sigma_1^1$-complete set embodies all the information content of every other $\Sigma_1^1$-definable set. One interesting characteristic of these sets is that any defining formula for a $\Sigma_1^1$-complete set must contain a set quantifier. Thus, no formula containing only quantifiers on numbers can possibly define a $\Sigma_1^1$-complete set.

An infinite graph $G$ has a *Hamiltonian path* if there is a sequence of vertices $v_0, v_1, v_2, \ldots$ such that each vertex of $G$ appears exactly once in the list and $v_j v_{j+1}$ is an edge of $G$ for each $j \in \mathbb{N}$. An index for a computable graph is the code for the characteristic function of the edge and vertex sets for the graph. Using all this terminology, we can state the following theorem of Harel [80].

**Theorem 3.43.** *The set of indices of computable graphs with Hamiltonian paths is $\Sigma_1^1$-complete.*

For a similar problem, suppose we weight the edges of an infinite graph with rational numbers that sum to 1. Such a graph may or may not have a minimal spanning tree. Is the set of indices of computable graphs with minimal spanning trees a $\Sigma_1^1$-complete set? There is a related reverse mathematics result for directed graphs [29], but the proof does not appear to adapt to the undirected case.

Lists of Theorems and Trees

There are a number of interesting Ramsey-style theorems that are worthy of exploration. Here is a list, with pointers to some good references.

- Van der Waerden's Theorem (and the related Szemerédi's Theorem) [71]
- Hindman's Theorem (and the related Folkman's Theorem) [71]
- Milliken's Theorem [106]
- Galvin–Prikry Theorem [61]
- Erdős–Rado Theorem [88]

Other good prospects for further study include special and regular Aronszajn trees, Suslin trees, and Kurepa trees, all of which appear in [88] and [101].

# 3.10   References

*His was a name to conjure with in certain circles.*
— E. Wallace, *The Just Men of Cordova*

In this section we will name a few authors whose books will be helpful to those interested in the further study of infinite sets, graphs, and combinatorics. Many good introductory texts on axiomatic set theory are available. The books of Enderton [41], Moschovakis [108], and Roitman [127] are all accessible and remarkably distinct. For a more technical approach, Drake [37] and Levy [101] are good choices. Jech's encyclopedic text [88] is an invaluable reference (and a good read).

Several books give extended treatments of specific topics in this chapter. Devlin's book [35] gives a comprehensive coverage of constructible sets. Large cardinals and partition cardinals play a central role in Drake's text [37]. *The Axiom of Choice* is the title and subject of another nice book by Jech [87]. For a detailed treatment of cardinal and ordinal arithmetic, it is hard to beat Sierpiński's old gem [132].

Much of the historical content of the chapter was gleaned from van Heijenoort's anthology [142] and Kanamori's insightful article [89]. For the final word in matters of logic, one can consult Kleene's blue bible [92] or the more accessible text of Mendelson [104]. Finally, other treatments of infinite graphs and combinatorics can be found in the books of Ore [112] and Diestel [36].

# References

[1] M. Aigner, *Combinatorial Theory*, Springer-Verlag, New York, 1979.

[2] M. Aigner, *Graph Theory, A Development from the 4-Color Problem*, BCS Associates, Moscow, ID, 1987.

[3] I. Anderson, *Perfect matchings of a graph*, J. Combin. Theory Ser. B **10** (1971), 183–186.

[4] G.E. Andrews, *The Theory of Partitions*, Cambridge Univ. Press, New York, 1984.

[5] T.M. Apostol, *Mathematical Analysis*, 2nd ed., Addison-Wesley, Reading, MA, 1974.

[6] K. Appel and W. Haken, *Every planar map is four-colorable*, Illinois J. Math. **21** (1977), 429–567.

[7] J. Baumgartner, *Ineffability properties of cardinals I*, in A. Hajnal, R. Rado, and V. Sós, eds., *Infinite and Finite Sets*, Colloq. Math. Soc. János Bolyai, North Holland, Amsterdam, 1975, 109–136.

[8] J. Baumgartner, *Ineffability properties of cardinals II*, in R. Butts and J. Hintikka, eds., *Logic, Foundations of Mathematics, and Computability Theory*, International Congress of Logic, Methodology, and Philosophy of Science, D. Reidel, Dordrecht, Boston, 1977, 87–106.

[9] E.F. Beckenbach, ed., *Applied Combinatorial Mathematics*, John Wiley & Sons, New York, 1964. Reprinted by Krieger, Melbourne, FL, 1981.

[10] L.W. Beineke and R.J. Wilson, eds., *Graph Connections*, Oxford University Press, New York, 1997.

[11] C. Berge, *Two Theorems in Graph Theory*, Proc. Nat. Acad. Sci. U. S. A. **43** (1957), 842–844.

[12] N.L. Biggs, E.K. Lloyd, and R.J. Wilson, *Graph Theory 1736–1936*, Clarendon Press, Oxford, 1976.

[13] G.D. Birkhoff and D.C. Lewis, *Chromatic Polynomials*, Trans. Amer. Math. Soc. **60** (1946) 355–451.

[14]  G. Boolos and R. Jeffrey, *Computability and Logic,* 3rd ed., Cambridge University Press, New York, 1989.

[15]  E. Borel, *Leçons sur la theorie des fonctions,* Gauthier-Villars et fils, Paris, 1898.

[16]  J.G. Broida and S.G. Williamson, *A Comprehensive Introduction to Linear Algebra,* Addison-Wesley, Redwood City, CA, 1989.

[17]  R.L. Brooks, *On colouring the nodes of a network,* Proc. Cambridge Phil. Soc. **37** (1941), 194–197.

[18]  N.G. de Bruijn, *Pólya's theory of counting,* in [9], 144–184.

[19]  W. Burnside, *Theory of Groups of Finite Order,* Cambridge Univ. Press, London, 1897. Second edition, Cambridge Univ. Press, London, 1911, reprinted by Dover, New York, 1955.

[20]  S.A. Burr, *Generalized Ramsey Theory for Graphs—A Survey,* in *Graphs and Combinatorics,* Springer (1974), 52–75.

[21]  G. Cantor, *Beiträge zur Begründung der transfiniten Mengenlehre,* Math. Ann. **46** (1895), 481–512. Translated in [23].

[22]  G. Cantor, *Beiträge zur Begründung der transfiniten Mengenlehre II,* Math. Ann. **49** (1897), 207–246. Translated in [23].

[23]  G. Cantor, *Contributions to the founding of the theory of transfinite numbers,* translated by P. Jourdain, Dover, New York, 1955.

[24]  A. Cayley, *A theorem on trees,* Quart. J. Math. **23** (1889), 276–378.

[25]  G. Chartrand and L. Lesniak, *Graphs & Digraphs,* 3rd ed., Chapman & Hall, London, 1996.

[26]  V. Chvátal, *Tree-Complete Ramsey Numbers,* J. Graph Theory **1** (1977), 93.

[27]  V. Chvátal and F. Harary, *Generalized Ramsey Theory for Graphs, III. Small Off-Diagonal Numbers,* Pacific J. Math. **41** (1972), 335–345.

[28]  J. Clark and D.A. Holton, *A First Look at Graph Theory,* World Scientific, Singapore, 1991.

[29]  P. Clote and J. Hirst, *Reverse mathematics of some topics from algorithmic graph theory,* Fund. Math. **157** (1998) 1–13.

[30]  P. Cohen, *The independence of the Continuum Hypothesis, I,* Proc. Nat. Acad. Sci. U.S.A. **50** (1963), 1143–1148.

[31]  P. Cohen, *The independence of the Continuum Hypothesis, II,* Proc. Nat. Acad. Sci. U.S.A. **51** (1964), 105–110.

[32]  L. Comtet, *Advanced Combinatorics,* D. Reidel, Dordrecht, Boston, 1974.

[33]  J.H. Conway and R.K. Guy, *The Book of Numbers,* Springer-Verlag, New York, 1996.

[34]  G.P. Csicsery, director, *N is a Number,* documentary film, 1993.

[35]  K. Devlin, *Constructability,* Springer-Verlag, Berlin Heidleberg, 1984.

[36]  R. Diestel, *Graph Decompositions: A Study in Infinite Graph Theory,* Oxford Claredon Press, New York, 1990.

[37]  F. Drake, *Set Theory: An Introduction to Large Cardinals,* North-Holland, Amsterdam, 1974.

[38]  G.A. Dirac and S. Schuster, *A theorem of Kuratowski,* Nederl. Akad. Wetensch. Proc. Ser. A **57** (1954), 343–348.

[39]  A.W.F. Edwards, *Pascal's Arithmetical Triangle,* Griffin, London, 1987.

[40]  E. Egerváry, *Matrixok Kombinatórius Tulajdonságairól,* Mat. Fiz. Lapok **38** (1931), 16–28.

[41]  H. Enderton, *Elements of Set Theory,* Academic Press, Orlando, 1977.

[42]  P. Erdős, *Some remarks on the theory of graphs*, Bull. Amer. Math. Soc. **53** (1947), 292–294.

[43]  P. Erdős, *Some of my favorite problems and results*, in [70], vol. I, 47–67.

[44]  P. Erdős and G. Szekeres, *A combinatorial problem in geometry*, Compositio Math. **2** (1935), 463–470.

[45]  P. Erdős and A. Tarski, *On families of mutually exclusive sets*, Ann. of Math. **44** (1943), 315–329.

[46]  P. Erdős and A. Tarski, *On some problems involving inaccessible cardinals*, in Y. Bar-Hillel, E. Poznansik, M. Rabin, and A. Robinson, eds., *Essays on the Foundations of Mathematics*, North Holland, Amsterdam, 1962, 50–82.

[47]  L. Euler, *Solutio problematis ad geometriam situs pertinentis*, Comment. Academiae Sci. I. Petropolitanae **8** (1736), 128–140.

[48]  L. Euler, *Demonstratio Nonnullarum Insignium Proprietatum Quibus Solida Hedris Planis Inclusa Sunt Praedita*, Novi Comm. Acad. Sci. Imp. Petropol **4** (1758), 140–160.

[49]  H. Eves, *In Mathematical Circles*, Prindle, Weber, and Schmidt, Boston, 1969.

[50]  I. Fáry, *On straight line representations of planar graphs*, Acta Sci. Math. **11** (1948), 229–233.

[51]  S. Feferman, *Does mathematics need new axioms?*, Amer. Math. Monthly, **106** (1999), 99–111.

[52]  L.R. Foulds, *Graph Theory Applications*, Springer-Verlag, New York, 1992.

[53]  A.A. Fraenkel, *Über die Zermelosche Begründung der Mengenlehre*, Jahresber. Deutsch. Math.-Verein. 30, 2nd section (1921) 97–98.

[54]  A.A. Fraenkel, *Axiomatische Begründung der transfiniten Kardinalzahlen I*, Math. Z. **13** (1922), 153–188.

[55]  A.A. Fraenkel, *Zu den Grundlagen der Cantor–Zermeloschen Mengenlehre*, Math. Ann. **86** (1922), 230–237.

[56]  A.A. Fraenkel, *Der Begriff "definit" und die Unabhängigkeit des Auswahlsaxioms*, Sitzungsberichte der Preussischen Akademie der Wissenschaften, Physikalisch-mathematische Klasse (1922), 253–257. Translated in [142] 284–289.

[57]  G. Frege, *Ueber die Begriffsschrift des Herrn Peano und meine eigene*, Berichte über die Verhandlungen der Königlich Sächsischen Gesellschaft der Wissenschaften zu Leipzig, Mathematisch-physikalische Klasse **48** (1896), 361–378. "*Summum bonum*" quotation translated in [142] on page 2.

[58]  H. Friedman, *Finite functions and the necessary use of large cardinals*, Ann. of Math., **148** (1998), 803–893.

[59]  F.G. Frobenius, *Über die Congruenz nach einem aus zwei endlichen Gruppen gebildeten Doppelmodul*, J. Reine Angew. Math. (Crelle's J.) **101** (1887), 273–299. Reprinted in *Gesammelte Abhandlungen*, vol. 2, Springer-Verlag, Berlin, 1968, 304–330.

[60]  D. Gale and L.S. Shapley, *College admissions and the stability of marriage*, Amer. Math. Monthly **69** (1962), 9–15.

[61]  F. Galvin and K. Prikry, *Borel sets and Ramsey's theorem*, J. Symbolic Logic **38** (1973), 193–198.

[62]  W. Gasarch, *A survey of recursive combinatorics,* in Y. Ershov, S. Goncharov, A. Nerode, and J. Remmel eds., *Handbook of Recursive Mathematics, Volume 2: Recursive Algebra, Analysis and Combinatorics*, North Holland, Amsterdam, 1998, 1041–1176.

[63] I. Gessel and G.-C. Rota, *Classic Papers in Combinatorics*, Birkhäuser Boston, 1987.

[64] K. Gödel, *Über formal unentscheidbare Sätze der* Principia Mathematica *und verwandter Systeme I*, Monatsh. Math. **38** (1931), 173–198. Reprinted and translated in [66], 144–195.

[65] K. Gödel, *The consistency of the Axiom of Choice and the Generalized Continuum Hypothesis*, Proc. Nat. Acad. Sci. U.S.A. **24** (1938), 556–557. Reprinted in [66], 26–27.

[66] K. Gödel, *Collected Works, Volume 1*, S. Feferman et al. eds., Oxford University Press, New York, 1990.

[67] R.J. Gould, *Graph Theory*, Benjamin/Cummings, Menlo Park, 1988.

[68] R.L. Graham, M. Grötschel, and L. Lovász, eds., *Handbook of Combinatorics* (two volumes), Elsevier Science B. V., Amsterdam, and MIT Press, Cambridge, 1995.

[69] R.L. Graham, D.E. Knuth, and O. Patashnik, *Concrete Mathematics: A Foundation for Computer Science*, Addison-Wesley, Reading, 1989.

[70] R.L. Graham and J. Nešetřil, eds., *The Mathematics of Paul Erdős* (two volumes), Springer-Verlag, Berlin, 1997.

[71] R.L. Graham, B.L. Rothschild, and J.H. Spencer, *Ramsey Theory*, 2nd ed., John Wiley & Sons, New York, 1990.

[72] J. Grantham, *There are infinitely many Perrin pseudoprimes*, preprint, 1997.

[73] A. Granville, *Zaphod Beeblebrox's brain and the fifty-ninth row of Pascal's triangle*, Amer. Math. Monthly **99** (1992), 318–331. Correction, ibid. **104** (1997), 848–851.

[74] J. Gross and J. Yellen, *Graph Theory and its Applications*, CRC Press, Boca Raton, FL, 1999.

[75] D. Gusfield and R.W. Irving, *The Stable Marriage Problem: Structure and Algorithms*, MIT Press, Cambridge, 1989.

[76] M. Hall, *Combinatorial Theory*, 2nd ed., John Wiley & Sons, New York, 1983.

[77] P. Hall, *On representation of subsets*, J. London Math. Soc. **10** (1935), 26–30.

[78] F. Harary, *Graph Theory*, Addison-Wesley, Reading, MA, 1969.

[79] F. Harary and E.M. Palmer, *Graphical Enumeration*, Academic Press, New York, 1973.

[80] D. Harel, *Hamiltonian paths in infinite graphs*, Israel J. Math. **76** (1991), 317–336.

[81] F. Hartogs, *Über das Problem der Wohlordnung*, Math. Ann. **76** (1915), 436–443.

[82] P.J. Heawood, *Map-colour theorem*, Quart. J. Math. **24** (1890), 332–339.

[83] C. Henrion, *Properties of subtle cardinals*, J. Symbolic Logic **52** (1987), 1005–1019.

[84] D. Hilbert, *Über das Undendliche*, Math. Ann. **95** (1926), 161–190. Translated in [142], 367–392.

[85] P. Hinman, *Recursion-theoretic Hierarchies*, Springer-Verlag, Berlin New York, 1977.

[86] J. Hirst, *Marriage theorems and reverse mathematics*, in W. Sieg, ed., *Logic and Computation*, Contemp. Math. **106**, American Mathematical Society, Providence, 1990, 181–196.

[87] T. Jech, *The Axiom of Choice*, North-Holland, Amsterdam, 1973.

[88] T. Jech, *Set Theory*, 2nd ed., Springer-Verlag, Berlin, 1997.

[89] H. Kanamori, *The mathematical development of set theory from Cantor to Cohen*, Bull. Symbolic Logic **2** (1996), 1–71.

[90] A.B. Kempe, *On the geographical problem of the four colors*, Amer. J. Math. **2** (1879), 193–200.

[91]  G. Kirchhoff, *Über der Auflösung der Gleichungen, auf welche man bei der Unter-suchung der linearen Verteilung galvanischer Ströme gefürt wird*, Ann. Phys. Chem. **72** (1847), 497–508.

[92]  S. Kleene, *Introduction to Metamathematics*, North Holland, Amsterdam, 1971.

[93]  D.E. Knuth, *Stable Marriage and Its Relation to Other Combinatorial Problems: An Introduction to the Mathematical Analysis of Algorithms*, CRM Proc. Lecture Notes **10**, Amer. Math. Soc., Providence, 1997.

[94]  D. König, *Über eine Schlussweise aus dem Endlichen ins Undendliche*, Acta Litt. Acad. Sci. Hung. (Szeged) **3** (1927), 121–130.

[95]  D. König, *Graphen und Matrizen*, Math. Fiz. Lapok **38** (1931), 116–119.

[96]  J. König, *Sur la theorie des ensembles*, C. R. Acad. Paris **143** (1906), 110–112.

[97]  J.B. Kruskal Jr., *On the shortest spanning subtree of a graph and the traveling salesman problem*, Proc. Am. Math. Soc. **7** (1956), 48–50.

[98]  K. Kuratowski, *Sur la notion del l'ordre dans la théorie des ensembles*, Fund. Math. **2** (1921), 161–171.

[99]  K. Kuratowski, *Sur le problème des courbes gauches en topologie*, Fund. Math. **15** (1930), 271–283.

[100]  N.J. Lennes, *On the foundation of the theory of sets* (abstract), Bull. Amer. Math. Soc. **28** (1922), 300.

[101]  A. Levy, *Basic Set Theory*, Springer-Verlag, Berlin, 1979.

[102]  L. Lovász, *Three short proofs in graph theory*, J. Combin Theory Ser. B **19** (1975), 269–271.

[103]  E. Lucas, *Théorie des fonctions numériques simplement périodiques*, Amer. J. Math. **1** (1878), 184–240, 289–321.

[104]  E. Mendelson, *Introduction to Mathematical Logic*, 3rd ed., Wadsworth and Brooks, Monterey, 1987.

[105]  K. Menger, *Zur allgemenen Kurventheorie*, Fund. Math. **10** (1927), 95–115.

[106]  K. Milliken, *Ramsey's theorem with sums or unions*, J. Combin. Theory Ser. A **18** (1975), 276–290.

[107]  D. Mirimanoff, *Les antinomies de Russell et de Burali-Forti et le problème fondamental de la théorie des ensembles*, Enseign. Math. **19** (1917), 37–52.

[108]  Y. Moschovakis, *Notes on Set Theory*, Springer-Verlag, New York, 1994.

[109]  P.M. Neumann, *A lemma that is not Burnside's*, Math. Sci. **4** (1979), 133–141.

[110]  J.R. Newman (ed.), *The World of Mathematics*, Simon and Schuster, New York, 1956.

[111]  A. Nijenhuis and H.S. Wilf, *Combinatorial Algorithms*, Academic Press, New York, 1975.

[112]  O. Ore, *Theory of Graphs*, Amer. Math. Soc., Providence, 1962.

[113]  G. Peano, *Super theorema de Cantor–Bernstein*, Rend. Circ. Mat. Palermo **21** (1906), 136–143.

[114]  J. Petersen, *Die Theorie der regulären Graphen* Acta Math. **15** (1891), 193–220.

[115]  M. Petkovšek, H.S. Wilf, and D. Zeilberger, *A = B*, A.K. Peters, Wellesley, MA, 1996.

[116]  G. Pólya, *Kombinatorische Anzahlbestimmungen für Gruppen, Graphen, und chemische Verbindungen*, Acta Math. **68** (1937), 145–254. Translated in [118].

[117]  G. Pólya, *On picture-writing*, Amer. Math. Monthly **63** (1956), 689–697. Reprinted in [63], 249–257.

[118]  G. Pólya and R.C. Read, *Combinatorial Enumeration of Groups, Graphs, and Chemical Compounds*, Springer-Verlag, New York, 1987.

216     References

[119] G. Pólya, R.E. Tarjan, and D.R. Woods, *Notes on Introductory Combinatorics*, Birkhäuser Boston, 1983.

[120] H. Prüfer, *Beweis eines Satzes über Permutationen*, Arch. Math. Phys. **27** (1918), 742–744.

[121] F.P. Ramsey, *On a problem of formal logic*, Proc. London Math. Soc. **30** (1930), 264–286. Reprinted in [63], 2–24.

[122] J. Riordan, *An Introduction to Combinatorial Analysis*, John Wiley & Sons, New York, 1958.

[123] J. Riordan, *Combinatorial Identities*, John Wiley & Sons, New York, 1968. Reprinted by Krieger, New York, 1979.

[124] F.R. Roberts, *Applied Combinatorics*, Prentice-Hall, Englewood Cliffs, 1984.

[125] N. Robertson, D.P. Sanders, P.D. Seymour, and R. Thomas, *The four-colour theorem*, J. Combin. Theory Ser. B **70** (1997), 2–44.

[126] H. Rogers, Jr., *Theory of Recursive Functions and Effective Computability*, McGraw-Hill, New York, 1967.

[127] J. Roitman, *Introduction to Modern Set Theory*, John Wiley & Sons, New York, 1990.

[128] J. Rosser, *Extensions of some theorems of Gödel and Church*, J. Symbolic Logic **1** (1936), 87–91.

[129] G.-C. Rota, ed., *Studies in Combinatorics*, Mathematical Association of America, 1978.

[130] B. Russell, *On some difficulties in the theory of transfinite numbers and order types*, Proc. London Math. Soc. Series 2 **4** (1906), 29–53. Reprinted in Russell's *Essays in Analysis*, D. Lackey, ed., George Braziller, New York, 1973.

[131] J. Schmerl, *Recursive colorings of graphs*, Canad. J. Math. **32** (1980), 821–830.

[132] W. Sierpiński, *Cardinal and Ordinal Numbers*, Państwowe Wydawnictwo Naukowe, Warszawa, 1958.

[133] S. Simpson, *Subsystems of Second Order Arithmetic*, Springer-Verlag, Berlin, Heidelberg, 1999.

[134] T. Skolem, *Einige Bemerkungen zur axiomatischen Begründung der Mengenlehre*, Matematikerkongressen i Helsingfors den 4–7 Juli 1922, Den femte skandinaviska matematikerkongressen, Redogörelse, Akademiska-Bokhandeln, Helsinki, 217–232. Translated in [142], 290–301.

[135] T. Skolem, *Two remarks on set theory*, Math. Scand. **5** (1957), 40–46. Reprinted in Skolem's *Selected works in logic,* J. Fenstad, ed., Universitetsforlaget, Oslo, 1970.

[136] N.J.A. Sloane and S. Plouffe, *The Encyclopedia of Integer Sequences*, Academic Press, San Diego, 1995.

[137] R. Soare, *Recursively Enumerable Sets and Degrees,* Springer-Verlag, Berlin, Heidelberg, 1987.

[138] R.P. Stanley, *Enumerative Combinatorics* (two volumes), Cambridge Univ. Press, New York, 1997 (vol. 1), 1999 (vol. 2).

[139] R. Thomas, *An Update on the Four-Color Theorem*, Notices Amer. Math. Soc. **45** 7 (1998), 848–859.

[140] A. Tucker, *Applied Combinatorics*, 3rd ed., John Wiley & Sons, New York, 1994.

[141] W.T. Tutte, *The factorization of linear graphs*, J. London Math. Soc. **22** (1947), 107–111.

[142] J. van Heijenoort, *From Frege to Gödel*, Harvard University Press, Cambridge, 1967.

[143] J.H. van Lint and R.M. Wilson, *A Course in Combinatorics*, Cambridge Univ. Press, New York, 1992.

[144] J. von Neumann, *Eine Axiomatisierung der Mengenlehre*, J. Reine Angew. Math. (Crelle's Journal) **154** (1925), 34–56. Translated in [142], 393–413.

[145] K. Wagner, *Bemerkungen zum Vierfarbenproblem*, Jahresber. Deutsch. Math.-Verein. **46** (1936), 21–22.

[146] D.B. West, *Introduction to Graph Theory*, Prentice Hall, Upper Saddle River, NJ, 1996.

[147] N. Wiener, *A simplification of logic of relations*, Proc. Cambridge Phil. Soc. **17** (1912-1914), 387–390.

[148] H.S. Wilf, *Generatingfunctionology*, 2nd ed., Academic Press, San Diego, 1994.

[149] E.M. Wright, *Burnside's lemma: A historical note*, J. Combin. Theory Ser. B **30** (1981), 89–90.

[150] E. Zermelo, *Beweis, daß jede Menge wohlgeordnet werden kann*, Math. Ann. **59** (1904), 514–516. Translated in [142], 139–141.

[151] E. Zermelo, *Neuer Beweis für die Möglichkeit einer Wohlordnung*, Math. Ann. **65** (1908), 107–128. Translated in [142], 183–198.

[152] E. Zermelo, *Untersuchungen über die Grundlagen der Mengenlehre I*, Math. Ann. **65** (1908), 261–281. Translated in [142], 199–215.

# Index

**Hilton/Holton/Pedersen:** Mathematical Reflections: In a Room with Many Mirrors.

**Iooss/Joseph:** Elementary Stability and Bifurcation Theory. Second edition.

**Isaac:** The Pleasures of Probability. *Readings in Mathematics.*

**James:** Topological and Uniform Spaces.

**Jänich:** Linear Algebra.

**Jänich:** Topology.

**Kemeny/Snell:** Finite Markov Chains.

**Kinsey:** Topology of Surfaces.

**Klambauer:** Aspects of Calculus.

**Lang:** A First Course in Calculus. Fifth edition.

**Lang:** Calculus of Several Variables. Third edition.

**Lang:** Introduction to Linear Algebra. Second edition.

**Lang:** Linear Algebra. Third edition.

**Lang:** Undergraduate Algebra. Second edition.

**Lang:** Undergraduate Analysis.

**Lax/Burstein/Lax:** Calculus with Applications and Computing. Volume 1.

**LeCuyer:** College Mathematics with APL.

**Lidl/Pilz:** Applied Abstract Algebra. Second edition.

**Logan:** Applied Partial Differential Equations.

**Macki-Strauss:** Introduction to Optimal Control Theory.

**Malitz:** Introduction to Mathematical Logic.

**Marsden/Weinstein:** Calculus I, II, III. Second edition.

**Martin:** The Foundations of Geometry and the Non-Euclidean Plane.

**Martin:** Geometric Constructions.

**Martin:** Transformation Geometry: An Introduction to Symmetry.

**Millman/Parker:** Geometry: A Metric Approach with Models. Second edition.

**Moschovakis:** Notes on Set Theory.

**Owen:** A First Course in the Mathematical Foundations of Thermodynamics.

**Palka:** An Introduction to Complex Function Theory.

**Pedrick:** A First Course in Analysis.

**Peressini/Sullivan/Uhl:** The Mathematics of Nonlinear Programming.

**Prenowitz/Jantosciak:** Join Geometries.

**Priestley:** Calculus: A Liberal Art. Second edition.

**Protter/Morrey:** A First Course in Real Analysis. Second edition.

**Protter/Morrey:** Intermediate Calculus. Second edition.

**Roman:** An Introduction to Coding and Information Theory.

**Ross:** Elementary Analysis: The Theory of Calculus.

**Samuel:** Projective Geometry. *Readings in Mathematics.*

**Scharlau/Opolka:** From Fermat to Minkowski.

**Schiff:** The Laplace Transform: Theory and Applications.

**Sethuraman:** Rings, Fields, and Vector Spaces: An Approach to Geometric Constructability.

**Sigler:** Algebra.

**Silverman/Tate:** Rational Points on Elliptic Curves.

**Simmonds:** A Brief on Tensor Analysis. Second edition.

**Singer:** Geometry: Plane and Fancy.

**Singer/Thorpe:** Lecture Notes on Elementary Topology and Geometry.

**Smith:** Linear Algebra. Third edition.

**Smith:** Primer of Modern Analysis. Second edition.

**Stanton/White:** Constructive Combinatorics.

**Stillwell:** Elements of Algebra: Geometry, Numbers, Equations.

**Stillwell:** Mathematics and Its History.

**Stillwell:** Numbers and Geometry. *Readings in Mathematics.*

**Strayer:** Linear Programming and Its Applications.

## Undergraduate Texts in Mathematics